Advances in
Heterocyclic
Chemistry

Volume 13

*Editorial Advisory Board*

A. Albert
A. T. Balaban
J. Gut
J. M. Lagowski
J. H. Ridd
Yu. N. Sheinker
H. A. Staab
M. Tišler

Advances in

# HETEROCYCLIC CHEMISTRY

*Edited by*

A. R. KATRITZKY

A. J. BOULTON

*School of Chemical Sciences*
*University of East Anglia*
*Norwich, England*

Volume 13

Academic Press · New York and London · 1971

COPYRIGHT © 1971, BY ACADEMIC PRESS, INC.
ALL RIGHTS RESERVED
NO PART OF THIS BOOK MAY BE REPRODUCED IN ANY FORM,
BY PHOTOSTAT, MICROFILM, RETRIEVAL SYSTEM, OR ANY
OTHER MEANS, WITHOUT WRITTEN PERMISSION FROM
THE PUBLISHERS.


ACADEMIC PRESS, INC.
111 Fifth Avenue, New York, New York 10003


*United Kingdom Edition published by*
ACADEMIC PRESS, INC. (LONDON) LTD.
24/28 Oval Road, London NW1 7DD


LIBRARY OF CONGRESS CATALOG CARD NUMBER: 62-13037


PRINTED IN THE UNITED STATES OF AMERICA

# Contents

CONTRIBUTORS . . . . . . . . . . vii

PREFACE . . . . . . . . . . . ix

## Heterocyclic Ferrocenes

F. D. POPP AND E. B. MOYNAHAN

I. Introduction . . . . . . . . . . 1
II. Ferrocenes Containing a Heterocyclic Ring Fused to the Ferrocene . 2
III. Ferrocenyl Heterocyclic Compounds . . . . . 13
IV. Other Heterocyclic Metallocenes . . . . . 41
V. Compounds of Potential Medicinal Interest . . . 44

## Synthesis and Reactions of 1-Azirines

FRANK W. FOWLER

I. Introduction . . . . . . . . . . 45
II. 2-Azirines . . . . . . . . . . 46
III. Synthesis of 1-Azirines . . . . . . . . 48
IV. Reactions of 1-Azirines . . . . . . . 63
V. Summary . . . . . . . . . . 76

## Electronic Aspects of Purine Tautomerism

BERNARD PULLMAN AND ALBERTE PULLMAN

I. Introduction . . . . . . . . . . 77
II. The Biological Importance of Purine Tautomerism . . . 79
III. Quantum Mechanical Methods on Investigation . . . 85
IV. The Four Prototropic Tautomers of Purine . . . 100
V. Amine–Imine Tautomerism in Adenines . . . . 111
VI. Lactam–Lactim Tautomerism in Hydroxypurines . . . 122
VII. The Fine Structure of Hydroxypurines . . . . 127
VIII. Guanine, with a Special Discussion of Its N(8)H Tautomer . 138
IX. Tautomerism in 8-Azapurines and the Problem of the N(8)H Tautomer . . . . . . . . . . 142
X. The Thione–Thiol Tautomerism in Mercaptopurines and the Fine Structure of Thiopurines . . . . . . 145
XI. The N(7)H⇌N(9)H Tautomerism and the Crystal Structure of Purines . . . . . . . . . . 150
XII. Conclusion . . . . . . . . . . 156

## 1,6,6a$S^{IV}$-Trithiapentalenes and Related Structures

NOËL LOZAC'H

|  |  |
|---|---|
| I. Introduction and Nomenclature | 162 |
| II. Preparation of α-(1,2-Dithiol-3-ylidene)carbonyl Compounds | 167 |
| III. Preparation of 1,6,6a$S^{IV}$-Trithiapentalenes | 195 |
| IV. Other Heterocycles Derived from 1,6,6a$S^{IV}$-Trithiapentalenes | 195 |
| V. Extended Structures | 204 |
| VI. Theoretical Studies and Physical Properties | 207 |

## Electrophilic Substitutions of Five-Membered Rings

GIANLORENZO MARINO

|  |  |
|---|---|
| I. Introduction | 235 |
| II. Kinetic Studies and Mechanism | 243 |
| III. Relative Reactivity of Rings and Ring Positions | 263 |
| IV. Effects of Substituents | 293 |

## Recent Developments in Phenanthridine Chemistry

B. R. T. KEENE AND P. TISSINGTON

|  |  |
|---|---|
| I. Introduction and Nomenclature | 316 |
| II. General Methods on Synthesis | 317 |
| III. Physical Properties | 369 |
| IV. Reactions of the Phenanthridine Nucleus | 378 |
| V. The Properties of Functional Groups | 402 |
| AUTHOR INDEX | 415 |
| CUMULATIVE INDEX OF TITLES | 437 |

# Contributors

Numbers in parentheses indicate the pages on which the authors' contributions begin.

FRANK W. FOWLER, *Chemistry Department, State University of New York at Stony Brook, Stony Brook, New York* (45)

B. R. T. KEENE, *Medway and Maidstone College of Technology, Kent, England* (315)

NOËL LOZAC'H, *Ecole Nationale Supérieure de Chimie de Caen, Caen, France* (161)

GIANLORENZO MARINO, *Istituto di Chimica Organica, Università di Perugia, Perugia, Italy* (235)

E. B. MOYNAHAN,* *Clarkson College of Technology, Potsdam, New York* (1)

F. D. POPP, *Clarkson College of Technology, Potsdam, New York* (1)

ALBERTE PULLMAN, *Institut de Biologie Physico-Chimique, Paris, France* (77)

BERNARD PULLMAN, *Institut de Biologie Physico-Chimique, Paris, France* (77)

P. TISSINGTON, *Ciba-Geigy Chemicals Ltd., Grimsby, England* (315)

---

* Present address: GAF Corporation, Binghamton, New York.

# Preface

Volume 13 of this serial publication comprises six chapters of which four deal with general accounts of compound classes: 1-azirines (F. W. Fowler), phenanthridines (B. R. T. Keene and P. Tissington), trithiapentalenes (N. Lozac'h), and heterocyclic ferrocenes (F. D. Popp and E. B. Moynahan). The other two chapters are concerned with particular aspects of the chemistry of groups of heterocycles: the tautomerism of purines (B. Pullman and A. Pullman) and quantitative aspects of the electrophilic substitution reactions of five-membered rings (G. Marino).

A. R. KATRITZKY
A. J. BOULTON

Advances in
Heterocyclic
Chemistry

Volume 13

# Heterocyclic Ferrocenes

F. D. POPP AND E. B. MOYNAHAN*

*Clarkson College of Technology, Potsdam, New York*

|     |     |     |
| --- | --- | --- |
| I. | Introduction | 1 |
| II. | Ferrocenes Containing a Heterocyclic Ring Fused to the Ferrocene | 2 |
|  | A. Ferroceno Heterocyclic Compounds | 2 |
|  | B. Bridged Heterocyclic Compounds | 7 |
| III. | Ferrocenyl Heterocyclic Compounds | 13 |
|  | A. Nitrogen-Containing Rings | 13 |
|  | B. Nitrogen and Other Heteroatoms in Ring | 27 |
|  | C. Oxygen-Containing Rings | 31 |
|  | D. Sulfur-Containing Rings | 37 |
| IV. | Other Heterocyclic Metallocenes | 41 |
| V. | Compounds of Potential Medicinal Interest | 44 |

## I. Introduction

Since the discovery of dicyclopentadienyliron (ferrocene) in 1951[1, 2] an immense body of literature has appeared concerning the chemistry of ferrocenes and related compounds. Although a large number of reviews[3] have appeared, they have given only brief treatment to ferrocenes containing heterocyclic systems. It is the purpose of this review to report on those ferrocenes which contain heterocyclic systems either fused to the ferrocene or as a substituent on the ferrocene.

The literature has been surveyed through mid-1970 and it is believed that all papers dealing specifically with heterocyclic ferrocenes, as well as most papers where such compounds were only incidental to the main theme of the paper, are included in this review.

---

* *Present address:* GAF Corporation, Binghamton, New York.

[1] T. J. Kealy and P. L. Pauson, *Nature (London)* **168**, 1039 (1951).
[2] S. A. Miller, J. A. Tebboth, and J. F. Tremaine, *J. Chem. Soc.* 632 (1952).
[3] See D. E. Bublitz and K. L. Rinehart, Jr., *Org. React.* **17**, 1 (1969) for a recent review and references to 41 other reviews.

## II. Ferrocenes Containing a Heterocyclic Ring Fused to the Ferrocene

Although ferrocene is theoretically capable of forming three types of cyclized structures (**1**, **2**, and **3**), only examples of **1** and **3** have been reported, with a heteroatom in the chain connecting the 1 and 2 carbons of one five-membered ring and in the bridge between the 1- and 1'-positions of ferrocene. The number of compounds of these two types are relatively few as compared with those in which ferrocene is a substituent on the heterocyclic compound (Section III). Relatively little has been reported regarding the chemistry of these compounds, despite the obvious potential of comparing the physical and chemical properties of ferroceno heterocyclic compounds (**1**) with the corresponding benzo compounds.

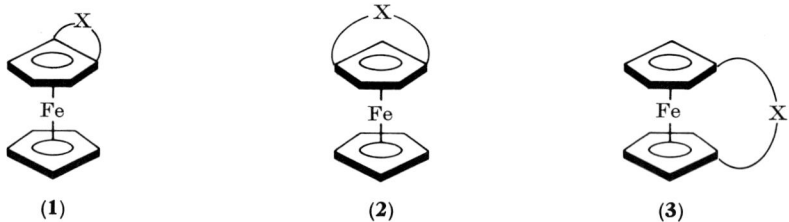

(**1**)         (**2**)         (**3**)

### A. Ferroceno Heterocyclic Compounds

#### 1. *Nitrogen-Containing Ferroceno Heterocyclic Compounds*

The first reported ferroceno heterocyclic compound of the type **1** was $N$-methyltetrahydropyridoferrocene (**4**). This compound was obtained from an anomalous Leuckart–Eschweiler–Clark reaction of 2-aminomethylferrocene (**5**; R = H), formaldehyde, and formic acid,[4] although the structure **4** was not immediately recognized for this compound.[5-8] The heterocyclic compound **4** can be used as an entry to 1,2-disubstituted ferrocenes since on reaction with potassium amide in liquid ammonia[7] or with potassium hydroxide in methanol[9] its

---

[4] D. Lednicer, J. K. Lindsay, and C. R. Hauser, *J. Org. Chem.* **23**, 653 (1958).
[5] J. M. Osgerby and P. L. Pauson, *Chem. Ind. (London)* 196 (1958).
[6] J. M. Osgerby and P. L. Pauson, *Chem. Ind. (London)* 1144 (1958).
[7] D. Lednicer and C. R. Hauser, *J. Org. Chem.* **24**, 43 (1959).
[8] J. M. Osgerby and P. L. Pauson, *J. Chem. Soc.* 4600 (1961).
[9] K. Schlögl, M. Fried, and H. Falk, *Monatsh. Chem.* **95**, 576 (1964).

methiodide gives **6**. Reaction of **5** (R = CH$_3$) with formaldehyde gave **7**, which could be converted into **4** by action of phosphoric acid.[6,8]

Bischler–Napieralski ring closure of **5** (R = CHO) with phosphorus oxychloride[6,8,9] and of **5** (R = COCH$_3$) with phosphorus oxychloride[6,8] or polyphosphate ester[10] gave **8** (R = H and R = CH$_3$, respectively). Reduction of **8** (R = H) with lithium aluminum hydride

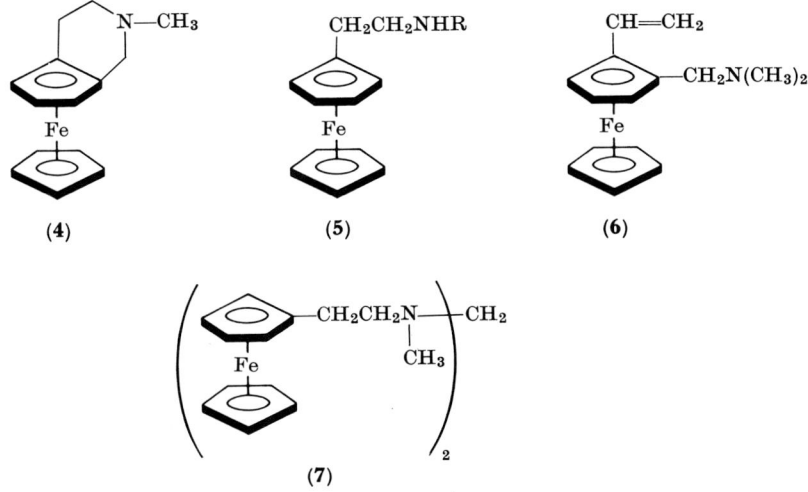

followed by methylation with dimethyl sulfate gave **4**.[6,8] Reduction of **8** (R = CH$_3$) was also carried out[8] to give **9** (R = CH$_3$) which could not be dehydrogenated to an isoquinoline analog under a wide range of conditions.[8,10] Although **9** (R = H) was not characterized in the conversion of **8** (R = H) to **4**, it was isolated by reduction of **8** (R = H) with lithium aluminum hydride or catalytically.[9] This same compound (**9**; R = H) was also obtained by treatment of **10** with hydrochloric acid[9] or by treatment of **11** with hydrochloric acid followed by hydrogenation.[10] If the hydrogenation step was omitted in this latter procedure, **12** could be isolated as its hydrochloride salt.[10] An attempt to cyclize, with hydrochloric acid, the homolog of **10** containing one less carbon atom led to polymer formation.[9] The amine **9** (R = H) has been resolved through crystallization of its (−)*o,o'*-dinitrodiphenic acid salt.

[10] F. D. Popp and E. B. Moynahan, unpublished results.

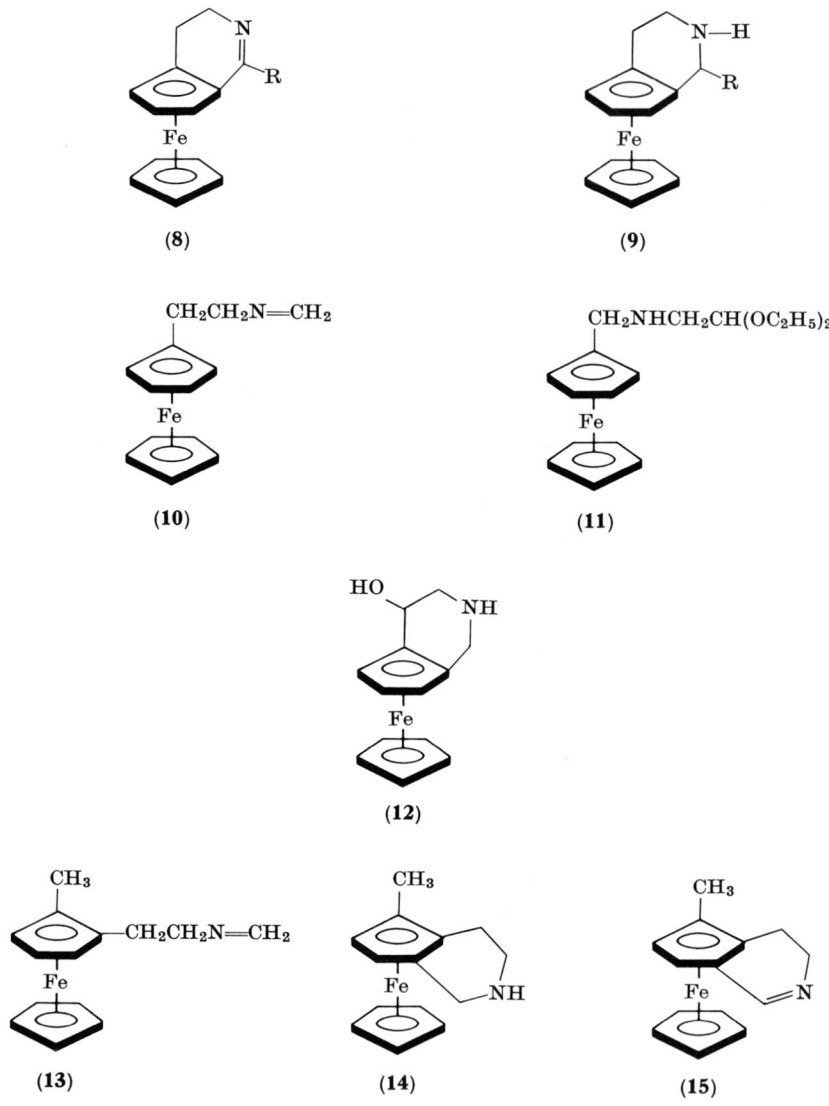

The 1,2-disubstituted ferrocene (13) on treatment with hydrochloric acid gives 14 which can be oxidized with manganese dioxide to give 15.[9]

The Schmidt reaction of 1,2-(α-ketotetramethylene)ferrocene (16) with hydrazoic acid and sulfuric acid in benzene[9] or the Beckmann rearrangement of **17** with *p*-toluenesulfonyl chloride[11] gave the lactam (**18**). In the case of the Schmidt reaction the optically active (+)-ketone was used and gave (−)-**18** as well as the (−)-tetrazole (**19**).[9] Reduction of (−)-**19** with lithium aluminum hydride gave the (−)-homolog of **9** (R = H), which was oxidized with manganese dioxide to give the (−)-homolog of **8** (R = H).[9]

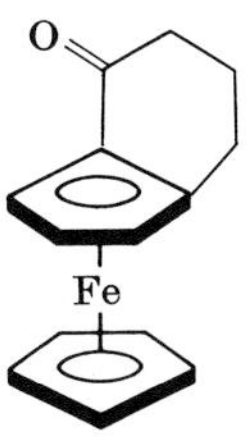

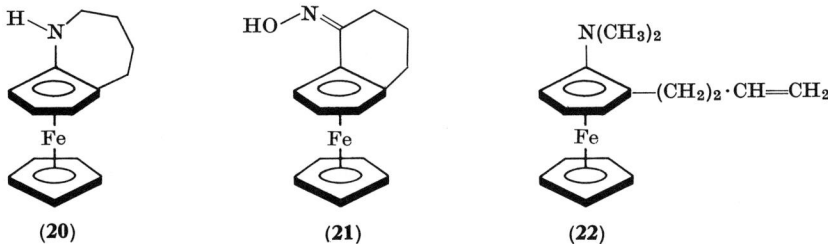

(16) 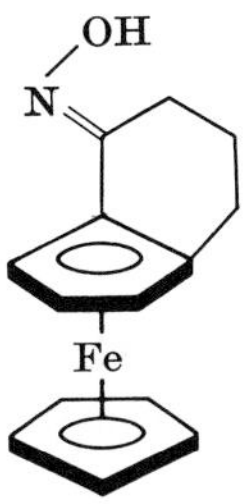  (17) 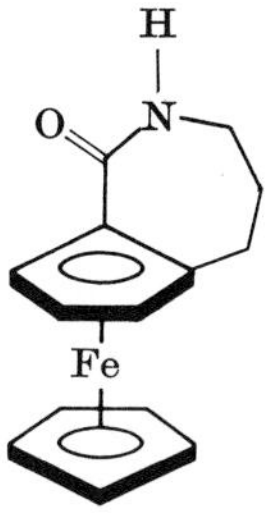  (18) 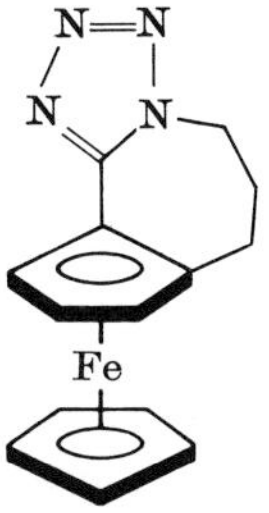  (19)

Treatment of the oxime (**17**) with lithium aluminum hydride–aluminum chloride led to a rearrangement of the ferrocene residue to give **20**.[11] The oxime (**21**) was unchanged under these conditions. Treatment of the quaternary methiodide of **20** with base led to the ring-opened product (**22**).

(20)  (21)  (22)

## 2. Oxygen-Containing Ferroceno Heterocyclic Compounds

The activity in the field of ferroceno heterocyclic compounds of the type **1** has been even less in the case of oxygen-containing compounds than it was with nitrogen-containing compounds. As might be anticipated, the first two examples of this class of compounds were an anhydride and a lactone. Thus treatment of ferrocene-1,2-dicarboxylic

---

[11] K. Schlögl and H. Mechtler, *Monatsh. Chem.* **97**, 150 (1966).

acid with $N,N'$-dicyclohexylcarbodiimide gave **23**[12] and the action of heat or acid on the appropriate hydroxy acid gave **24**.[13]

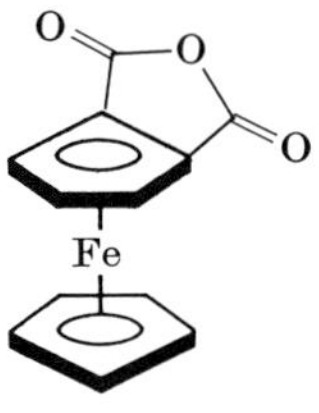

(23) 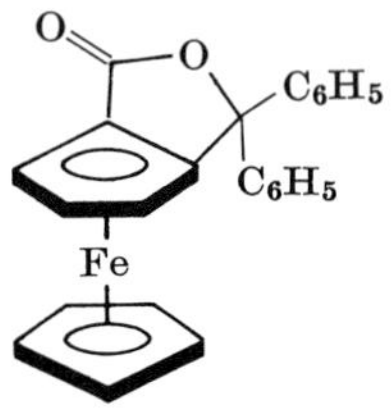  (24)

The cyclic ether (**25**) has been prepared via the cyclization of **26** in monoglyme with potassium amide.[14, 15] The magnetic nonequivalence of the methylene group protons in **25** have been compared with other 1,2-disubstituted ferrocenes.[16]

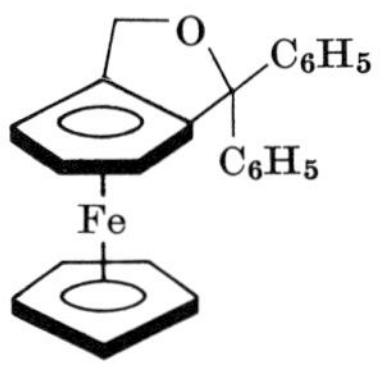

(25) 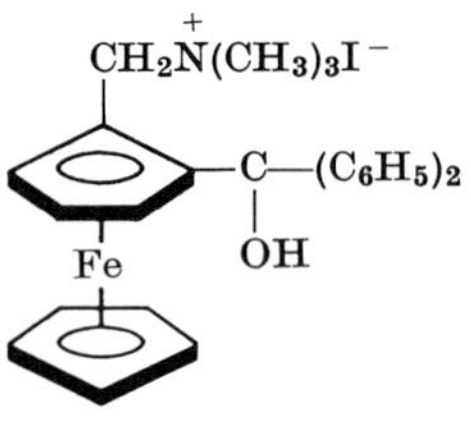  (26)

3. *Miscellaneous Ferroceno Heterocyclic Compounds*

Treatment of the acid chloride of $S$-ferrocenylmethylthioglycolic acid with stannic chloride in methylene chloride at $-70°$ gave 1,2-(2-thia-4-oxotetramethylene)ferrocene (**27**).[17] Reaction of **27** with Raney nickel led to ring opening and gave a mixture of 1-methyl-2-ethyl-

[12] J. H. Richards and T. J. Curphey, *Chem. Ind.* (*London*) 1456 (1956).
[13] R. A. Benkeser, W. P. Fitzgerald, and M. S. Melzer, *J. Org. Chem.* **26**, 2569 (1961).
[14] D. W. Slocum, B. W. Rockett, and C. R. Hauser, *Chem. Ind.* (*London*) 1831 (1964).
[15] D. W. Slocum, B. W. Rockett, and C. R. Hauser, *J. Amer. Chem. Soc.* **87**, 1241 (1965).
[16] P. Smith, J. J. McLeskey, III, and D. W. Slocum, *J. Org. Chem.* **30**, 4356 (1965).
[17] A. N. Nesmeyanov, E. G. Perevalova, L. I. Leonteva, and Yu. A. Ustynyuk, *Izv. Akad. Nauk. SSSR, Ser. Khim.* 1882 (1965); *Chem. Abstr.* **64**, 2123 (1966).

ferrocene and 1-methyl-2-acetylferrocene. Reduction of **27** with lithium aluminum hydride gave the corresponding alcohol (**28**), whereas use of aluminum chloride with the lithium aluminum hydride gave 30% methylferrocene, 5% **28**, and 40% 1,2-(2-thiatetramethylene)ferrocene (**29**).[17]

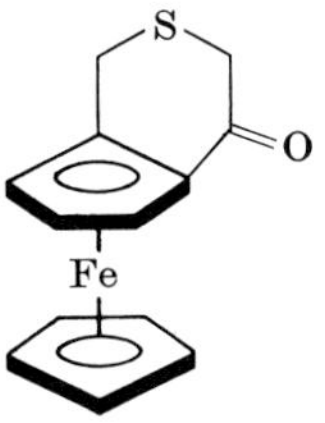

(**27**) 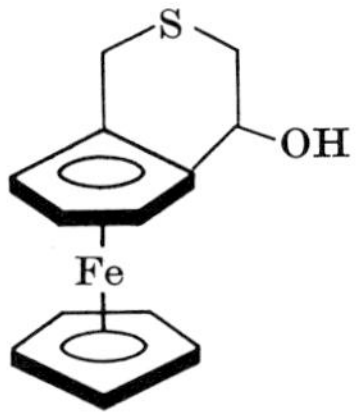   (**28**) 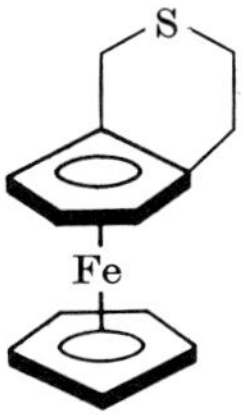   (**29**)

## B. Bridged Heterocyclic Compounds

### 1. *Compounds with Nitrogen in the Bridge*

Reaction of the heteroannular diol (**30**) with phenylisocyanate or *p*-methoxyphenylisocyanate at room temperature gave the expected urethans, but use of these isocyanates at elevated temperature or use of *p*-nitrophenylisocyanate at room or elevated temperature gave the bridged amine (**31**; R = H, OCH$_3$, and NO$_2$).[18] The NMR spectrum of **31** shows a triplet for the ferrocene ring protons.[19] This triplet is attributed to the direct inductive influence on the α-ring proton of the bridge nitrogen atom.

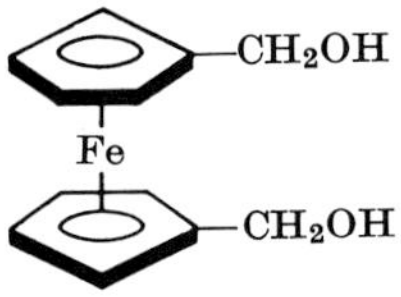

(**30**) 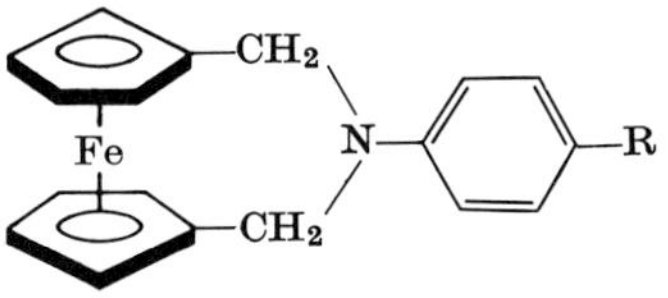   (**31**)

Reduction of the monooxime of 1,1'-diacetylferrocene with lithium aluminum hydride–aluminum chloride gave a mixture of products which apparently included the bridged amine (**32**).[11] A compound,

[18] H. J. Lorkowski and P. Kieselack, *Chem. Ber.* **99**, 3619 (1966).
[19] H. J. Lorkowski, G. Engelhart, P. Kieselack, and H. Jancke, *J. Organometal. Chem.* **7**, 523 (1967).

whose analysis was in approximate agreement with the bridged structure (33), was obtained from the reaction of o-phenylenediamine and ferrocene-1,1'-dicarboxaldehyde.[20]

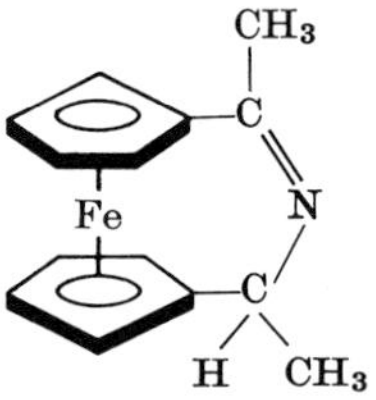

(32)

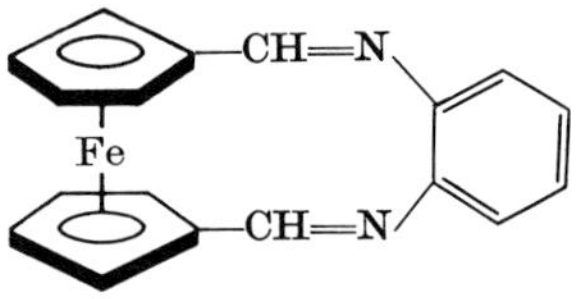

(33)

### 2. Compounds with Oxygen in the Bridge

Just as homoannular ferrocenedicarboxylic acid leads to a cyclic anhydride (23), so does the heteroannular ferrocenedicarboxylic acid. Thus treatment of the dichloride of ferrocene-1,1'-dicarboxylic acid with water in chloroform–pyridine gave the anhydride (34).[21] Treatment of 34 with ammonium hydroxide led to ring opening and formation of the diamide and the monoamide–monoacid.[21]

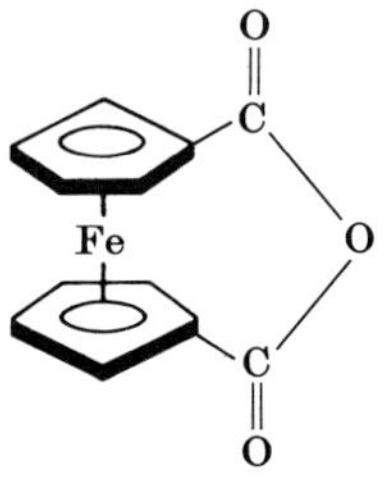

(34)

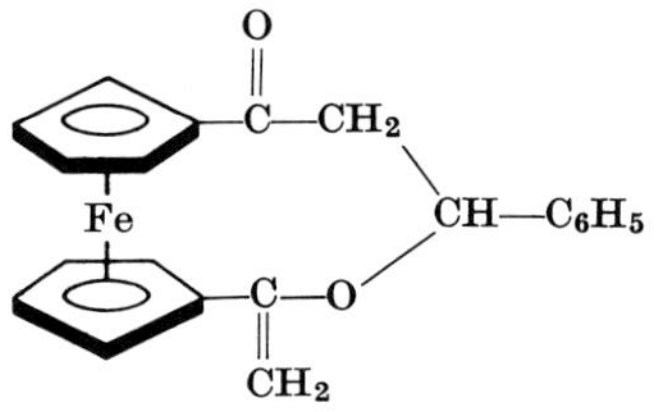

(35)

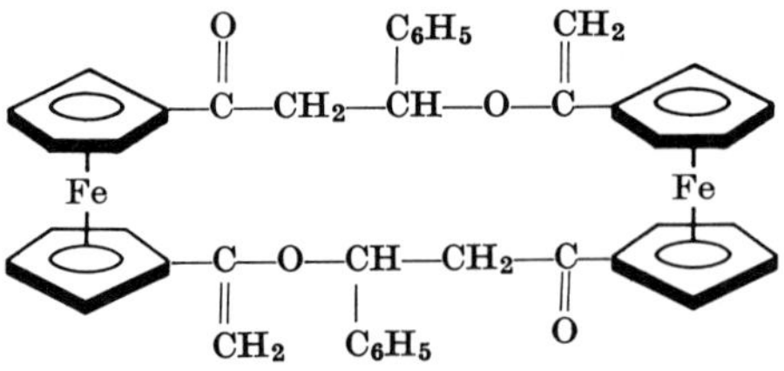

(36)

[20] J. M. Osgerby and P. L. Pauson, *J. Chem. Soc.* 4604 (1961).
[21] A. N. Nesmeyanov and O. A. Reutov, *Dokl. Akad. Nauk SSSR* **120**, 1267 (1958); *Chem. Abstr.* **53**, 1292 (1959).

Reaction of 1,1'-diacetylferrocene and benzaldehyde with 5% aqueous–ethanolic sodium hydroxide gave the expected dibenzal derivative and a yellow compound believed to be **35** or **36**.[22]

The remaining heterocyclic compounds with oxygen in the bridge are ethers of the types **37** and **38**. The parent compound (**37**; R = R' = H) has been obtained by treatment of 1,1'-di(hydroxymethyl)ferrocene with *p*-toluenesulfonyl chloride in refluxing benzene[23] or with 10% hydrochloric acid in ether.[24] The ring protons appear as a singlet in the NMR spectrum of **37** (R = R' = H).[23] In a similar manner other diols have been dehydrated with *p*-toluenesulfonyl chloride, acid alumina, dilute hydrochloric acid, dilute acetic acid, and silica to give **37** (R = R' = $CH_3$[25–32]; R = R' = $C_6H_5$[25–27,]

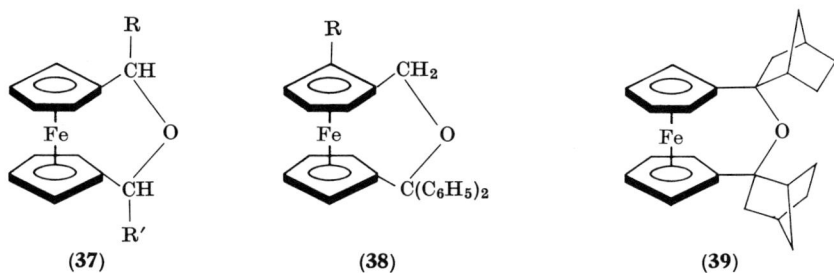

(37)  (38)  (39)

---

[22] T. A. Mashburn, Jr., C. E. Cain, and C. R. Hauser, *J. Org. Chem.* **25**, 1982 (1960).

[23] K. L. Rinehart, Jr., A. K. Frerichs, P. A. Kittle, L. F. Westman, D. H. Gustafson, R. L. Pruett, and J. E. McMahon, *J. Amer. Chem. Soc.* **82**, 4111 (1960).

[24] A. N. Nesmeyanov, S. S. Churanov, Yu. A. Ustynyuk, and E. G. Perevalova, *Izv. Akad. Nauk SSSR, Ser. Khim.* 1648 (1966); *Chem. Abstr.* **66**, 65613j (1967).

[25] P. L. Pauson, M. A. Sandhu, and W. E. Watts, *J. Chem. Soc. C* 251 (1966).

[26] K. Schlögl and A. Mohar, *Naturwissenschaften* **48**, 376 (1961).

[27] K. Schlögl and A. Mohar, *Monatsh. Chem.* **92**, 219 (1961).

[28] T. A. Mashburn, Jr. and C. R. Hauser, *J. Org. Chem.* **26**, 1671 (1961).

[29] E. C. Winslow and E. W. Brewster, *J. Org. Chem.* **26**, 2982 (1961).

[30] K. Yamakawa, H. Ochi, and K. Arakawa, *Chem. Pharm. Bull.* **11**, 905 (1963).

[31] K. Yamakawa and M. Hisatome, *Tetrahedron Lett.* 2827 (1967).

[32] J. T. Suh, U.S. Patent 3,408,376 (1968); *Chem. Abstr.* **71**, 50229t (1969).

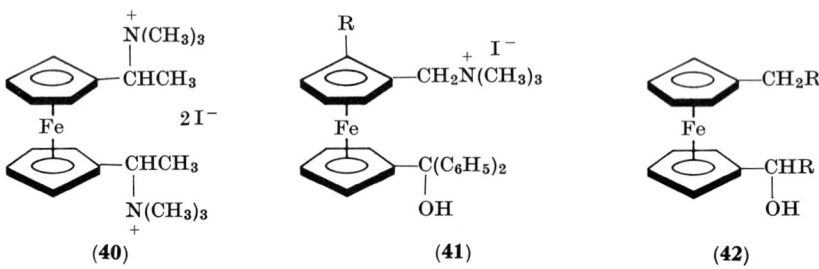

(40)            (41)            (42)

[31-35]; $R = R' = n\text{-}C_5H_{11}$[32,33]; $R = CH_3$, $R' = C_6H_5$[32]; $R = CH_3$, $R' = C_2H_5$[32,33]; and $R = CH_3$, $R' = H$[36]). In a similar manner the appropriate diol with p-toluenesulfonyl chloride gave **39**.[37] The cyclic ether (**37**; $R = R' = CH_3$) has also been obtained[25,38] by hydrolysis of the dimethiodide (**40**). The methiodide (**41**) has been cyclized to the ether (**38**; $R = H$) by action of sodamide in dimethylformamide.[39] With **41** [$R = C(C_6H_5)_2OH$] the ether [**38**; $R = C(C_6H_5)_2OH$] was obtained[39] rather than an ether of the type **25**.

The ethers (**37**; $R = R' = CH_3$ or $C_6H_5$) can be converted into low yields of diols by prolonged treatment with acidic alumina.[31] Treatment of **37** ($R = R' = H$ or $CH_3$) with lithium aluminum hydride–aluminum chloride gave 1,1'-dimethylferrocene and 1,1'-diethylferrocene, respectively.[40] Action of lithium in tetrahydrofuran on **37** ($R = R' = H$)[24] or ($R = R' = CH_3$ or $C_6H_5$)[34] led after hydrolysis to ring-opened products of the type **42** ($R = H$, $CH_3$, and $C_6H_5$, respectively). In the case of **37** ($R = R' = C_6H_5$) some 1,1'-dibenzylferrocene and other products were obtained in addition to **42** ($R = C_6H_5$).[34] The visible and ultraviolet absorption spectra of **37** and **38** have been reported[41] and compared with other ferrocene derivatives.

[33] J. E. Robertson, U.S. Patent 3,377,248 (1968); *Chem. Abstr.* **69**, 96878r (1968).
[34] A. N. Nesmeyanov, E. G. Perevalova, Yu. A. Ustynyuk, N. S. Prozorova, T. I. Tatashina, and S. S. Churanov, *Izv. Akad. Nauk SSSR, Ser. Khim.* 1646 (1966); *Chem. Abstr.* **66**, 65614k (1967).
[35] A. N. Nesmeyanov, I. I. Kritskaya, and T. V. Antipina, *Izv. Akad. Nauk SSSR, Otd. Khim. Nauk* 1777 (1962); *Chem. Abstr.* **58**, 7971 (1963).
[36] K. Schlögl, M. Peterlik, and H. Seiler, *Monatsh. Chem.* **93**, 1309 (1962).
[37] M. J. A. Habib and W. E. Watts, *J. Chem. Soc. C* 1469 (1969).
[38] G. R. Knox, J. D. Munro, P. L. Pauson, G. H. Smith, and W. E. Watts, *J. Chem. Soc.* 4319 (1961).
[39] E. S. Bolton, P. L. Pauson, M. A. Sandhu, and W. E. Watts, *J. Chem. Soc. C* 2260 (1969).
[40] K. Schlögl, A. Mohar, and M. Peterlik, *Monatsh. Chem.* **92**, 921 (1961).
[41] T. H. Barr and W. E. Watts, *J. Organometal. Chem.* **15**, 177 (1968).

The separation of stereoisomers of **37** (R = R' = CH$_3$ or C$_6$H$_5$) has been reported[25] and a detailed discussion of the stereochemistry of the system has appeared.[31]

### 3. *Miscellaneous Bridged Heterocyclic Compounds*

Photolysis of ferrocenylsulfonyl azide in hydrocarbon solvents using 3500 Å radiation yields the novel bridged ferrocene derivative (**43**).[42] Methylation of **43** with dimethyl sulfate and sodium hydroxide gave the *N*-methyl derivative. The trithia system (**44**) has been reported[43] from the reaction of 1,1'-dilithioferrocene with sulfur.

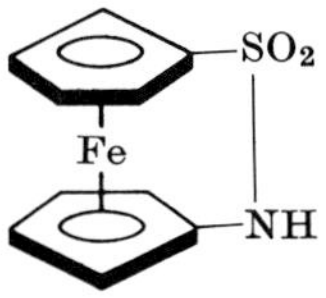

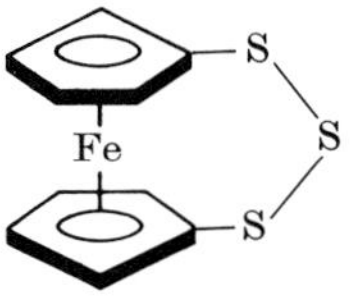

A number of compounds containing silicon–oxygen bridges have been prepared.[44, 45] For example, acid-catalyzed hydrolysis of **45** or reaction of **46** with butyllithium and ferrous chloride gave **47**.[44] Reaction of chlorosilanes and 1,1'-dilithioferrocene can lead to compounds such as **48** and **49**.[46] Although **48** is stable at 300°[46] the

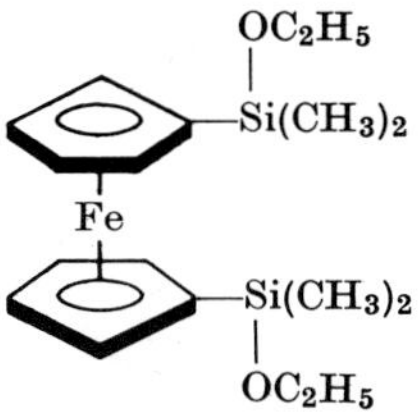

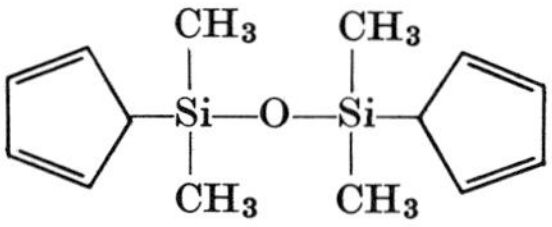

---

[42] R. A. Abramovitch, C. I. Azogu, and R. G. Sutherland, *Chem. Commun.* 1439 (1969).
[43] A. Davison and J. C. Smart, *J. Organometal. Chem.* **19**, P7 (1969).
[44] R. L. Schaaf, P. T. Kan, and C. T. Lenk, *J. Org. Chem.* **26**, 1790 (1961).
[45] M. Kumada, K. Mimura, M. Ishikawa, and K. Shiina, *Tetrahedron Lett.* 83 (1965).
[46] M. Kumada, H. Tsunemi, and S. Iwasaki, *J. Organometal. Chem.* **10**, 111 (1967).

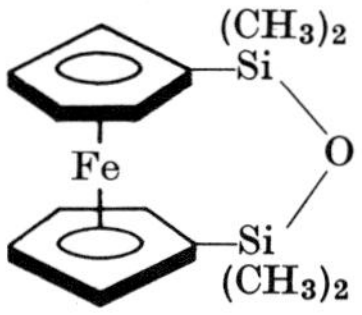

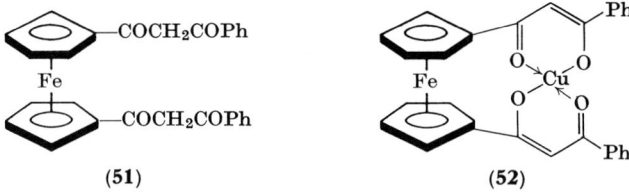

bridged silicon compound (**49**; R = H or CH₃) has been reported[47] to undergo a thermal rearrangement to **50**. Although they may be considered beyond the scope of this review, it should be noted that metals have been used to bridge the gap between groups attached to the ferrocene rings. For example, **51** gives the copper chelate (**52**).[48]

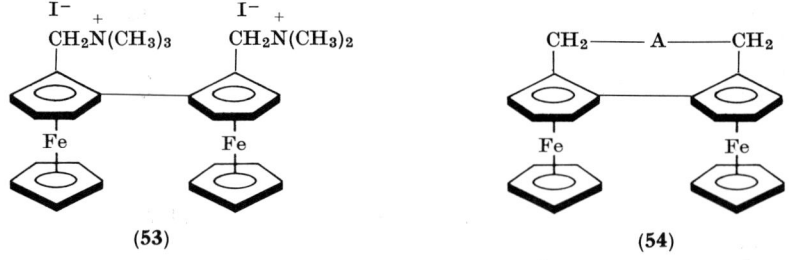

The dimethiodide (**53**) has been used to prepare a number of bridged structures of the type **54**.[49] Thus **53** with aqueous acid or base gave the ether (**54**; A = O), while the sulfur (**54**; A = S) and nitrogen (**54**; A = NCH₂C₆H₅) analogs were prepared by reaction of **53** with sodium sulfide and benzylamine, respectively.[49]

A series of compounds with sulfur in the bridge have been prepared[50]

[47] M. Kumada, M. Ogura, H. Tsunemi, and M. Ishikawa, *Chem. Commun.* 207 (1969).
[48] C. R. Hauser and C. E. Cain, *J. Org. Chem.* **23**, 1142 (1958).
[49] G. Marr, R. E. Moore, and B. W. Rockett, *Tetrahedron* **25**, 3477 (1969).
[50] J. T. Suh, U.S. Patent 3,382,267 (1968); *Chem. Abstr.* **69**, 59378g (1968).

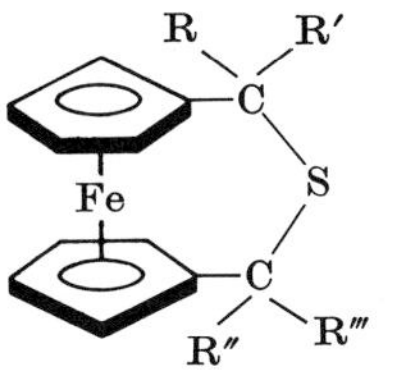

(55)

by reaction of the appropriate diol with hydrogen sulfide in the presence of a mineral acid. These products (55) are analogous to the ethers (37) noted earlier.

## III. Ferrocenyl Heterocyclic Compounds

Compounds containing a heteronitrogen are discussed in Section III,A; those with a nitrogen and another heteroatom are discussed in Section III,B; those with a heterooxygen or an oxygen and another heteroatom (other than nitrogen) are considered in Section III,C; and those with a heterosulfur (but no nitrogen or oxygen) are found in Section III,D. If more than one heterocyclic ring is present in a compound, only the ring closest to the ferrocene is used to decide on the appropriate section. Compounds which contain the heterocyclic ring attached directly to the ferrocene are discussed first in each section.

### A. Nitrogen-Containing Rings

A number of azetidinones have been prepared, with ferrocene attached to carbon (56) and to nitrogen (57).[51] These azetidinones were prepared by a Reformatsky-type reaction of aldimines with zinc and ethyl bromoacetate.

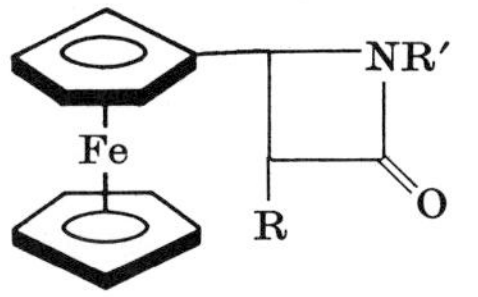

(56) 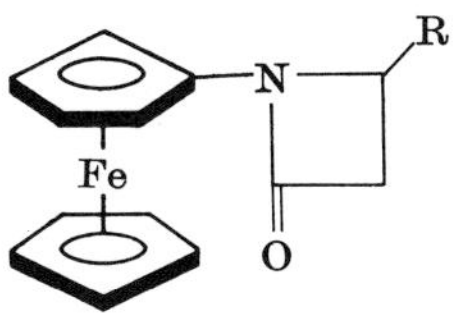   (57)

[51] E. Cuingnet, D. Poulain, and M. Tarterat-Adalberon, *Bull. Soc. Chim. Fr.* 514 (1969).

Some 3-ferrocenyl- (58) and 5-ferrocenylpyrazolines (59) have been prepared by the action of hydrazine or phenylhydrazine on $\alpha,\beta$-unsaturated ketones[52] or by the action of phenylhydrazines on Mannich bases.[53,54] The infrared,[52,53] NMR,[52] and mass spectra of some of these pyrazolines are discussed. A variety of $\beta$-diketones have been reacted with hydrazine[55-59] to yield pyrazoles of the type 60, including one in which R was another ferrocenyl group.[57] pH values have been measured for methanolic solutions of salts of some of these pyrazoles.[59] In a similar manner bisdiketones gave rise to bispyrazoles (61).[48,60] The parent 3-ferrocenylpyrazole (60; R = H) has been prepared by the action of hydrazine on a variety of ferrocene derivatives,[61-63] while the parent bis compound (61; R = H) has been prepared from the bisketoaldehyde.[62] The potentiometric titration of 60 (R = H) and other ferrocenes in aqueous acetic acid with potassium dichromate has been reported[64] to involve a one-electron change.

(58)

(59)

[52] F. D. Popp and E. B. Moynahan, *J. Heterocycl. Chem.* **7**, 351 (1970).
[53] C. R. Hauser, R. L. Pruett, and T. A. Mashburn, Jr., *J. Org. Chem.* **26**, 1800 (1961).
[54] M. Furdik, P. Elecko, and S. Kovac, *Chem. Zvesti* **19**, 371 (1965); *Chem. Abstr.* **63**, 7041 (1965).
[55] C. R. Hauser and J. K. Lindsay, *J. Org. Chem.* **22**, 482 (1957).
[56] I. I. Grandberg and A. N. Kost, *Zh. Obshch. Khim.* **32**, 3025 (1962); *Chem. Abstr.* **58**, 8881 (1963).
[57] L. Wolf and H. Hennig, *Z. Chem.* **3**, 469 (1963).
[58] L. Wolf and H. Hennig, *Z. Anorg. Allgem. Chem.* **341**, 1 (1965).
[59] I. I. Grandberg, *Zh. Obshch. Khim.* **32**, 3029 (1962); *Chem. Abstr.* **58**, 8882 (1963).
[60] C. E. Cain, T. A. Mashburn, Jr., and C. R. Hauser, *J. Org. Chem.* **26**, 1030 (1961).
[61] K. Schlögl and A. Mohar, *Monatsh. Chem.* **93**, 861 (1962).
[62] K. Schlögl and H. Egger, *Monatsh. Chem.* **94**, 1054 (1963).
[63] K. Schlögl and W. Steyrer, *Monatsh. Chem.* **96**, 1520 (1965).
[64] M. Peterlik and K. Schlögl, *Z. Anal. Chem.* **195**, 113 (1963).

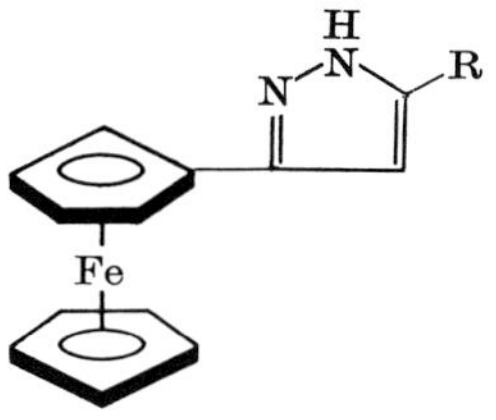

(60)

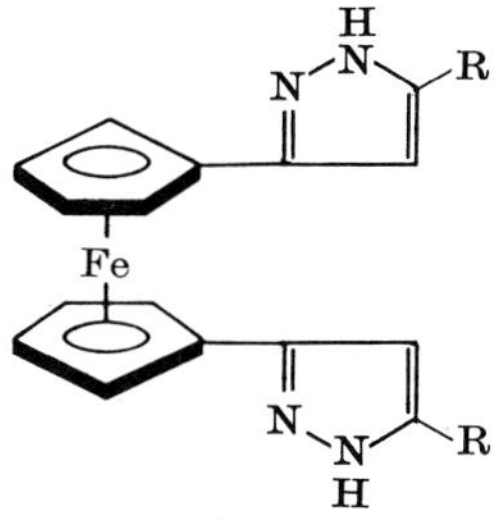

(61)

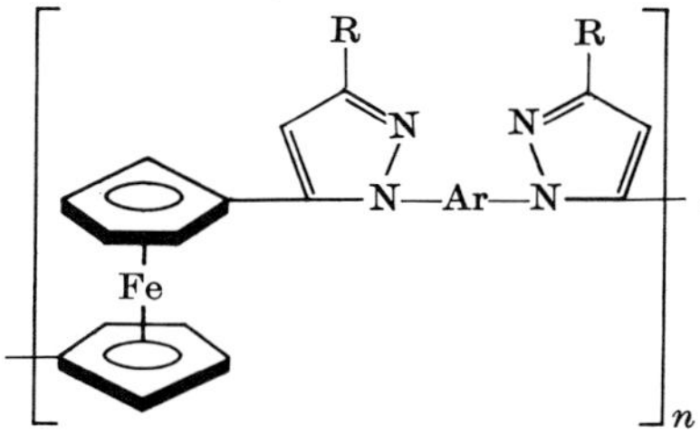

(62)

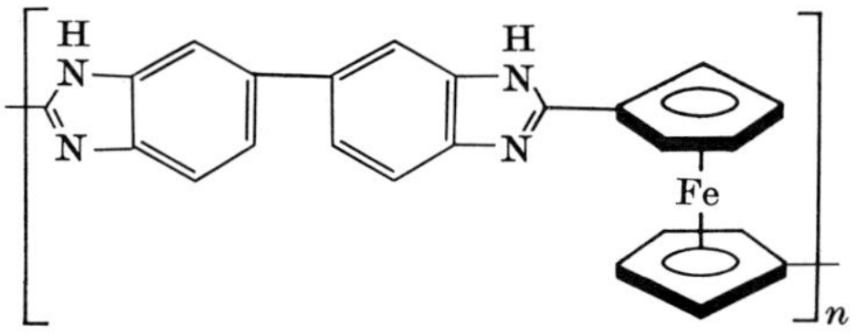

(63)

Polypyrazoles of the type **62** have been prepared.[65] These polymers are formed from the low-temperature solution condensation of 1,1'-bis(diketo)ferrocenes with aromatic or pseudoaromatic dihydrazines to give polyhydrazones which are transformed thermally to the polypyrazoles. Polybenzimadazole (**63**) has been prepared[66] from 1,1'-ferrocenedicarboxylic acid and 3,3'-diaminobenzidine.

Reaction of phenylhydrazine with the β-keto ester (**64**) gives the pyrazolone (**65**).[55] The hydantoin (**66**) has been obtained by the action of potassium cyanide and ammonium carbonate on ferrocenecarboxaldehyde.[67-69] The potentiometric titration[64] of hydantoin (**67**) has been reported.

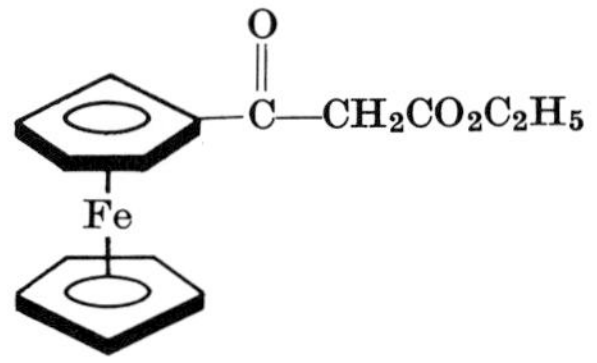

(64)

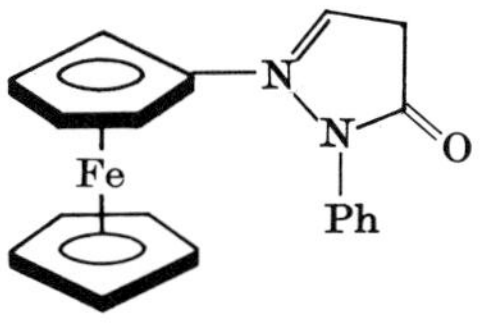

(65)

[65] E. W. Neuse, *Macromolecules* **1**, 171 (1968).
[66] L. Plummer and C. S. Marvel, *J. Polym. Sci. Part A* **2**, 2559 (1964).
[67] P. J. Graham, R. V. Lindsey, G. W. Parshall, M. L. Peterson, and G. M. Whitman, *J. Amer. Chem. Soc.* **79**, 3416 (1957).
[68] K. Schlögl, *Monatsh. Chem.* **88**. 601 (1957).
[69] B. Loev and M. Flores, *J. Org. Chem.* **26**, 3595 (1961).

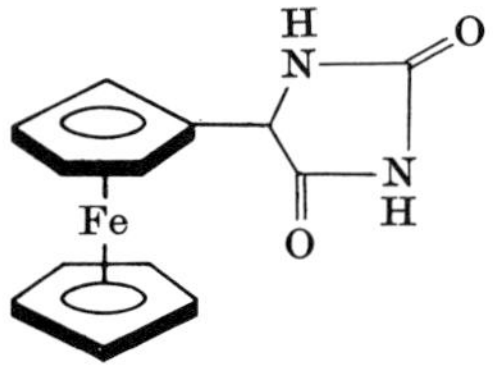

(66) 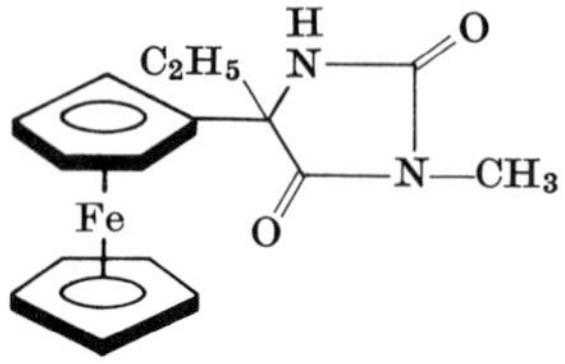     (67)

The reaction of ferrocenylazide and dimethyl *exo-cis*-3,6-endoxo-$\Delta^4$-tetrahydrophthalate gives the ferrocenyl compound **68**.[70] Reaction of cyanoferrocene with trimethylsilyl azide and aluminum chloride gave ferrocenyltetrazole (**69**) as the sole organometallic product.[71] Alkylation of **69** with methyl iodide affords two *N*-methyl derivatives. The major product is assigned the structure 2-methyl-5-ferrocenyltetrazole, while the minor product is assumed to be the 1-methyl isomer.

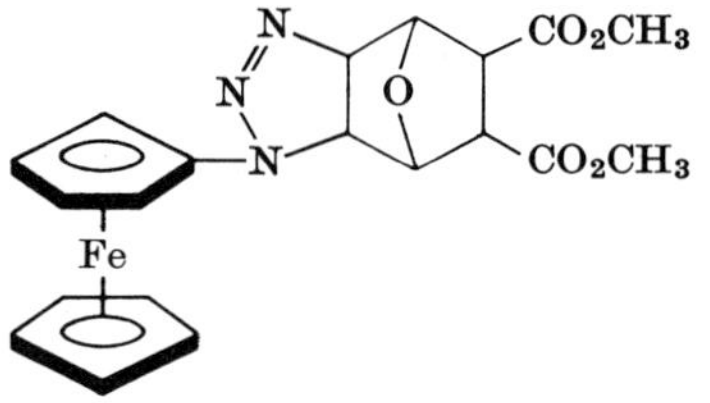

(68) 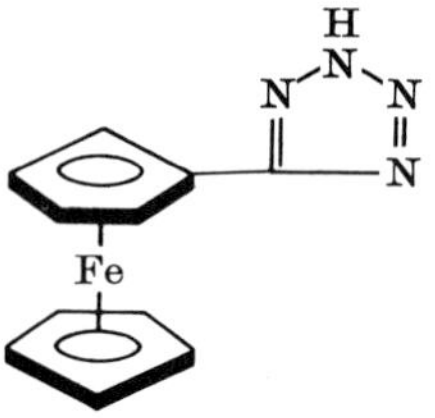     (69)

A number of reactions similar to those noted above have been used to prepare six-membered rings. The ketoaldehyde used to prepare **60** (R = H) reacts with urea to give 4-ferrocenyl-2-pyrimidone (**70**).[62] The appropriate $\gamma$-diketone[72] and $\gamma$-keto acids[73] have reacted with hydrazine to give **71** and **72** (R = $CH_3$ or $C_6H_5$), respectively. The Michael addition of cyanoacetamide to **73** in the presence of base gave the compounds **74–76**.[74]* Ferrocenylphenylglyoxal and *o*-phenylene-

---

* Structure **76** is as given in *Chem. Abstr.*

[70] A. N. Nesmeyanov, V. N. Drozd, and V. A. Sazonova, *Dokl. Akad. Nauk SSSR* **150**, 321 (1963); *Chem. Abstr.* **59**, 5196 (1963).

[71] S. S. Washburne and W. R. Peterson, Jr., *J. Organometal. Chem.* **21**, 427 (1970).

[72] N. Sugiyama and T. Teitei, *Bull. Chem. Soc. Japan* **35**, 1423 (1962).

[73] J. Tirouflet, B. Gautheron, and R. Dabard, *Bull. Soc. Chim. Fr.* 96 (1965).

[74] M. Furdik and S. Toma, *Chem. Zvesti* **20**, 3 (1966); *Chem. Abstr.* **66**, 105041 (1967).

diamine yielded 2-ferrocenyl-3-phenylquinoxaline (**77**).[75] Condensation of ferrocenecarboxaldehyde with 1,8-diaminonaphthalene gave the 2,3-dihydro-1*H*-perimidine (**78**) which was dehydrogenated to **79**.[76]

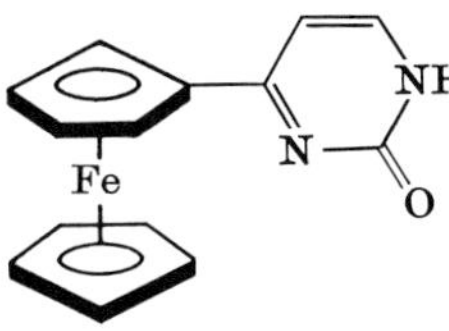

(**70**) 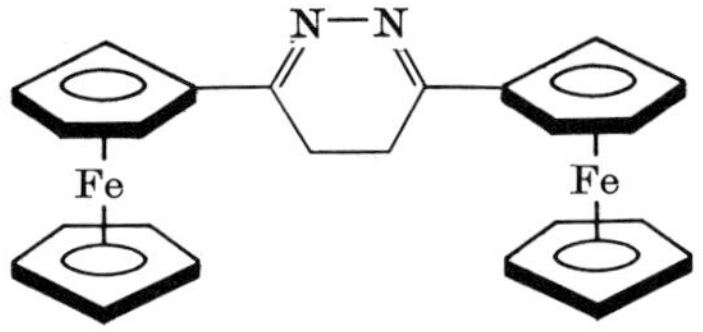  (**71**)

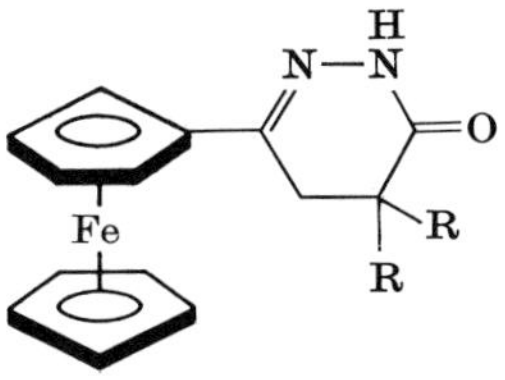

(**72**) 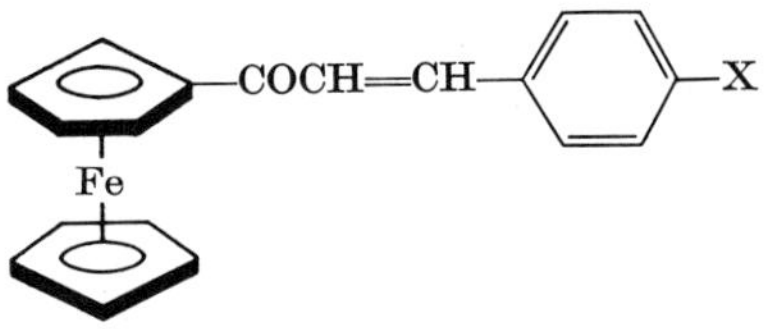  (**73**)

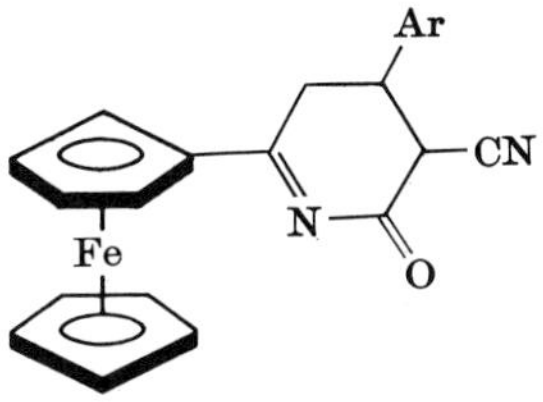

(**74**) 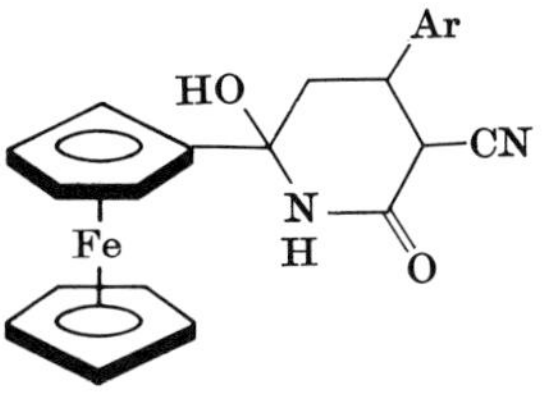  (**75**) 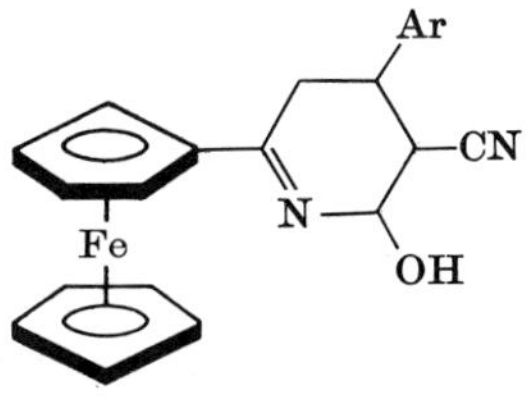  (**76**)

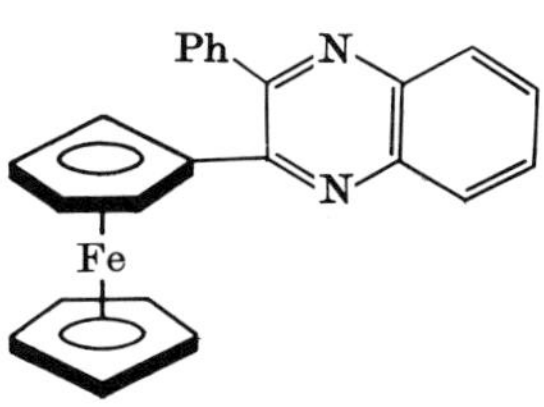

(**77**) 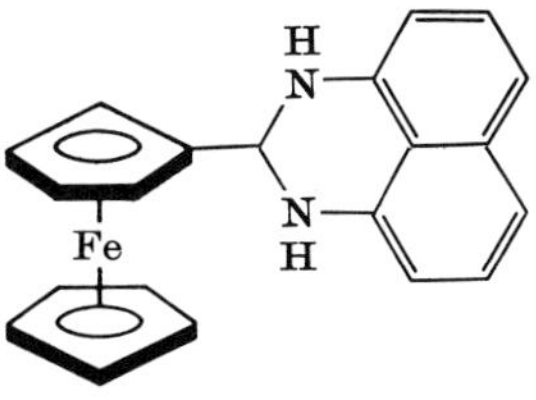  (**78**) 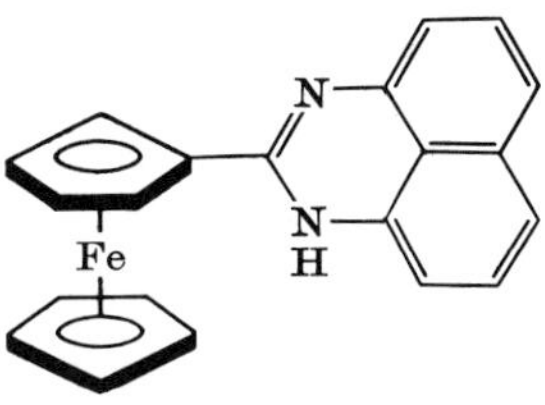  (**79**)

Reaction of pyrrolemagnesium bromide with bromoferrocene and cuprous bromide gave a mixture of 15% 2-pyrrylferrocene (**80**) and

[75] M. D. Rausch and A. Siegel, *J. Org. Chem.* **33**, 4545 (1968).
[76] F. D. Popp and E. B. Moynahan, *J. Heterocycl. Chem.* **7**, 739 (1970).

11% 3-isomer (**81**),[77] whereas use of sodiopyrrole and bromoferrocene with cuprous bromide gave[78] the *N*-ferrocenyl compound **82** and biferrocene. Under similar conditions potassium tetra-1-indolylborate gave **83**[78] and the indole Grignard reagent gave a mixture of **83** and **84**.[77] Lithium piperidide and chloroferrocene in the presence of butyllithium gave a number of products including 1-ferrocenyl-piperidine (**85**).[79]

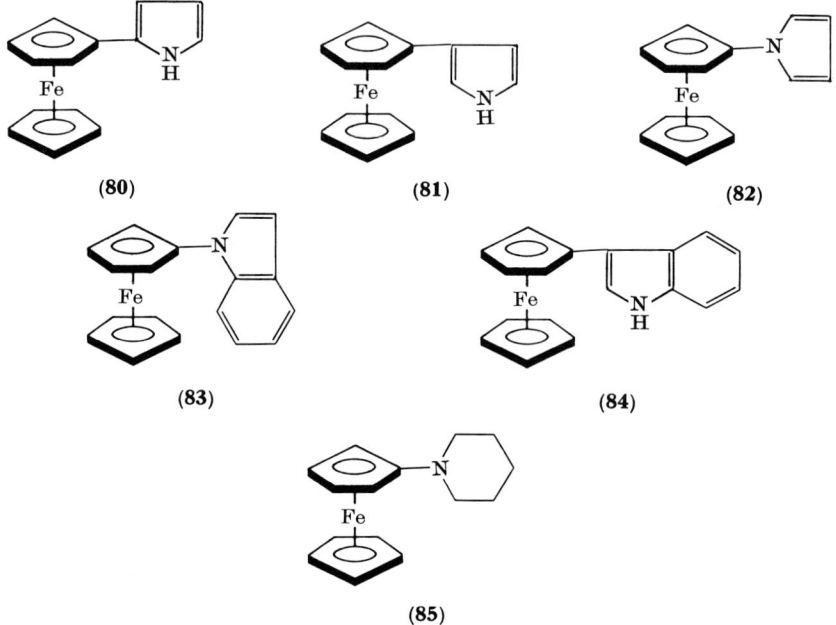

(80)   (81)   (82)
(83)   (84)
(85)

Reaction of the lithio derivative of ferrocene with pyridine gave α-pyridylferrocene (**86**) together with a small amount of **87** (from dilithioferrocene formed during the preparation of lithioferrocene).[80, 81] A small yield of **86** was also obtained[80] when 1,1'-ferro-

[77] A. N. Nesmeyanov, V. A. Sazonova, and V. N. Drozd, *Dokl. Akad. Nauk SSSR*, **165**, 575 (1965); *Chem. Abstr.* **64**, 6687 (1966).
[78] A. N. Nesmeyanov, V. A. Sazonova, and V. N. Drozd, *Dokl. Akad. Nauk SSSR* **154**, 158 (1964); *Chem. Abstr.* **60**, 9309 (1964).
[79] J. W. Huffman, L. H. Keith, and R. L. Asbury, *J. Org. Chem.* **30**, 1600 (1965).
[80] A. N. Nesmeyanov, V. A. Sazonova, and A. V. Gerasimenko, *Dokl. Akad. Nauk SSSR* **147**, 634 (1962); *Chem. Abstr.* **58**, 9133 (1963).
[81] K. Schlögl and M. Fried, *Monatsh. Chem.* **94**, 537 (1963).

cenylenediboronic acid and copper carbonate were heated in pyridine. Irradiation of the methiodide of **86** gave an 84% yield of N-methyl-2-cyclopentadienylene-1,2-dihydropyridine and some cyclopentadiene.[82] Catalytic hydrogenation of **86** and **87** gave the corresponding piperidyl derivatives.[81] The lithio derivative of methylferrocene and pyridine gave a mixture of 1-α-pyridyl-1′-methylferrocene, 1-α-pyridyl-2-methylferrocene, and 1-α-pyridyl-3-methylferrocene,[81] while lithioferrocene and quinoline gave **88** and **89**.[81, 83] As with **86**, irradiation of the methiodide of **88** gave N-methyl-2-cyclopentadienylidene-1,2-dihydroquinoline.[83]

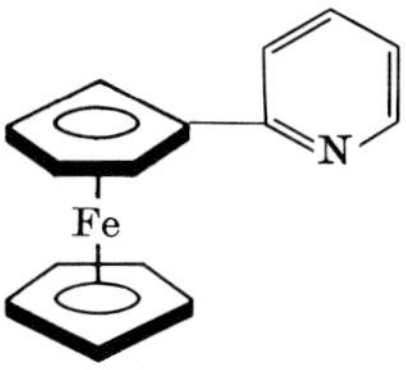

(86)

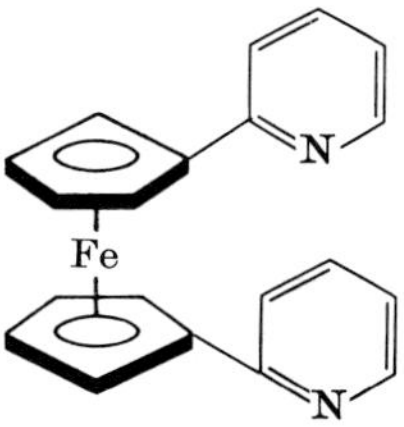

(87)

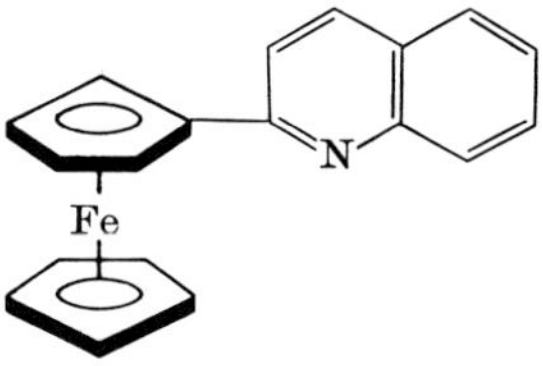

(88)

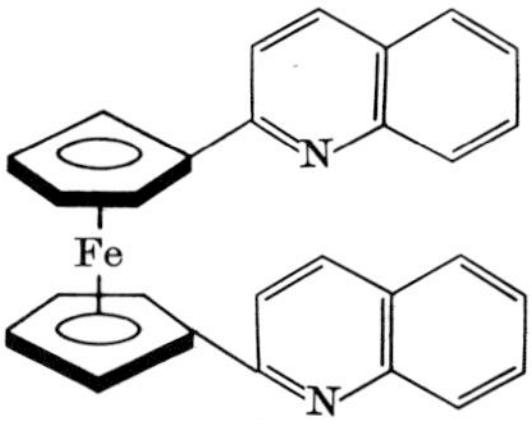

(89)

The reaction of **86** with n-butyllithium gave some reaction in the pyridine ring, but resulted mainly in lithiation of the 2-position of the ferrocene.[84, 85] This lithiated intermediate has proven useful for the synthesis of some 2-substituted (2-pyridyl)ferrocenes.[86]

[82] A. N. Nesmeyanov, V. A. Sazonova, A. V. Gerasimenko, and N. S. Sazonova, *Dokl. Akad. Nauk SSSR* **149**, 1354 (1963); *Chem. Abstr.* **59**, 3460 (1963).
[83] A. N. Nesmeyanov, V. A. Sazonova, V. I. Romanenko, N. A. Rodionova, and G. P. Zolnikova, *Dokl. Akad. Nauk SSSR* **155**, 1130 (1964); *Chem. Abstr.* **61**, 1891 (1964).
[84] D. J. Booth and B. W. Rockett, *Tetrahedron Lett.* 1483 (1967).
[85] D. J. Booth and B. W. Rockett, *J. Chem. Soc. C* 656 (1968).
[86] D. J. Booth, G. Marr, B. W. Rockett, and A. Rushworth, *J. Chem. Soc. C* 2701 (1969).

Treatment of ferrocene with the diazonium salt of 3-aminopyridine gave **90** together with some of the dipyridylferrocene.[81]

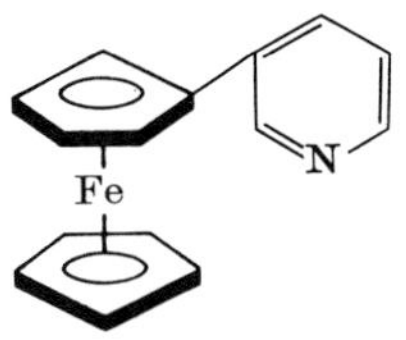

(90) 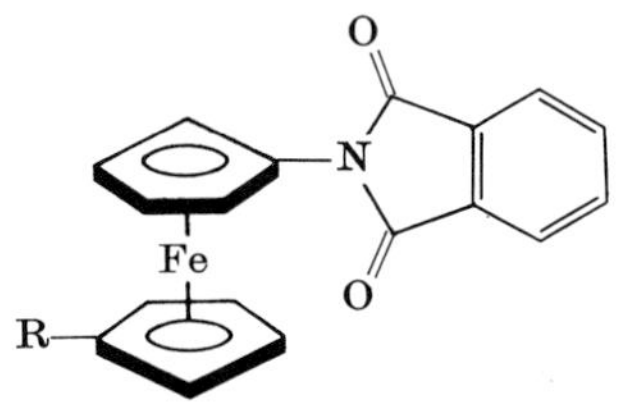     (91)

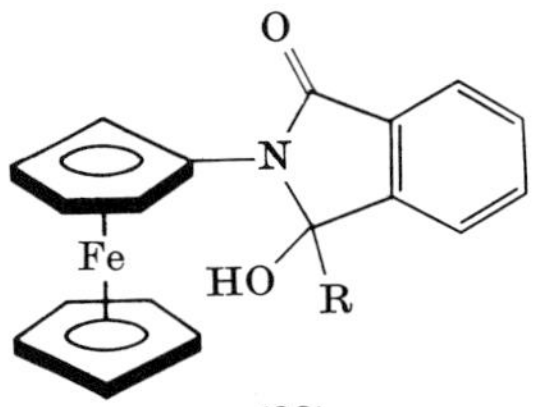

(92) 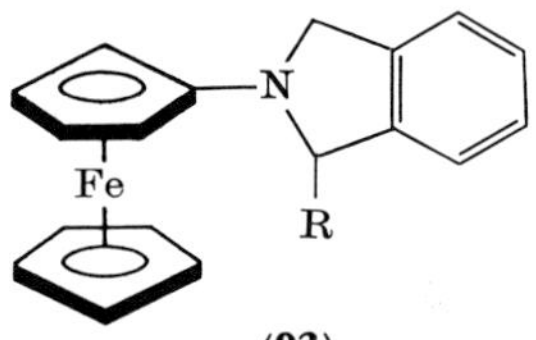     (93)

N-Ferrocenylphthalimide (**91**; R = H) has been prepared by the reaction of chloromercuriferrocene and copper phthalimide[87] or better by reaction of bromoferrocene and copper phthalimide at 140°.[88,89] Reaction of **91** (R = H) with hydrazine gives ferrocenylamine.[88,89] In a similar manner 1′-bromo-1-ethylferrocene gave **91** (R = $C_2H_5$) which can also be converted into the corresponding aminoferrocene.[90] Reaction of **91** (R = H) with various Grignard reagents gave **92** which was reduced by lithium aluminum hydride to **93**.[91] Photolysis of the methiodide of **93** (R = $CH_3$) gave cyclopentadiene and a 63% yield of 1,2-dimethylisoindolinium cyclopentadienylide, isolated as its tetraphenylborate.[91]

[87] V. A. Nefedov and M. N. Nefedova, *Zh. Obshch. Khim.* **36**, 122 (1966); *Chem. Abstr.* **64**, 14215 (1966).
[88] A. N. Nesmeyanov, V. A. Sazonova, and V. N. Drozd, *Dokl. Akad. Nauk SSSR* **130**, 1030 (1960); *Chem. Abstr.* **54**, 12089 (1960).
[89] A. N. Nesmeyanov, V. A. Sazonova, and V. N. Drozd, *Chem. Ber.* **93**, 2717 (1960).
[90] A. N. Nesmeyanov, V. A. Sazonova, and V. N. Drozd, *Dokl. Akad. Nauk SSSR* **137**, 102 (1961); *Chem. Abstr.* **55**, 21081 (1961).
[91] A. N. Nesmeyanov, V. N. Postnov, V. A. Sazonova, and V. A. Dobryak, *Izv. Akad. Nauk SSSR, Ser. Khim.* 2372 (1968); *Chem. Abstr.* **70**, 29031s (1969).

The Diels–Alder reaction of $N$-ferrocenylmaleimide (**94**) with a variety of dienes gave bicyclic adducts, such as **95**, which was obtained from furan.[92] All these adducts were endo because the exo position is sterically hindered by the ferrocenyl group on the imide nitrogen.

(**94**)          (**95**)

As might be expected, ferrocenecarboxaldehyde has been a favorite starting material for the preparation of a wide variety of heterocyclic ferrocenes. Compounds of the type **96** have been prepared from the reaction of 1-amino-2-ethyl-3,3-diphenylpyrrolidine,[93] 1-hydrazinophthalazine,[94] 3-aminocarbazole,[76] 3-(2-aminoethyl)indole,[76] 2-hydrazino-4-hydroxy-6-methylpyrimidine,[76] indole-3-acetic acid hydrazide,[76] 2-hydrazinopyridine,[67,76] 2-hydrazinoquinoline,[76] and isonicotinic acid hydrazide[76] with ferrocenecarboxaldehyde. Reaction of ferrocenecarboxaldehyde with 2-acetylpyrrole[95,96] and 2-acetylpyridine[96,97] gave **97**, whereas use of acetylferrocene with 2-formylpyrrole,[95,96] 2-, 3-, and 4-formylpyridine,[95,96] 2-formyl-4- and 2-formyl-5-nitropyrrole,[95,96] and 2-formylquinoline[96] gave the isomeric chalcones (**98**). A polarographic study of some of the compounds of the type **98** has appeared.[98] Ferrocenecarboxaldehyde also reacts with barbituric acid to give **99**,[67] with 1,2-dimethylpyridinium iodide to give **100**,[99] with a series of substituted

---

[92] M. Furdik, S. Toma, and J. Suchy, *Chem. Zvesti* **17**, 21 (1963); *Chem. Abstr.* **59**, 10116 (1963).
[93] J. W. Cusic and P. Yonan, U.S. Patent 3,395,144 (1968); *Chem. Abstr.* **69**, 96460y (1968).
[94] F. D. Popp and E. B. Moynahan, *J. Med. Chem.* **13**, 1020 (1970).
[95] J. Tirouflet and J. Boichard, *C. R. Acad. Sci.* **250**, 1861 (1960).
[96] J. Boichard, J. Monin, and J. Tirouflet, *Bull. Chim. Soc. Fr.* 851 (1963).
[97] J. Boichard and J. Tirouflet, *C. R. Acad. Sci.* **251**, 1394 (1960).
[98] J. Tirouflet, E. Laviron, J. Metzger, and J. Boichard, *Collect. Czech. Chem. Commun.* **25**, 3277 (1960).
[99] G. D. Broadhead, J. M. Osgerby, and P. L. Pauson, *J. Chem. Soc.* 650 (1958).

pyrroles to give methine dyes of the type **101**,[100] with 1,3,3-trimethyl-2-methylindolenine to give a methine dye,[101] with homophthalimides to give **102**, [96, 97] with 2-aminopyridine in the presence of phosphoryl chloride, followed by reduction with borohydride, to give **103**,[102] and with nitroalkanes and piperidine to give **104**.[103] A compound similar to **104** was also obtained with pyrrolidine.[102] The methine dyes (**101**) could also be obtained by condensation of ferrocene with pyrrolecarboxaldehydes.[100] Use of ferrocenecarboxaldehyde with the anion of Reissert compounds[104] derived from quinoline and isoquinoline

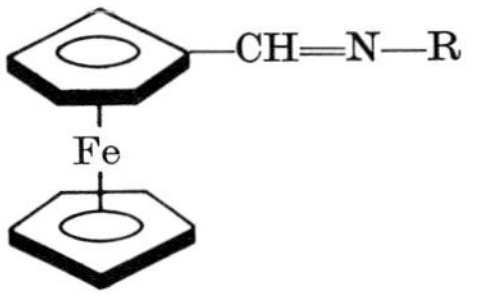

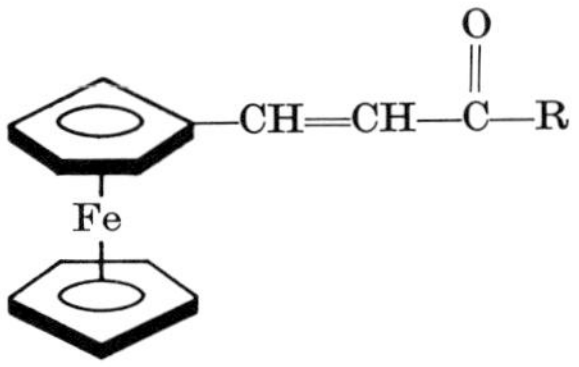

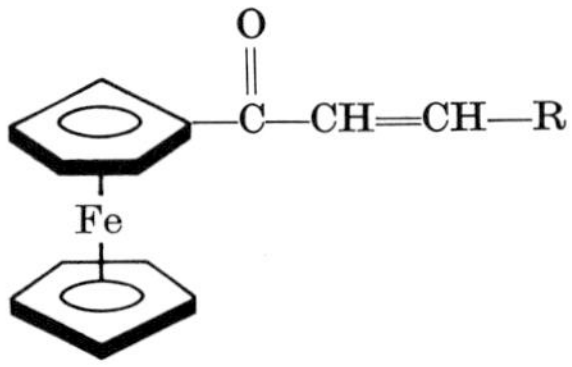

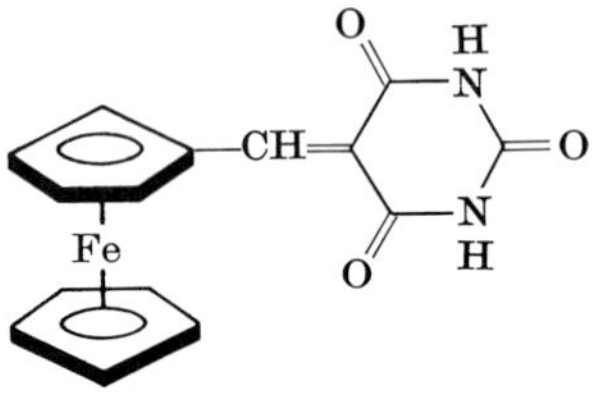

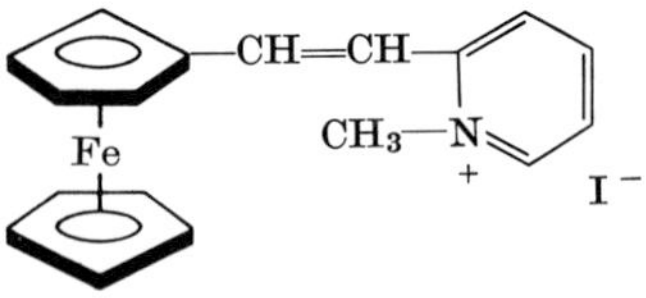

[100] A. Treibs and R. Zimmer-Galler, *Chem. Ber.* **93**, 2539 (1960).
[101] A. I. Titov, E. S. Lisitsyna, and M. R. Shemtova, *Dokl. Akad. Nauk SSSR* **130**, 341 (1960); *Chem. Abstr.* **54**, 10986 (1960).
[102] I. K. Barben, *J. Chem. Soc.* 1827 (1961).
[103] M. Shiga, H. Kono, I. Motoyama, and K. Hata, *Bull. Chem. Soc. Japan* **41**, 1897 (1968).
[104] F. D. Popp, *Advan. Heterocycl. Chem.* **9**, 1 (1968).

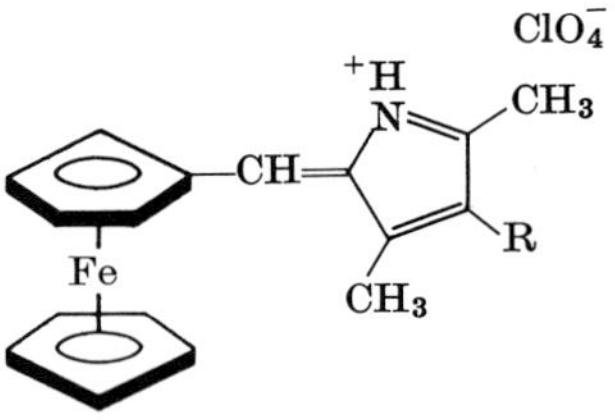

gave **105** and **106**, respectively.[76] Oxidation of **106** with manganese dioxide gave the corresponding ketone.[76]

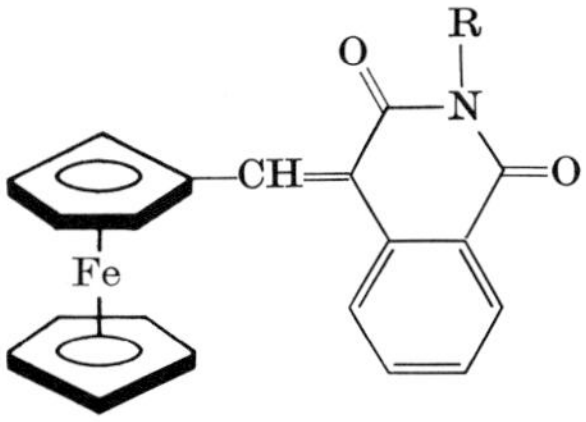

(102)

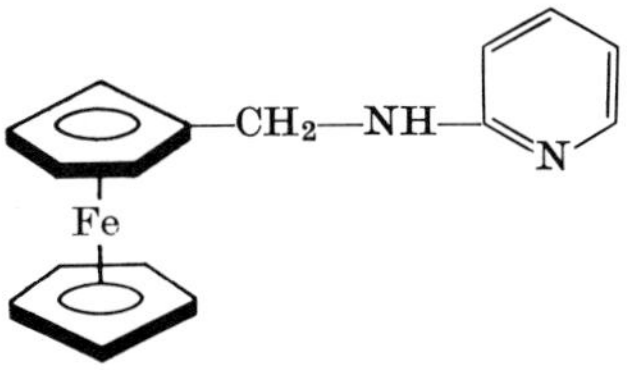

(103)

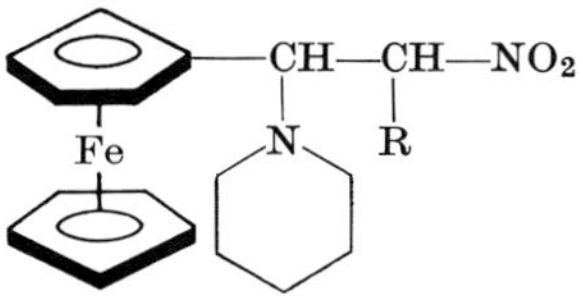

(104)

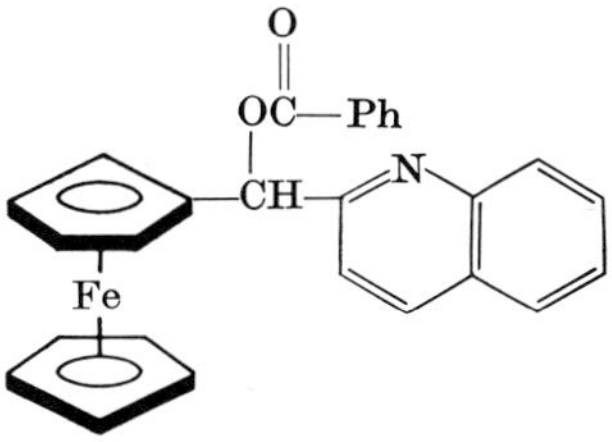

(105)

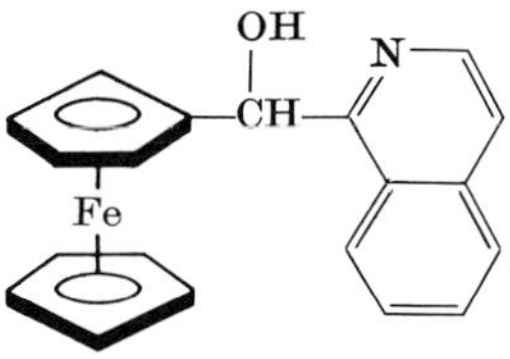

(106)

Reaction of the quaternary ammonium compound **107** with piperidine,[105] pyrrolidine,[106–107] and potassium phthalimide[105] gave compounds of the type **108–110**. Reaction of **110** with hydrazine gave (ferrocenylmethyl)amine.[105] Amine **108** was also obtained by reaction of ferrocene with methylenebispiperidine in the presence of phosphoric

[105] A. N. Nesmeyanov, E. G. Perevalova, L. S. Shilovtseva, and V. D. Tyurin, *Izv. Akad. Nauk SSSR, Otd. Khim. Nauk* 1997 (1962); *Chem. Abstr.* **58**, 9132 (1963).
[106] T. I. Bieber and M. Y. Dorsett, *J. Org. Chem.* **29**, 2028 (1964).
[107] M. Hadlington, B. W. Rockett, and A. Nelhans, *J. Chem. Soc. C* 1436 (1967).

acid,[108] while **109** was also obtained in addition to 2-(ferrocenylmethyl)cyclohexanone from **107** and N-cyclohexenylpyrrolidine in wet acetonitrile.[106] Metallation of **108** and **109** with n-butyllithium is a route to 1,2-disubstituted ferrocenes.[107] The asymmetric lithiation of 1-ferrocenylmethyl-2-methylpiperidine has been discussed.[108a]

The piperidine derivative (**111**) has been reported[109] from the reaction of piperidine and **112** in acetic acid with sodium tetraphenylborate. The piperidine derivative (**113**) arises from a displacement of

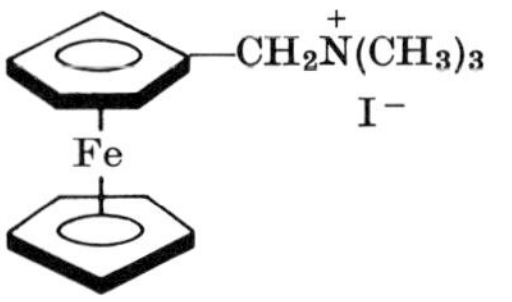

(**107**) Ferrocene-CH$_2$N$^+$(CH$_3$)$_3$ I$^-$

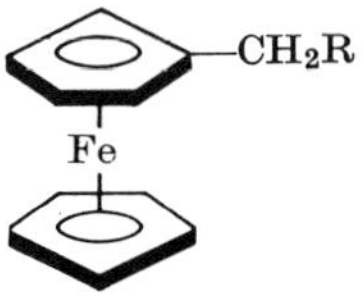

(**108**) R = —N(piperidine)

(**109**) R = —N(pyrrolidine)

(**110**) R = —N(phthalimide)

Ferrocene-CH$_2$R

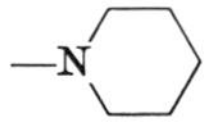

(**111**) Ferrocene-CH(N-piperidinyl)—CH=CPh$_2$

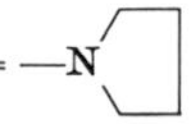

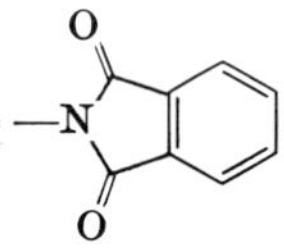

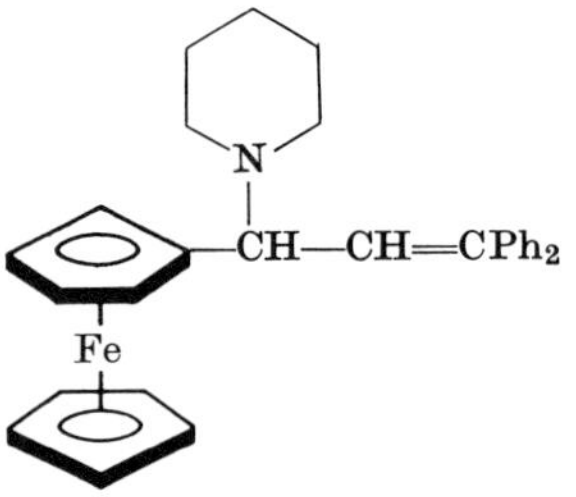

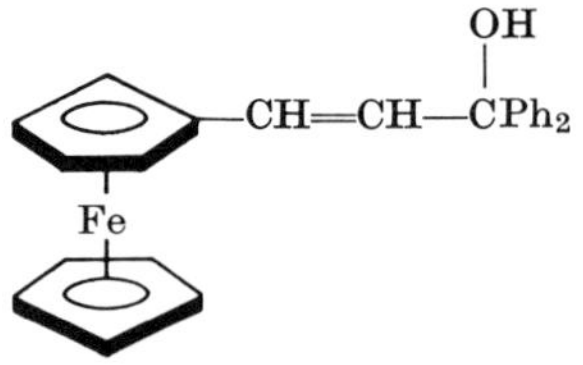

(**112**) Ferrocene-CH=CH—C(OH)Ph$_2$

---

[108] J. M. Osgerby and P. L. Pauson, *J. Chem. Soc.* 656 (1958).
[108a] T. Aratani, T. Gonda, and H. Nozaki, *Tetrahedron Lett.* 2265 (1969); G. Gokel, P. Hoffmann, H. Kleimann, H. Klusacek, D. Marquarding, and I. Ugi, *Tetrahedron Lett.* 1771 (1970).
[109] A. N. Nesmeyanov, V. A. Sazonova, V. N. Postnov, I. F. Leshcheva, and O. P. Yurchenko, *Dokl. Akad. Nauk SSSR* **189**, 555 (1969); *Chem. Abstr.* **72**, 67076j (1970).

chlorine by piperidine.[102] A number of substituted ferrocenylcarbinols in pyridine with tosyl chloride gave **114**.[110]

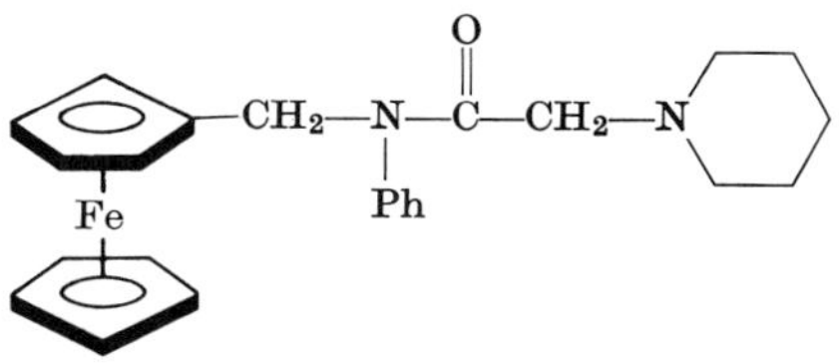

(113)

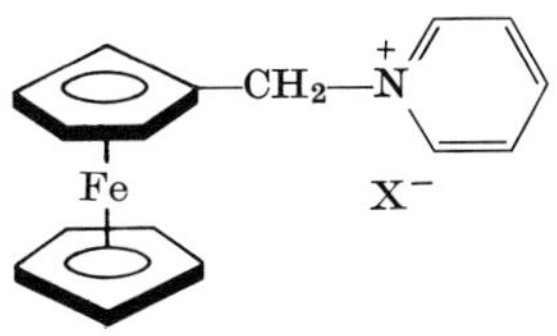

(114)

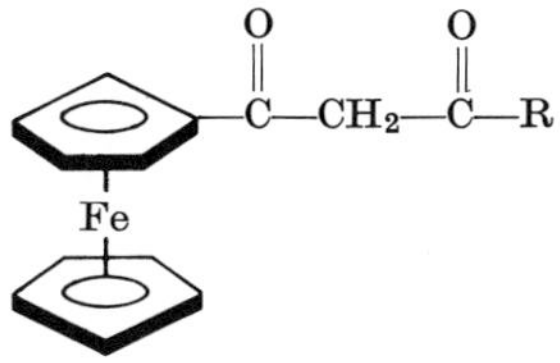

(115)

Diketones of the type **115** have been prepared in which R is 2-, 3-, or 4-pyridyl.[57, 58] As was noted earlier, these compounds have been converted into pyrazoles. The $pK_a$ and $pK_b$[111] and color reactions[58] of these compounds have been investigated. The ferrocenyl compounds (**115**) have a lower acidity and a higher basicity than the corresponding phenyl compounds.[111]

The hydantoins (**116** and **117**) have been prepared from DL-$\beta$-ferrocenyl-$\alpha$-alanine and $N$-ferrocenylmethyl-$N$-($\alpha$-carbethoxybenzyl)urea, respectively.[68] The Friedel–Crafts reaction of ferrocene and 2-pyrroylchloride gave the ketone (**118**).[112, 113] The polarographic half-wave potentials of **118** and other related ferrocenes were measured[113] and it was concluded that the ferrocene radical was more electropositive than the phenyl radical.

A number of amino acid derivatives of the type **119** were prepared[114] for ORD and CD studies. Among the amino acids used to prepare **119** were tryptophan, proline, and hydroxyproline.[114]

---

[110] A. N. Nesmeyanov, E. G. Perevalova, and M. D. Reshetova, *Izv. Akad. Nauk SSSR, Ser. Khim.* 335 (1966); *Chem. Abstr.* **64**, 17634 (1966).
[111] L. Wolf, H. Hennig, and I. P. Sereda, *J. Prakt. Chem.* **32**, 105 (1966).
[112] R. Dabard and B. Gautheron, *C. R. Acad. Sci.* **254**, 2014 (1962).
[113] J. Tirouflet, R. Dabard, and E. Laviron, *Bull. Soc. Chim. Fr.* 1655 (1963).
[114] H. Falk, C. Krasa, and K. Schlögl, *Monatsh. Chem.* **100**, 1552 (1969).

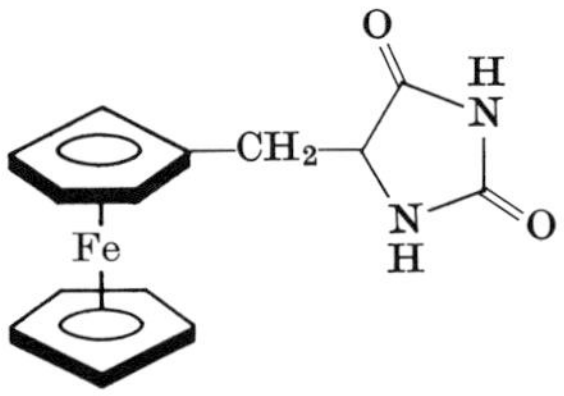

(116) 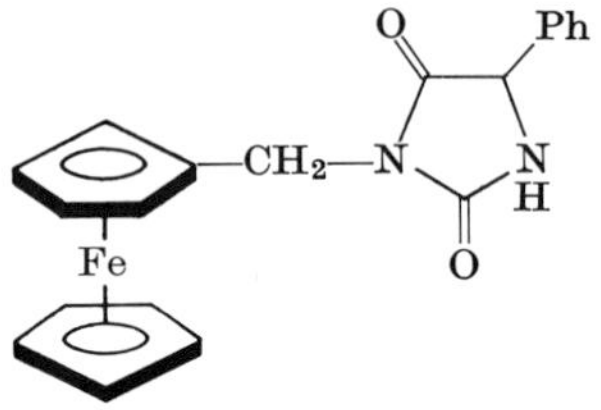 (117) 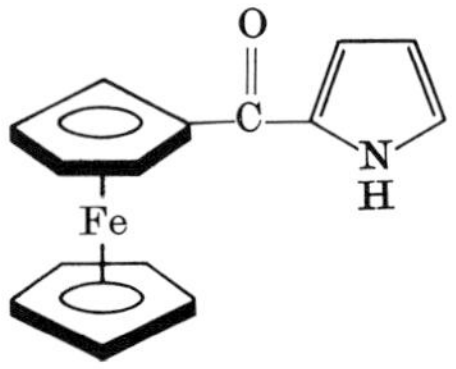 (118) 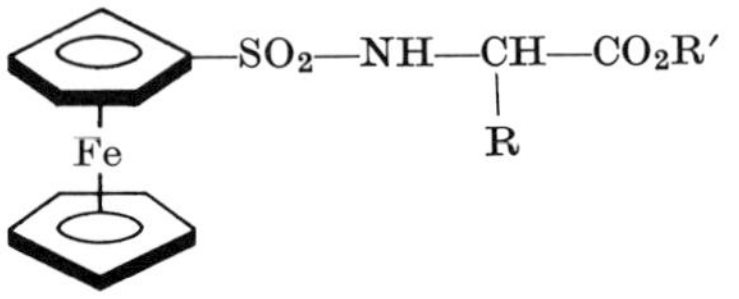 (119)

A series of compounds of the type **120** was prepared by the condensation of ω-ferrocenyl aliphatic acids with aromatic o-diamines.[115] The Diels–Alder reaction of **121** (R = H or $CH_3$) with various N-substituted maleimides gave a series of compounds of the type **122**.[116] Only the endo isomers were formed. A number of additional nitrogen-

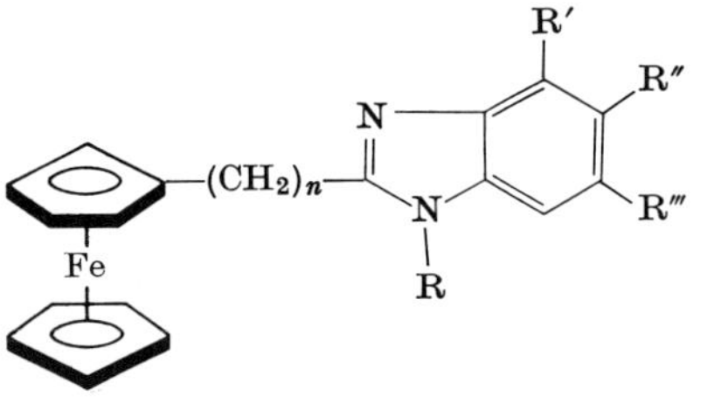

(120) 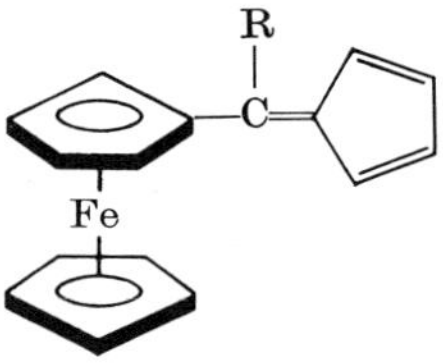 (121) 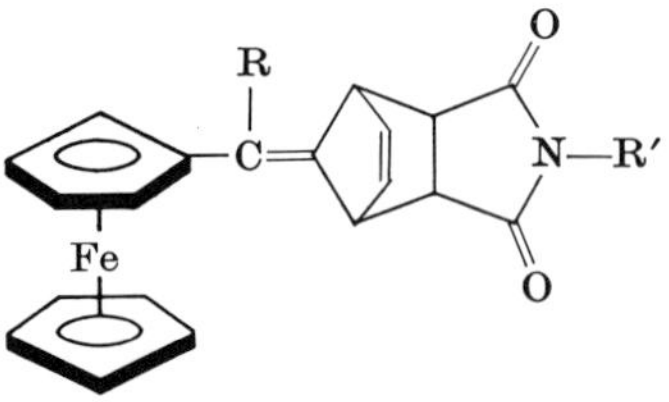 (122)

[115] D. Heydenhauss and H. Schubert, *Z. Chem.* **4**, 459 (1964).
[116] M. Furdik, S. Toma, J. Suchy, and M. Dzurilla, *Chem. Zvesti* **16**, 719 (1962); *Chem. Abstr.* **59**, 3955 (1963).

containing ferrocenyl heterocyclic compounds have been reported.[116a]

## B. Nitrogen and Other Heteroatoms in Ring

Reaction of the acetylenic ketone (**123**) with hydroxylamine gave two isomeric isoxazoles.[61] These were probably **124** and the 3-ferrocenyl isomer. The reaction of $N$-ferrocenoylglycine with benzaldehyde and sodium acetate in acetic anhydride gave the oxazolone (**125**)[68]

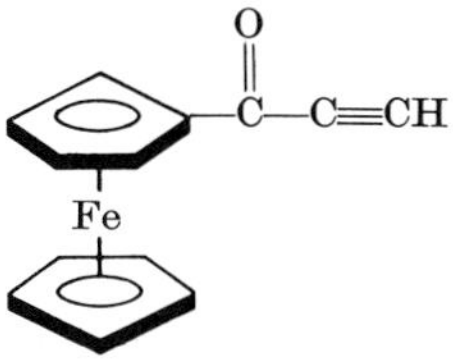

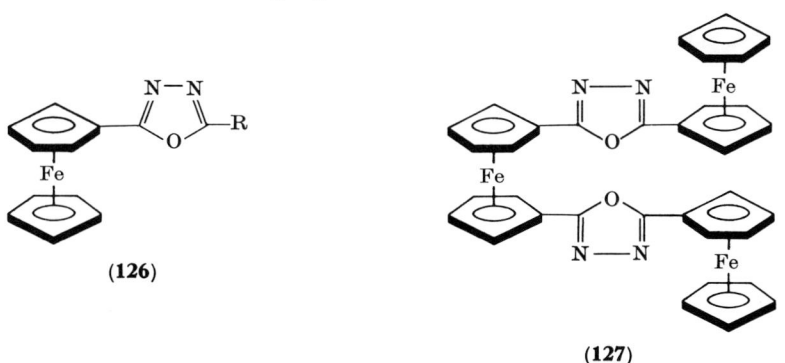

(123) 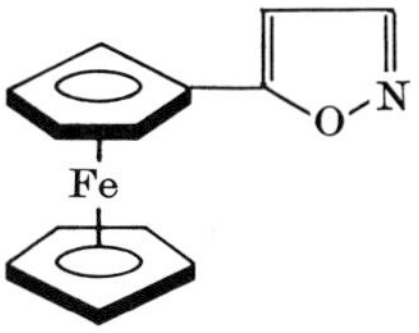            (124) 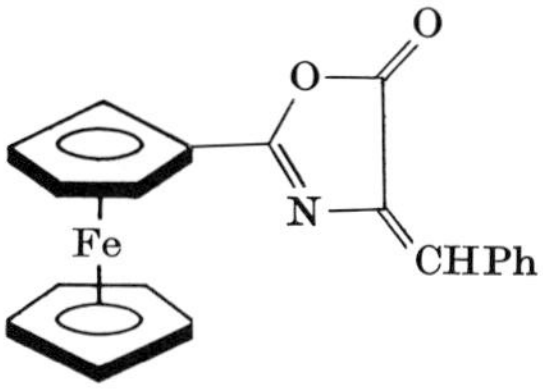            (125)

A number of ferrocene-containing 1,3,4-oxadiazoles have been prepared. The reaction of ferrocenecarboxhydrazide with triethyl orthoformate gave three products, the principal one being 2-ferrocenyl-1,3,4-oxadiazole (**126**; R = H). [76] 2-Phenyl-5-ferrocenyl-1,3,4-oxadiazole (**126**; R = $C_6H_5$) and 2,5-diferrocenyl-1,3,4-oxadiazole (**126**; R = ferrocenyl) were prepared from the appropriate 1,2-diacylhydrazines.[117] A number of other ferrocenyloxadiazoles, including the bis compound **127**, were prepared in a similar manner.[117] The oxadi-

(126)            (127)

---

[116a] M. Lacan and V. Rapic, *Croat. Chem. Acta* **42**, 411 (1970); T. Aratani, T. Gonda, and M. Nozaki, *Tetrahedron* **26**, 5453 (1970); R. A. Abramovitch, C. I. Azogu, and R. G. Sutherland, *Chem. Commun.* **134** (1971); A. N. Nesmeyanov, V. A. Sazonova, and V. E. Ferorov, *Izv. Akad. Nauk. SSSR, Ser. Khim.* 2133 (1970); *Chem. Abstr.* **74**, 31837k (1971).

[117] H. J. Lorkowski, R. Pannier, and A. Wende, *J. Prakt. Chem.* **35**, 149 (1967).

azole derivative (**128**) was obtained from the reaction of ferrocenecarboxylic acid chloride and 1,4-bis(tetrazol-5-yl)benzene, whereas ferrocene-1,1'-dicarboxylic acid chloride and this same benzene derivative gave a ferrocene-containing polyoxadiazole.[117]

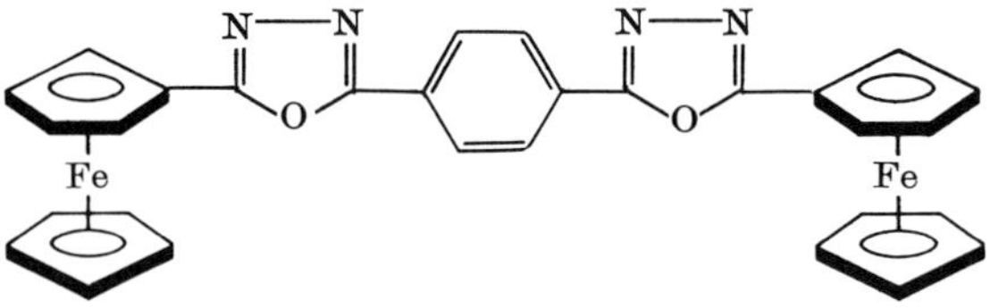

(**128**)

The reaction of ferrocenecarboxaldehyde with hippuric acid[68, 108, 118] and cycloserine[94] gave **129** and **130**, respectively. Hydrolysis of the oxazolone (**129**) gave α-benzoylamino-β-ferrocenylacrylic acid,[68, 108, 118] whereas reaction of **129** with sodium and ethanol converted it to β-ferrocenyl-α-cyclohexylformamidopropionic acid.[108]

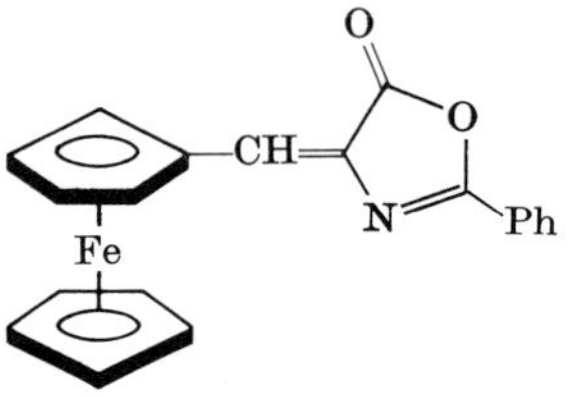

(**129**) 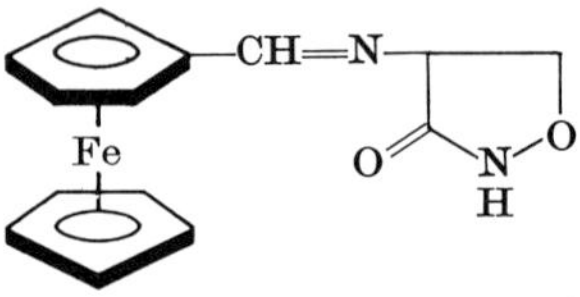   (**130**)

The remaining ferrocenyl heterocyclic compounds containing nitrogen and oxygen are all derivatives of morpholine. Use of acetylferrocene in the Willgerodt reaction gave the thiomorpholide (**131**; X = S) which could be hydrolyzed by base to **131** (X = O).[67] Reaction of morpholine with the quaternary ammonium compound **107**[105] and with ferrocenecarboxylic acid chloride[119] gave **132** and **133**, respectively. Morpholine analogs of **104** have been prepared from ferrocenecarboxaldehyde, nitroalkanes, and morpholine.[103, 120]

[118] G. D. Broadhead, J. M. Osgerby, and P. L. Pauson, *Chem. Ind.* (*London*) 209 (1957).
[119] E. M. Acton and R. M. Silverstein, *J. Org. Chem.* **24**, 1487 (1959).
[120] H. Kono, M. Shiga, I. Motyama, and K. Hata, *Bull. Chem. Soc. Japan* **42**, 3267 (1969).

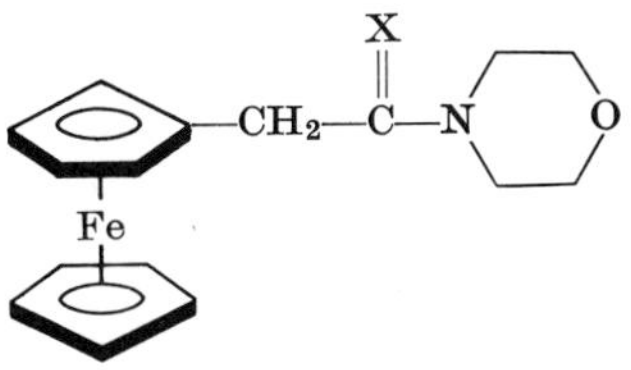

(131)

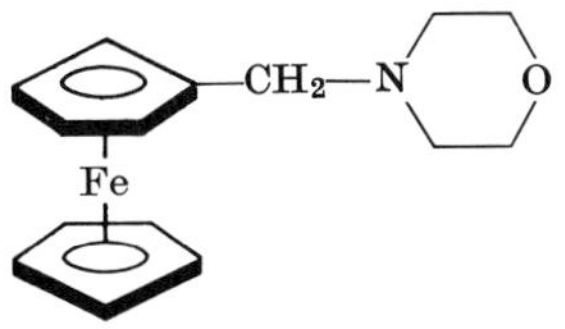

(132)

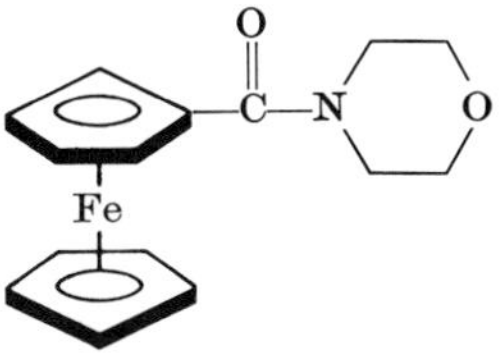

(133)

2-Aminothiazole reacts with both ferrocenecarboxaldehyde[96, 97] and ferrocenecarboxylic acid chloride[119] to give the expected imine and amide, respectively. Ferrocenecarboxylic acid chloride also reacts with phenothiazine to give both N- and C-acetylation.[119] Ferrocenecarboxaldehyde and rhodanine gave **134**,[67] while reaction of the thiosemicarbazone of ferrocenecarboxaldehyde with chloroacetic acid in acetic acid gave **135**.[94]

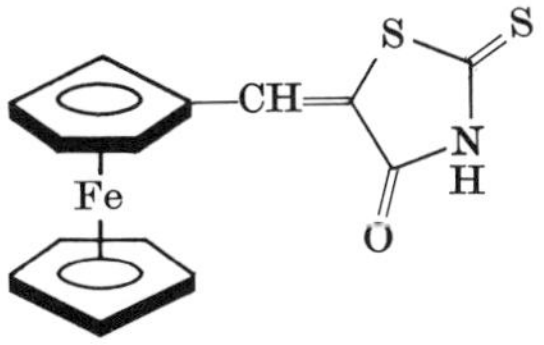

(134)

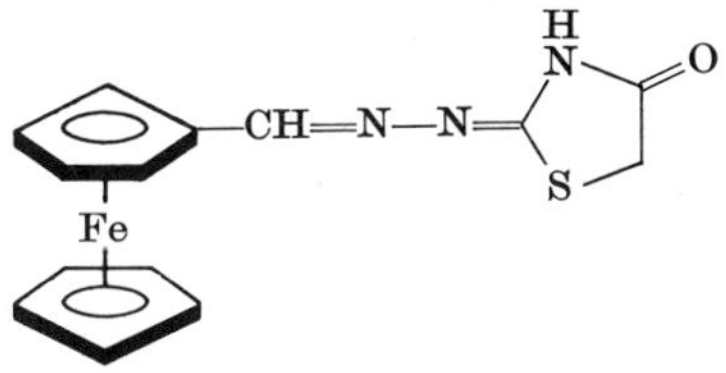

(135)

Reaction of (2-methyl-5-)- and (2-methyl-6-benzothiazolyl)-dimethyltriazene or 2-methyl-5- and 2-methyl-6-benzothiazolyl-diazonium salts with ferrocene gave **136** (R = ferrocenyl, R′ = H; R = H, R′ = ferrocenyl, respectively).[121] Use of the diazonium salts also led to the isolation of some 1,1′-disubstituted ferrocenes in this reaction. The compounds **136** were used to prepare thiacyanines. Reaction of ferrocenecarboxaldehyde with the ethiodide of 2-methyl-benzothiazole followed by treatment with potassium iodide gave **137**.[121]

[121] I. K. Ushenko, K. D. Zhikhareva, and F. Z. Rodova, *Zh. Obshch. Khim.* **33**, 798 (1963); *Chem. Abstr.* **59**, 10268 (1963).

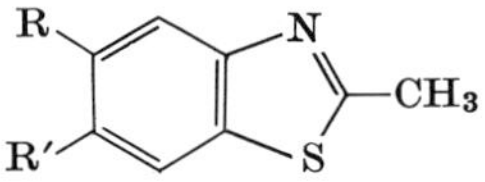

Condensation of ferroceneboronic acid and o-phenylenediamine gave 2-ferrocenylborabenzimidazoline (**138**; R = H).[122] The corresponding chloro (**138**; R = Cl) and bromo (**138**, R = Br) analogs were similarly prepared. Treatment of **138** with aqueous–alcoholic hydrogen chloride gave back the ferroceneboronic acids. Use of 1,1′-diboronate esters of diboronic acids with o-phenylenediamine or 3,3′-diaminobenzidine gave **139** and the polybenzborimidazoline (**140**), respectively.[123]

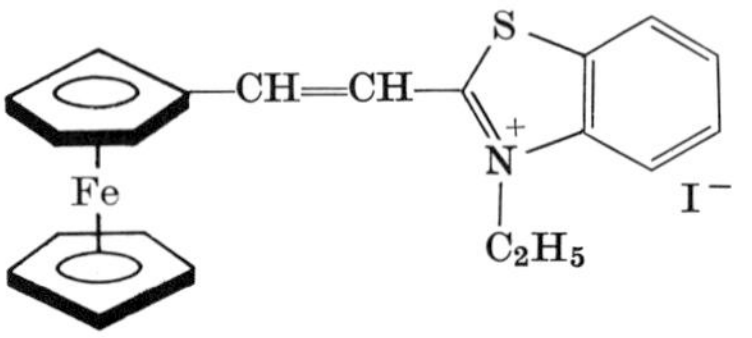

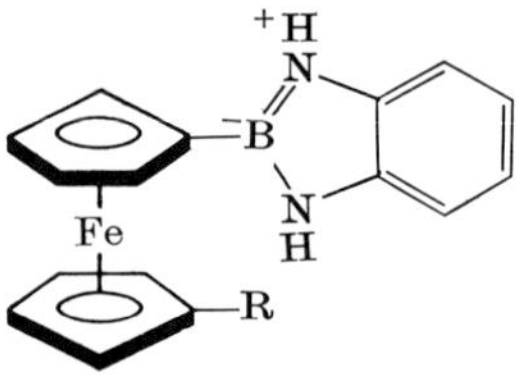

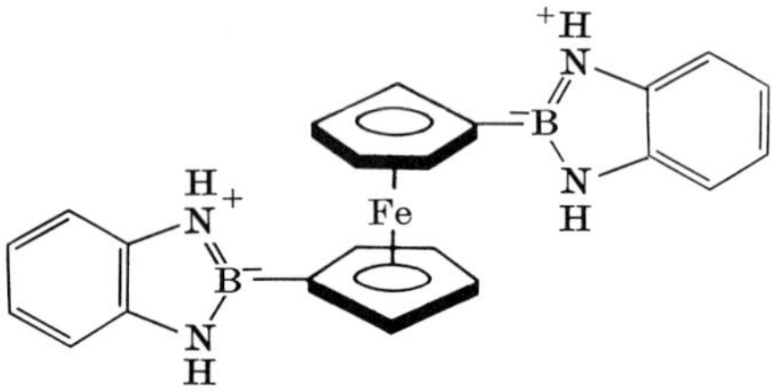

---

[122] A. N. Nesmeyanov, V. A. Sazonova, N. S. Sazonova, and R. I. Komarova, *Izv. Akad. Nauk SSSR, Ser. Khim.* 1352 (1969); *Chem. Abstr.* **71**, 91618n (1969).
[123] J. E. Mulvaney, J. J. Bloomfield, and C. S. Marvel, *J. Polym. Sci.* **62**, 59 (1962).

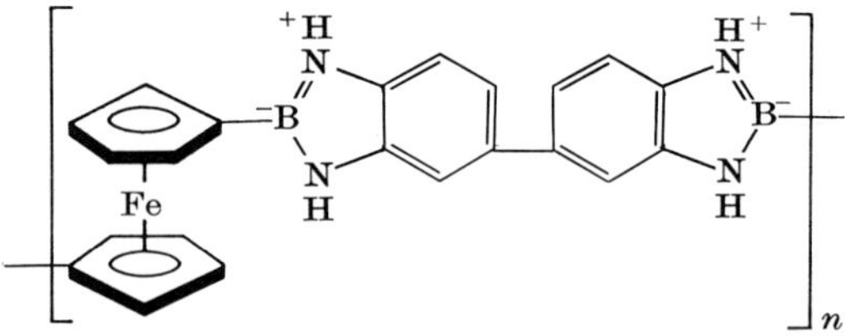

Whether or not chelates can be considered as true cyclic compounds is open to question, but the formation of **141**,[124] **142**,[125] and **143**[126] might be noted.

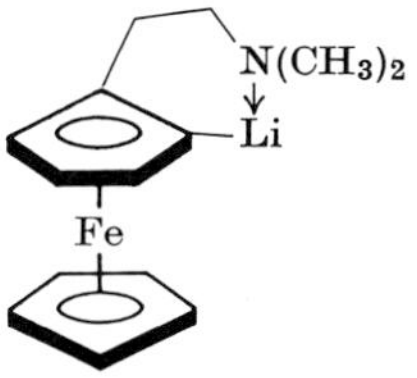

**(141)**

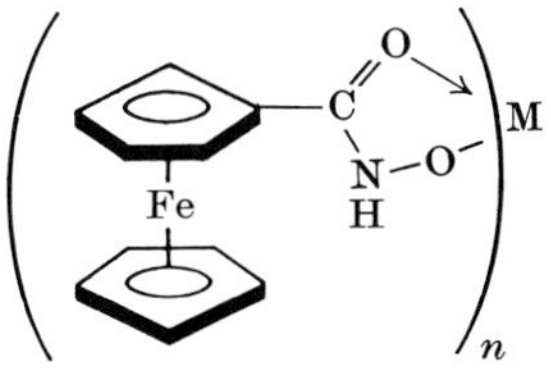

**(142)**
M = Fe, $n = 3$
M = Cu, $n = 2$

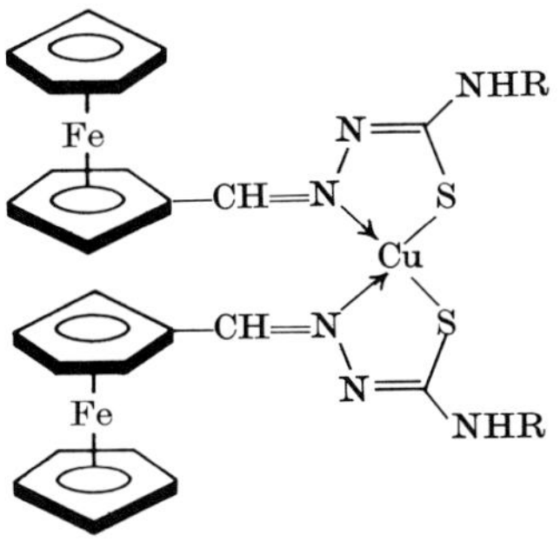

**(143)**

## C. Oxygen-Containing Rings

Ferrocenecarboxaldehyde has been converted into the cyclic acetal (**144**) which is easily hydrolyzed back to the aldehyde.[99] A number of cyclic lactones have been reported. Thus, reduction of β-ferrocenoylpropionic acid with sodium borohydride gave the butyrolactone (**145**)[127,128] which was smoothly hydrogenated to γ-ferrocenylbutyric

---

[124] D. W. Slocum, T. R. Engelmann, and C. A. Jennings, *Aust. J. Chem.* **21**, 2319 (1968).
[125] T. P. Vishnyakova, I. A. Golubeva, and L. M. Timofeeva, *Zh. Obshch. Khim.* **39**, 2534 (1969); *Chem. Abstr.* **72**, 79196m (1970).
[126] D. M. Wiles and T. Suprunchuk, *Can. J. Chem.* **46**, 1865 (1968).
[127] J. W. Huffman and D. J. Rabb, *J. Org. Chem.* **26**, 3588 (1951).
[128] N. Sugiyama, H. Suzuki, Y. Shioura, and T. Teitei, *Bull. Chem. Soc. Japan* **35**, 767 (1962).

acid.[127] The butyrolactone (146) is formed together with two other products when ferrocene and succinyl chloride react in a Friedel–Crafts acylation.[128] Catalytic reduction of bis(3-carboxypropionyl)-ferrocene or its diester gave the dilactone (147).[67] Reaction of the appropriate γ-keto acid with trifluoroacetic anhydride gave the ethylenic lactone (148).[73] The lactones (149; R = $C_6H_5$, R' = H and R = H, R' = $C_6H_5$) were also obtained by dehydration of γ-keto acids.[129] Finally, dehydration of the keto acid (150) gives α-ferrocenylisocoumarin (151).[130]

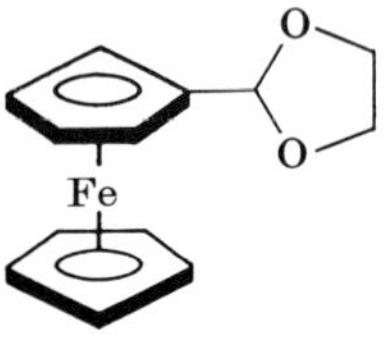

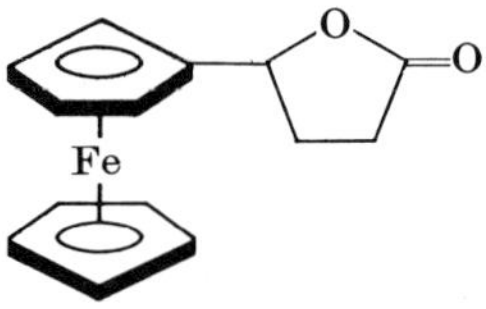

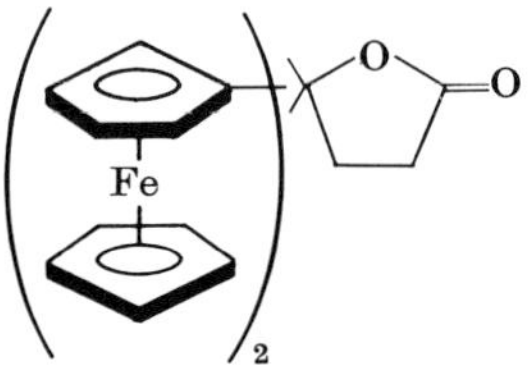

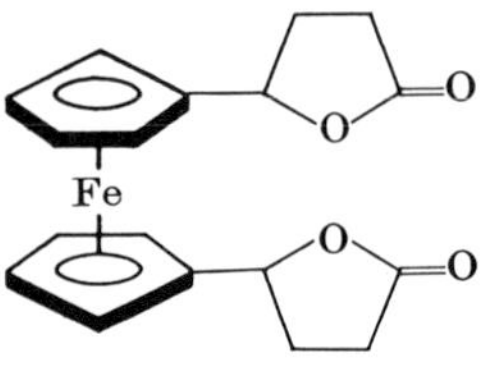

When o-carboxybenzoylferrocene was treated with acetic anhydride and sulfuric acid, the diphthalide (152) was obtained,[131] whereas acetic anhydride at higher temperatures gave 153. Treatment of 152 with base gave 154 and use of methanol or ethanol converted 153 to 155.[131] Treatment with acid converted 153 and 155 into 152.

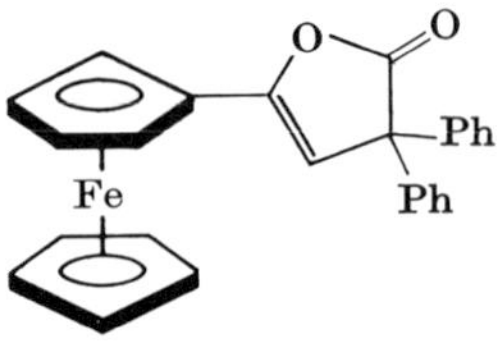

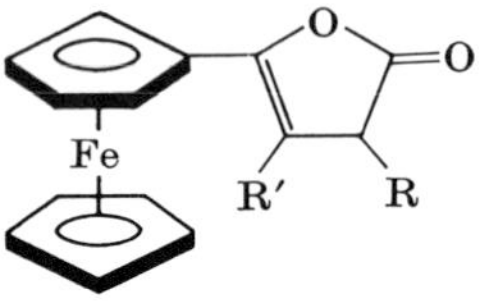

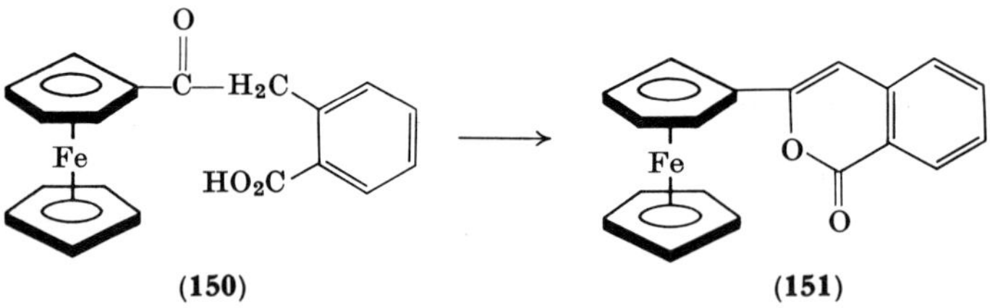

[129] J. Tirouflet, R. Daburd, and B. Gautheron, *C. R. Acad. Sci.* **256**, 1315 (1963).
[130] J. Boichard, *C. R. Acad. Sci.* **253**, 2702 (1961).
[131] A. N. Nesmeyanov, V. D. Vilchevskaya, and N. S. Kochetkova, *Dokl. Akad. Nauk SSSR* **138**, 390 (1961); *Chem. Abstr.* **55**, 21080 (1961).

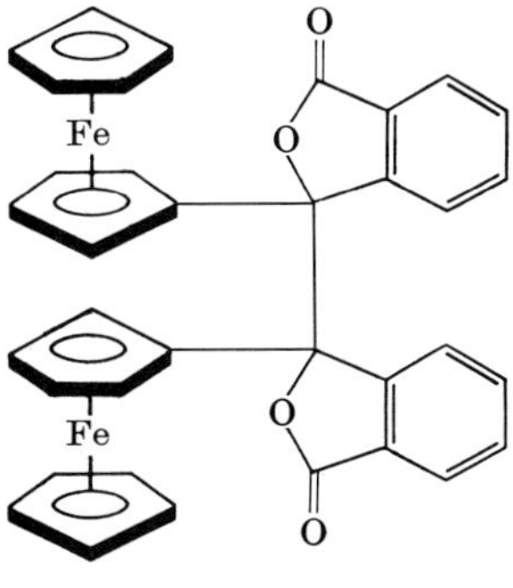

The compound **156** can be readily transformed to 2-ferrocenyl-4-chromanone **(157)** by the action of heat in the presence of sodium acetate.[96,97] Ferrocenecarboxaldehyde, chloroacetonitrile, and sodium acetate react at −60° to give β-ferrocenylglycidonitrile **(158)**.[132] A number of 1,4-glycols of the type **159** have been converted to tetrahydrofurans **(160)** by dehydration with heat or acid.[133] The diferrocenyl analog **(161)** has also been prepared[128] by the action of acid on the appropriate diol. The synthesis and stereochemistry of the cis and trans forms of **161** and the synthesis of the corresponding 2,5-diferrocenylfuran have been reported.[134] The reaction of o-iodophenol and ferrocenylethynylcopper gave 2-ferrocenylbenzofuran **(162)**.[135]

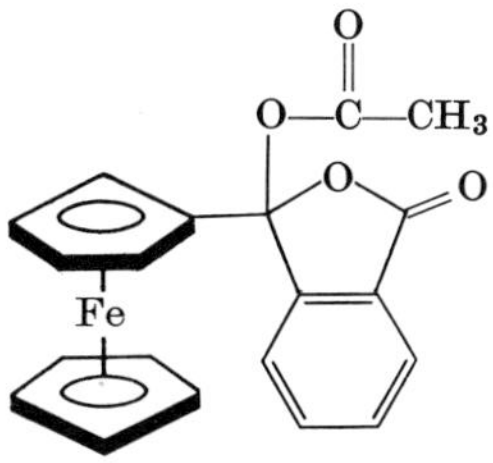

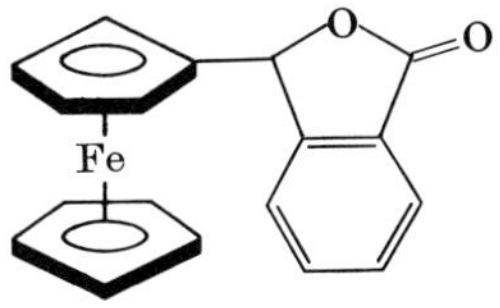

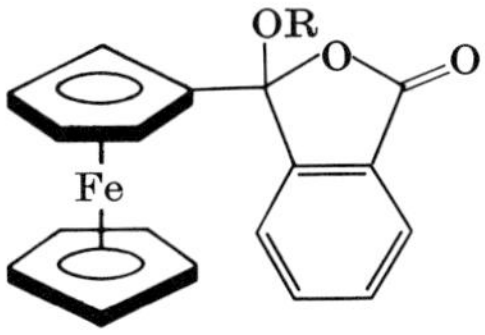

---

[132] A. A. Koridze, *Zh. Obshch. Khim.* **39**, 1649 (1969); *Chem. Abstr.* **71**, 91615j (1969).
[133] I. M. Gverdtsiteli and L. P. Asatiani, *Soobshch. Akad. Nauk Gruz. SSR* **51**, 585 (1968); *Chem. Abstr.* **70**, 96900n (1969).
[134] K. Yamakawa and M. Moroe, *Tetrahedron* **24**, 3615 (1968).
[135] M. D. Rausch and A. Siegel, *J. Org. Chem.* **34**, 1974 (1969).

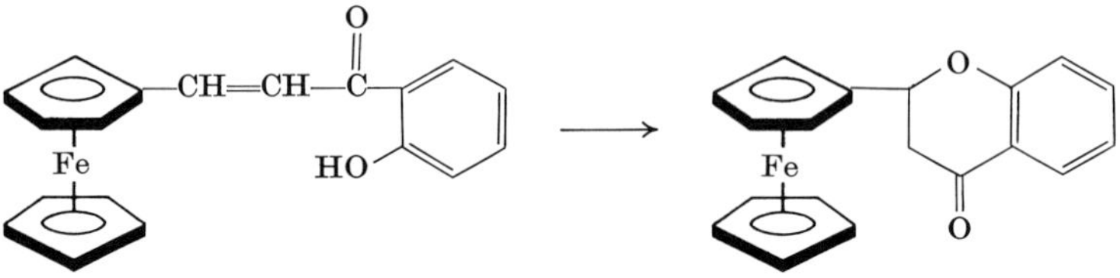

(156) (157)

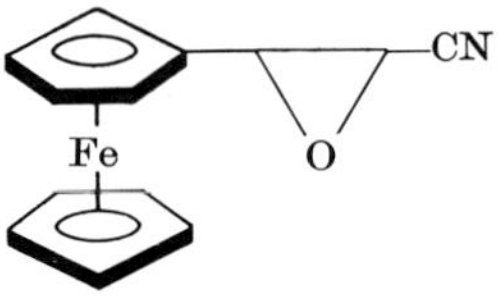

(158) 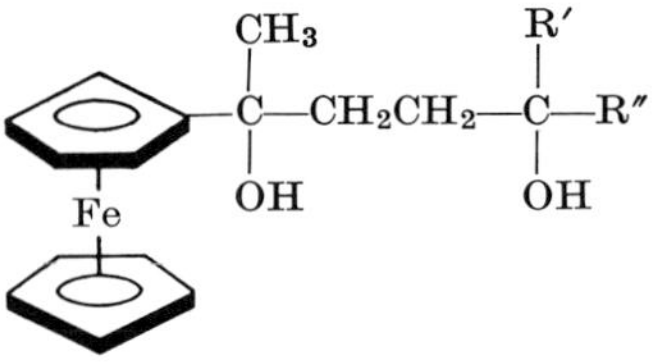 (159)

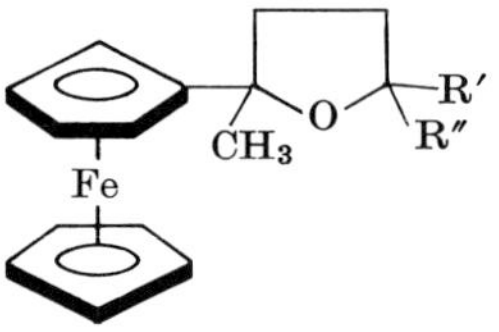

(160) 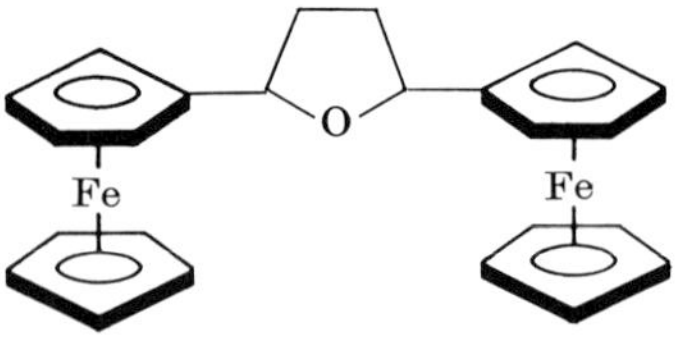 (161)

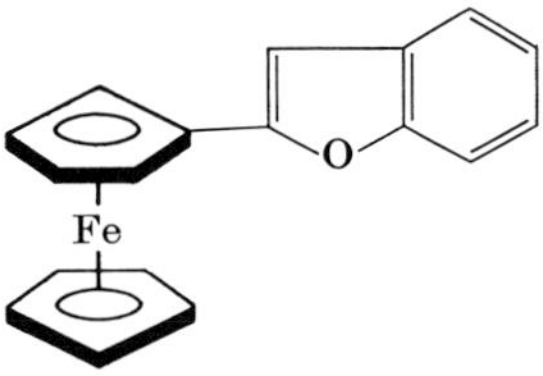

(162)

Ferrocene,[136] ferrocenecarboxaldehyde,[137] and hydroxymethyl- and hydroxyethylferrocene[138] have all been used to prepare various ferrocenylcarbohydrates. Ferrocenecarboxylic acid chloride and 1,3-$O$-methyleneglycerol[139, 140] gave the ferrocenecarboxylate (163).

[136] W. Treibs, *Naturwissenschaften* **52**, 496 (1965).
[137] A. N. DeBelder, E. J. Bourne, and J. B. Pridham, *J. Chem. Soc.* 5486 (1964).
[138] A. N. DeBelder, E. J. Bourne, and J. B. Pridham, *J. Chem. Soc.* 4464 (1961).
[139] J. S. Brimacombe, A. B. Foster, and A. H. Haines, *J. Chem. Soc.* 2582 (1960).
[140] N. Baggett, A. B. Foster, A. H. Haines, and M. Stacey, *J. Chem. Soc.* 3528 (1960).

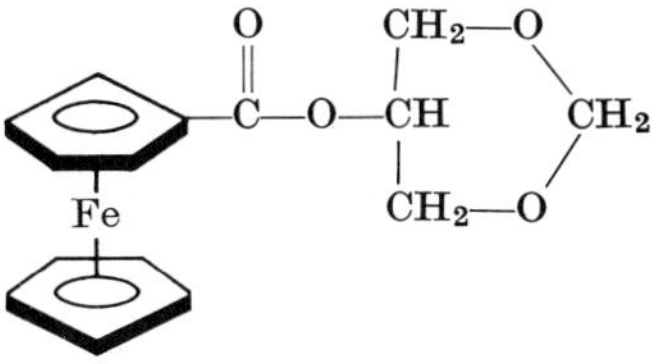

(163) 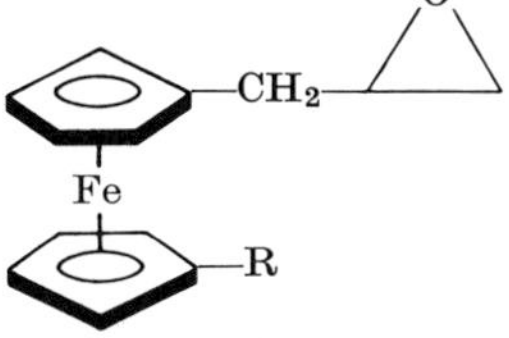   (164)

A number of glycidylferrocenes [**164**; R = H, CH$_2$CH(OH)CH$_2$Cl, or CH$_2$CHOCH$_2$] have been prepared by the addition of ferrocenyllithium to epichlorohydrin followed by cyclization of the chlorohydrin obtained.[141] Treatment of **164** (R = CH$_2$CHOCH$_2$) with ethoxide gave a small amount of **164** [R = CH$_2$CH(OH)CH$_2$OC$_2$H$_5$] and an 81% recovery of starting material.[141]

Reaction of the acid chloride of furan-2-carboxylic acid with ferrocene in a Friedel–Crafts acylation gave the expected ketone.[112, 113] As noted earlier for the corresponding ketone with a pyrrole group, polarographic studies have been carried out.[113] The diketone (**165**) was prepared in the usual manner.[56, 57] The reaction of the appropriate quaternary ammonium compound and furfuryl alcohol gave the ether (**166**).[142] Reaction of acetylferrocene with furfural[52, 95, 96, 143] and 5-methyl-2-furaldehyde[52] gave chalcones (**167**), whereas compounds containing the isomeric skeleton (**168**) were obtained from the reaction

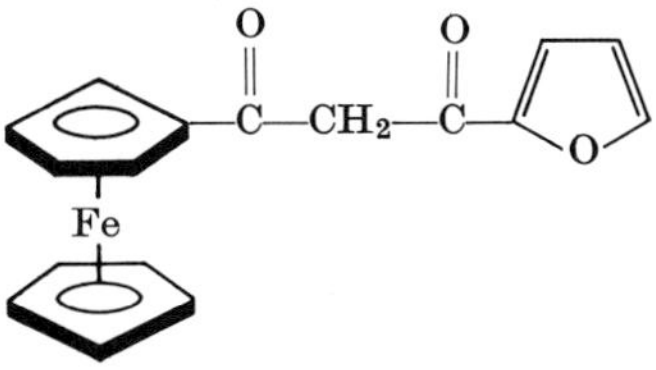

(165) 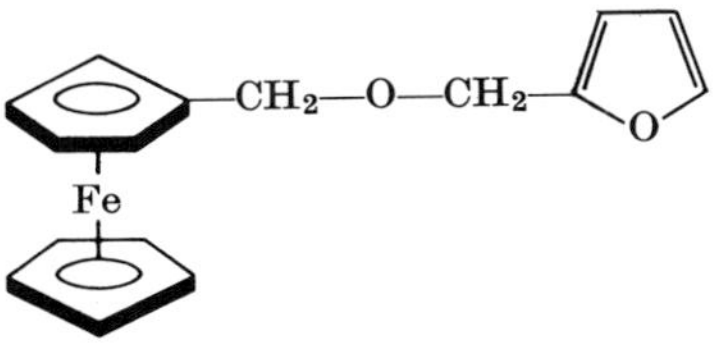   (166)

---

[141] H. Watanabe, I. Motoyama, and K. Hata, *Bull. Chem. Soc. Japan* **39**, 784 (1966).

[142] A. N. Nesmeyanov, E. G. Perevalova, Yu. A. Ustynyuk, and L. S. Shilovtseva, *Izv. Akad. Nauk SSSR, Otdel. Khim. Nauk* 554 (1960); *Chem. Abstr.* **54**, 22540 (1960).

[143] M. Furdik, P. Elecko, S. Toma, and J. Suchy, *Chem. Zvesti* **14**, 501 (1960); *Chem. Abstr.* **55**, 16508 (1961).

of ferrocenecarboxaldehyde with 2-acetylfuran and 2-acetyl-5-nitrofuran.[52] The compound **167** (R = H) was studied polarographically,[98] and **167** and **168** were treated with hydrazine and phenylhydrazine to give pyrazolines.[52] 1,1'-Diacetylferrocene has been reported to react with furfural to give **169**[144] and **170**[145] or a resin.[145]

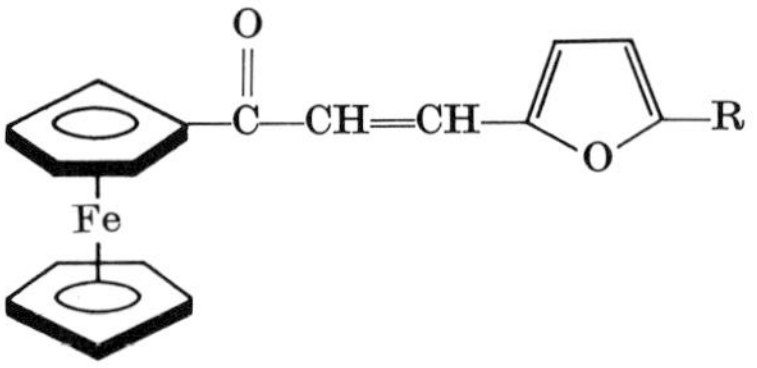

(167)

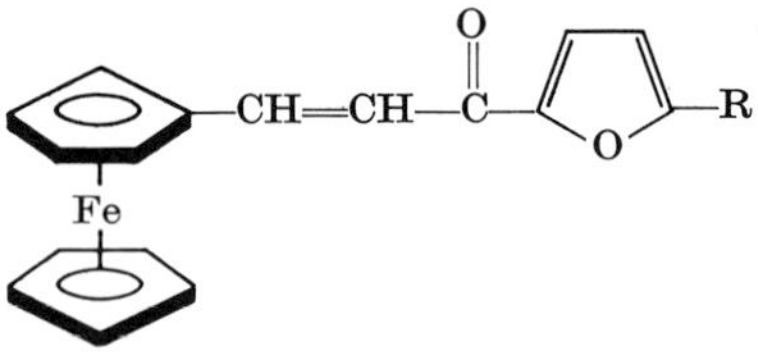

(168)

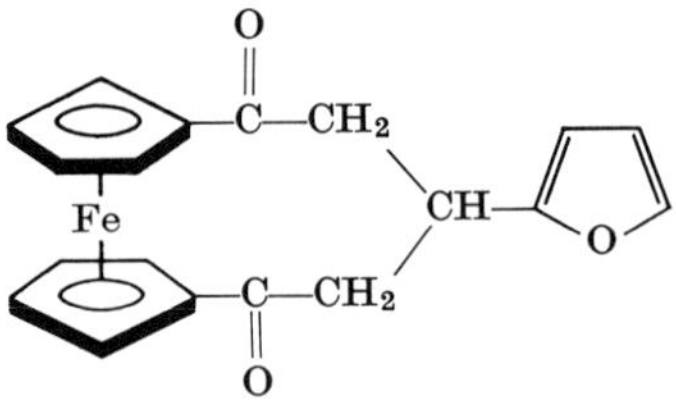

(169)

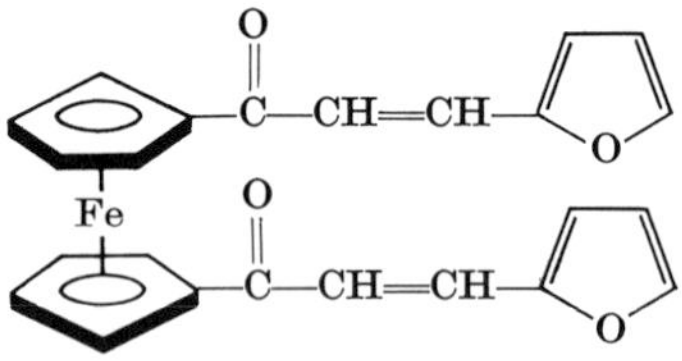

(170)

Condensation of ferrocenecarboxyhydrazide with 5-nitro-2-furaldehyde or 5-nitro-2-acetylfuran gave **171**, and the same hydrazide with ethyl 5-nitro-2-furimidate hydrochloride gave **172**.[52] The reaction of β-ferrocenylnitroethylene and furfural with sodium methoxide gave **173**.[146]

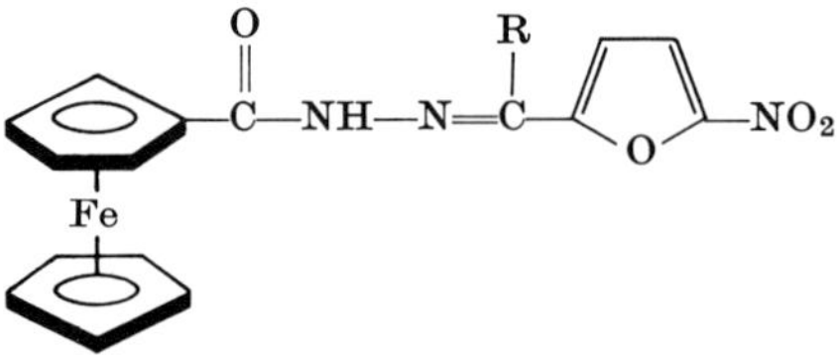

(171)

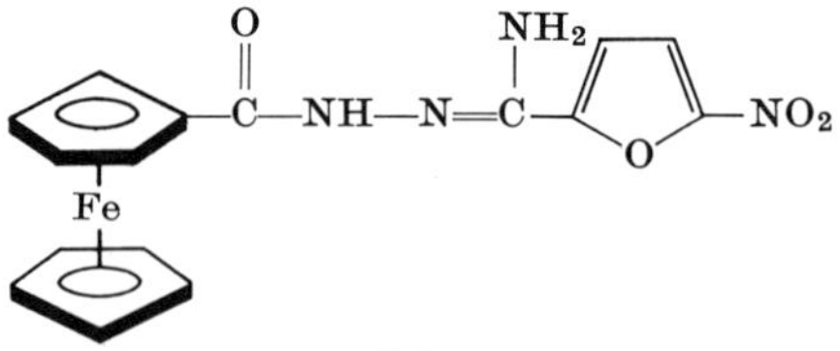

(172)

[144] M. Furdik, S. Toma, J. Suchy, and P. Elecko, *Chem. Zvesti* **15**, 45 (1961); *Chem. Abstr.* **55**, 18692 (1961).

[145] V. N. Kotrelev, S. P. Kalinina, and G. I. Kuznetsova, *Plast. Massy* **3**, 24 (1961); *Chem. Abstr.* **55**, 25341 (1961).

[146] M. Shiga, H. Kono, I. Motoyama, and K. Hata, *Bull. Chem. Soc. Japan* **42**, 798 (1969).

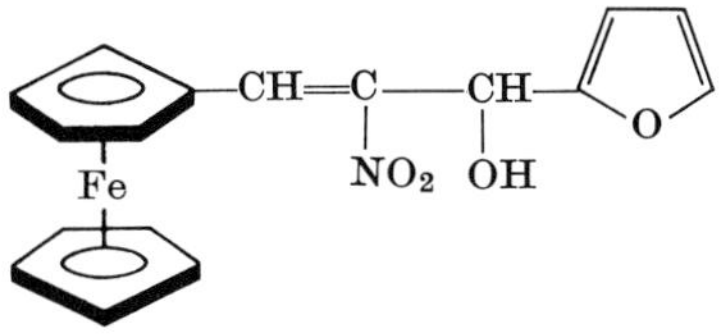

(173) 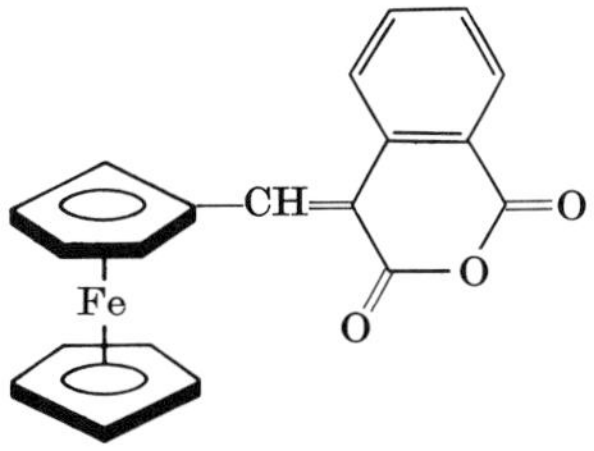     (174)

The Diels–Alder reaction of **121** with maleic anhydride proceeds as noted earlier for maleimides.[116] Ferrocenecarboxaldehyde and homophthalic anhydride have been condensed to give **174**.[96, 97]

A number of ferrocenyl chelates containing oxygen and copper,[48, 55, 57, 147] as well as other metals,[57, 147, 147a] have been reported.

### D. SULFUR-CONTAINING RINGS

All the ferrocenyl sulfur heterocyclic compounds contain the thiophene ring system. Reaction of bromoferrocene with potassium tetra-α-thienylborate[78] or the thiophene Grignard reagent[77] in the presence of cuprous bromide gave **175** (R = H). A number of additional ferrocenylthiophenes (**175**; R = $CH_2OH$, $C_6H_5$, or ferrocenyl) have been prepared by closing the thiophene ring in the final step of the synthesis.[148] Reaction of **175** (R = $CH_2OH$) with manganese dioxide gave the corresponding formyl compound (**175**; R = CHO) and reduction with lithium aluminum hydride gave the methyl compound (**175**; R = $CH_3$).[148] A study of the absolute configuration and circular

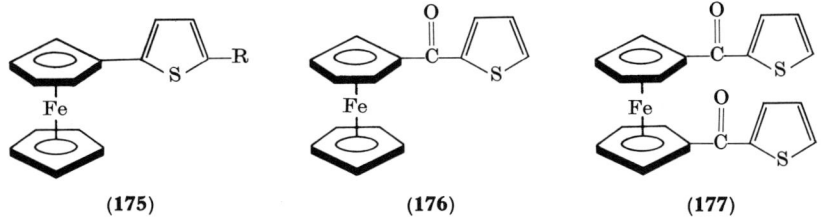

(175)     (176)     (177)

[147] Y. M. Paushkin, T. P. Vishnyakova, I. D. Vlasova, and F. F. Machus, *Zh. Obshch. Khim.* **39**, 2379 (1969).
[147a] I. Pavlik, J. Klikorka, K. Handlar, *Z. Chem.* **8**, 390 (1968).
[148] H. Egger and K. Schlögl, *Monatsh. Chem.* **95**, 1750 (1964).

(178)    (179)

dichroism of 2-(α-methylferrocenyl)-5-substituted thiophenes has appeared.[148a]

Use of thienyl chloride in the Friedel–Crafts acylation of ferrocene gave the expected ketone (**176**),[112, 113, 149] as well as the diketone (**177**).[149] The oxidation potential[150] of **176** and the comparison of the polarographic behavior[113] of **176** with other compounds have been reported. Reduction of **176** with lithium aluminum hydride gave the alcohol (**178**), while reduction of **176** or **178** with lithium aluminum hydride–aluminum chloride gave **179**. Reduction of **176** with sodium and ethanol in xylene gave **179** and ferrocenylbutylcarbinol, whereas Raney nickel and ethanol gave n-pentylferrocene from **179**. With the exception of the sodium and alcohol reduction, the diketone (**177**) also underwent a similar series of reactions.[149] The reduction of **177** with sodium and ethanol gave the bismethylene compound and **180**. The action of Raney nickel on **177** gave 1,1′-divalerylferrocene.[149] The Friedel–Crafts acylation of **179** with one mole of acetyl chloride–aluminum chloride gave a mixture of **181** and **182**, while the use of four moles of acetyl chloride–aluminum chloride gave **183**.[149] Compound **181** was reduced with lithium aluminum hydride to 5-ethyl-2-thenylferrocene which was also obtained through reduction of **184**. The diketone (**184**) was prepared by Friedel–Crafts acylation of ferrocene with 5-acetyl-2-thiophenecarboxylic acid chloride. Treatment of **181** with Raney nickel gave (6-hydroxy-1-heptyl)ferrocene.[149] Reaction of the bismethylene analog of **177** with acetyl chloride and aluminum chloride gave **185**.[149] Treatment of the oxime of **176** with lithium aluminum hydride gave, in addition to the expected amine (**186**), the rearranged product **187**.[11] The chemistry of **176** and related compounds has been studied in more detail than any other heterocyclic ferrocene system.

[148a] H. Falk, H. Lehner, and K. Schlögl, *Monatsh. Chem.* **101**, 967 (1970).
[149] K. Schlögl and H. Pelousek, *Ann.* **651**, 1 (1962).
[150] K. Schlögl and M. Peterlik, *Monatsh. Chem.* **93**, 1328 (1962).

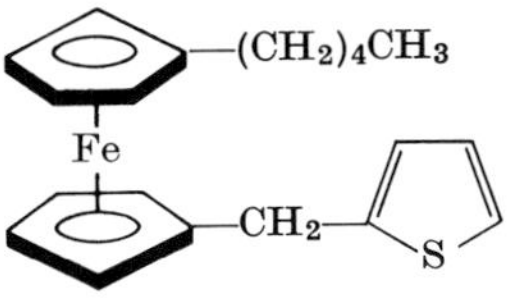

(180) 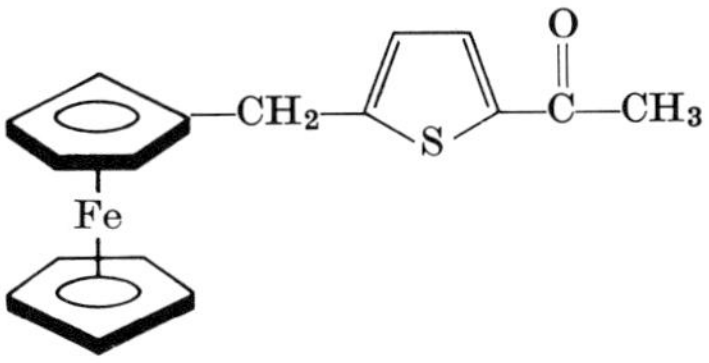 (181) 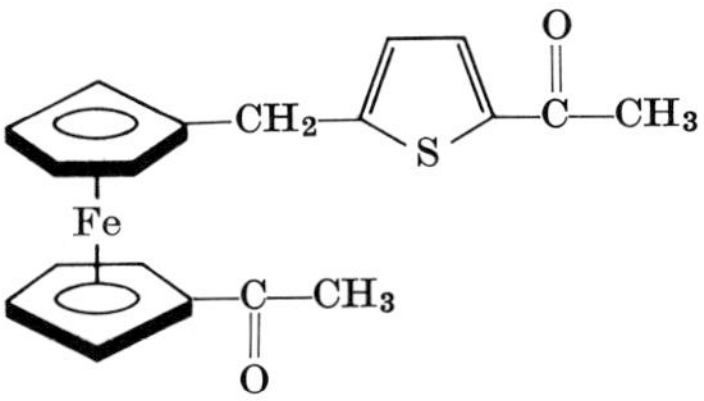 (182) 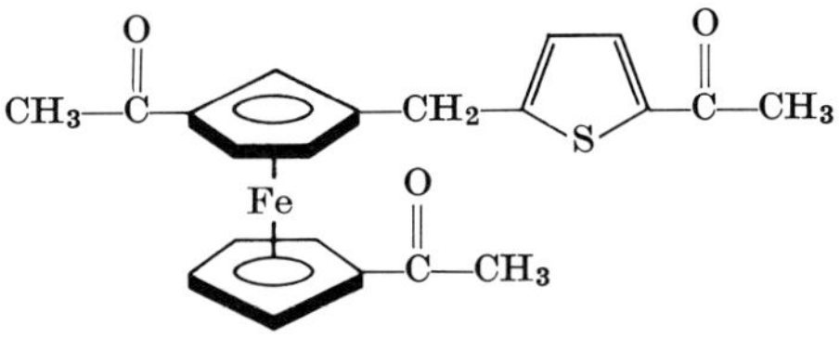 (183) 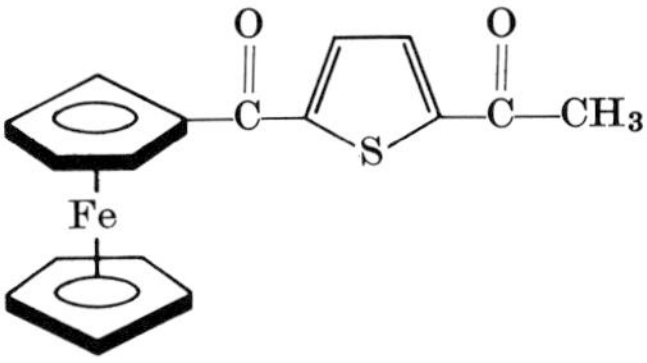 (184) 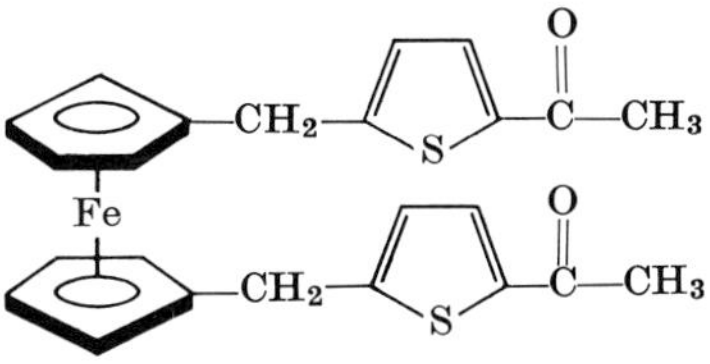 (185) 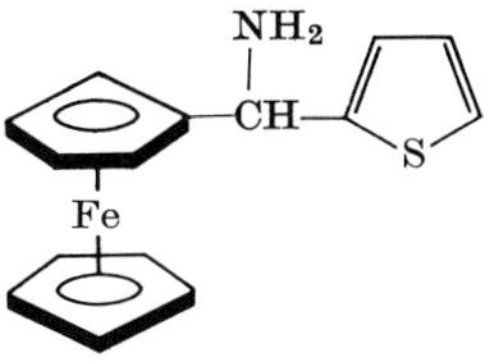 (186) 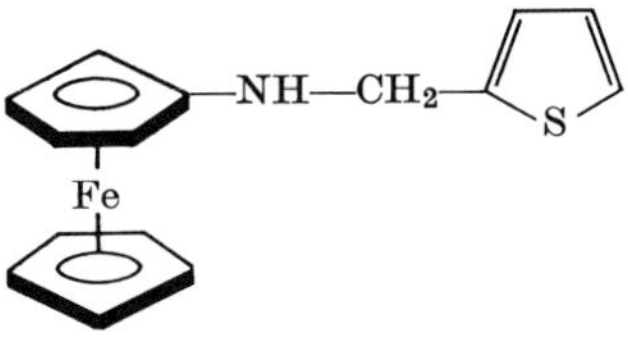 (187)

Reaction of ferrocenylamine and α-thiophenealdehyde gave the expected imine (188).[96] As has been the case with the nitrogen and oxygen heterocyclic ferrocenes, acetylferrocene[95, 96, 98] and ferrocenecarboxaldehyde[95, 96] have been used to prepare 189 and 190, respectively. A chalcone of the general type 190 has also been prepared from 3-acetylthianaphthene.[96, 97] The diketone (191) has also been prepared by methods noted earlier.[56, 57] Reaction of the thiophene Grignard reagent with the appropriate ketone gave 192.[151]

[151] J. Boichard and J. Tirouflet, *C. R. Acad. Sci.* 253, 1337 (1961).

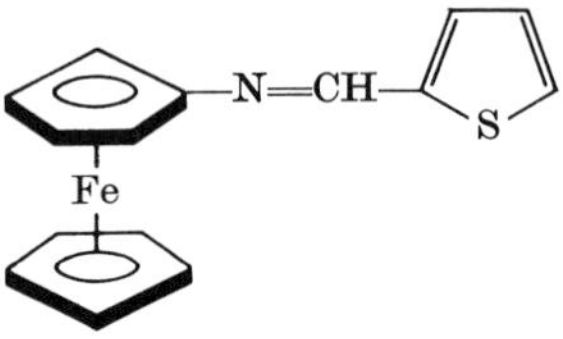

(188) 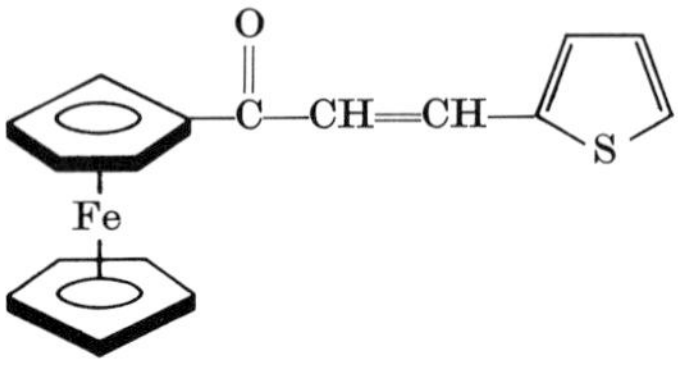 (189) 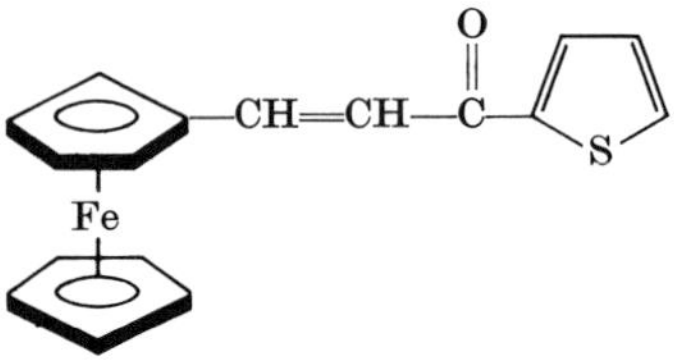 (190) 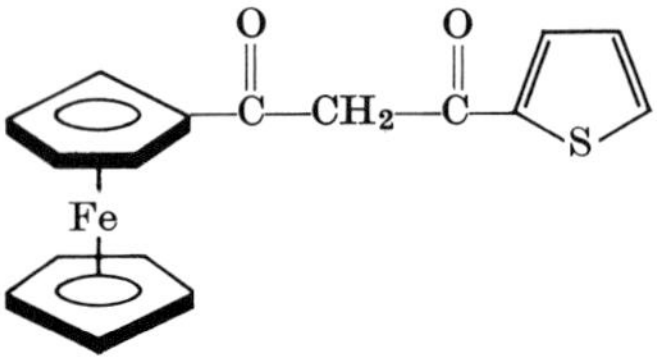 (191) 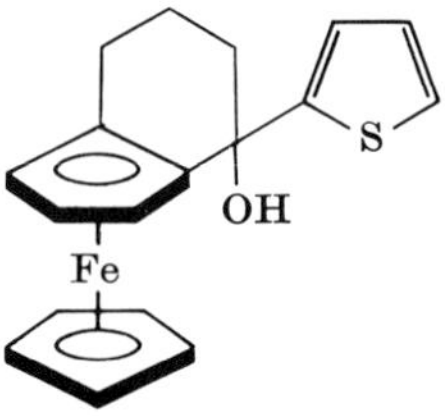 (192)

Friedel–Crafts acylation of ferrocene gave the ketone (193) which was oxidized rapidly on alumina columns to the diketone.[63] The action of phosphorus oxychloride–dimethylformamide on 193 gave 194 and then 195, which could then be converted into the ferrocenyl-acetylene (196).[63]

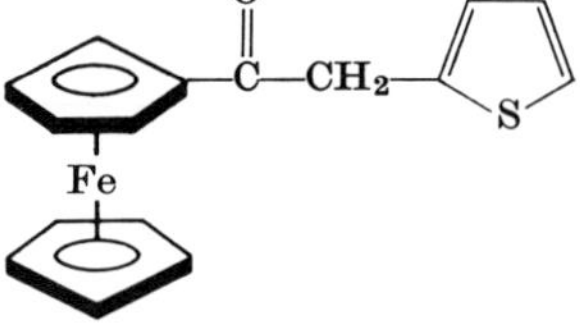

(193) 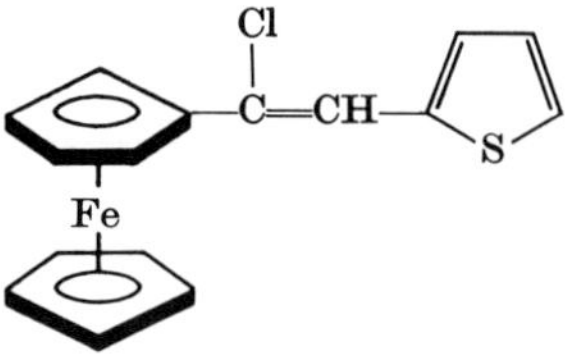 (194)

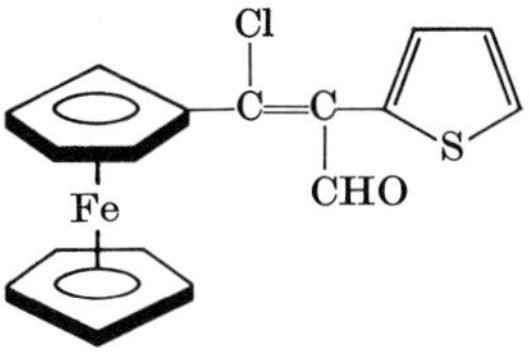

(195)

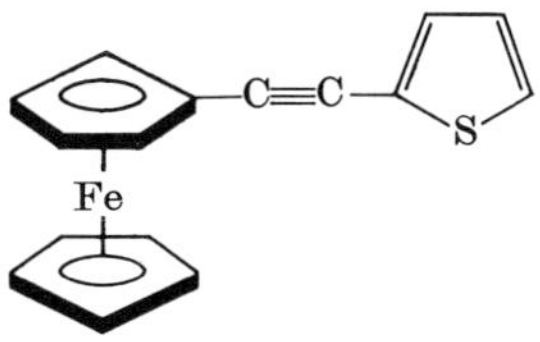

(196)

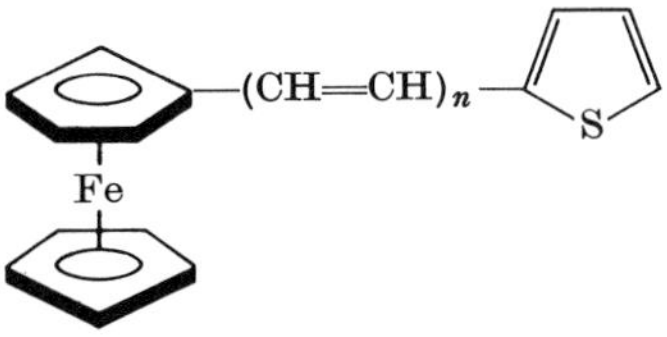

(197)

α-Ferrocenyl-ω-(2-thienyl)polyenes (**197**; $n = 1$–3) have been prepared[152, 153] by the Wittig reaction from (ferrocenylmethyl)- or (2-thienylmethyl)triphenylphosphonium salts and the appropriate aldehyde in ethanol with lithium.

## IV. Other Heterocyclic Metallocenes

The compounds in this section can be divided into two categories: those in which the heterocyclic group is a substituent on the metallocene, such as heterocyclic compounds containing cyclopentadienylmanganese tricarbonyl, and those in which a heteroatom is in a ring of the metallocene, such as in azaferrocene.

Bromoacetylcyclopentadienylmanganese tricarbonyl has been converted into **198**, **199**, and **200**.[154] Action of hydrazine on the appropriate diketone gave **201**.[155] Friedel–Crafts acylation of cyclopentadienylmanganese tricarbonyl gave **202** (X = O or S).[113] The half-wave potentials of **202** were measured polarographically and compared with ferrocene and benzene analogs.[113]

---

[152] K. Schlögl and H. Egger, *Angew. Chem.* **75**, 1123 (1963).
[153] K. Schlögl and H. Egger, *Ann.* **676**, 76 (1964).
[154] A. N. Nesmeyanov, I. B. Zlotina, N. E. Kolobova, and K. N. Anisimov, *Izv. Akad. Nauk SSSR, Ser. Khim.* 2127 (1968); *Chem. Abstr.* **70**, 29030r (1969).
[155] E. O. Fischer and K. Plesske, *Chem. Ber.* **92**, 2841 (1959).

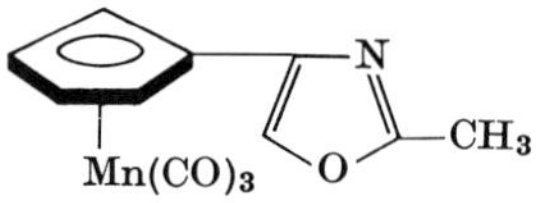

**(198)** — Mn(CO)₃-Cp with oxazole-CH₃

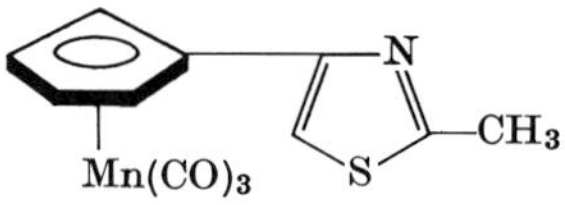

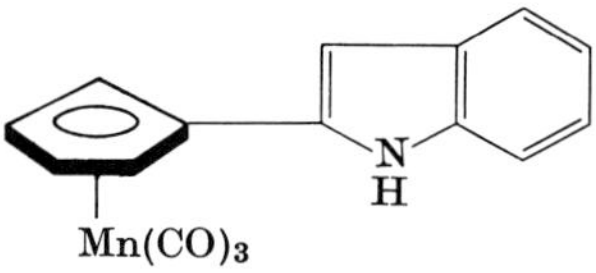

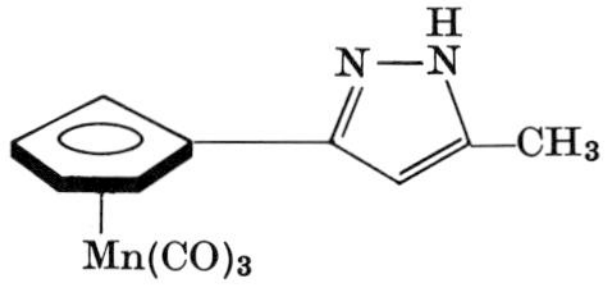

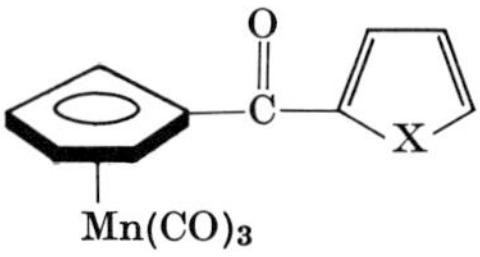

**(199)** — Mn(CO)₃-Cp with thiazole-CH₃

**(200)** — Mn(CO)₃-Cp with indole

**(201)** — Mn(CO)₃-Cp with pyrazole-CH₃

**(202)** — Mn(CO)₃-Cp with C(O)-furan-X

Cyclopentadienylmanganese tricarbonylcarboxylic acid chloride has been reacted with quinoline and hydrogen cyanide to give the Reissert compound (203).[51] Analogous compounds have also been prepared from 2-methyl- and 3-methylcyclopentadienylmanganese tricarbonylcarboxylic acid chloride.[156] Hydrolysis of these Reissert compounds gave the metallocenealdehydes.[51, 156] Reaction between zinc, ethyl bromoacetate, and 204 (R = C₆H₅ or ferrocenyl) gave the azetidinones (205).[51]

A number of chelates containing oxygen–metal rings have also been prepared in this series.[147, 155]

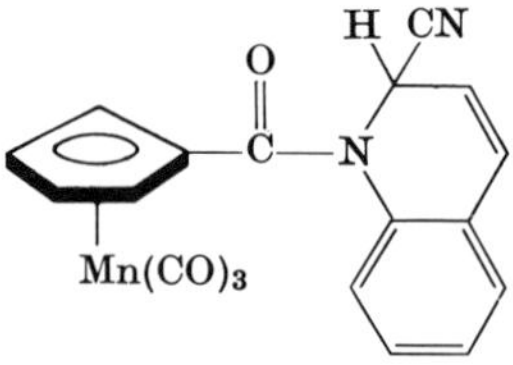

**(203)**

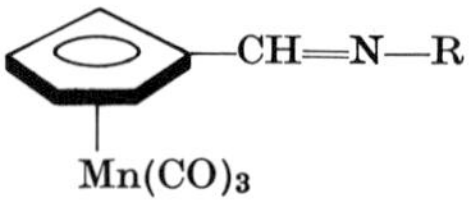

**(204)**

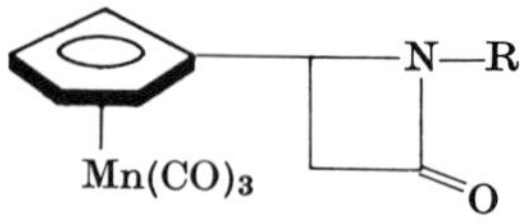

**(205)**

[156] E. Cuingnet and M. Adalberon, *C. R. Acad. Sci.* **258**, 3053 (1964).

Sec. IV.]  HETEROCYCLIC FERROCENES  43

Azaferrocene (**206**) has been prepared by the reaction of sodium pyrrole, cyclopentadienylsodium, and ferrous chloride,[157] or better by the reaction of potassium pyrrole with cyclopentadienyliron dicarbonyl iodide.[158] It has been found[159] that dioxane is a better solvent than benzene for this latter reaction. This reaction has also been applied to 2,4- and 2,5-dimethylpyrrole,[158] 3-acetyl-2-methyl- and 3-acetyl-2,4-dimethylpyrrole,[160] and 2-methylpyrrole.[159] The 2-methylazaferrocene from this latter pyrrole has been resolved with (−)6,6′-dinitrodiphenic acid.[159] A compound (**207**) believed to be an intermediate in the formation of azaferrocenes has been isolated.[161]

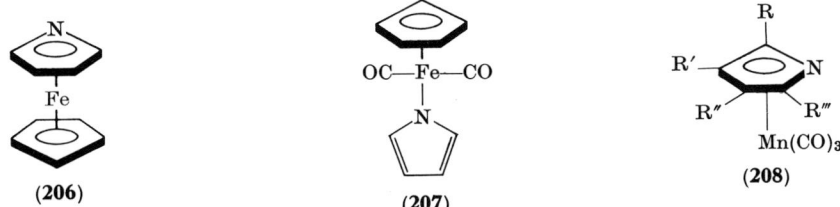

(**206**)   (**207**)   (**208**)

Aza analogs of cyclopentadienylmanganese tricarbonyl have also been prepared.[158, 160, 162] Application of this sequence to pyrazole, imidazole, and 1,2,4-triazole, however, led not to aza analogs but to coordination polymers.[163] A mixed manganese and chromium carbonyl complex of 2-benzylpyrrole has been prepared.[163a]

The azametallocenes, such as **206** and **208**, are less stable than their cyclopentadienyl analogs.[158] The p$K_a$ for azaferrocene (**206**) is similar to that of quinoline, whereas the manganese compound was a much weaker base.[158] Azaferrocene yields a picrate and a methiodide, but, as might be expected, it does not undergo electrophilic substitution.[158] The mass spectra of azametallocenes have been discussed.[163b] The preparation of cyclopentadienyliodobis(imidazole)cobalt(III)iodide has been reported.[163c]

[157] R. B. King and M. B. Bisnette, *Inorg. Chem.* **3**, 796 (1964).
[158] K. K. Joshi, P. L. Pauson, A. R. Qazi, and W. H. Stubbs, *J. Organometal. Chem.* **1**, 471 (1964).
[159] K. Bauer, H. Falk, and K. Schlögl, *Angew. Chem.* **81**, 150 (1969).
[160] P. L. Pauson, A. R. Qazi, and B. W. Rockett, *J. Organometal. Chem.* **7**, 325 (1967).
[161] P. L. Pauson and A. R. Qazi, *J. Organometal. Chem.* **7**, 321 (1967).
[162] K. K. Joshi and P. L. Pauson, *Proc. Chem. Soc.* 326 (1962).
[163] F. Seel and V. Sperber, *Angew. Chem. Int. Ed.* **7**, 70 (1968).
[163a] K. J. Coleman, C. S. Davies, and N. J. Gogan, *Chem. Commun.* 1414 (1970).
[163b] R. B. King, *Appl. Spectrosc.* **23**, 148 (1968); F. Seel and V. Sperber, *J. Organometal. Chem.* **14**, 405 (1968); R. B. King and A. Efraty, *Org. Mass Spectrom.* **3**, 1227 (1970).
[163c] D. J. O'Sullivan and F. J. Lalor, *J. Organometal. Chem.* **25**, C80 (1970).

## V. Compounds of Potential Medicinal Interest

For the sake of completeness we have included in this section other reports concerning ferrocene and its derivatives, whether or not a heterocyclic system is involved.

Ferrocene itself and a variety of ferrocene derivatives have been studied as potential hematinic agents.[32, 33, 50, 164, 165] They apparently have some value in this area. Ferrocene derivatives, however, were of no value in supplying iron to plants.[166]

A nitrogen mustard containing ferrocene did not exhibit any antineoplastic activity.[167] A variety of ferrocene-containing nitrofurans did not exhibit any significant antibacterial, antifungal, or antiparasitic activity[52]; these same activities were also absent in a series of steroids, sulfonamides, antibiotics, and other miscellaneous compounds which contained a ferrocene system.[94] The hydrazone (209) was among a series of compounds reported to have antiulcer activity.[93] A series of thiosemicarbazones of ferrocenecarboxaldehyde and their copper complexes were ineffective as potential fungicides.[126] Ferrocene analogs of amphetamine and diphenylhydantoin showed very weak anorexic and central nervous system stimulant activity and no central nervous system depressant activity, respectively.[69]

(209)

A variety of ferrocene derivatives have been used for the labeling of immunoproteins and related functions.[168–174]

[164] J. L. Madinaveitia, *Brit. J. Pharmacol.* **24**, 352 (1965).
[165] Imperial Chemical Industries Ltd., Fr. Addn. 79,755 to Fr. Patent 1,305,312 (1963); *Chem. Abstr.* **59**, 11567 (1963).
[166] E. R. Page, *Nature* **212**, 640 (1966).
[167] F. D. Popp, S. Roth, and J. Kirby, *J. Med. Chem.* **6**, 83 (1963).
[168] T. J. Gill, III and L. T. Mann, Jr., *J. Immunol.* **96**, 906 (1966).
[169] H. Franz, *Naturwissenschaften* **54**, 339 (1967).
[170] H. Franz, *Z. Chem.* **7**, 427 (1967).
[171] L. T. Mann, Jr., *J. Label. Compounds* **3**, 87 (1967).
[172] H. Falk, M. Peterlik, and K. Schlögl, *Monatsh. Chem.* **100**, 787 (1969).
[173] H. Franz and G. Scheuner, *Histochemie* **16**, 159 (1968).
[174] M. Peterlik, *Monatsh. Chem.* **98**, 2133 (1967).

# Synthesis and Reactions of 1-Azirines

FRANK W. FOWLER

*Chemistry Department, State University of New York at Stony Brook, Stony Brook, New York*

|  |  |  |
|---|---|---|
| I. Introduction | . . . . . . . . . | 45 |
| II. 2-Azirines | . . . . . . . . . | 46 |
| III. Synthesis of 1-Azirines | . . . . . . . | 48 |
|     A. Neber Reaction and Related Reactions | . . . . | 48 |
|     B. The Preparation and Decomposition of Vinyl Azides | . . | 51 |
|     C. Photolysis and Pyrolysis of Isoxazoles | . . . . | 60 |
|     D. Miscellaneous 1-Azirine Synthesis | . . . . . | 62 |
| IV. Reactions of 1-Azirines | . . . . . . . | 63 |
|     A. Electrophilic Reagents | . . . . . . . | 63 |
|     B. Nucleophilic Reagents | . . . . . . . | 69 |
|     C. Cycloadditions | . . . . . . . . | 74 |
| V. Summary | . . . . . . . . . | 76 |

## I. Introduction

Azirine is the term used to describe a three-membered heterocycle containing one nitrogen atom and one double bond. There are two isomeric azirines (**1** and **2**) which have been designated by *Chemical Abstracts* and "The Ring Index,"[1] as 1*H*- and 2*H*-azirine, respectively. Due to some confusion in the literature using the "indicated hydrogen" nomenclature 2- and 1-azirine have been suggested as terms to describe **1** and **2**. Since many of the recent articles and texts have used the latter designations they will be used in this review.

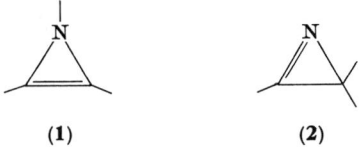

(1)         (2)

[1] A. M. Patterson, "The Ring Index." Amer. Chem. Soc. Washington, D.C., (1960).

## II. 2-Azirines

Although the 1-azirine ring system is well known, to date, there are no authentic examples of 2-azirines in spite of several attempts directed toward their synthesis.

The 2-azirine ring system is of theoretical interest since it is a cyclic conjugated structure containing $4\pi$ electrons and is predicted by Hückel's rule not to be stabilized by cyclic delocalization. Electronically it is analogous to cyclobutadiene.

The decomposition of a *vic*-triazole would appear to offer a direct route to the 2-azirine ring system. However, photolysis of triazole (**3**) does not give the 2-azirine (**4**) but indole (**5**) and ketenimine (**6**).[2]

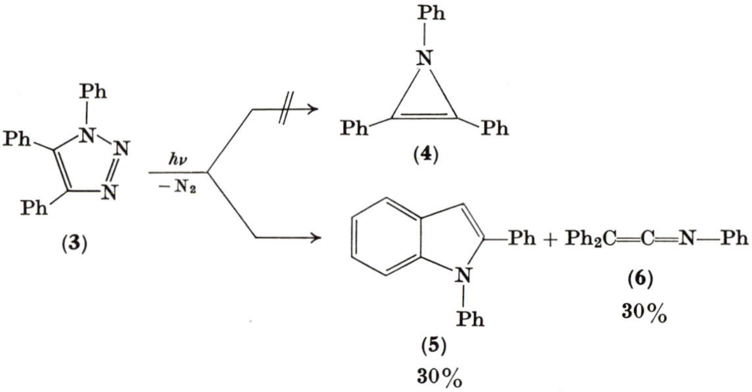

Treatment of triazole (**7**) with acetic anhydride was reported to give the 2-azirine (**8**),[3] but further investigation of this reaction by Rees *et al.* has shown that the correct structure is the oxazole (**9**).[4]

Since strongly electron-withdrawing groups at the 1-position have a stabilizing effect on $1H$-azepines, precursor **10** was thought to have a fair probability of producing a stable 2-azirine. However, dehydrohalogenation of chloroaziridine (**10**) with a variety of bases also failed to give the 2-azirine ring system (**11**).[5]

---

[2] E. M. Burgess, R. Carithers, and L. McCullagh, *J. Amer. Chem. Soc.* **90**, 1923 (1968).
[3] S. Yamada, T. Mizoguchi, and A. Ayata, *J. Pharm. Soc. Japan* **77**, 452 (1957); *Chem. Abstr.* **51**, 146986 (1957).
[4] D. J. Anderson, T. L. Gilchrist, and C. W. Rees, *Chem. Commun.* 147 (1969).
[5] F. W. Fowler and A. Hassner, *J. Amer. Chem. Soc.* **90**, 2875 (1968).

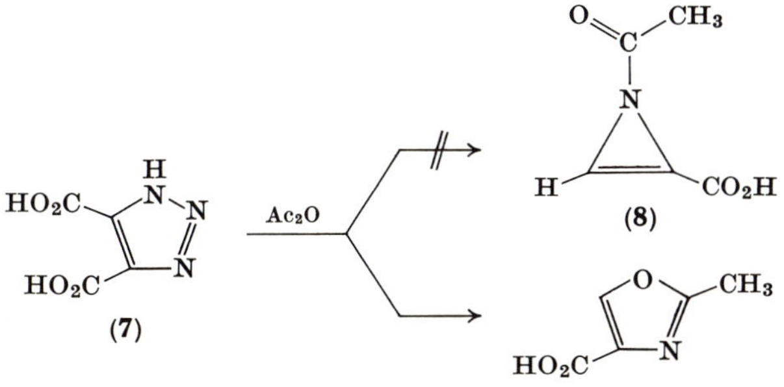

The addition of a nitrene to an acetylene would appear to be another direct route to 2-azirines. The reaction of ethyl azidoformate (**12**) with either diphenyl- or diethylacetylene (**15**) produces mainly the oxazole (**16**).[6,7]

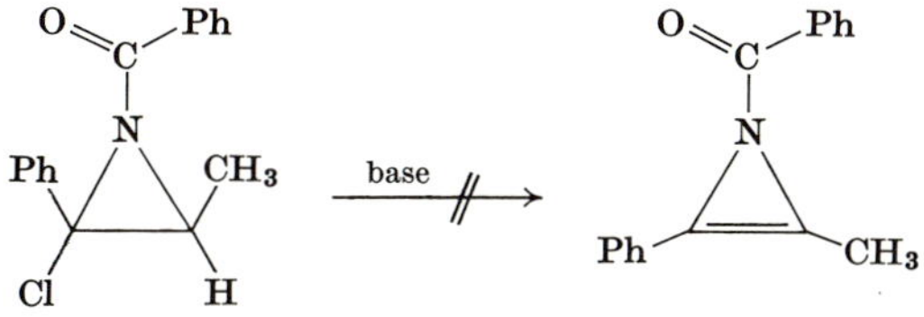

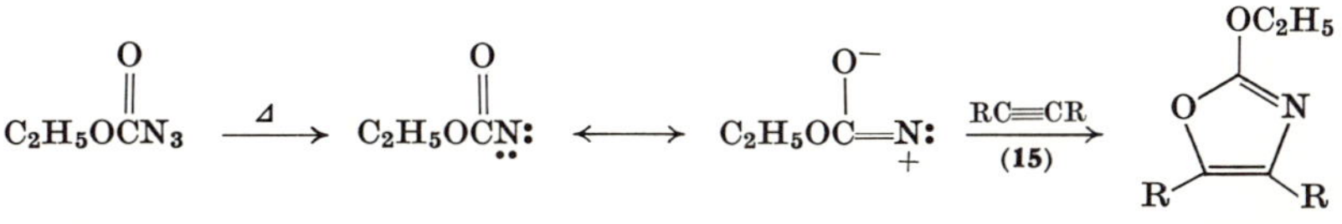

However, the recently reported reaction of phthalimidonitrene with acetylenes probably offers the clearest example of the 2-azirine ring system actually being a reaction intermediate.[4] The oxidation of *N*-aminophthalimide (**17**) with lead tetraacetate produces nitrene (**18**) which in the presence of 3-hexyne gives 1-azirine (**20**). This product is most easily rationalized in terms of 2-azirine (**19**) as an unstable intermediate which undergoes a 1,3-sigmatropic shift of $NR_2$ giving the product (**20**).

[6] R. Huisgen and H. Blaschke, *Chem. Ber.* **98**, 2985 (1965).
[7] J. Meinwald and D. H. Aue, *J. Amer. Chem. Soc.* **88**, 2849 (1966).

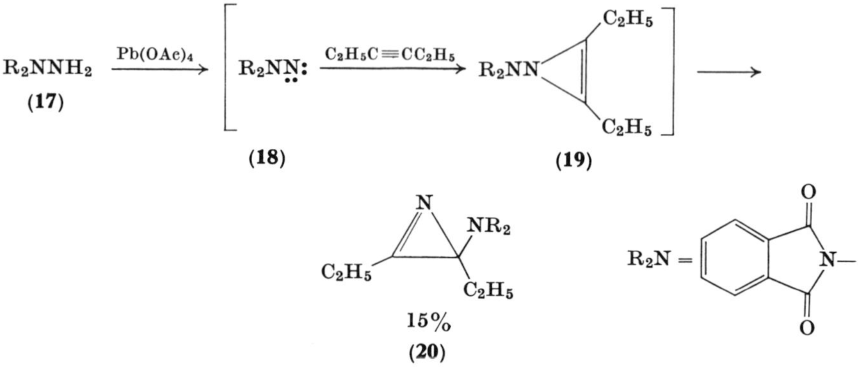

$R_2NNH_2$ $\xrightarrow{Pb(OAc)_4}$ [$R_2N\ddot{N}$: $\xrightarrow{C_2H_5C\equiv CC_2H_5}$ $R_2NN\overset{C_2H_5}{\underset{C_2H_5}{\triangleleft}}$] $\longrightarrow$

(17)                      (18)                      (19)

$C_2H_5\overset{N}{\underset{C_2H_5}{\triangleleft}}NR_2$      $R_2N = \text{(phthalimido)}$

15%
(20)

Since 1-azirines are well known, it does not seem reasonable to attribute the nonexistence of 2-azirines exclusively to strain energy. An unfavorable electronic situation is most likely responsible for their alleged instability. As mentioned previously, the 2-azirine ring system is a cyclic conjugated necessarily planar $\pi$ system containing four $\pi$ electrons and Hückel's rule would not predict it to be stabilized by delocalization. In fact, HMO theory predicts that delocalization results in a less stable $\pi$ system than the open chain analog.[5] Such systems are predicted to be relatively unstable and have recently been designated as antiaromatic.[8,9]

### III. Synthesis of 1-Azirines

#### A. Neber Reaction and Related Reactions

Neber and co-workers undoubtedly synthesized the first authentic 1-azirine in 1932 while studying the reaction of oxime $p$-toluenesulfonates with base to give aminoketones. Neber observed that while trying to prepare the $p$-toluenesulfonate derivatives of oximes (21), pyridine was a sufficiently strong base to convert the intermediate $p$-toluenesulfonates (22) into the aziridines (23).[10,11] Azirines (24) could then be prepared by treating 23 with sodium carbonate.

---

[8] R. Breslow, *Angew. Chem. Int. Ed. Engl.* **7**, 565 (1968).
[9] R. Breslow, J. Brown, and J. J. Gajewski, *J. Amer. Chem. Soc.* **89**, 4383 (1967).
[10] P. W. Neber and A. Burgard, *Ann.* **493**, 281 (1932).
[11] P. W. Neber and G. Huh, *Ann.* **515**, 283 (1935).

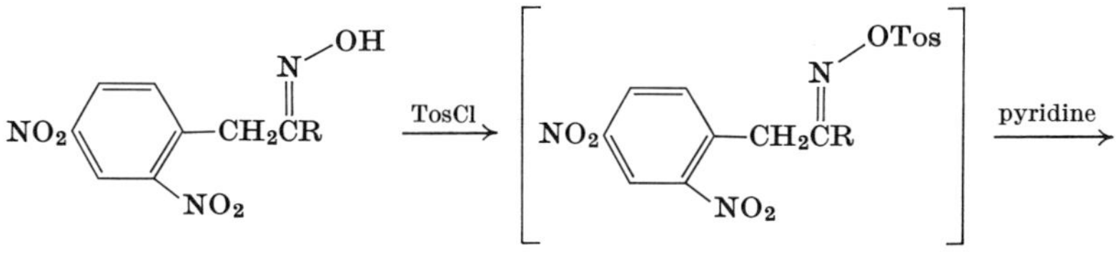

Sec. III. A.]  SYNTHESIS AND REACTIONS OF 1-AZIRINES  49

(21) → [ (22) ] pyridine →

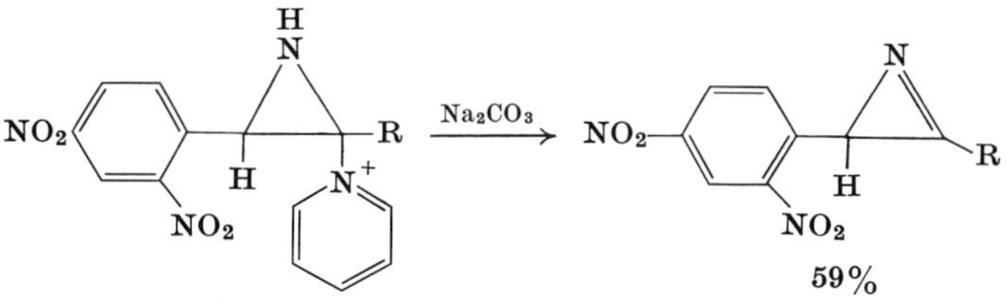

(23) → Na₂CO₃ → (24) 59%

R = CH₃ or Ph

Because of the high degree of instability and reactivity that might be expected for the 1-azirine structure, Cram and Hatch reinvestigated, but confirmed Neber's original assignments.[12]

The Neber reaction generally does not lend itself as a useful method for the synthesis of 1-azirines. The above reaction is probably successful because of the strong electron-withdrawing substituent which increases the acidity of the α-hydrogens and allows the ring closure to occur under mild conditions. Nevertheless, a few azirines have been prepared by related Neber reactions.

Smith and Most[13] developed a modified Neber reaction for the synthesis of aminoketones which was used successfully by Parcell for the synthesis of 3,3-dimethyl-2-phenyl-1-azirine (26).[14]

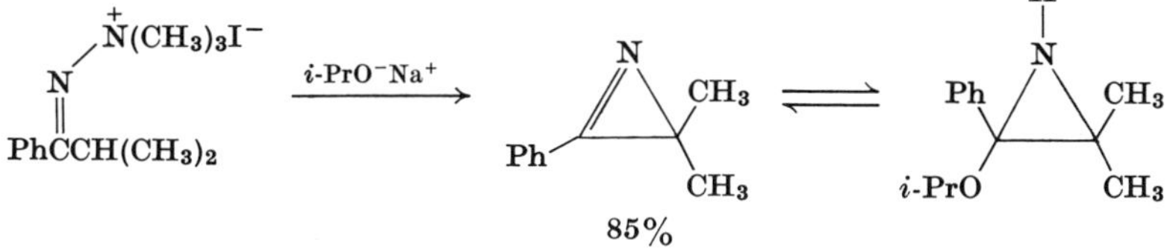

(25) $\xrightarrow{i\text{-PrO}^-\text{Na}^+}$ (26) 85% ⇌ (27)

[12] D. J. Cram and M. J. Hatch, *J. Amer. Chem. Soc.* **75**, 33 (1953).
[13] P. A. S. Smith and E. E. Most, *J. Org. Chem.* **22**, 358 (1957).
[14] R. F. Parcell, *Chem. Ind. (London)* 1396 (1963).

Morrow *et al.* applied this reaction to the synthesis of the steroidal spiroazirine (**29**) and discovered that higher yields can be achieved if the dimethylsulfinyl carbanion is substituted for sodium isopropoxide as the base.[15,16] Unfortunately, this method lacks generality.

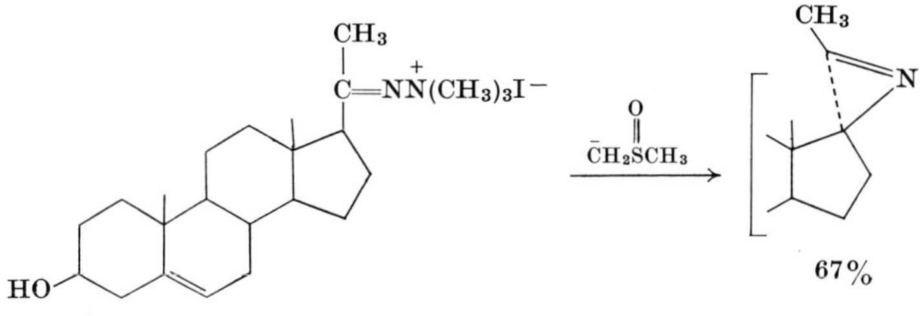

(**28**) → (**29**) 67%

An attempt by Sato[17] to prepare 2-phenyl-1-azirine from the dimethylhydrazone methiodide (**30**) resulted in the formation of 2,4-diphenylpyrrole (**31**). Although the 1-azirine may be an intermediate (see Section IV, B, 2) in this reaction, its presence could not be detected.

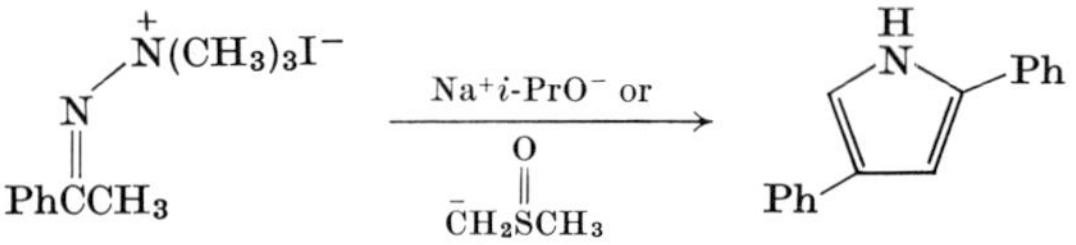

(**30**) → (**31**)

Apparently, this method is affected by the type of hydrogen that is available on the α-carbon. As Sato and previous workers discovered, this reaction proceeds well if the α-hydrogen is tertiary. Thus, spiro-1-azirine (**33**) is prepared in 80% yield when **32** is treated with sodium isopropoxide.

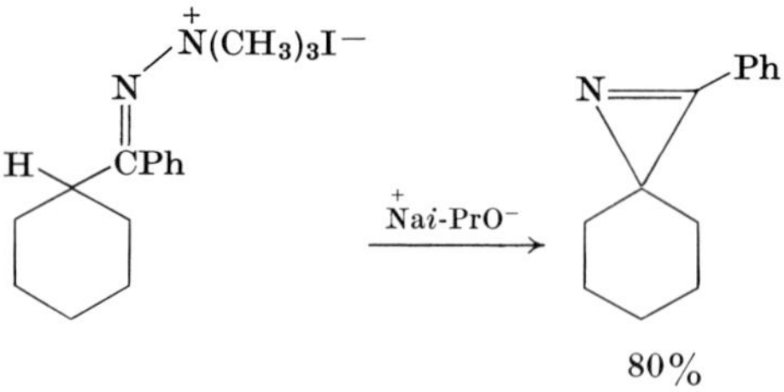

(**32**) → (**33**) 80%

[15] D. F. Morrow, M. E. Butler, and E. C. Y. Huang, *J. Org. Chem.* **30**, 579 (1965).
[16] D. F. Morrow and M. E. Butler, *J. Heterocycl. Chem.* **1**, 53 (1964).
[17] S. Sato, *Bull. Chem. Soc. (Japan)* **41**, 1440 (1968).

The reaction of **34**, which contains a secondary hydrogen alpha to the hydrazone, with sodium isopropoxide gave a mixture of alkoxyaziridines (**35** and **36**). On heating, **35** and **36** partially eliminated isopropanol to give the 1-azirine (**37**).

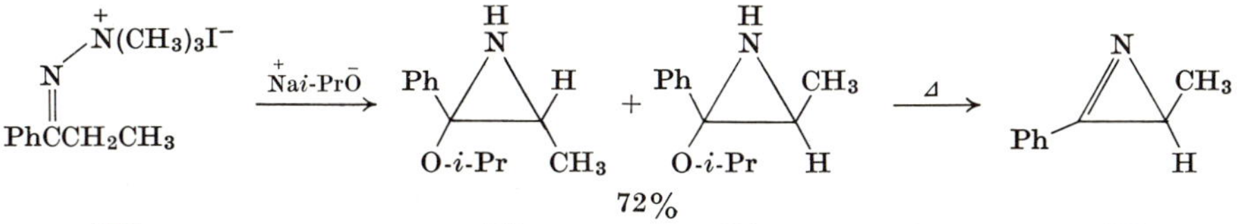

$$\underset{(34)}{\underset{PhCCH_2CH_3}{\overset{\overset{+}{N}(CH_3)_3I^-}{N}}} \xrightarrow{Na\textit{i}-Pr\bar{O}} \underset{(35)}{\underset{O\text{-}\textit{i}\text{-}Pr\ \ CH_3}{Ph\triangle H}} + \underset{(36)}{\underset{O\text{-}\textit{i}\text{-}Pr\ \ H}{Ph\triangle CH_3}} \xrightarrow{\Delta} \underset{(37)}{\underset{Ph\ \ \ \ \ \ H}{\triangle CH_3}}$$

72%

The use of potassium *t*-butoxide as the base directly produced a low yield of 1-azirine contaminated with propiophenone. However, Nair has reported that azirine (**37**) can be prepared from hydrazone (**34**) in 63% yield by using the dimethylsulfinyl carbanion as the base.[18]

Even if the α-hydrogen is tertiary, this does not always ensure success of the reaction. Treatment of **38** with sodium isopropoxide

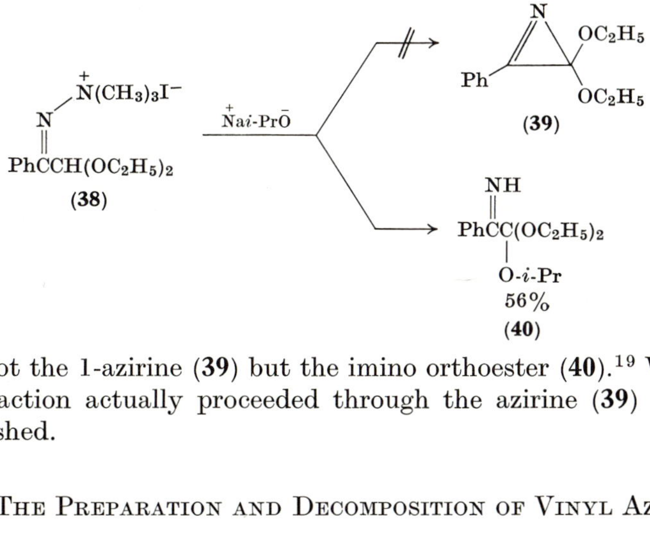

gave not the 1-azirine (**39**) but the imino orthoester (**40**).[19] Whether this reaction actually proceeded through the azirine (**39**) was not established.

### B. THE PREPARATION AND DECOMPOSITION OF VINYL AZIDES

Smolinsky developed the first general synthesis of 1-azirines by the vapor phase pyrolysis of vinyl azides (**41**).[20, 21] In addition to 50–60%

[18] V. Nair, *J. Org. Chem.* **33**, 2121 (1968).
[19] K. R. Henery-Logan and T. L. Fridinger, *J. Amer. Chem. Soc.* **89**, 5724 (1967).
[20] G. Smolinsky, *J. Org. Chem.* **27**, 3557 (1962).
[21] G. Smolinsky, *J. Amer. Chem. Soc.* **83**, 4483 (1961).

yields of the 1-azirines (**42**), small amounts (5–6%) of the ketenimines (**43**) are also formed. The ketenimine results from a migration of the group alpha to the azido function by a Curtius-type rearrangement.

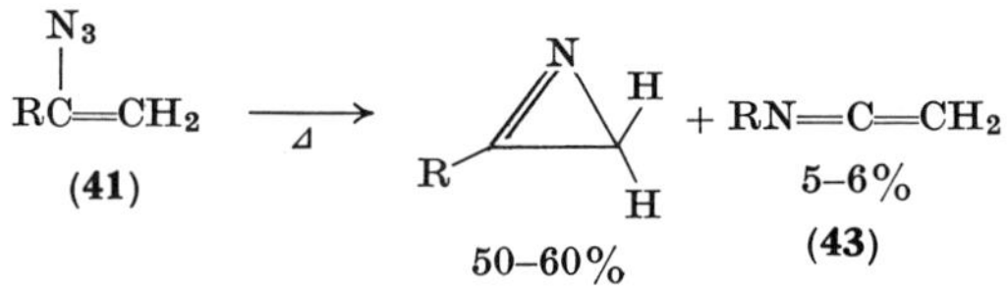

$$\underset{(41)}{\overset{N_3}{\underset{|}{RC}}{=}CH_2} \xrightarrow{\Delta} \underset{\underset{(42)}{50-60\%}}{R\overset{N}{\triangle}{\overset{H}{H}}} + \underset{\underset{(43)}{5-6\%}}{RN{=}C{=}CH_2}$$

(a) R = Ph
(b) R = *n*-Bu
(c) R = *p*-CH₃

Irradiation of vinyl azides also produces 1-azirines. In some instances it may be the method of choice since the reaction can be carried out at low temperature and little polymerization of the azirines occurs.

Three plausible mechanisms[20] have been postulated for the thermal decomposition of vinyl azides (**41**) to 1-azirines (**47**). The azide can either (1) lose nitrogen to give the vinyl nitrene which then cyclizes to the 1-azirine, (2) lose nitrogen with simultaneous ring closure (**45**), or (3) first cyclize to triazole (**46**) which then loses nitrogen. To date little experimental evidence exists to favor any of these alternatives.

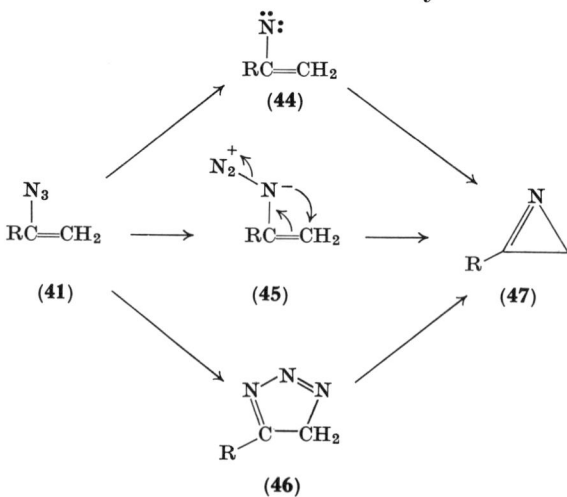

1-Azirines are usually vile-smelling compounds which cause skin irritation. They are prone to polymerization unless kept cold under an inert atmosphere. The infrared absorption of the carbon–nitrogen double bond in 2-substituted unconjugated 1-azirines normally occurs at ca. 1770 cm$^{-1}$.[22] The NMR spectra of 1-azirines are consistent with the strained nature of this heterocycle. The $^{13}$C—H coupling constant for the methylene hydrogens in 2-phenyl-1-azirine (**42a**) is 178 Hz which indicates 36% $s$ character of the exocyclic carbon orbitals at C-3.[23] This value can be compared to 32% obtained for ethylenimine and is consistent with the smaller endocyclic interorbital angle caused by the presence of the carbon–nitrogen double bond.

**(42a)**   **(48)**

Hydrogens Ha and Hb in **48** are both alpha to an imine bond and hydrogens Hb are attached to a carbon atom bearing an electronegative nitrogen. From these considerations only, Hb should occur downfield with respect to Ha. However, hydrogens Hb occur upfield 1.76 ppm from Ha. Also, the methylene hydrogens of 1-azirines occur slightly upfield from the corresponding aziridines. The shielding of hydrogens attached to saturated carbon in three-membered rings has been attributed to diamagnetic ring currents in these three-membered rings which are apparently greater in 1-azirines than aziridines.[22]

Azirines substituted at the 2-position are most stable and are easily prepared. Azirines fused to rings containing less than eight atoms are unknown, although their existence as unstable intermediates has been inferred.[22] For example, the photolysis of 1-azido-3,4-dihydronaphthalene (**49**) in inert solvents gives only polymeric substances, whereas the photolysis in methanol containing sodium methoxide gives aminoketal (**51**) suggesting the intermediacy of azirine (**50**).

**(49)**   **(50)**   **(51)**

[22] F. W. Fowler and A. Hassner, *J. Amer. Chem. Soc.* **90**, 2869 (1968).
[23] A. Hassner and F. W. Fowler, *J. Amer. Chem. Soc.* **90**, 2875 (1968).

It should be noted that the decomposition of a number of β-azidovinyl ketones gives the isoxazole rather than the azirine.[24–28] Although it is unlikely that the azirine is an intermediate in these reactions this point has not been established conclusively. In contrast, β-azido esters behave normally, giving the azirines.[22]

Photolysis of 5,5-dimethyl-3-azido-2-cyclohexene-1-one gave only the azepine (**52**).[26] This product probably arises by a Curtius-type rearrangement to give a cyclic ketenimine which is hydrolyzed to the product **52**.

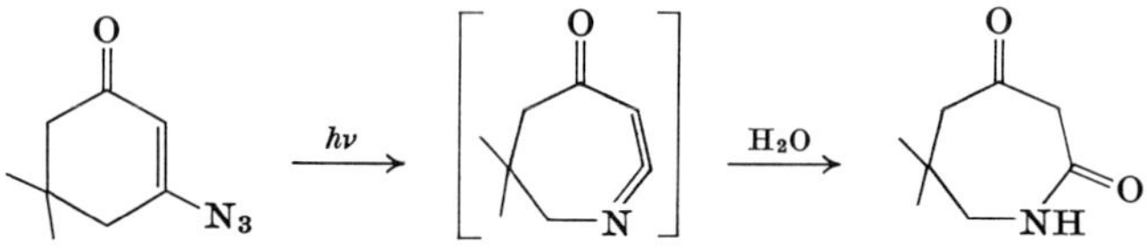

The thermal decomposition of terminal vinyl azides was originally believed to give only nitriles or, in some cases, indoles. The presence of 3-monosubstituted 1-azirines, however, has been inferred in the photolytic decomposition of some terminal azides.[22, 29, 30] The nitrile is thought to arise in a similar manner to the ketenimine by an analogous Curtius-type rearrangement. The ketenimine (**53**) derived from terminal azides is unstable and rearranges to the nitrile (**54**).

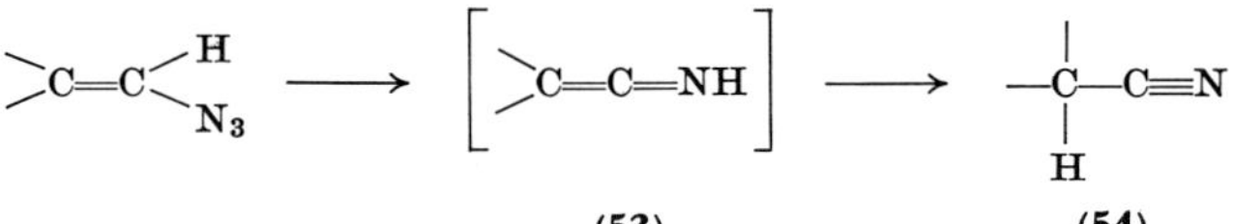

Recently Isomura *et al.* have shown that both photolysis and pyrolysis of vinyl azides (**55**) can lead to 1-azirines (**56**).[27, 28] Pyrolysis

---

[24] J. S. Meek and J. S. Fowler, *J. Org. Chem.* **33**, 3418 (1968).
[25] S. Sato[26] has reported that photolysis of β-azidovinyl phenyl ketone gives benzoylacetonitrile and not 3-benzoyl-1-azirine.
[26] S. Sato, *Bull. Chem. Soc. Japan* **41**, 2524 (1968).
[27] K. Isomura, M. Okada, and H. Taniguchi, *Tetrahedron Lett.* 4073 (1969).
[28] K. Isomura, S. Kobayashi, and H. Taniguchi, *Tetrahedron Lett.* 3499 (1968).
[29] J. H. Boyer, W. E. Krueger, and G. J. Mikole, *J. Amer. Chem. Soc.* **89**, 5504 (1967).
[30] G. Smolinsky and C. A. Pryde, *J. Org. Chem.* **33**, 2411 (1968).

of the 1-azirines leads to similar products that were originally formulated as arising directly from the vinyl azide. Photolysis of either *cis*-

$$\underset{(55)}{\overset{R}{\underset{R'}{\diagdown}}C=C\overset{H}{\underset{N_3}{\diagup}}} \xrightarrow{h\nu \text{ or } \Delta} \underset{(56)}{\overset{N}{\underset{R'}{\triangle}}\overset{}{\underset{H}{}}}$$

or *trans*-β-azidostyrene (**57**) with 3650 Å light at −50° in a nitrogen atmosphere gives only 3-phenyl-1-azirine (**58**). Pyrolysis of this azirine in boiling *n*-hexadecane results in an equimolar mixture of indole (**59**) and phenylacetonitrile (**60**).

$$\underset{\underset{(57)}{cis \text{ or } trans}}{\text{PhCH}=\text{CHN}_3} \xrightarrow[-50°]{h\nu} \underset{(58)}{\text{Ph}\overset{N}{\triangle}\text{H}} \xrightarrow[287°]{n\text{-}C_{16}H_{34}} \underset{(59)}{\text{[indole]}} + \underset{\underset{1:1}{(60)}}{\text{PhCH}_2\text{C}\equiv\text{N}}$$

These results suggest the intriguing possibility that the azirine may be in thermal equilibrium with the vinyl nitrene. The vinyl nitrene can then react further to give the thermodynamically more stable nitrile or, in some cases, the indole. However, the intermediacy of the vinyl nitrene in the 1-azirine pyrolysis reactions has not been determined.

$$\overset{N}{\triangle} \rightleftharpoons \overset{:\ddot{N}}{\diagup}\diagdown \longrightarrow \text{Products}$$

These 2-unsubstituted 1-azirines, compared with those substituted at the 2-position, are very unstable, polymerizing rapidly at room temperature in the presence of oxygen. The carbon–nitrogen double bond of the 2-unsubstituted 1-azirines absorbs at ca. 100 cm$^{-1}$ lower energy than the 2-alkyl-substituted 1-azirines. This result is consistent with the corresponding cyclopropanes. The aldimine hydrogen is also anomalous in that its NMR absorption occurs at about 0τ. This same result also occurs with the corresponding cyclopropanes and is attributed to the 2-hydrogen being situated in the deshielding zone of the 1-azirine.[27]

Carrying out the reaction at low temperature can be particularly advantageous in the preparation of the 2-unsubstituted 1-azirines, since the azirines are thermally unstable. The major by-product is usually the isomeric nitrile.

Bauer and Hafner have prepared the thermally unstable 1-azirine (**62**) from vinyl azide (**61**) which is also unsubstituted at the 2-position.[31] This azirine readily loses hydrogen cyanide to give the carbene (see Section IV,C).

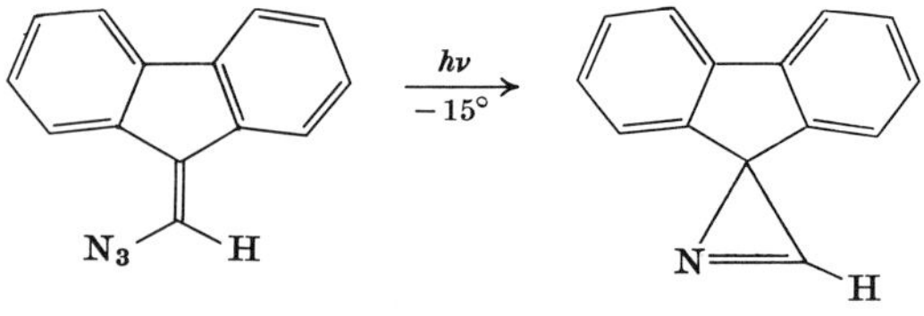

The thermal decomposition of perfluoro-2-azidopropene (**64**), prepared from perfluoro propene (**63**), was originally reported to give azetine (**65**). However, two groups[32, 33] independently observed that perfluoroazirine (**66**) is the product of this reaction.

This azirine is unstable with respect to azirine (**69**) and is converted

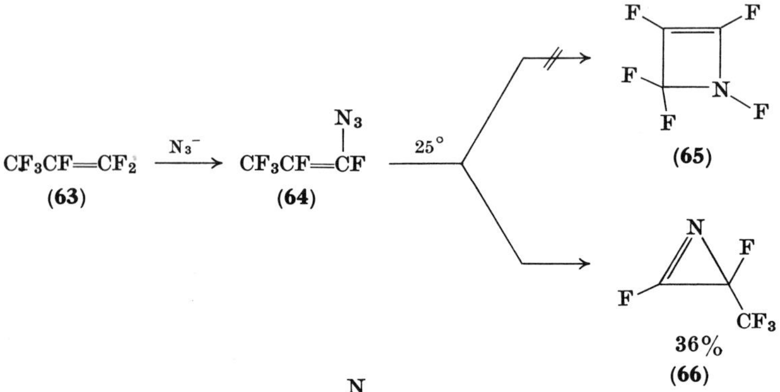

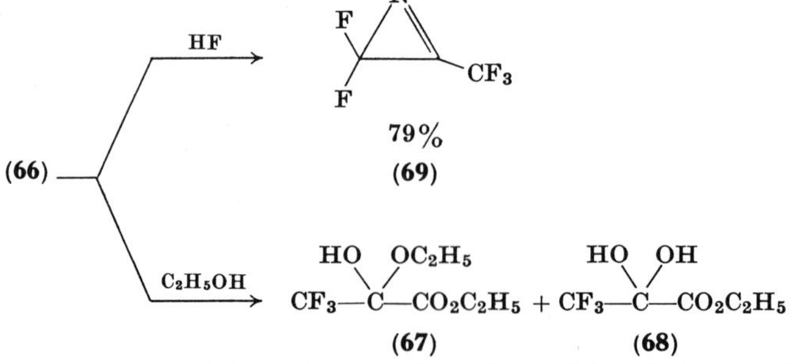

[31] W. Bauer and K. Hafner, *Angew. Chem. Int. Ed. Engl.* **8**, 772 (1969).
[32] C. S. Cleaver and C. G. Krespan, *J. Amer. Chem. Soc.* **87**, 3716 (1965).
[33] R. E. Banks and G. J. Moore, *J. Chem. Soc. C* 2304 (1966).

into its isomer in the presence of catalytic amounts of hydrogen fluoride. Azirine (**66**) is also more prone to polymerization than azirine (**69**) which is consistent with the hydrogen-substituted azirines previously discussed. Ethanolysis of azirine (**66**) results in destruction of the three-membered ring giving ethyl-2-ethoxy-3,3,3-trifluoro-2-hydroxy- and 3,3,3-trifluoro-2,2-dihydroxypropionate (**67** and **68**).[34]

The decomposition of vinyl azides offers an excellent method for the preparation of 1-azirines. The main limitation of this method is the availability of the prerequisite vinyl azide, which can, in some instances, be a severe one.

The reagent iodine azide offers an excellent route to vinyl azides.[35, 36] Iodine azide adds to many olefinic compounds to give β-iodoazides. Elimination of hydrogen iodide with base usually occurs preferentially in the direction of the azide function to give the vinyl azide. The direction of iodine azide addition is consistent with electrophilic attack of $I^+$ to give a cyclic iodonium ion which is opened by azide ion. Usually from a given olefin only one vinyl azide is obtained. Thus, 1-hexene (**70**) leads to 2-azidohexene (**41b**) rather than the isomeric 1-azidohexene. A conjugated olefin such as methyl cinnamate (**71**) gives methyl β-azidocinnamate (**72**).

$$n\text{-BuCH}=CH_2 \xrightarrow[\text{(2) base}]{\text{(1) IN}_3} n\text{-BuC}=CH_2 \quad 78\%$$
$$\qquad\qquad\qquad\qquad\qquad\quad |$$
$$\qquad\qquad\qquad\qquad\qquad\; N_3$$
(**70**)       (**41b**)

$$PhCH=CHCO_2CH_3 \xrightarrow[\text{(2) base}]{\text{(1) IN}_3} PhC=CHCO_2CH_3$$
$$\qquad\qquad\qquad\qquad\qquad\qquad\qquad |$$
$$\qquad\qquad\qquad\qquad\qquad\qquad\quad N_3$$
(**71**)       (**72**)

Frequently, steric effects also play an important role in determining the orientation of the azido function; *t*-butylethylene (**73**) leads to vinyl azide (**74**) rather than the azide (**75**) expected from electronic considerations.

$$t\text{-BuCH}=CH_2 \xrightarrow[\text{(2) base}]{\text{(1) IN}_3} t\text{-BuCH}=CHN_3 \qquad t\text{-BuC}=CH_2$$
$$\qquad\qquad\qquad\qquad\qquad\quad 53\% \qquad\qquad\qquad\; |$$
$$\qquad\qquad\qquad\qquad\qquad\qquad\qquad\qquad\qquad\; N_3$$
(**73**)     (**74**)     (**75**)

[34] R. E. Banks, D. Berry, and G. J. Moore, *J. Chem. Soc. C* 2598 (1969).
[35] A. Hassner and L. A. Levy, *J. Amer. Chem. Soc.* **87**, 4203 (1965).
[36] F. W. Fowler, A. Hassner, and L. A. Levy, *J. Amer. Chem. Soc.* **89**, 2077 (1967).

The free radical addition of bromine azide has been complementary to iodine azide for the synthesis of some vinyl azides. Using bromine azide β-azidostyrene (**57**) has been prepared from styrene (**76**) through bromoazide (**77**).[37]

$$\text{PhCH=CH}_2 \xrightarrow{\text{BrN}_3} \underset{\text{(77)}}{\text{Ph}\overset{\text{Br}}{\underset{|}{\text{C}}}\text{H}\overset{\text{N}_3}{\underset{|}{\text{C}}}\text{H}_2} \xrightarrow{\text{base}} \underset{\text{(57)}}{\text{PhCH=CHN}_3}$$
(**76**)

Vinyl azides have been prepared by treating epoxides (**78**) with azide ion to give azido alcohols (**79**) which are then dehydrated to the vinyl azide.[30] This method also complements the iodine azide method since the epoxide route usually gives the isomeric vinyl azide. β-Hydroxy azides (**81**) can also be prepared by the sodium borohydride reduction of α-azidoketones (**80**).

$$\underset{\text{(78)}}{\text{Ph}_2\overset{\text{O}}{\overset{|}{\triangle}}\text{H}_2} \xrightarrow[\text{CH}_3\text{OH}]{\text{NaN}_3} \underset{\text{(79)}}{\text{Ph}_2\overset{\text{OH}}{\underset{|}{\text{C}}}-\overset{}{\underset{|}{\text{CH}_2}}\atop{\text{N}_3}} \xrightarrow[\text{DMF-Py-SO}_2]{\text{CH}_3\text{SO}_2\text{Cl}} \underset{50\%}{\text{Ph}_2\text{C=CHN}_3}$$

$$\underset{\text{(80)}}{\text{Ph}\overset{\text{O}}{\overset{\|}{\text{C}}}\text{CH}_2\text{N}_3} \xrightarrow{\text{NaBH}_4} \underset{\text{(81)}}{\text{Ph}\overset{\text{OH}}{\underset{|}{\text{C}}}\text{HCH}_2\text{N}_3}$$

The displacement of activated vinyl halides and sulfinates has also been used for the synthesis of some vinyl azides (**82**).[24, 30, 38, 39]

$$\underset{\text{Cl}}{\overset{\text{H}}{\diagdown}}\text{C=CH}\overset{\text{O}}{\overset{\|}{\text{C}}}\text{Ph} \xrightarrow{\text{NaN}_3} \underset{\text{N}_3}{\overset{\text{H}}{\diagdown}}\text{C=CH}\overset{\text{O}}{\overset{\|}{\text{C}}}\text{Ph}$$
$$80\text{–}90\%$$
(**82**)

Harvey and Ratts have discovered an interesting vinyl azide synthesis from allenic esters.[40] Treatment of allenes (**83a** and **b**) with

[37] A. Hassner and F. P. Boerwinkle, *J. Amer. Chem. Soc.* **90**, 216 (1968).
[38] A. N. Nesmeyanov and M. I. Rybinskaya, *Izv. Akad. Nauk SSSR, Otd. Khim. Nauk* 816 (1962); *Chem. Abstr.* **58**, 3408d (1963).
[39] S. Maiorana, *Ann. Chim.* **56**, 1531 (1966).
[40] G. R. Harvey and K. W. Ratts, *J. Org. Chem.* **31**, 3907 (1966).

sodium azide in tetrahydrofuran (THF)–water gave good yields (70 and 74%) of the vinyl azides (**84a** and **b**). Terminal disubstituted allenic esters or amides failed to give the vinyl azide or any definite product. The fate of terminal monosubstituted allenic esters upon treatment with azide ion was not reported.

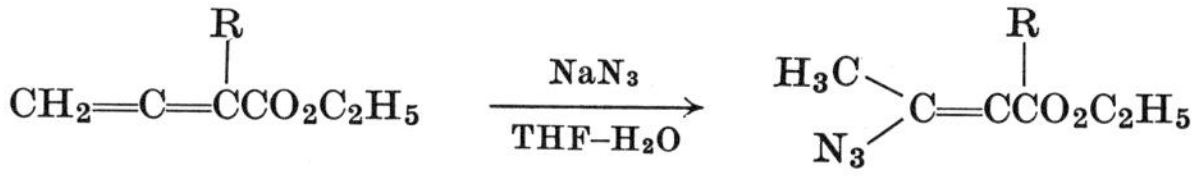

$$CH_2=C(R)-CCO_2C_2H_5 \xrightarrow{NaN_3, THF-H_2O} \underset{N_3}{\overset{H_3C}{>}}C=C(R)CO_2C_2H_5$$

(83)   (84)

(a) R = H
(b) R = CH$_3$

Related to the above method is the addition of hydrazoic acid to conjugated acetylenes (**85**) which has been applied to the synthesis of the vinyl azide (**86**).[41]

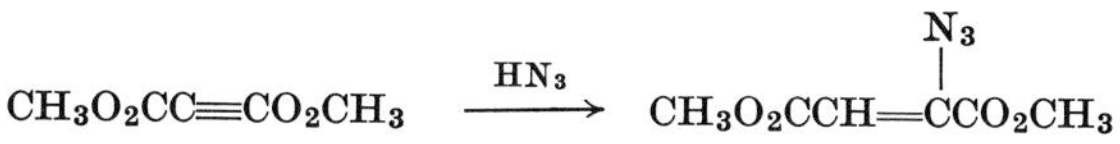

$$CH_3O_2CC\equiv CO_2CH_3 \xrightarrow{HN_3} CH_3O_2CCH=C(N_3)CO_2CH_3$$

(85)   (86)

The base-catalyzed condensation of α-azido esters and ketones with aromatic aldehydes has recently been developed as a new vinyl azide synthesis.[42,43] The yields range from moderate to excellent in some cases. The thermal decomposition of ethyl α-azidocinnamate (**87**) in xylene gives only 2-ethoxycarbonylindole (**88**).[44] The unstable 2-ethoxycarbonyl-3-phenyl-1-azirine could be detected if the thermolysis was carried out at a lower temperature. This fact indicates that the 1-azirine is probably an intermediate leading to the indole, although the intermediacy of the vinyl nitrene could not be established. This result is similar to that observed by Isomura *et al.* on the pyrolysis of terminal vinyl azides.[27,28]

---

[41] V. G. Ostroverkhov and E. A. Shilov, *Ukr. Khim. Zh.* **23**, 615 (1957); *Chem. Abstr.* **52**, 7828d (1958).

[42] D. Knittel, H. Hemetsberger, and H. Weidman, *Monatsh. Chem.* **101**, 157 (1970).

[43] H. Hemetsberger, D. Knittel, and H. Weidman, *Monatsh. Chem.* **100**, 599 (1969).

[44] H. Hemetsberger, D. Knittel, and H. Weidman, *Monatsh. Chem.* **101**, 161 (1970).

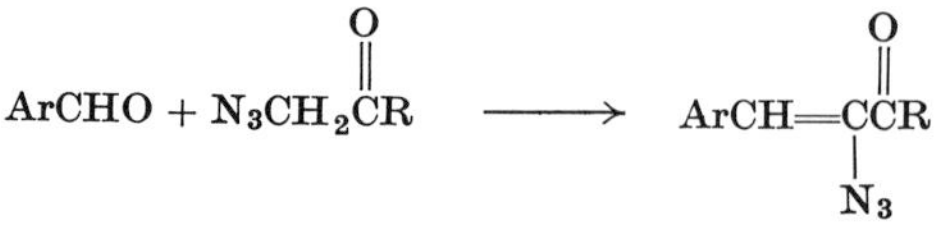

$$\text{ArCHO} + \text{N}_3\text{CH}_2\overset{\text{O}}{\overset{\|}{\text{C}}}\text{R} \longrightarrow \text{ArCH}=\overset{\text{O}}{\overset{\|}{\text{C}}}\text{CR}$$
$$\underset{\text{N}_3}{}$$

(a) $R = OC_2H_5$
(b) $R = Ph$

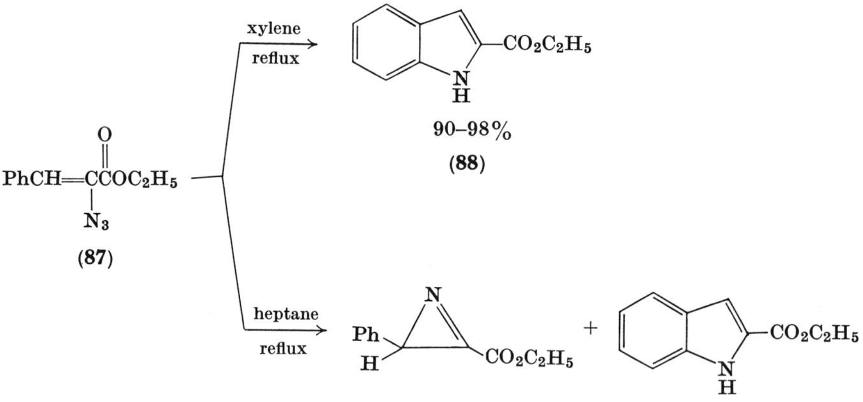

PhCH=CCOC$_2$H$_5$ (87), with N$_3$ substituent:

- xylene reflux → indole-2-CO$_2$C$_2$H$_5$ (88), 90–98%
- heptane reflux → 2-Ph-3-CO$_2$C$_2$H$_5$-azirine (80%) + indole-2-CO$_2$C$_2$H$_5$ (20%)

J. H. Boyer et al. have observed interesting products on storage of α-azidostyrene.[45] In addition to 2-phenyl-1-azirine, 3,6-diphenyl-pyridazine (89) and 2,5-diphenylpyrrole (90) were also formed. The mechanistic details of this reaction were not elucidated.

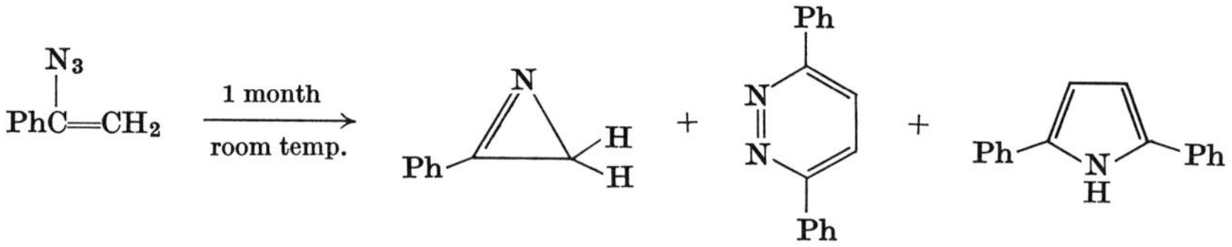

$$\text{PhC}(\text{N}_3)=\text{CH}_2 \xrightarrow[\text{room temp.}]{1 \text{ month}} \text{2-Ph-azirine} + \text{3,6-diphenylpyridazine (89)} + \text{2,5-diphenylpyrrole (90)}$$

10%    7% (89)    32% (90)

## C. Photolysis and Pyrolysis of Isoxazoles

Ullman and Singh[46, 47] first showed that 1-azirines could be prepared from isoxazoles. While studying the photochemical rearrange-

[45] J. H. Boyer, W. E. Krueger and R. Modler, *Tetrahedron Lett.* 5979 (1968).
[46] B. Singh and E. F. Ullman, *J. Amer. Chem. Soc.* **89**, 6911 (1967).
[47] E. F. Ullman and B. Singh, *J. Amer. Chem. Soc.* **88**, 1844 (1966).

ment of 3,5-diphenylisoxazole to 2,5-diphenyloxazole, they observed the formation of an oily intermediate which proved to be 2-phenyl-3-benzoyl-1-azirine (**92**). This reaction shows a fascinating wavelength dependence. Irradiation at 2537 Å resulted in up to an 82% conversion into azirine (**92**), whereas irradiation of the azirine at wavelengths greater than 3000 Å gave the isoxazole (**91**).

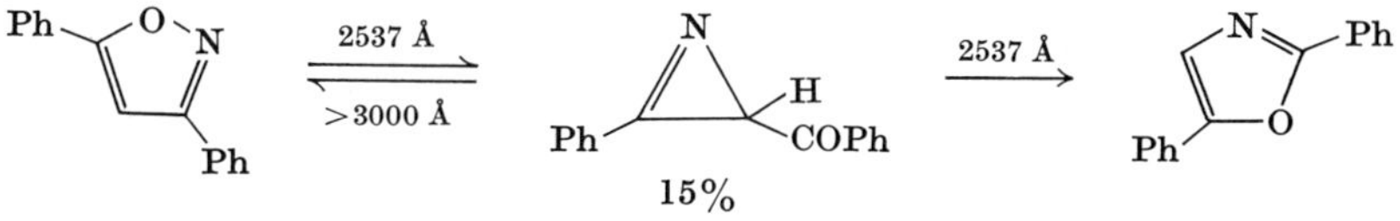

Kurtz and Shechter[48] have observed similarly that irradiation of 3,4,5-triphenylisoxazole gives $N$-phenylbenzoylphenylketenimine, 3-benzoyl-2,3-diphenyl-1-azirine, and 1,4,5-triphenyloxazole. The formation of the azirine and ketenimine from this reaction and from the decomposition of vinyl azides would suggest a similarity in mechanism.

Singh and Ullman[46] have observed that pyrolysis of azirine (**92**) in nonhydroxylic solvents gives a 30% yield of the isoxazole (**91**). This result should be compared with the work of Nishiwaki et al.,[49] who observed that the pyrolysis of 5-alkoxy-substituted isoxazoles (**94**) gives the 1-azirines (**95**). Possibly at high temperatures there is a thermal equilibrium between the isoxazole and 1-azirine and the success of the azirine synthesis is due to the selective removal of the azirine at high temperatures. Although several 1-azirines have been prepared in moderate yield by the above method, the high temperatures needed to bring about the reaction results also in extensive decomposition.

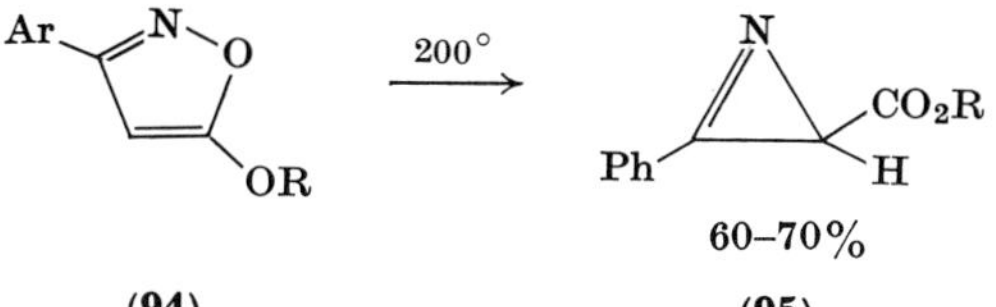

[48] D. W. Kurtz and H. Shechter, *Chem. Commun.* 689 (1966).
[49] T. Nishiwaki, T. Kitimura and A. Nakano, *Tetrahedron* **26**, 453 (1970).

## D. Miscellaneous 1-Azirine Synthesis

Two research groups have independently observed that the pyrolysis of 4,5-dihydro-1,2,5-oxazaphospholes can be a useful synthesis of 1-azirines (**99**) in some cases.[50–53] The oxazaphospholes (**98**) are readily prepared by the cycloaddition of nitrile oxides (**96**) to alkylidenephosphoranes (**97**).

(**96**) (**97**) (**98**) (**99**)

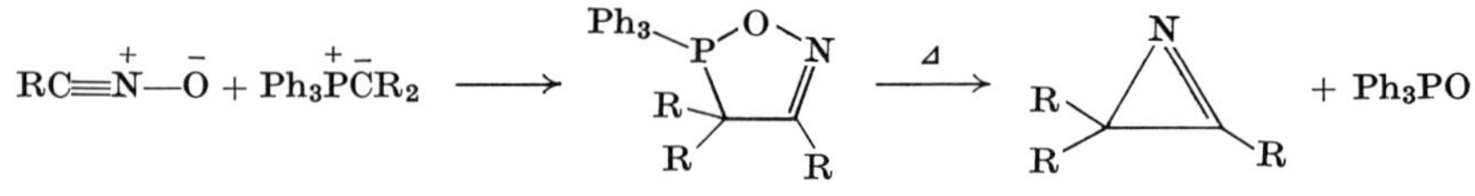

The reaction of the cyclopropyl ylide (**100**) with benzonitrile oxide gave the oxazaphosphole (**101**) in 61% yield.[51] Pyrolysis of **101** at 100°–110° (0.1 torr) gave an 84% yield of the relatively stable spiroazirine (**102**).

(**100**) (**101**) (**102**)

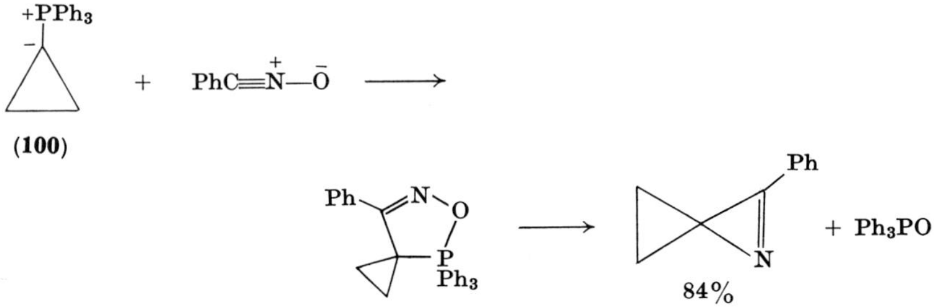

Electron-withdrawing groups such as carbomethoxy on the phosphorus ylide (**103**) result in very unstable oxazaphospholes and

(**103**) (**104**)

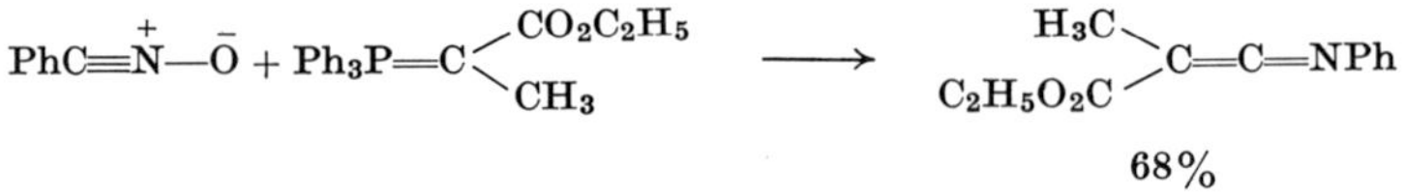

[50] H. J. Bestmann and R. Kunstmann, *Angew. Chem. Int. Ed. Engl.* **5**, 1039 (1966).
[51] H. J. Bestmann and R. Kunstmann, *Chem. Ber.* **102**, 1816 (1969).
[52] R. Huisgen and J. Wulff, *Tetrahedron Lett.* 917 (1967).
[53] R. Huisgen and J. Wulff, *Chem. Ber.* **102**, 1833 (1969).

suppress 1-azirine formation in favor of the ketenimine (**104**). If electron-withdrawing groups are on the nitrile oxide, the oxazaphosphole is still unstable, but the 1-azirines (**106**) are produced.

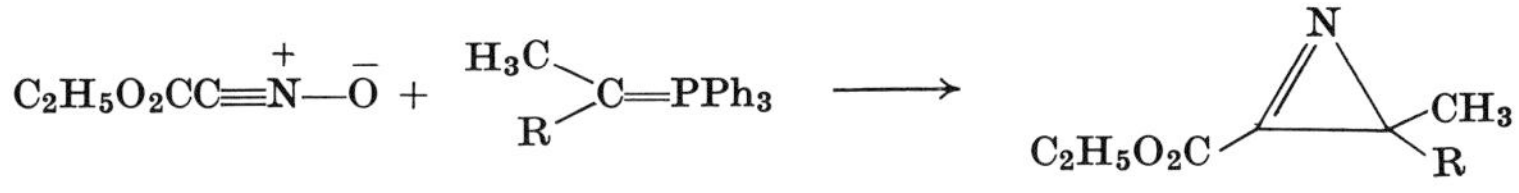

$$C_2H_5O_2CC\equiv\overset{+}{N}-\overset{-}{O} + \underset{R}{\overset{H_3C}{>}}C=PPh_3 \longrightarrow \text{(106)}$$

(**105**) (**106**)

(a) R = $CO_2C_2H_5$ (27%)
(b) R = Ph (18%)

There has been a brief mention of 2-phenylazirine being prepared by the reaction of dimethyl oxosulfonium methylide (**108**) and benzonitrile (**107**).[54] Due to the simplicity of this approach this method deserves further study.

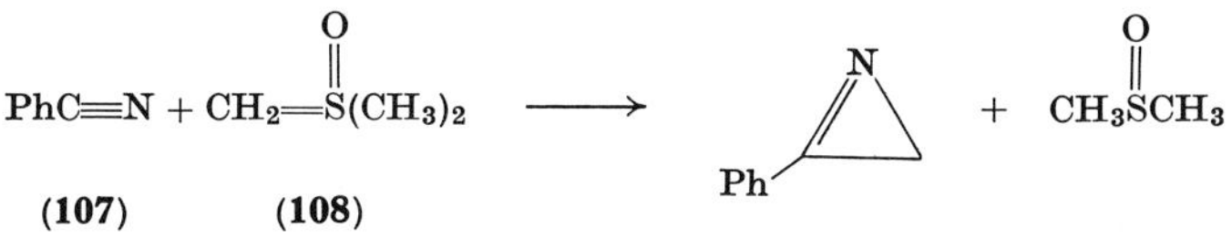

$$PhC\equiv N + CH_2=\overset{O}{\underset{\|}{S}}(CH_3)_2 \longrightarrow \underset{Ph}{\triangle}^N + CH_3\overset{O}{\underset{\|}{S}}CH_3$$

(**107**) (**108**)

## IV. Reactions of 1-Azirines

### A. Electrophilic Reagents

#### 1. *Acids*

Due to the large degree of *s* character in the orbital containing the lone pair of electrons on the nitrogen, 1-azirines are very nonbasic. 2-Phenyl-3-methyl-1-azirine is insoluble in 10% hydrochloric acid, although it does dissolve in 37%. Neutralization of this solution after 5 minutes gives a 47% recovery of the 1-azirine, demonstrating that the three-membered ring is not destroyed rapidly even in strong acidic aqueous solutions.[23]

Leonard and Zwanenburg[55] have treated azirine (**26**) with anhydrous perchloric acid and studied the reactions of the protonated azirine with acetone and acetonitrile. It is believed that protonated azirine ring **109** opens to cation **110** which then adds to the carbon–oxygen double bond or carbon–nitrogen triple bond to give the observed products (**111** and **112**).

[54] H. Koenig, H. Metzger, and K. Seelert, 100 (*Hundert*) *Jahre BASF aus Forsch.* 49 (1965); *Chem. Abstr.* **64**, 17409 (1966).
[55] N. J. Leonard and B. Zwanenburg, *J. Amer. Chem. Soc.* **89**, 4456 (1967).

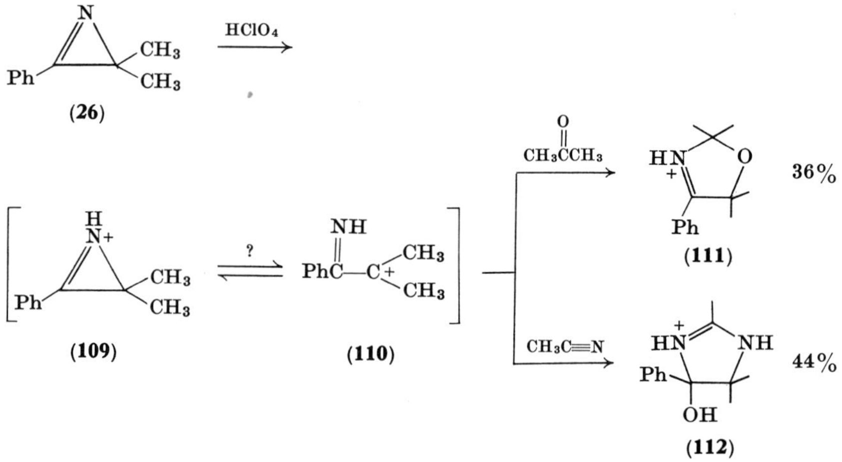

Treatment of azirine (**26**) with anilinium perchlorate resulted in destruction of the three-membered ring giving α-ammonioisobutyrophenone anil perchlorate (**116**). The reactants have been postulated to proceed through aziridine (**113**), which rearranges through intermediates **114** and **115**.[56]

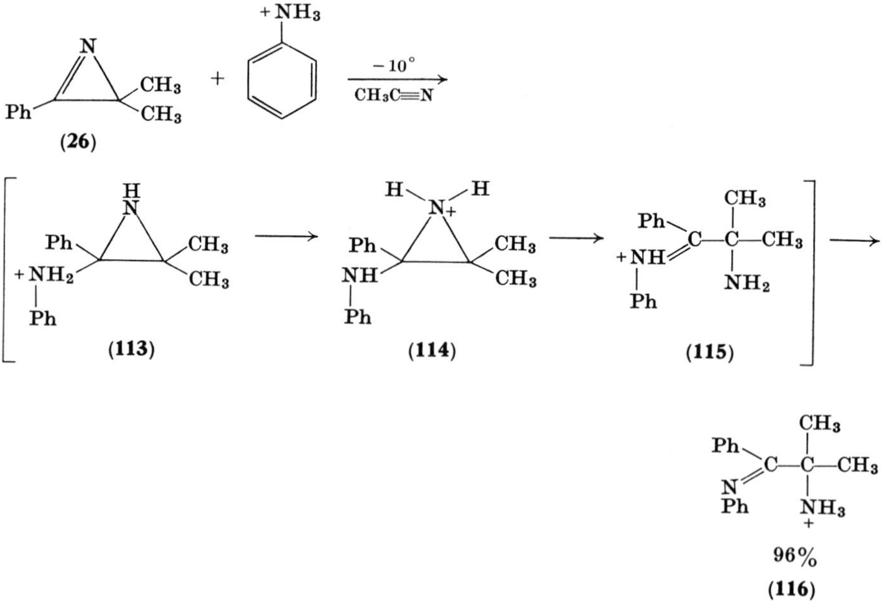

---

[56] N. J. Leonard, E. F. Muth, and V. Nair, *J. Org. Chem.* **33**, 827 (1967).

The treatment of azirine (**92**) with hydrazine perchlorate to give aminopyrazole (**118**)[46] probably follows a mechanism similar to that for the reaction of anilinium perchlorate with azirine (**26**). However, now the intermediate **117** can cyclize to the pyrazole.

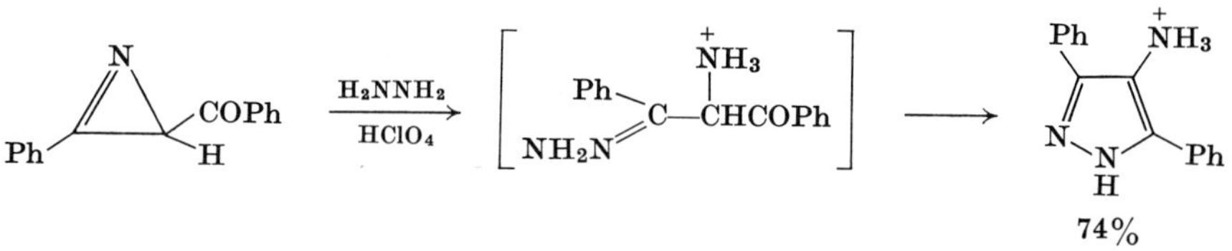

$$\text{(92)} \xrightarrow[\text{HClO}_4]{\text{H}_2\text{NNH}_2} \text{(117)} \longrightarrow \text{(118)} \quad 74\%$$

The reaction of 2-phenyl-1-azirine with benzoic acid gave N-benzoylphenacylamine (**120**). Aziridine (**119**), which can rearrange to **120**, is believed to be an intermediate in this reaction.[57]

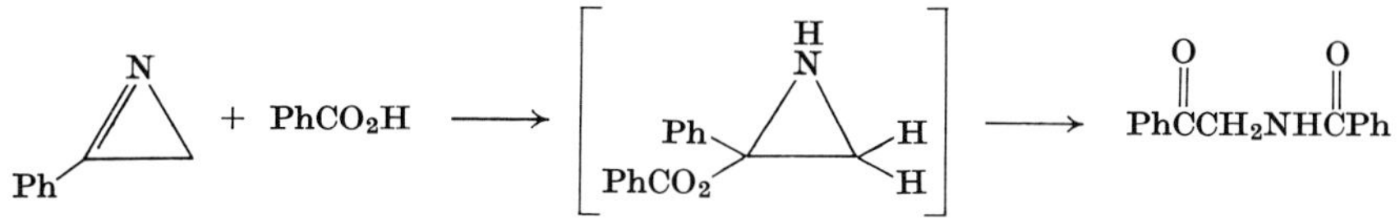

$$\text{(42a)} + \text{PhCO}_2\text{H} \longrightarrow \text{(119)} \longrightarrow \text{PhCOCH}_2\text{NHCPh} \;\;\text{(120)}$$

The reaction of thiobenzoic acid takes a slightly different course, although addition of the acid to the imine bond giving (**121**) is a likely initial step. Further reaction with thiobenzoic acid produces the final product (**122**).[57]

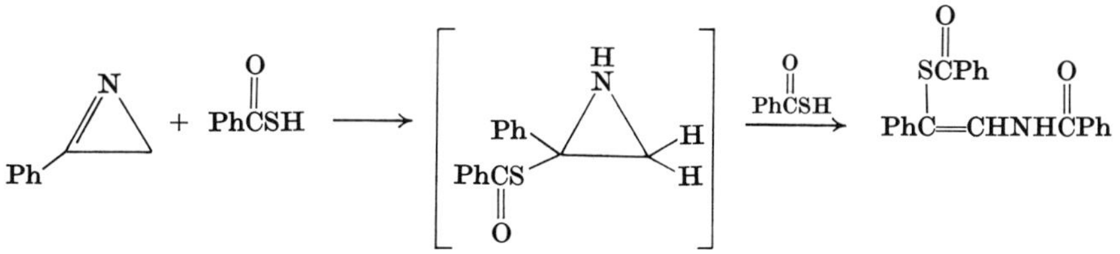

$$\text{(42a)} + \text{PhCSH} \longrightarrow \text{(121)} \xrightarrow{\text{PhCSH}} \text{(122)}$$

The addition of some acids to 1-azirines does not result in destruction of the three-membered ring. Leonard and Zwanenberg obtained aziridine (**123**) when 1-azirine (**26**) was treated with pyridinium perchlorate in pyridine.[55] The structure of a similar product (**23**), prepared from Neber's azirine, was proposed by Cram and Hatch.[12]

---

[57] S. Sato, H. Kato and M. Ohta, *Bull. Chem. Soc. (Japan)* **40**, 2938 (1967).

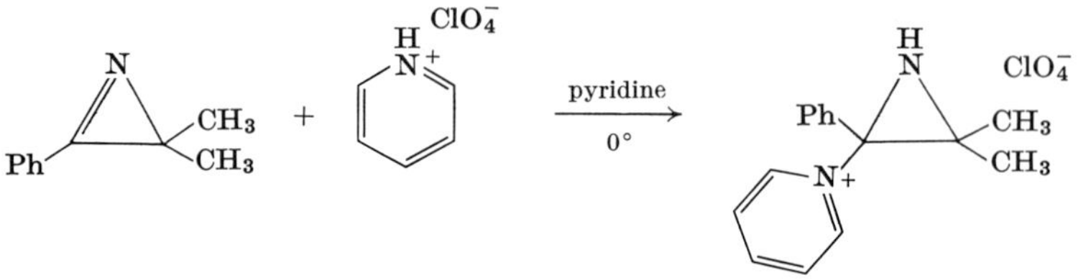

[Structural formulas: (26) + pyridine → (123) at 0°]

Meek and Fowler[58] observed that the addition of *p*-toluenesulfinic acid occurs readily to azirines (**124a** and **b**) to give the sulfonyl aziridines (**125a** and **b**), respectively.

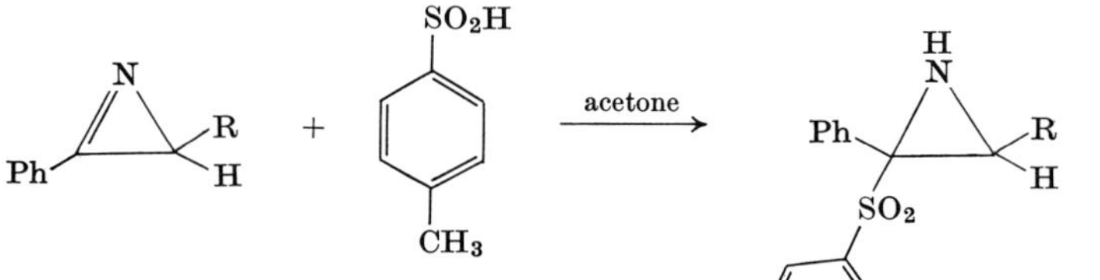

[Structural formulas: (124) + p-toluenesulfinic acid, acetone → (125)]

(**124**)
(a) R = $CH_3$
(b) R = Ph

One of the best known reactions of 1-azirines is the acid/catalyzed hydrolysis to aminoketones. Since the Neber reaction also accomplishes this same synthetic end, this reaction may appear to have little practical value. This is not the situation because with the Neber reaction there is no control over the aminoketone that will be obtained from a given ketone. For example, when oxime (**127**) derived from benzyl methyl ketone (**126**) is subjected to the Neber reaction aminoketone **128** is obtained.[59] The amino function is substituted for the most acidic α-hydrogen. The isomeric aminoketone (**132**) that could not be prepared by the Neber reaction can be formed by the hydrolysis of 1-azirine (**131**). The synthesis of this 1-azirine has been accomplished from allyl benzene (**129**) through vinyl azide (**130**) using iodine azide.[22]

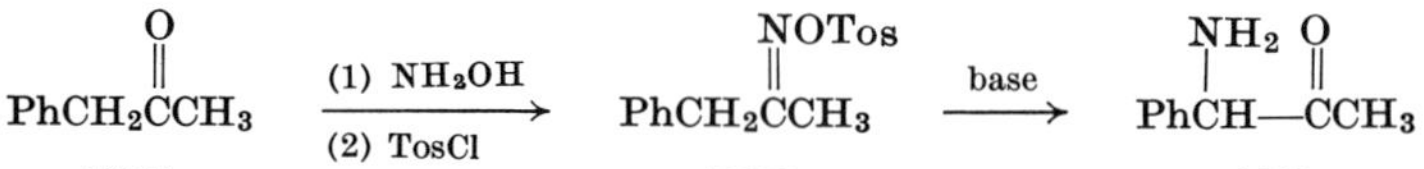

$PhCH_2CCH_3$ (126) $\xrightarrow{(1)\ NH_2OH}{(2)\ TosCl}$ $PhCH_2CCH_3$ with NOTos (127) $\xrightarrow{base}$ $PhCH-CCH_3$ with $NH_2$ and O (128)

[58] J. S. Meek and J. S. Fowler, *J. Org. Chem.* **33**, 985 (1968).
[59] C. O'Brien, *Chem. Rev.* **64**, 81 (1964).

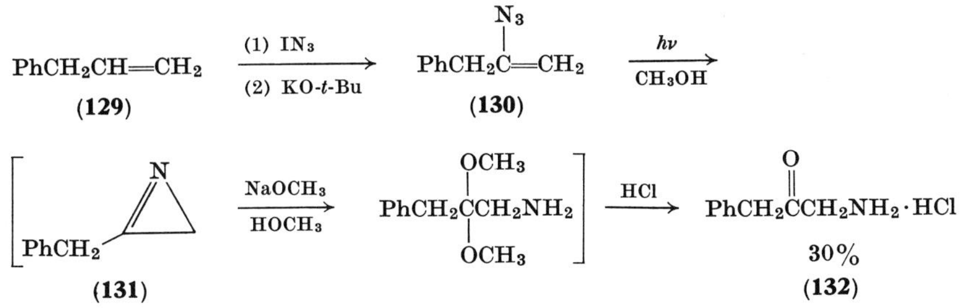

PhCH₂CH=CH₂ $\xrightarrow[\text{(2) KO-}t\text{-Bu}]{\text{(1) IN}_3}$ PhCH₂C(N₃)=CH₂ $\xrightarrow[\text{CH}_3\text{OH}]{h\nu}$
(129)                                  (130)

[PhCH₂—azirine] (131) $\xrightarrow[\text{HOCH}_3]{\text{NaOCH}_3}$ [PhCH₂C(OCH₃)₂CH₂NH₂] $\xrightarrow{\text{HCl}}$ PhCH₂COCH₂NH₂·HCl
                                                                           30%
                                                                           (132)

Hydrolysis of 2-unsubstituted 1-azirines is a potentially valuable method for the preparation of aminoaldehydes. The application of the Neber reaction to oxime tosylates derived from aldehydes does not give the aminoaldehydes.[59] Instead, nitriles are produced by elimination.

The frequently reported dimerization of 1-azirines to 3,6-dihydropyrazines probably involves initial hydrolysis to the aminoketone which then undergoes a cyclodehydration to the dihydropyrazine, rather than a direct dimerization of the 1-azirine. Catalytic amounts of water absorbed on the acid surface of the glass are probably responsible for this reaction.

## 2. Acid Chlorides and Anhydrides

1-Azirines react with carboxylic acid chlorides in benzene to give the $N$-benzoyl-2-chloroaziridines (133) in good yield.[23, 60] These aziridines are unstable and are converted in polar solvents or by heating into a mixture of the oxazole and dichloroamide (134).

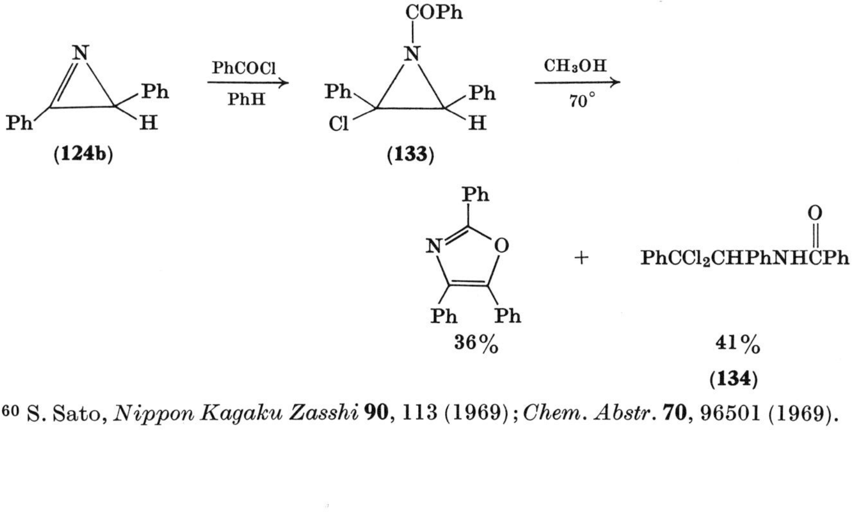

(124b) $\xrightarrow[\text{PhH}]{\text{PhCOCl}}$ (133) $\xrightarrow[70°]{\text{CH}_3\text{OH}}$

2,4,5-triphenyloxazole (36%) + PhCCl₂CHPhNHCOPh (41%)
                                                                          (134)

[60] S. Sato, *Nippon Kagaku Zasshi* 90, 113 (1969); *Chem. Abstr.* 70, 96501 (1969).

Sato et al. observed that the reaction of 2-phenyl-1-azirine with acid chlorides and anhydrides in the presence of triethylamine gives the oxazole directly.[57] They have reported that azirines also react with carboxylic anhydrides to give oxazoles (136).[57] Aziridine (135) is suggested as a likely intermediate in the reaction of acetic anhydride with azirine (42a), since carrying out the reaction at lower temperature and for a shorter reaction time gave a compound to which they assigned structure 135.

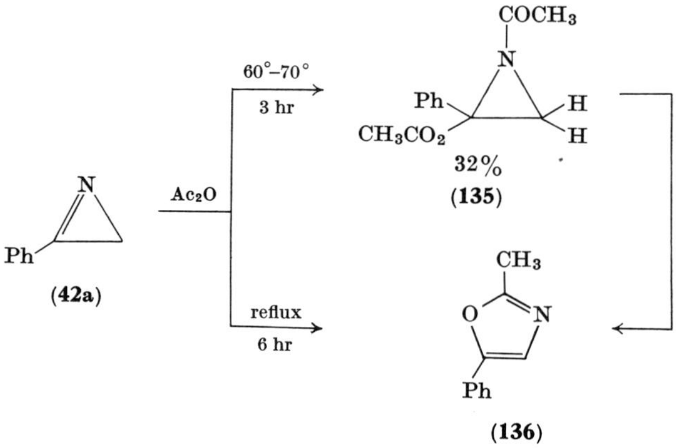

The reaction of azirine (42a) with phthalic anhydride does not lead to the oxazole.[57] Rather, the ketoamide (138) is produced, presumably through hydrolysis of intermediate 137.

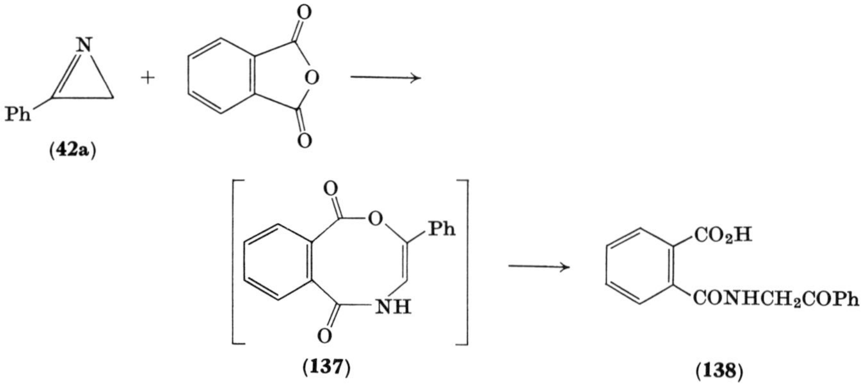

Predictably, 1,2,5-triphenylimidazole (**139**) is produced when *N*-phenylbenzimidoyl chloride is treated with 2-phenyl-1-azirine.

(**42a**) (**139**)

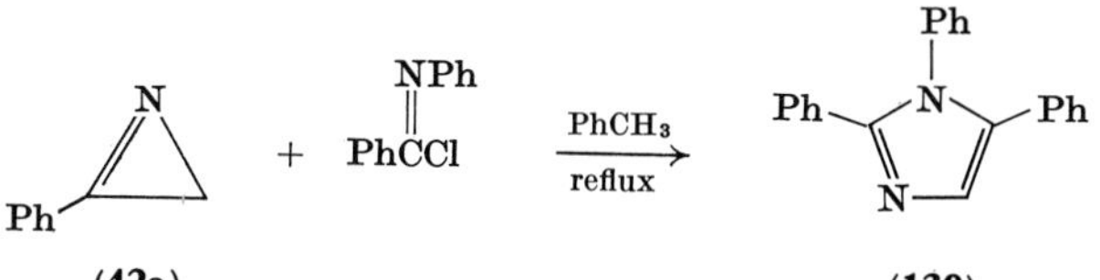

The reaction of 2-phenyl-3-methyl-1-azirine with benzenesulfonyl chloride gives a mixture of sulfonamides (**141** and **142**). It is likely that the chloroaziridine (**140**) is an intermediate in this reaction, since it is known that *N*-sulfonylaziridines rearrange to vinyl sulfonamides.[61] Attempts to isolate the chloroaziridine (**140**) were unsuccessful.

(**124a**) (**140**)

$PhCCl{=}CCH_3NHSO_2Ph + PhCCl{=}CCH_3N(SO_2Ph)_2$

(**141**) (**142**)

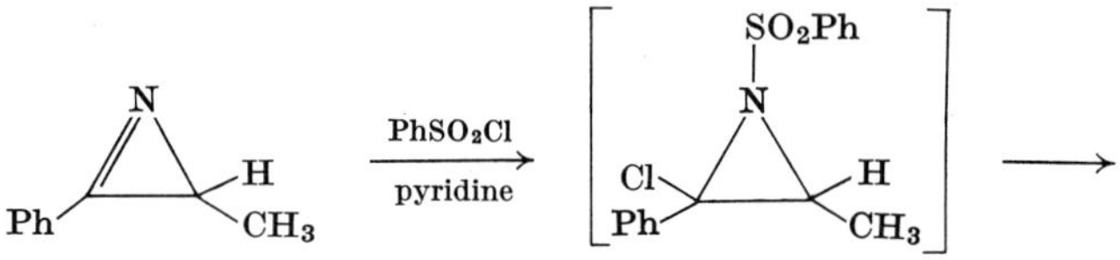

## B. Nucleophilic Reagents

### 1. *Metal Hydrides*

Several 1-azirines have been reduced to the aziridines (**143**) with lithium aluminum hydride. Although this reaction has been used

(**124**) (**143**)

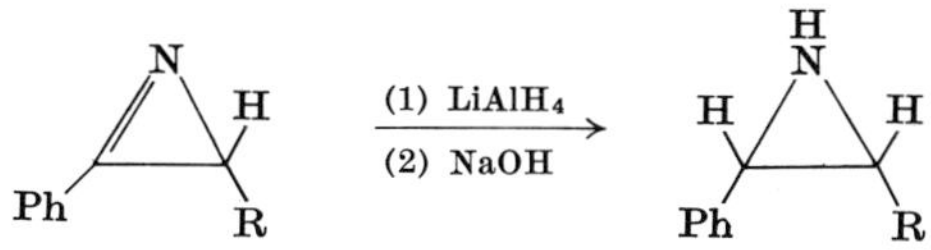

(**a**) R = $CH_3$ (97%)
(**b**) R = Ph (85%)

[61] O. C. Dermer and G. E. Ham, "Ethylenimine and Other Aziridines." Academic Press, New York, 1969.

mainly as a method for proof of structure[27] it can be a useful aziridine synthesis. The reaction usually proceeds in high yield and is stereospecific.[20, 21] With the substituted 1-azirines that have been studied, the approach of hydride occurs exclusively from the side opposite the group at position three giving *cis*-aziridines. This can be a useful synthesis since preparation of the *cis*-aziridine usually requires the thermodynamically less stable *cis*-olefin as a precursor. The prerequisite azirine can be prepared from either the *cis,trans*- or a mixture of *cis*- and *trans*-olefin using iodine azide.

The reaction of 1-azirines with sodium borohydride has not been studied in detail, but one case was believed also to give the aziridine initially which ring-opened under the acidic reaction conditions.[12]

## 2. *Carbanions*

Grignard reagents have been shown to react with 1-azirines to give aziridines.[23, 62] Furthermore, the attack of the Grignard reagent undoubtedly occurs stereospecifically from the less sterically hindered side of the azirine.[63] This is analogous to the stereospecific reduction of azirines to aziridines by lithium aluminum hydride.

Azirines have also been shown to be likely intermediates in the Campbell aziridine synthesis.[62] This reaction is analogous to the Neber reaction in which the intermediate azirine is hydrolyzed to the aminoketone. Here the proposed azirine intermediate prepared from oxime (**144**) reacts with a Grignard reagent to give the aziridine (**145**).

PhCCH$_3$ + 2C$_2$H$_5$MgBr ⟶ [azirine intermediate with Ph] $\xrightarrow[(2)\ NH_4Cl]{(1)\ C_2H_5MgBr}$ aziridine product

(**144**)　　　　　　(**41a**)　　　　　　(**145**)
　　　　　　　　　　　　　　　　40–54%

The reaction of azirines with Grignard reagents is an anomalous reaction of imines. Normally an α-hydrogen is abstracted to give the enamine anion which is unreactive toward further attack of the Grignard reagent.[64] The enamine derived from a 1-azirine is a 2-azirine (**146**); it is an unknown and probably unstable compound (Section I,A).

[62] S. Eguchi and Y. Ishii, *Bull. Chem. Soc. (Japan)* **36**, 1434 (1963).
[63] A. Laurent and A. Muller, *Tetrahedron Lett.* 759 (1969).
[64] R. W. Layer, *Chem. Rev.* **63**, 489 (1963).

(42a) (146)

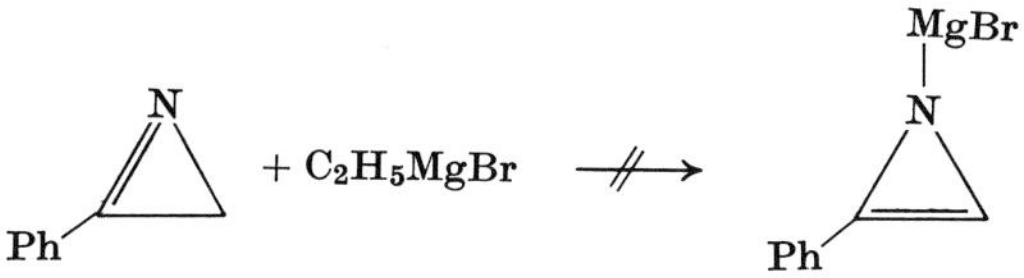

2-Phenyl-1-azirine (**42a**) reacts with acetophenone in the presence of the dimethylsulfinyl carbanion to give 2,4-diphenylpyrrole (**150**).[65] This reaction probably involves initial attack of the enolate anion on the carbon–nitrogen double dond, to give intermediate **147** and **148** which ring-opens to 149 and loses hydroxide ion giving pyrrole (**150**). A similar reaction of ethyl benzoylacetate with 2-phenyl-1-azirine yields 3-benzoyl-4-phenyl-2-oxopyrroline (**151**).

(42a) (147)

(148) (149) (150)

(42a) (151)

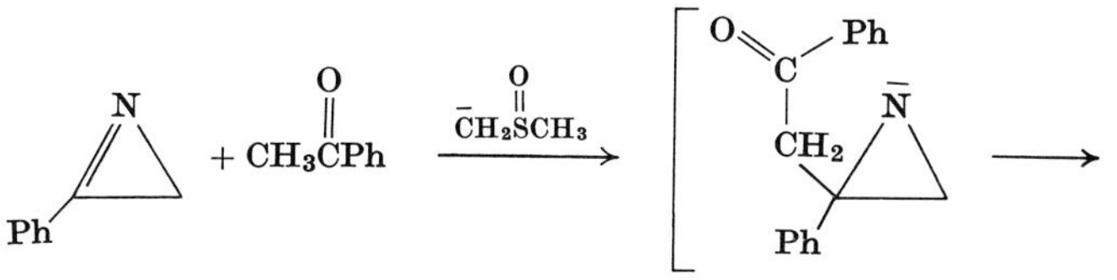

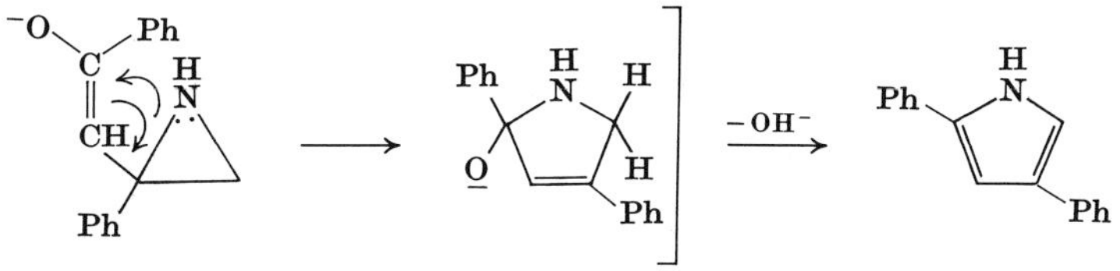

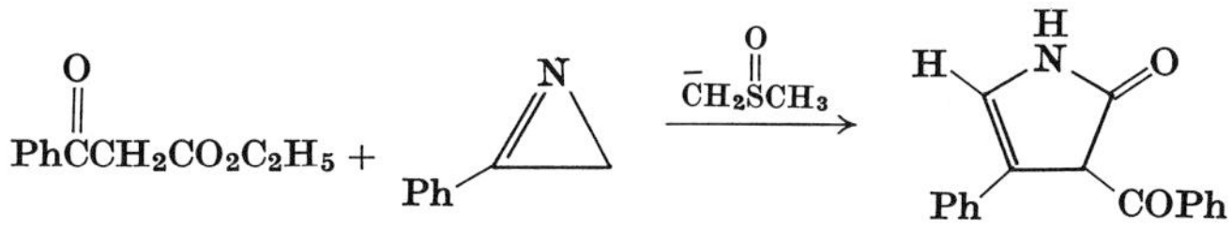

[65] S. Sato, H. Kato and M. Ohta, *Bull. Chem. Soc.* (*Japan*) **40**, 2936 (1967).

Benzyl cyanide under similar conditions reacts with 2-phenyl-1-azirine to give 3,4-diphenyl-2-oxo-5-iminopyrroline (**153**). The iminopyrroline (**152**) is probably produced initially and is air-oxidized to the final product.

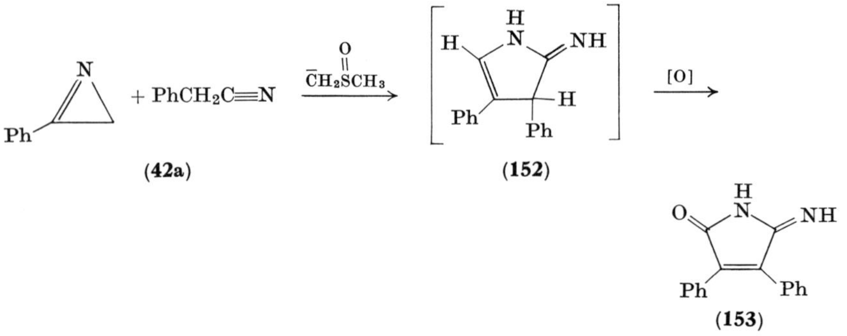

### 3. *Amines*

Smolinsky and Feuer have observed that treatment of 2-phenyl-1-azirine with aniline followed by mild acid hydrolysis gives mainly benzanilide, in addition to smaller amounts of 2,5-diphenylpyrazine and 3,4-dianilino-1,2,5-triphenylpyrrole.[66] These products have been rationalized as initially proceeding by nucleophilic attack of the amine at the imine carbon atom to give intermediate **154**. No effort to isolate the 2-phenyl-2-anilinoaziridine (**155**) or (**156**) was made.

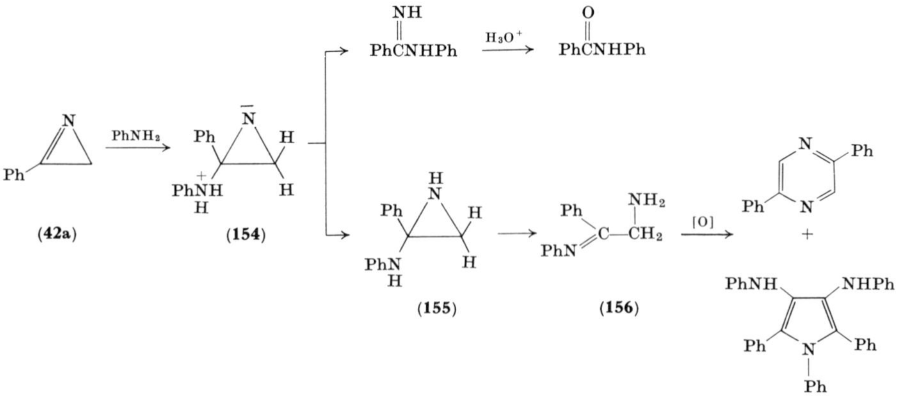

[66] G. Smolinsky and B. Feuer, *J. Org. Chem.* **31**, 1423 (1966).

## 4. Alcohols

Aminoketone acetals and alkoxyaziridines can be prepared by treating 1-azirines with methanol containing a catalytic amount of sodium methoxide. The first product is the alkoxyaziridine (**157**), which has been isolated in some cases. Further treatment with methanol and sodium methoxide produces the aminoketone acetal (**158**).

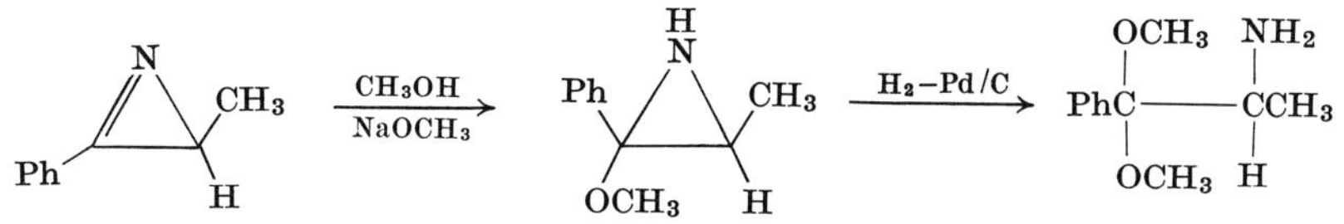

If the aminoketone or its acetal derivative are the desired compounds, then the best procedure is to carry out the photolysis of the vinyl azide in methanol containing sodium methoxide. Attempts to first isolate the 1-azirine result in decreased yields due to polymerization of the unstable 1-azirine ring system.[22]

Oxazole (**93**) can be obtained from the azirine (**92**) when it is heated in weakly alkaline methanol.[46] The reaction presumably involves initial attack of the alcohol on the imine bond to give the intermediates **159** and **160**.

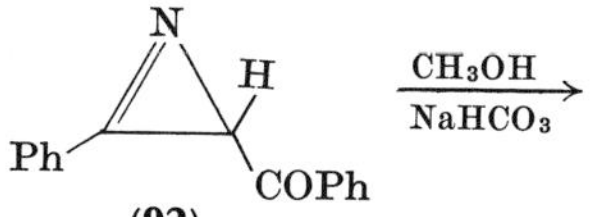

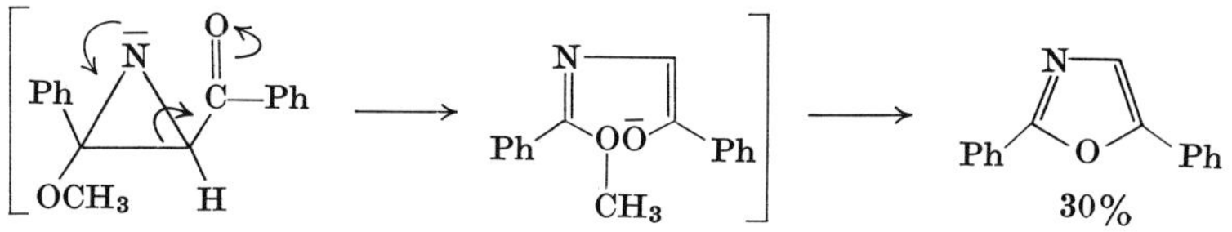

Probably one of the most interesting reactions related to nucleophilic addition was reported by Hortmann and Robertson.[67] They

[67] A. G. Hortmann and D. A. Robertson, *J. Amer. Chem. Soc.* **89**, 5974 (1967).

observed that when 2-phenyl-1-azirine is treated with dimethylsulfonium methylide in tetrahydrofuran (THF) the first example of the 1-azabicyclo[1.1.0]butane ring system (**161**) was produced.

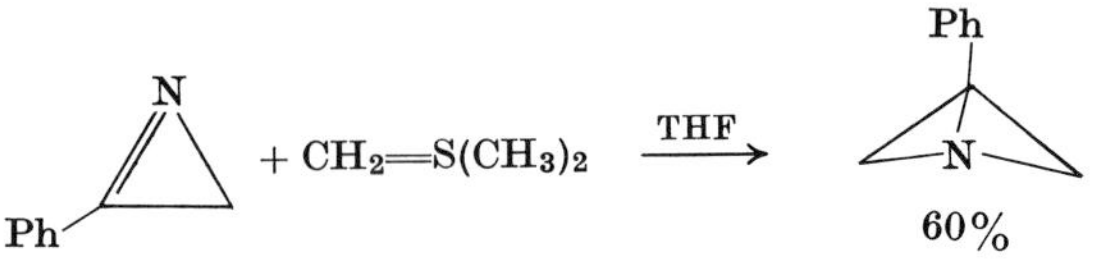

$$\underset{(\mathbf{42a})}{\text{Ph-azirine}} + CH_2\!=\!S(CH_3)_2 \xrightarrow{\text{THF}} \underset{\underset{60\%}{(\mathbf{161})}}{\text{Ph-azabicyclobutane}}$$

## C. CYCLOADDITIONS

Although cycloadditions to the carbon–nitrogen double bond of 1-azirine are potentially valuable routes for preparing unusual heterocyclic compounds, there are few reported examples of this type of reaction.

Logothetis first reported that azirine (**24a**) reacts with diazomethane to produce the allyl azide (**163**).[68] This reaction is postulated to proceed by a 1,3-dipolar cycloaddition to form the triazoline (**162**) which then undergoes a valence tautomerization to the allyl azide (**163**).

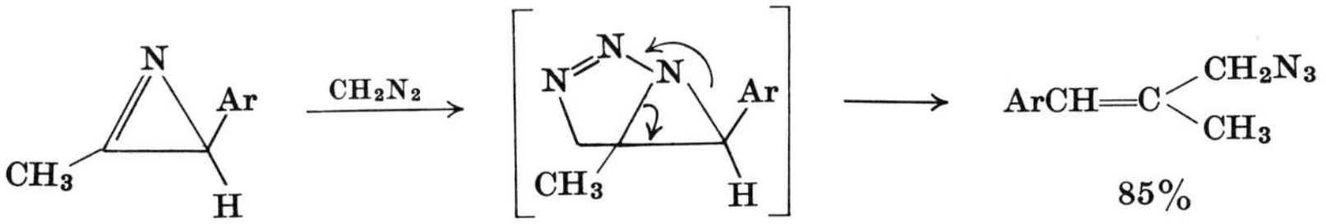

$$\underset{(\mathbf{24a})}{\text{CH}_3\text{-azirine-Ar}} \xrightarrow{\text{CH}_2\text{N}_2} \underset{(\mathbf{162})}{\left[\text{triazoline}\right]} \longrightarrow \underset{\underset{85\%}{(\mathbf{163})}}{\text{ArCH}\!=\!\text{C}\!\!\begin{array}{c}\text{CH}_2\text{N}_3\\\text{CH}_3\end{array}}$$

Nair has recently observed a similar reaction with 2-phenyl-3-methyl- and 2-phenyl-3,3-dimethyl-1-azirine.[69]

The photolysis of α-azidostyrene in benzene using filtered light is reported to give the azabicyclo[2.1.0]pentane (**164**).[70] This product

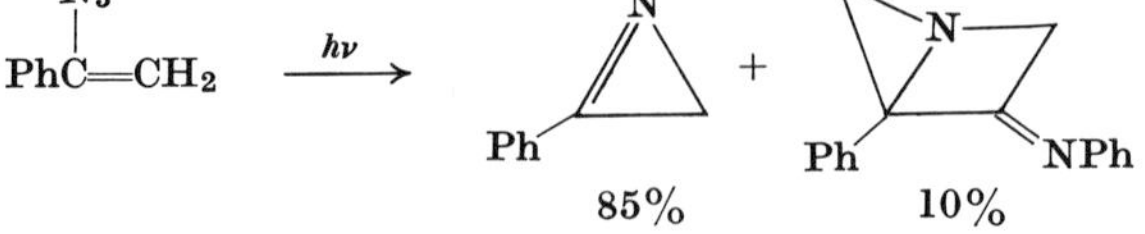

$$\underset{(\mathbf{41a})}{\text{PhC}(\text{N}_3)\!=\!\text{CH}_2} \xrightarrow{h\nu} \underset{\underset{85\%}{(\mathbf{42a})}}{\text{Ph-azirine}} + \underset{\underset{10\%}{(\mathbf{164})}}{\text{Ph-azabicyclopentane-NPh}}$$

[68] A. L. Logothetis, *J. Org. Chem.* **29**, 3049 (1964).
[69] V. Nair, *J. Org. Chem.* **33**, 2121 (1968).
[70] F. P. Woerner, H. Reimlinger, and D. R. Arnold, *Angew. Chem. Int. Ed. Engl.* **7**, 130 (1968).

is believed to arise by a photochemically induced cycloaddition between 2-phenyl-1-azirine and *N*-phenylketenimine. The ketenimine is probably produced as a primary photoproduct from the vinyl azide.

Catalytic hydrogenation (palladium or Raney nickel catalyst) surprisingly results in reduction of the carbon–nitrogen single bond rather than the double bond.[4, 12, 40] The imines, or possibly enamines, are usually not isolated and their existence has only been inferred in most instances. Harvey and Ratts have shown that this reaction with azirine (**165**) does not proceed first to the aziridine which is then reduced to **166**, since aziridine (**167**) is inert to hydrogen and palladium on carbon.[40]

(**165**) (**166**) (**167**)

(**24a**) (**168**)

Ar = 2,4-dinitrophenyl

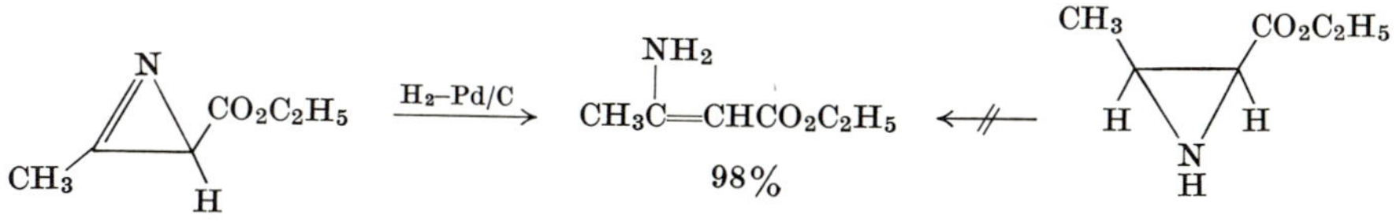

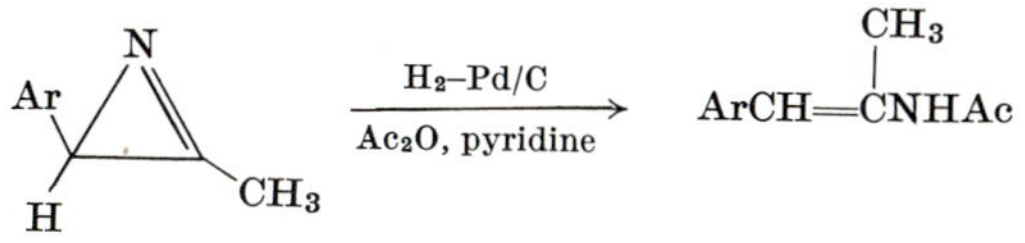

Interestingly, Morrow *et al.* have reported that the steroidal 1-azirine (**29**) is hydrogenated to the aziridine (**169**) using platinum oxide.[15, 16]

(**29**) (**169**)

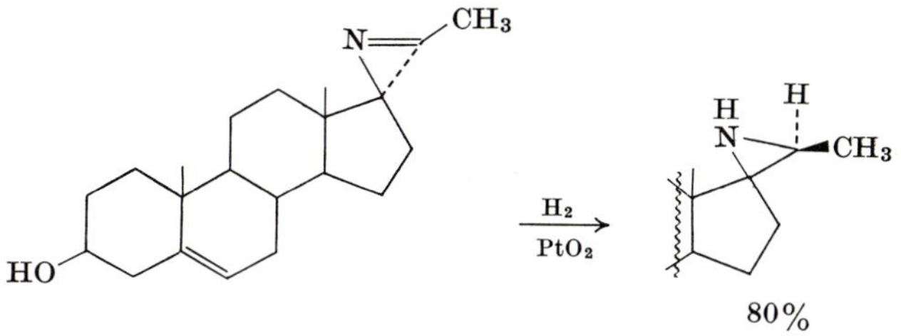

Two groups have independently observed an interesting product when the vinyl azide (**61**) is thermolyzed.[30, 31] Bauer and Hafner have isolated the azirine (**62**) by low-temperature photolysis of **61** and have shown that thermolysis of this azirine also leads to **170**.[31] Apparently, elimination of hydrogen cyanide occurs to give 9-fluorenylidene carbene (**171**), which then reacts with azirine (**62**) to give the final product.

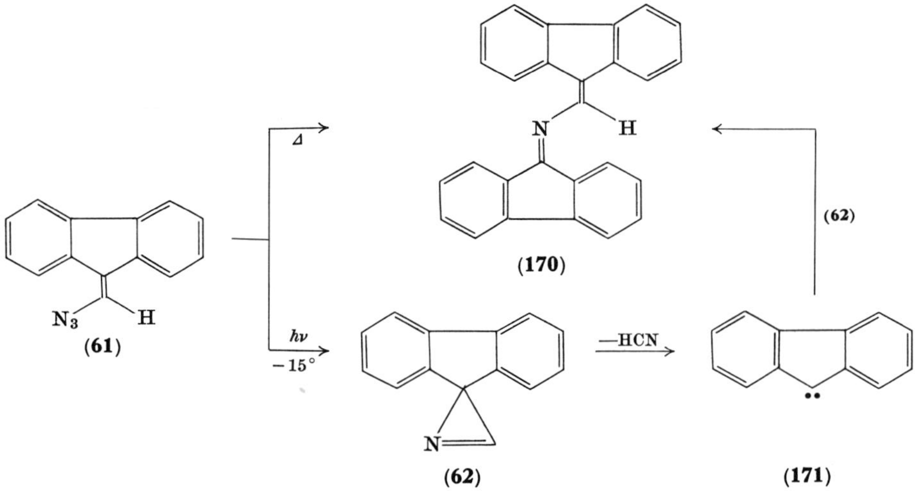

## V. Summary

It is clear from this review that our knowledge of the azirine ring system is still in its infancy. For example, there are no authentic derivatives of the 2-azirine ring system known. The parent 1-azirine ring system and 1-azirines fused to five- or six-membered rings are unknown. Fundamental physical parameters such as the strain energy, bond lengths, and bond angles remain to be determined for the 1-azirine ring system.

Although a considerable number of 1-azirines have been prepared in a very short time, discovery of new and refinement of old synthetic techniques are clearly needed. Finally, the 1-azirine ring system is potentially a very valuable starting point for the preparation of new and unusual heterocyclic compounds. Already a number of very interesting aziridines have been prepared and 1-azirines were key intermediates for the preparation of the 1-azabicyclo[2.1.0]pentane and 1-azabicyclo[1.1.0]butane ring systems.

# Electronic Aspects of Purine Tautomerism*

BERNARD PULLMAN AND ALBERTE PULLMAN

*Institut de Biologie Physico-Chimique, Paris, France*

|      |                                                                                    |      |
|------|------------------------------------------------------------------------------------|------|
| I.   | Introduction                                                                       | 77   |
| II.  | The Biological Importance of Purine Tautomerism                                    | 79   |
| III. | Quantum Mechanical Methods of Investigation                                        | 85   |
|      | A. Generalities                                                                    | 85   |
|      | B. The Self-Consistent Field Method                                                | 87   |
|      | C. The ZDO Approximation                                                           | 92   |
|      | D. The Parametrization of the Pariser–Parr–Pople Approximation for $\pi$ Electrons | 93   |
|      | E. The Simple Representation of the $\sigma$ Bonds                                 | 96   |
|      | F. The CNDO Approximation                                                          | 98   |
| IV.  | The Four Prototropic Tautomers of Purine                                           | 100  |
| V.   | Amine–Imine Tautomerism in Adenines                                                | 111  |
| VI.  | Lactam–Lactim Tautomerism in Hydroxypurines                                        | 122  |
| VII. | The Fine Structure of Hydroxypurines                                               | 127  |
| VIII.| Guanine, with a Special Discussion of Its N(3)H Tautomer                           | 138  |
| IX.  | Tautomerism in 8-Azapurines and the Problem of the N(8)H Tautomer                  | 142  |
| X.   | The Thione–Thiol Tautomerism in Mercaptopurines and the Fine Structure of Thiopurines | 145  |
| XI.  | The N(7)H $\rightleftarrows$ N(9)H Tautomerism and the Crystal Structure of Purines | 150  |
| XII. | Conclusion                                                                         | 156  |

## I. Introduction

Because of the multiplicity of possible forms, the tautomerism of purines offers a challenging field of investigation for physical and quantum chemists. The following principal types of tautomeric transformation liable to occur in the most significant group of purines, namely, those of biological interest, can be considered.

* This work was supported by the RCP 173 of the Centre National de la Recherche Scientifique and grant No. CR 66-236 of the Institut National de la Santé et de la Recherche Médicale.

1. *The prototropic tautomerism* corresponding to the displacement of the proton among the four available ring nitrogens. Thus purine is generally represented in the form in which a hydrogen atom is attached to N-9. We shall call such a form the N(9)H tautomer (**1**). It is obvious, however, that structures may be considered in which the proton is attached to the other nitrogens of the molecule, yielding the tautomers N(7)H, (**2**); N(3)H, (**3**); and N(1)H, (**4**). In azapurines

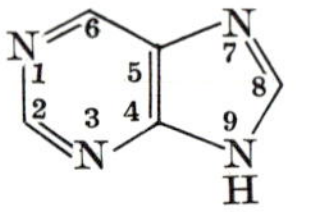

N(9)H tautomer of purine
(**1**)

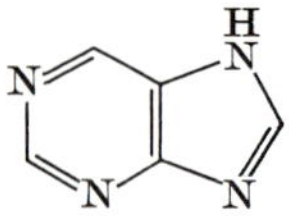

N(7)H tautomer of purine
(**2**)

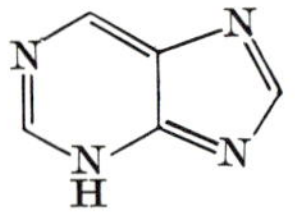

N(3)H tautomer of purine
(**3**)

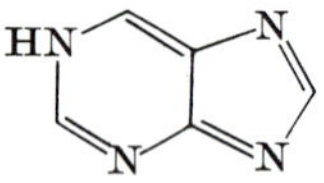

N(1)H tautomer of purine
(**4**)

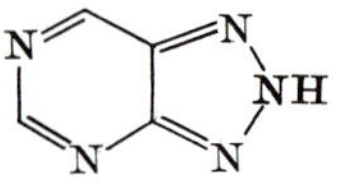

(**5**)

a supplementary form of the same type may be considered corresponding to the fixation of the proton on the supplementary nitrogen [e.g., the N(8)H tautomer of 8-azapurine (**5**)]. For the sake of convenience we shall frequently refer to this tautomerism as involving the "imidazole proton" of purines.

2. *The amine–imine tautomerism*, liable to occur in aminopurines which may be illustrated in the particularly important case of adenine by the structures **6** and **7**.

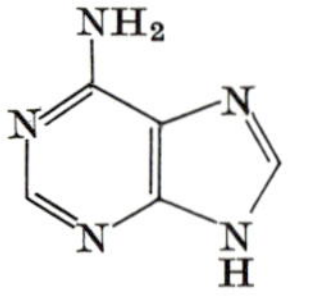

(**6**)

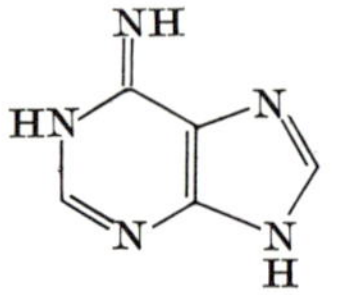

(**7**)

3. *The lactam–lactim tautomerism* of hydroxypurines illustrated, for example, for hypoxanthine by the structures **8** and **9**.

In the related mercapto derivatives this corresponds to the *thione–thiol* tautomerism, illustrated by structures **10** and **11** for the important antitumor agent 6-mercaptopurine.

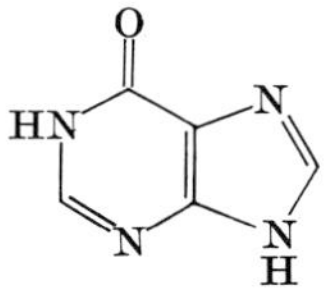

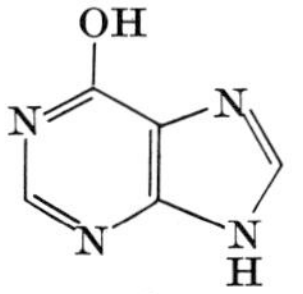

The amine–imine and the lactam–lactim tautomerisms may, of course, be coupled with the shift of the "imidazole proton" on the purine skeleton. In polysubstituted derivatives, e.g., in 8-azaguanine, all the different types of tautomerisms may be intermingled.

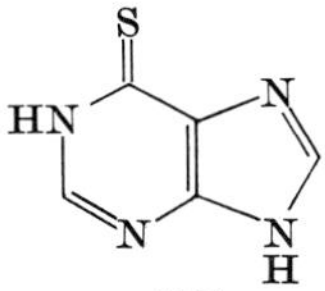

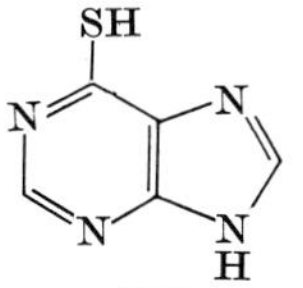

It is surprising to observe to what a large extent the compounds of this series are frequently depicted in arbitrary and erroneous forms in otherwise excellent papers and textbooks—the more so as the utilization of the appropriate tautomeric forms is, as will be seen in Section II, of fundamental importance for the appropriate presentation and understanding of biological structures and phenomena in which these compounds play a decisive role. During recent years much work has been carried out, both experimentally by a wide variety of techniques and theoretically with the help of the refined methods of quantum chemistry, in order to ascertain the relative importance of the different tautomeric forms and to determine their essential electronic characteristics. The present chapter is devoted to a summary of this work.

## II. The Biological Importance of Purine Tautomerism

No better illustration of this importance can probably be given than is described in the narrative of Jim Watson about the discovery of the purine–pyrimidine base-pairing scheme and thus of the fundamental structure of deoxyribonucleic acid (DNA).[1] Although he took his formulas for the bases from a very excellent textbook of biochemistry "so as to be sure to have the correct structures," he happened to stumble on the incorrect ones for guanine and thymine. There were strong odds that this should have happened. For obscure reasons there seemed (and still seems frequently) to be a persistent tendency in the literature to depict these two molecules in the incorrect *lactim* (hydroxy) form. It is possible that this tendency springs from the fact that such a form is the appropriate one for phenol and that this situation leads most people to believe that such should

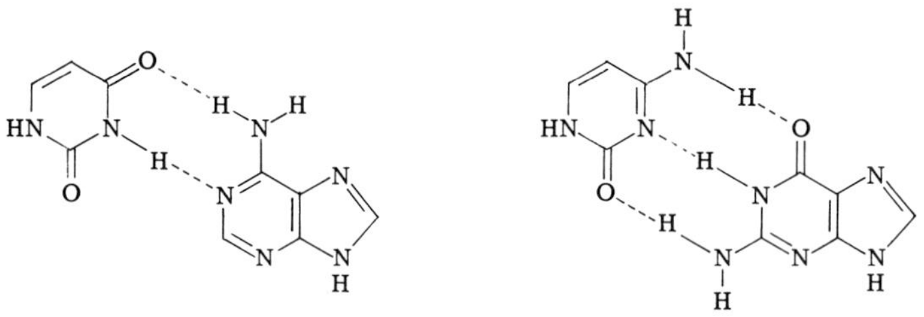

FIG. 1. The Watson–Crick base pairing in DNA.

be the general case of hydroxy derivatives of conjugated hydrocarbons and heterocycles. Whatever the reason, this is, as we shall see shortly, an erroneous belief and in the case of Watson it prevented him for some time from finding a base-pairing scheme which would agree with the X-rays pattern for DNA. It is only after Jerry Donohue, who happened to be in Cambridge at that time, had drawn his attention to the incorrectness of the formulas he was using and had pointed to the probable *lactam* (oxo) structure of the two bases that Watson found the Watson–Crick complementary pairing and thus solved one of the most important mysteries in biology.

The above-mentioned purine–pyrimidine base-pairing scheme consists, as it is well known, of hydrogen bonding between specific, *complementary* base pairs, namely, adenine–thymine [or uracil in ribonucleic acid (RNA)] and guanine–cytosine (Fig. 1) (shorthand notations: A–T or A–U and G–C). The specificity of the bonding concerns both this exclusiveness and the steric arrangement which is

[1] J. D. Watson, "The Double Helix." Atheneum, New York, 1968.

as depicted in Fig. 1. It is easy to see that the existence of this complementarity necessitates the simultaneous presence of the bases in definite tautomeric forms, namely, the amine and lactam forms, as it is only with such "complementary" forms that the appropriate hydrogen bonds may be formed.

There is abundant evidence, obtained with a large variety of experimental techniques, such as infrared spectroscopy,[2-18] Raman spectroscopy,[19a, b] ultraviolet spectroscopy,[17, 20-23] nuclear magnetic resonance spectroscopy,[10, 24-27] measurements of ionization con-

---

[2] H. T. Miles, *Biochim. Biophys. Acta* **22**, 247 (1956).
[3] H. T. Miles, *Biochim. Biophys. Acta* **27**, 46 (1958).
[4] H. T. Miles, *Biochim. Biophys. Acta* **30**, 324 (1958).
[5] H. T. Miles, *Biochim. Biophys. Acta* **35**, 274 (1959).
[6] H. T. Miles, *Proc. nat. Acad. Sci. U.S.* **47**, 791 (1961).
[7] C. L. Angell, *J. Chem. Soc.* 504 (1961).
[8] T. Shimanouchi, M. Tsuboi, and Y. Kyogoku, *Advan. Phys. Chem.* **7**, 435 (1964).
[9] D. Brown and S. Mason, *J. Chem. Soc.* 682 (1957).
[10] H. T. Miles, F. B. Howard, and J. Frazier, *Science* **142**, 1458 (1963).
[11] F. B. Howard and H. T. Miles, *Biochem. Biophys. Res. Commun.* **15**, 18 (1964).
[12] S. F. Mason, *in* "Chemistry and Biology of Purines," Ciba Found. Symp., p. 60. Churchill, London, 1957.
[13] E. A. Blout and M. Fields, *Science* **107**, 252 (1948).
[14] E. R. Blout and M. Fields, *J. Biol. Chem.* **178**, 335 (1949).
[15] E. R. Blout and M. Fields, *J. Amer. Chem. Soc.* **72**, 479 (1950).
[16] R. D. B. Fraser, *Progr. Biophys. Biophys. Chem.* **3**, 47 (1953).
[17] B. C. Pal and C. A. Horton, *J. Chem. Soc.* 400 (1964).
[18] C. H. Willets, J. C. Decius, K. L. Dille, and B. E. Christensen, *J. Amer. Chem. Soc.* **77**, 2569 (1955).
[19a] R. C. Lord and G. J. Thomas, *Spectrochim. Acta* **A23**, 2551 (1967).
[19b] G. Madeiras and G. J. Thomas, *Biophys. Soc. Abstr. 14th Ann. Meeting* **10**, 31a (1970).
[20] S. F. Mason, *J. Chem. Soc.* 1253 (1959).
[21] D. J. Brown, E. Hoerger, and S. F. Mason, *J. Chem. Soc.* 4035 (1955).
[22] S. F. Mason, *J. Chem. Soc.* 2071 (1954).
[23] H. G. Mautner and G. Bergsson, *Acta Chem. Scand.* **17**, 1694 (1963).
[24] L. Gatlin and J. C. Davis, Jr. *J. Amer. Chem. Soc.* **84**, 4464 (1962).
[25] C. Jardetzky and O. Jardetzky, *J. Amer. Chem. Soc.* **82**, 222 (1960).
[26] J. P. Kokko, J. H. Goldstein, and L. Mandell, *J. Amer. Chem. Soc.* **83**, 2909 (1961).
[27] P. O. P. Ts'o, M. P. Schweizer, and D. P. Hollis, *Ann. N.Y. Acad. Sci.* **158**, 256 (1969).

stants,[28,29] and X-ray crystallography,[30-45] that these are the common tautomeric forms for the majority of biological purines and, in particular, for those involved in the nucleic acids, both in the solid state and in solution.

It is possible to specify this concept of complementarity in an explicit presentation.[46,47] Thus in classical chemistry, a "hydrogen bond" consists of a hydrogen atom which is simultaneously attracted to two electronegative atoms. In modern formulation it is a proton which is shared between two electron pairs belonging to the two electronegative atoms. Considering only the possible hydrogen bonds in the specific positions of the bases, determined by their linkage to the sugar–phosphate strands, one obtains the following self-explanatory shorthand presentation of the *proton–electron pair code* for the four bases involved which immediately indicates as the only possible combinations A–T and G–C.

$$A \begin{Bmatrix} :H & : \\ : & H: \\ & : \end{Bmatrix} T \qquad G \begin{Bmatrix} : & H: \\ :H & : \\ :H & : \end{Bmatrix} C$$

Associated with this scheme is, however, the observation that if a base happens to be in one of its rare tautomeric forms, *imine* for

[28] R. V. Wolfenden, *J. Mol. Biol.* **40**, 307 (1969).
[29] S. F. Mason, *J. Chem. Soc.* 674 (1958).
[30] R. F. Stewart and L. H. Jensen, *J. Chem. Phys.* **40**, 2071 (1964).
[31] H. M. Sobell and K. I. Tomita, *Acta Crystallog.* **17**, 126 (1964).
[32] J. Kraut and C. H. Jensen, *Acta Crystallog.* **16**, 79 (1963).
[33] S. T. Rao and M. Sundaralingam, *J. Amer. Chem. Soc.* **91**, 1210 (1969).
[34] D. G. Watson, D. J. Sutor, and P. Tollin, *Acta Crystallog.* **19**, 111 (1965).
[35] R. F. Bryan and K. I. Tomita, *Acta Crystallog.* **15**, 1179 (1962).
[36] M. Sundaralingam, *Acta Crystallog.* **21**, 495 (1966).
[37] H. Ringertz, *Acta Crystallog.* **20**, 397 (1966).
[38] W. M. Macintyre, P. Singh, and M. S. Werkema, *Biophys. J.* **5**, 697 (1965).
[39] J. Donohue, *Arch. Biochem. Biophys.* **128**, 591 (1968).
[40] H. Ringertz, Ph.D. Thesis, Stockholm (1969).
[41] C. E. Bugg, U. T. Thewalt, and R. E. Marsh, *Biochem. Biophys. Res. Commun.* **33**, 430 (1968).
[42] J. Sletlen, E. Sletlen, and L. H. Jensen, *Acta Crystallog.* **B24**, 1692 (1968).
[43] D. J. Sutor, *Acta Crystallog.* **11**, 453 (1958).
[44] D. J. Sutor, *Acta Crystallog.* **11**, 83 (1958).
[45] P. Tollin and A. R. L. Munns, *Nature* **222**, 1170 (1969).
[46] P. O. Löwdin, *Rev. Mod. Phys.* **35**, 724 (1963).
[47] P. O. Löwdin, in "Electronic Aspects of Biochemistry" (B. Pullman, ed.), p. 167. Academic Press, New York, 1964.

adenine and cytosine or *lactim* for guanine and thymine (denoted by A\*, T\*, G\*, and C\*), this could lead to a *miscoupling* of the bases.

Thus, the shorthand proton–electron pair codes would then be modified as follows:

$$A^* \begin{cases} : \\ :H \\ : \end{cases} \quad C^* \begin{cases} : \\ :H \\ : \end{cases} \quad G^* \begin{cases} :H \\ : \\ :H \end{cases} \quad T^* \begin{cases} :H \\ : \\ : \end{cases}$$

Therefore, as can be seen from Fig. 2, cytosine in its *imine* form would be able to hydrogen-bond to adenine in its common form (and vice

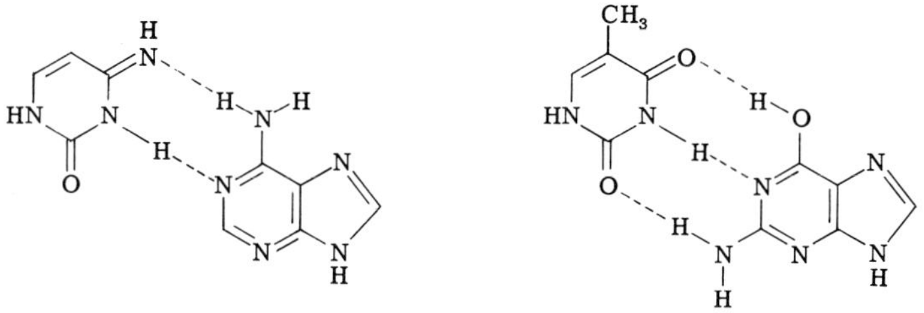

Cytosine (rare imino form)-
adenine (normal form)

Guanine (rare enol form)-
thymine (usual form)

Fig. 2. Examples of miscoupling of the bases of the nucleic acids.

versa) and guanine in its lactim form could couple with thymine in its usual form (and vice versa). As a result of such *miscouplings*, the original order of the arrangement of successive base pairs along the axis of the nucleic acid would be modified and the modification perpetuated during DNA replication, as indicated schematically in Fig. 3 (in which two cases are considered following the appearance of

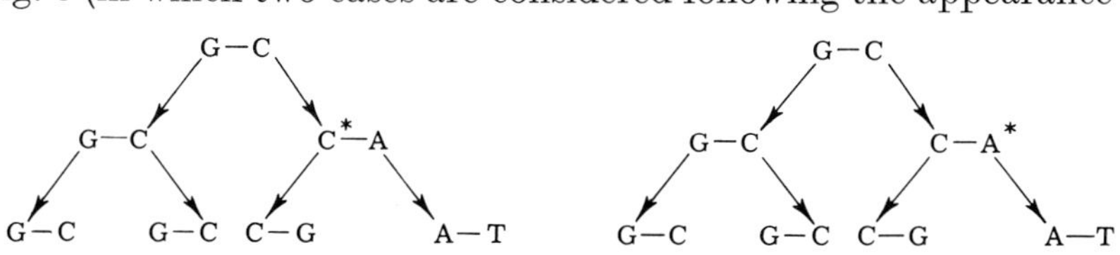

Fig. 3. The effect of rare tautomeric forms on the change in the base-pairing scheme and its perpetuation through DNA replication.

a rare form in the template chain or in the newly formed chain). The order of the complementary base pairs along the axis of DNA being most probably responsible for the genetic code of the species, any perturbation of this order represents by definition *a mutation*. In fact, the possible involvement of rare tautomeric forms as a cause of point mutations has already been recognized by Watson and Crick in their celebrated notes in *Nature* defining the structure of DNA [48, 49] (see also Ref. 50).

It may be observed that if both bases engaged in hydrogen bonding are *simultaneously* in their rare tautomeric forms, *imine* in adenine or cytosine and *lactim* in guanine or thymine, the original complementary coupling scheme may be preserved; it would simply consist of A*–T* and G*–C* instead of A–T and G–C. From the viewpoint of the base composition the two types of nucleic acids, the one formed only with the starred bases and the one formed only with the unstarred ones, would be undistinguishable. They could, of course, be distinguished by other properties. It may thus, for instance, be predicted on purely theoretical grounds [51] that while the stability of nucleic acids formed of unstarred bases should increase with their G–C content (which is, in fact, the case corresponding to the experimental situation), the reverse should be true for nucleic acids formed of starred bases. On the other hand, the possibility of the formation of A*–T* and G*–C* pairs through a double proton tunnelling in the A–T and G–C pairs has been considered as a possible mechanism for mutations.[46, 47]

The possible involvement of the rare tautomeric forms of purines (and pyrimidines) in the mechanism of mutagenesis, although representing, if verified, their probably most significant biological role, does not exhaust the possible biological manifestation of such forms. For example, a rare tautomeric form of purines, namely, N(7)H, is present at least in some (and possibly in all) natural or synthetic purine analogs of vitamin $B_{12}$ (analogs of the vitamin in which a purine replaces the 5,6-dimethylbenzimidazole ring).[52–56] The same form is also present in crystalline purine, 6-mercaptopurine,

---

[48] J. D. Watson and F. H. C. Crick, *Nature* **171**, 737 (1953).
[49] J. D. Watson and F. H. C. Crick, *Nature* **171**, 964 (1953).
[50] A. Pullman, *in* "Electronic Aspects of Biochemistry" (B. Pullman, ed.), p. 135. Academic Press, New York, 1964.
[51] B. Pullman and J. Caillet, *C. R. Acad. Sci.* **264**, 1900 (1967).
[52] K. Bernhauer, O. Muller and E. Muller, *Biochem. Z.* **335**, 37 (1961).
[53] S. K. Kon and J. Pawelkiewicz, *in* "Vitamin Metabolism," p. 115. Pergamon Press, London, 1960.
[54] E. Lester Smith, *in* "Vitamin $B_{12}$," p. 81. Methuen, London, 1960.
[55] J. A. Montgomery and H. J. Thomas, *J. Amer. Chem. Soc.* **85**, 2672 (1963).
[56] A. Veillard and B. Pullman, *J. Theoret. Biol.* **8**, 307 (1965).

and some hydroxypurines (*vide infra*). On the other hand, the role of the N(8)H form of some azapurines in a modified hydrogen bonding with a complementary base or the solvent has been considered in connection with their activity in cancer chemotherapy.

## III. Quantum Mechanical Methods of Investigation

### A. Generalities

A number of tautomeric forms of some fundamental purines have been studied quantum mechanically with the help of the standard, semiempirical Hückel approximation of the molecular orbital method.[57,58] Limited to $\pi$ electrons, these studies nevertheless enabled the determination of a number of electronic differences among the tautomers. Using some simplifying ideas, typical in $\pi$-electron calculations, information has also been obtained, within the limits of this approximation, concerning the *relative tendencies* of the bases to undergo a *given type* of tautomerization, and this information has been used successfully for the prediction of the preferential mutagenic sites in the nucleic acids.[50]

More recently the scope of the work has been very much extended and more refined approximations of the molecular orbital method have been employed. In the present account we shall essentially use the results obtained by two such procedures.

1. In the first place the Hückel approximation for the $\pi$ electrons has been replaced by a *self-consistent field (SCF) procedure*, generally in a semiempirical approximation of the Pariser–Parr–Pople type completed with some limited configuration interaction (PPP-CI method).[59,60] Second, the $\sigma$ skeleton of the molecules has been treated by the Del Re procedure[61] for saturated systems (which is the counterpart for the $\sigma$ electrons of the Hückel method for $\pi$ electrons) as refined for the $\sigma$ skeletons of conjugated heterocycles by

[57] B. Pullman and A. Pullman, "Quantum Biochemistry." Wiley (Interscience), New York, 1963.
[58] J. I. Fernandez-Alonso, *Advan. Chem. Phys.* **7**, 3 (1964).
[59] H. Berthod, C. Giessner-Prettre, and A. Pullman, *Theoret. Chim. Acta* **5**, 53 (1966).
[60] H. Berthod, C. Giessner-Prettre, and A. Pullman, *Int. J. Quant. Chem.* **1**, 123 (1967).
[61] G. Del Re, *in* "Electronic Aspects of Biochemistry" (B. Pullman, ed.), p. 221. Academic Press, New York, 1964.

Berthod and Pullman (DRBP method).[62] The representations of the π and σ systems obtained in this way are then added together to give an overall picture of the electronic structure (see Pullman and Pullman,[63] which contains also references to similar work by other authors).

2. It is obvious that this way of calculating separately and then adding together the distribution of the σ and π electrons suffers from the drawback of neglecting the fine aspects of their mutual interaction. Recently, methods have been proposed which remedy this defect by treating *simultaneously all the valence electrons*. Three of these methods are particularly prominent. These are the extended Hückel theory (EHT),[64,65] the iterative extended Hückel theory (IEHT),[66,67] and the so-called CNDO/2 method (the abbreviation standing for "complete neglect of differential overlap").[68-70] Whereas the first two procedures are extensions to all valence electrons of the basic Hückel procedure for π systems, the CNDO method is based on the SCF-MO scheme. It is this last procedure which has been most used, in particular, in our laboratory, for the investigation of purine tautomerism.[71-76] Its utilization has been of decisive importance because

[62] H. Berthod and A. Pullman, *J. Chim. Phys.* **62**, 942 (1965).
[63] A. Pullman and B. Pullman, *Advan. Quant. Chem.* **4**, 267 (1968).
[64] R. Hoffmann, *J. Chem. Phys.* **39**, 1397 (1963).
[65] R. Hoffmann, *J. Chem. Phys.* **40**, 2745 (1964).
[66] L. C. Cusachs and J. W. Reynolds, *J. Chem. Phys.* **43**, S160 (1965).
[67] D. G. Caroll, A. T. Armstrong, and S. P. McGlynn, *J. Chem. Phys.* **44**, 1865 (1966).
[68] J. A. Pople, D. P. Santry, and G. A. Segal, *J. Chem. Phys.* **43**, S129 (1965).
[69] J. A. Pople and G. A. Segal, *J. Chem. Phys.* **43**, S136 (1965).
[70] J. A. Pople and G. A. Segal, *J. Chem. Phys.* **44**, 3289 (1966).
[71] C. Giessner-Prettre and A. Pullman, *Theoret. Chim. Acta* **9**, 779 (1968).
[72] A. Pullman, *Int. J. Quant. Chem.* **2S**, 187 (1968).
[73] A. Pullman, *Ann. N. Y. Acad. Sci.* **158**, 65 (1969).
[74] B. Pullman and A. Pullman, *Progr. Nucl. Acid Res. Mol. Biol.* **9**, 327 (1969).
[75] A. Pullman, *in* "Quantum Aspects of Heterocyclic Compounds in Chemistry and Biochemistry" (E. D. Bergman and B. Pullman, eds.), p. 9. Israel Acad. Sci. Humanities, Jerusalem (distributed by Academic Press, New York), 1970.
[76] B. Pullman, *in* "Quantum Aspects of Heterocyclic Compounds in Chemistry and Biochemistry" (E. D. Bergman and B. Pullman, eds.), p. 292. Israel Acad. Sci. Humanities, Jerusalem (distributed by Academic Press, New York), 1970.

it enables a direct comparison of total molecular energies, and thus of the relative stabilities of *different* tautomeric forms to be made.

In spite of this essential advantage of the CNDO method, it must nevertheless be understood that the simultaneous utilization of this procedure and of the combined PPPCI-DRBP method is most useful. For a number of electronic properties of the conjugated heterocycles the PPPCI method is more appropriate than CNDO—in particular, the case for electronic transitions and other spectroscopic studies. Also, in cases in which both methods are known to give satisfactory results for some electronic indices and the corresponding physicochemical properties, e.g., dipole moments, their simultaneous utilization provides a useful mutual check. Finally, some properties, such as the ionization potentials, are probably in between the values indicated by the PPPCI method (too low) and those indicated by the CNDO method (too high).

The results based on the electronic structure of the purines presented here will thus correspond, in general, to computations carried out by the two types of method mentioned above. Before presenting these results and discussing them in conjunction with the available experimental data, we shall summarize the broad lines of the computational techniques.

## B. The Self-Consistent Field Method

The basic idea at the foundations of the *method of molecular orbitals* consists, as is well known, of *constructing the wave function of a polyelectronic system as a suitable combination of individual one-electron wave functions.*

The most "suitable" combination has been shown to be of the general form

$$\begin{vmatrix} a(1) & b(1) & c(1) \ldots \\ a(2) & b(2) & c(2) \ldots \\ \vdots & \vdots & \vdots \\ a(n) & b(n) & c(n) \ldots \end{vmatrix} \qquad (1)$$

a notation which stands for the determinant built on the n *individual wavefunctions* $a, b, c$, etc. Since each of those is a product of an *"orbital"* part

$$\phi(x, y, z) \qquad (2)$$

and a *spin function* α or β, the *total wave function* for an even number of electrons is written as:

$$\Psi = (n!)^{-1/2} \begin{vmatrix} \phi_1(1)\,\alpha(1) & \phi_1(1)\,\beta(1) & \phi_2(1)\,\alpha(1) & \phi_2(1)\,\beta(1) \dots \\ \phi_1(2)\,\alpha(2) & \phi_1(2)\,\beta(2) & \phi_2(2)\,\alpha(2) & \phi_2(2)\,\beta(2) \dots \\ \vdots & \vdots & \vdots & \vdots \\ \phi_1(n)\,\alpha(n) & \phi_1(n)\,\beta(n) & \phi_2(n)\,\alpha(n) & \phi_2(n)\,\beta(n) \dots \end{vmatrix} \quad (3)$$

Such a "Slater determinant," as it is often called, would, in fact, be the correct wave function for a system of noninteracting electrons. Electrons, however, do interact in real molecular systems. In order to obtain a more satisfactory representation, *the individual orbitals $\phi$ are determined so as to take into account the presence of the other electrons.* The best procedure which allows this determination is the "self-consistent field" method, whose main features are as follows. (a) One writes the exact *total* Hamiltonian for the system with explicit inclusion of electron interactions

$$H = \sum_\nu H(\nu) + \sum_{\mu<\nu} (1/r_{\mu\nu}) \quad (4)$$

where $H(\nu)$ is the Hamiltonian for *one* electron $\nu$ in the field of *all* the *bare* nuclei. (b) One expresses the *total* energy of the system by the standard quantum mechanical expression

$$\epsilon = \frac{\int \Psi^* H \Psi \, d\tau}{\int \Psi^* \Psi \, d\tau} \quad (5)$$

in terms of the individual orbitals $\phi$, by using the determinantal expression of $\Psi$. (c) One satisfies the variation principle for the energy. This is a standard procedure which is based on a fundamental theorem of quantum mechanics, namely, that the energy calculated by the above expression using an approximate wave function lies always higher than the exact energy. Thus if one uses an approximate wave function expressed in terms of certain parameters, minimization of the energy with respect to these parameters will yield the best possible energy value attainable with this form of $\Psi$. Carrying out this program yields the general "Fock" equations, one for each individual orbital $\phi$

$$F\phi_i = \epsilon_i \phi_i \quad (6)$$

where $F$ is an operator playing the role of an individual Hamiltonian and $\epsilon_i$ is the individual energy of one electron occupying the orbital $\phi_i$.

An essential characteristic of the Fock equations resides in the fact that each individual operator $F$ depends on all the orbitals which are occupied in the system (on account of the explicit inclusion of the interaction terms). Thus, each $\phi$ is given by an equation which depends on all the $\phi$'s. The way out of this difficulty is to choose arbitrarily a starting set of $\phi$'s, calculate the $F(\nu)$'s, solve the series of equations for a new set of $\phi$'s, and go over the same series of operations again and again until the $p$th set of $\phi$'s reproduce the $(p-1)$th set with good accuracy—hence the name "self-consistent" given to the procedure. The orbitals obtained in this fashion are, in principle, the best possible orbitals compatible with a determinantal $\Psi$.

One restriction must be made, however, about this last statement: a choice must be made of a starting set of $\phi$'s. Since it is impossible to guess *ab initio* the appropriate analytical form of a molecular orbital one must rely on a "reasonable" possibility. Thus the final orbitals are the best possible orbitals of *the form chosen*.

The classical choice of the starting orbitals is based on the following idea. Suppose that we deal with a chemical bond formed between two monovalent atoms A and B by the pairing of their valence electrons, one on A, the other on B. It is natural to assume that when one electron in the molecule is close to nucleus A, its molecular orbital will resemble the atomic orbital that it would occupy in A, and a similar situation would occur in the vicinity of B. This leads to the idea that *the molecular orbital may be approximated by a linear combination*

$$\phi = c_1 \chi_A + c_2 \chi_B \tag{7}$$

where the $\chi$'s are the atomic orbitals.

The idea can be extended to a polyatomic molecule and generalized so that *each molecular orbital in a molecule is a linear combination of all the atomic orbitals occupied by the electrons in the constituent atoms*.

$$\phi = \sum_r c_r \chi_r \tag{8}$$

This is the classical and general *LCAO approximation* (linear combination of atomic orbitals) of the molecular orbital method.

Given this form of molecular orbitals, the Fock equations yield a system of homogeneous linear equations in the $c_i$'s, *the Roothaan equations*[77]

$$\sum_q c_{iq}(F_{pq} - \epsilon S_{pq}) = 0 \tag{9}$$

where

$$S_{pq} = \int \chi_p^*(\nu) \chi_q(\nu) d\tau_\nu \tag{10}$$

is the *overlap integral*, and $F_{pq}$ is the $pq$ matrix element of the Fock operator

$$F_{pq} = \int \chi_p^*(\nu) F \chi_q(\nu) d\tau_\nu \tag{11}$$

All the necessary elements can be calculated in terms of integrals over the atomic orbitals $\chi$, integrals which involve either the nuclear attraction operators $H(\nu)$ or the interelectronic repulsions $1/r_{\mu\nu}$.

As is well known, a set of equations like Eq. (9) has nontrivial solution only if

$$|F_{pq} - \epsilon S_{pq}| = 0 \tag{12}$$

Solution of Eq. (12) yields the energy parameters $\epsilon$ and their replacement in the system of Eq. (9) yields the $c_{ir}$'s. But, the elements $F_{pq}$ depend on the coefficients; indeed, writing

$$F_{pq} = H_{pq} + G_{pq} \tag{13}$$

yields

$$H_{pq} = \int \chi_p^*(\nu) H(\nu) \chi_q(\nu) d\tau_\nu \tag{14}$$

and

$$G_{pq} = \sum_j \sum_{r,s} c_{jr} c_{js} [2(rs, pq) - (rp, qs)] \tag{15}$$

with

$$(pq, rs) = \iint \chi_p^*(\mu) \chi_q(\mu) 1/r_{\mu\nu} \chi_r^*(\nu) \chi_s(\nu) d\tau_\mu d\tau_\nu \tag{16}$$

[77] C. C. J. Roothaan, *Rev. Mod. Phys.* **23**, 69 (1951).

Thus, the practical way of solving the Roothaan equations is to choose an initial set of $c_r^0$'s, calculate the $F_{pq}$'s, solve the equations for a new set, and iterate again until *consistency* is attained.

In principle, this kind of scheme may be carried out for any molecule, with any number of electrons and any number of atomic orbitals $\chi$ in the LCAO basis set. The practical calculation, however, involves the tedious evaluation of a large number of integrals, a number which increases so rapidly with the number of electrons that, for large molecules, complete self-consistent field calculations are not really feasible on a large scale.

When dealing with *conjugated* molecules which are usually defined as molecules containing double bonds separated from each other by not more than one single bond, a classical and general simplification consists of treating their system of $\pi$ electrons alone in the field created by the nuclei *and* the so-called "$\sigma$ core." For this sake one writes a Hamiltonian *for the $\pi$ electrons*:

$$H_\pi = \sum_\nu \overset{\text{core}}{H}(\nu) + \sum_{\mu<\nu}(1/r_{\mu\nu}) \qquad (17)$$

defining a *core* which includes everything but the $\pi$ electrons. Provided $H$ is so defined, self-consistent procedure can be applied to the $\pi$ system alone and the *best $\pi$ orbitals* thus computed using the Fock equations appropriate for the system.

When the molecular orbitals are taken as a linear combination of the atomic $p_z$ orbitals only, the form taken by the equations amounts to solving a determinantal equation:

$$|F_{pq}^\pi - \epsilon S_{pq}| = 0 \qquad (18)$$

where $S_{pq} = \int \chi_p \chi_q d\tau$ is the overlap integral between two $p_z$ orbitals on atoms P and Q and $F_{pq} = \int \chi_p F \chi_q d\tau$ the corresponding $pq$ matrix element of the Fock operator.

Solving the equation in $\epsilon$ yields the possible values of the individual energies and for each $\epsilon_i$ the corresponding coefficients $c_{ir}$ of the atomic orbital $\chi_r$ in the molecular orbital $\phi_i$ are obtained by a system of linear equations

$$\sum_q c_{iq}(F_{pq}^\pi - \epsilon_i S_{pq}) = 0 \qquad (19)$$

The resolution of the equations yielding $\epsilon$ and the $c$'s is achieved by an iterative procedure of a "self-consistent" character as in the general method, since all the quantities involved in $F_{pq}$ can be calculated when the $\sigma$ core is assumed to have a definite configuration, in principle, as close as possible to the "exact" $\sigma$ core. In fact, this LCAO-SCF method for $\pi$ electrons only is quite time consuming and is essentially utilized in its simplified version, namely, the Pariser–Parr–Pople approximation.

## C. THE ZDO APPROXIMATION

The bottleneck in all self-consistent field calculations is the difficulty of calculating the integrals [Eq. (16)] over the atomic orbitals. Thus, reducing their number has been an imperative requirement and has led to the fundamental zero-differential overlap approximation (ZDO) which assumes

$$\chi_p(\mu)\chi_q(\mu) = 0 \tag{20}$$

for $p \neq q$. This hypothesis was initiated by Pariser and Parr[78] in the case of $\pi$ electrons and was later generalized by Pople et al.[68] for any pair of valence orbitals (complete neglect of differential overlap or CNDO). It is easily seen that such an assumption simplifies considerably both the SCF equations and the calculation of the matrix elements involved in them, since all overlap integrals vanish, and, moreover, among the $(pq,rs)$ integrals, only those of the $(pp,pp)$ or $(pp,qq)$ type remain.

The Roothaan LCAO-SCF equations in the ZDO approximation are thus simplified to

$$\sum_q c_{iq}(F_{pq} - \epsilon \delta_{pq}) = 0 \tag{21}$$

where $\delta_{pq}$ is the Kronecker $\delta$ and the $F_{pq}$'s involve interaction elements $G_{pq}$ which reduce to

$$G_{pq} = -\sum_j \sum_{p,q} c_{jp} c_{jq}(pp,qq) \tag{22}$$

and

$$G_{pp} = \sum_j \sum_r 2c_{ir}^2(rr,pp) \tag{23}$$

These simplifications are formally identical whether applied to the all-electron Eqs. (9) and (12) or to the $\pi$-electron Eqs. (18) and (19).

[78] R. Pariser and R. G. Parr, J. Chem. Phys. **21**, 466 (1953).

The ZDO hypothesis may seem a drastic assumption. It has been discussed thoroughly for the case of $\pi$ electrons by various authors [79] who have shown that the set of Eqs. (20) to (22) may be considered exact if the molecular orbitals $\phi$ are built as linear combinations of *orthogonal* "atomic" orbitals $\lambda$ and all the quantities calculated in terms of $\lambda$'s instead of the usual atomic orbitals $\chi$. Such a set of orthogonal $\lambda$'s can be obtained, for instance, by an appropriate unitary transformation of the usual $\chi$'s.[80] Unfortunately, in order to calculate the matrix elements over the $\lambda$'s, one goes back first to calculate integrals over the $\chi$'s which is precisely what one wants to avoid. Thus, *in practice*, when the ZDO assumption is adopted, the simplified equations are used, and some empirical corrections are introduced so as to make up for the "built-in" errors of the procedure.[81] It must be emphasized that the ZDO assumption is only one procedure. The limitation of the LCAO basis set is another. Further, there is the intrinsic error of the SCF method, namely, the use of doubly occupied orbitals which permit two electrons of opposite spins to be at the same place at the same time, thus ignoring at least in part what is called their "correlation." Finally, in the $\pi$-electron approximation, there will be the representation chosen for the $\sigma$ core which is generally taken as the sum of the atomic cores with the $\sigma$ electrons occupying atomic valence-state orbitals.

We shall see in more detail how these empirical corrections are made in the two cases of interest in this chapter, that is, the Pariser–Parr–Pople procedure and the CNDO method.

### D. The Parametrization of the Pariser–Parr–Pople Approximation for $\pi$ Electrons

It has been shown in Pariser and Parr's early articles [78, 82] that the main corrections were: (a) a strong reduction of the one-center Coulomb integrals $(pp|pp)$ with respect to their value calculated with atomic Slater orbitals. In practice, the value to adopt is the

---

[79] See R. G. Parr, *in* "The Quantum Theory of Molecular Electronic Structure." Benjamin, New York, 1963.
[80] P. O. Löwdin, *J. Chem. Phys.* **18**, 365 (1950).
[81] A. Pullman, *in* "Molecular Biophysics" (M. Weisbluth and B. Pullman, eds.), p. 81. Academic Press, New York, 1965.
[82] R. Pariser and R. G. Parr, *J. Chem. Phys.* **21**, 767 (1953).

difference between the atomic valence-state ionization potential and electron affinity [83]

$$(pp,pp) = I - A \qquad (24)$$

(b) a corresponding reduction of the two-center $(pp|qq)$ integrals, at least for nearest neighbors (*vide infra*); (c) an empirical choice of $H_{pq} = \beta_{pq}$ values so as to fit spectral data for reference compounds.

These are the main features of the Pariser–Parr approximation. There remains the choice of the $H_{pp}$ integrals, the classical expression for which is

$$\alpha_p = I_p - \sum_A (A|pp) - \sum_{q \neq p} (qq|pp) \qquad (25)$$

This expression is generally used as such with or without neglect of the second term and using as $I_p$ the negative of the valence-state ionization potential of atom $P$. Although such a simplification may be of little importance in hydrocarbons, where it seems to influence merely the absolute values of the molecular ionization potentials and where the atomic $\pi$-electron populations are close to unity, this is not the case for heteromolecules. Here, an empirical choice of $U_p$ values in place of the first two terms in Eq. (24) is better.[84]

A certain flexibility exists thus for the reduction of the $(pp,qq)$ integrals and the choice of the $H_{pq}$ and $U_p$ values. The calculations reported in this paper have used the parametrization adopted by Berthod, Giessner-Prettre, and Pullman[59] and determined by trial and error so as to reproduce in a satisfactory way as many ground state properties as possible in a series of reference compounds. More precisely this choice of the integral values can be summarized briefly as follows: (a) the one-center Coulomb integrals are given the values

$$\gamma_p^0 = (pp|pp) = I_p - A_p \qquad (26)$$

according to the above-mentioned Pariser relation,[83] where $I_p$ and $A_p$ are the valence-state ionization potential and electron affinity of atom $p$, which have been tabulated by Hinze and Jaffé[85] for the appropriate valence states. For $\pi$ lone-pairs like those of the pyrrolic nitrogen, Eq. (26) was also adopted. (b) Concerning the two-center

[83] R. Pariser, *J. Chem. Phys.* **21**, 568 (1953).
[84] A. Pullman and M. Rossi, *Biochim. Biophys. Acta* **88**, 211 (1964).
[85] J. Hinze and H. H. Jaffé, *J. Amer. Chem. Soc.* **84**, 540 (1962).

Coulomb integrals $\gamma_{pq} = (pp|qq)$, one adopts for all pairs $p$, $q$ of non-directly bonded atoms the theoretical values calculated with the usual Slater orbitals. For the Coulomb integrals for pairs of bonded atoms, they affect the results essentially through their difference with respect to the one-center value adopted. The best fit for this difference in carbon compounds yields a value of 6.9 eV for $\gamma_{CC}$ (1.39 Å) when $\gamma_C^0$ is equal to the $I-A$ value. This number would be obtained through the theoretical formulas by using Slater orbitals with an effective quantum number $\zeta'$ equal to 0.83 (instead of the Slater value of 1.625). For adjacent atoms, all the $\gamma_{CC}$ values have thus been calculated with this value of $\zeta'$. A similar procedure has been used for heteronuclear integrals [for details see the original papers[59]] and has yielded the set of $\zeta'$ listed in Table I. (c) The remaining parameters concern the one-center and two-center core integrals $\alpha_p$ and $\beta_{pq}$. Adopting as a starting point the $\beta_{CC}$ value, $-2.39$ eV, determined for benzene (C–C = 1.39 Å) and the $\beta_{CN}$ value of $-2.576$ eV adjusted for s-triazine[82] (C–N = 1.34 Å), $\beta_{CO}$, $\beta_{C-NH}$, and $\beta_{C-NH_2}$ have been found by trial and error for formaldehyde, pyrrole, and aniline. The distance dependence of $\beta_{pq}$ ($\beta = -K/R^6$) has been used.

TABLE I

Integral Values for the $\pi$ System

| $p$ | $\gamma_p^0$ (eV) | $\zeta'$ | $K_{(C-p)}$ | $U_p$ (eV) |
|---|---|---|---|---|
| C | 11.13 | 0.83 | 17.238 | $-9.5$ |
| N | 12.34 | 0.92 | 14.913 | $-11.8$ |
| NH | 12.34 | 0.92 | 15.195 | $-9.9$ |
| NH$_2$ | 12.34 | 0.92 | 11.579 | $-8.9$ |
| O | 15.23 | 1.135 | 8.573 | $-14.5$ |

There remains the choice of $U_p$ values. As a starting point we have adopted $U_C = -9.5$ eV, which is known to give satisfactory values for the ionization potentials in hydrocarbons[86] and determined $U_O$, $U_N$, $U_{NH}$, and $U_{NH_2}$ so as to reproduce the ionization potentials of formaldehyde, pyridine, pyrimidine, pyrrole, and aniline. Moreover,

[86] J. A. Pople, *Trans. Faraday Soc.* **49**, 1375 (1953).

in this trial-and-error procedure, we have attempted to reproduce at the same time the values of the dipole moments of the reference compounds, obtained by adding to the calculated $\pi$ component, the $\sigma$ component (obtained in the way described in the next section).

The set of integral values which gave the best overall agreement with as many ground state properties as possible in the series of reference compounds chosen is listed in Table I.

### E. The Simple Representation of the $\sigma$ Bonds

A very simple procedure was proposed by Del Re [87] for representing the $\sigma$ bonds in saturated compounds. Its essential features are as follows. (a) Treat each bond as a two-electron problem. (b) Describe each electron in the bond by a molecular orbital linear combination of

$$\phi = a\chi_a + b\chi_b \tag{27}$$

two atomic orbitals. (c) Use an effective one-electron Hückel-type Hamiltonian yielding the secular equation of second-order

$$\begin{vmatrix} H_{aa} - E & H_{ab} - ES_{ab} \\ H_{ab} - ES_{ab} & H_{bb} - E \end{vmatrix} = 0 \tag{28}$$

with

$$H_{aa} = \int \chi_a H \chi_a \, d\tau \tag{29}$$

$$H_{ab} = \int \chi_a H \chi_b \, d\tau \tag{30}$$

Solving for $E$ permits then the calculation of the coefficients by

$$\frac{b}{a} = \frac{H_{aa} - E}{H_{ab} - ES_{ab}} \tag{31}$$

(d) Neglect overlap, put

$$H_{aa} = \alpha_a = \alpha + \delta_a \beta \tag{32}$$

$$H_{ab} = \epsilon_{ab} \beta \tag{33}$$

in terms of two units $\alpha$ and $\beta$. (These units are $\alpha_H$ and $\epsilon_{CH}$, respectively.) (e) As an original feature of the procedure, one defines

$$\delta_a = \delta_a^0 + \sum_c \gamma_a(c) \delta_c^0 \tag{34}$$

[87] G. Del Re, *J. Chem. Soc.* 4031 (1958).

where $\delta_a^0$ is a characteristic of atom $a$, $c$ stands for all the atoms bound to atom $a$, $\gamma_a(c)$ is a proportionality factor which introduces what Del Re calls the inductive effect of atom $c$ on atom $a$. (f) The set of $\delta^0, \gamma$, and $\epsilon$ values were chosen by Del Re by successive approximations for reproducing as many properties as possible for a number of saturated molecules, imposing a few logical conditions on the parameters. In particular, the initial set of $\delta^0$ values obeyed a linear law in the *global* atomic electronegativities $x$.

$$\delta_a^0 = k\frac{x_a - x_H}{x_H} \tag{35}$$

(g) The preceding technique was adopted by Berthod and Pullman[62] so as to make allowance for the possible differences in valence state or hybridization ratio of the same atom. Thus, orbital electronegativities must be used as a guide for the choice of $\delta^0$ values. This leads, in particular, to the use of different $\sigma$-parameter values for saturated and conjugated molecules, a logical step in a procedure which uses linear combinations of valence orbitals. The final parameter values utilized are given in Table II.

TABLE II

$\sigma$-Parameter Values

|  | $\mu$ | $\delta_\mu^0$ | $\mu\nu$ | $\epsilon_{\mu\nu}$ |
|---|---|---|---|---|
| C: | 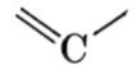 =C− | 0.12 | C—H | 1 |
| N: | 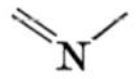 =N− | 0.38 | C=N | 0.7 |
| N: | 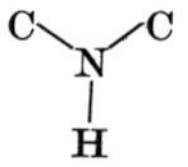 C−N(H)−C | 0.30 | C—N<br>N—H | 0.7<br>0.6 |
| N: | 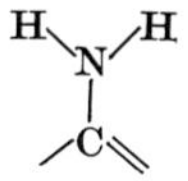 H−N(H)−C= | 0.24 | C—N<br>N—H | 1<br>0.45 |
| O: | O= | 0.28 | C=O | 0.7 |
| O: | —OH | 0.40 | C—OH<br>O—H | 0.95<br>0.45 |

## F. The CNDO Approximation

As already mentioned, the basic hypothesis of the CNDO method is the generalization of the ZDO approximation beyond the $\pi$-electron theory. More precisely, all the valence electrons of a molecule are treated explicitly in the LCAO-SCF framework previously described [Eqs. (9)–(12) of Section B]. The inner-shell electrons are included in the core and *all* the products

$$\chi_p(\mu)\chi_q(\mu) \tag{36}$$

between valence orbitals are neglected. Thus the equations to be solved are those of the ZDO approximation [Eq. (21)] and the matrix elements $G_{pq}$ and $G_{pp}$ are given by Eqs. (22) and (23) which we shall write now for convenience in Pople's usual notations: the Greek letter indices refer to atomic orbitals, the capital letter indices to atoms. Thus, Eq. (21) becomes

$$\sum_\nu c_{i\nu}(F_{\mu\nu} - \epsilon\delta_{\mu\nu}) = 0 \tag{37}$$

Using, moreover, the definition of the charge density and bond-order matrix

$$P_{\mu\nu} = 2\sum_i c_{i\mu}c_{i\nu} \tag{38}$$

where $i$ stands for the occupied molecular orbitals, one obtains

$$G_{\mu\mu} = \tfrac{1}{2}P_{\mu\mu}\gamma_{\mu\mu} + \sum_{\lambda\neq\mu}P_{\lambda\lambda}\gamma_{\mu\lambda} \tag{39}$$

$$G_{\mu\nu} = -\tfrac{1}{2}P_{\mu\nu}\gamma_{\mu\nu} \tag{40}$$

The next fundamental hypothesis in the CNDO procedure concerns the values of the Coulomb integrals $\gamma_{\mu\nu}$ which are all approximated as Coulomb integrals over $2s$ Slater atomic orbitals. If $\mu$ and $\nu$ are on the same atom A

$$\gamma_{\mu\nu} = \gamma_{2s_A 2s_A} = \gamma_{AA} \tag{41}$$

If not,

$$\gamma_{\mu\nu} = \gamma_{2s_A 2s_B} = \gamma_{AB} \tag{42}$$

This is a sort of mean value between $\gamma_{\sigma\sigma}$ and $\gamma_{\pi\pi}$ and this approximation ensures the invariance with respect to a rotation of the molecular axis,[68] an important requirement if the theory is to make any sense.*

Using this approximation and putting

$$P_{AA} = \sum_{\mu_A} P_{\mu\mu} \qquad (43)$$

the $G_{\mu\mu}$ matrix elements become

$$G_{\mu\mu} = (P_{AA} - \tfrac{1}{2}P_{\mu\mu})\gamma_{AA} + \sum_{B \neq A} P_{BB}\gamma_{AB} \qquad (44)$$

The third fundamental hypothesis of the CNDO approximation concerns the core matrix elements $H_{\mu\nu}$. These correspond to the $\alpha$ and $\beta$ parameters of the Pariser–Parr–Pople method, the core including here only the 1s electrons and the nuclei.

Using a partition of the core Hamiltonian into atom-centered fractions, the one-center–one-orbital core integral may be written

$$H_{\mu_A \mu_A} = U_{\mu_A \mu_A} + \sum_{B \neq A} \langle \mu_A V_B \mu_A \rangle$$

$$= U_{\mu_A \mu_A} - \sum_{B \neq A} V_{BA} \qquad (45)$$

where $V_B$ is the potential due to the $B$th core.

The values of all $H_{\mu_A \nu_A}$ for two different orbitals on the same atom are neglected.

Concerning the $H_{\mu_A \nu_B}$ values, they are assumed to be expressible as

$$H_{\mu_A \nu_B} = \tfrac{1}{2}(\beta_A^0 + \beta_B^0) S_{\mu\nu} \qquad (46)$$

where $S_{\mu\nu}$ is the overlap integral between orbitals $\mu$ and $\nu$ which is calculated using the appropriate atomic Slater orbitals. The *parameters proper* of the procedure are then the values of the $U_{\mu\mu}$'s and the $\beta^0$'s characteristic of each atom. A discussion of this choice is

* When hydrogen atoms are involved, 1s atomic orbitals with $\zeta = 1.2$ are used in the calculation of the corresponding $\gamma$ values.

detailed in the original papers.[68–70] In the version called CNDO/2, one approximates the $U_{\mu\mu}$'s as

$$U_{\mu\mu} = -\tfrac{1}{2}(I_\mu + A_\mu) - (Z_A - \tfrac{1}{2})\gamma_{AA} \qquad (47)$$

and the $V_{AB}$'s as

$$V_{AB} = Z_B \gamma_{AB} \qquad (48)$$

where $I_\mu$ and $A_\mu$ are the orbital ionization potential and electron affinity, and $Z_B$ the $B$th core charge.

Inside this framework, the $\beta^0$ values have been chosen so that the results fit as well as possible with those of nonempirical calculations for small molecules. Table III summarizes the numerical values used in CNDO/2.

TABLE III

Atomic Parameters for CNDO/2 (eV)

| Atom | $\tfrac{1}{2}(I+A)$ | | $\beta^0$ |
|---|---|---|---|
| | $s$ | $p$ | |
| H | 7.175 | — | −9 |
| C | 14.051 | 5.572 | −21 |
| N | 19.316 | 7.275 | −25 |
| O | 25.390 | 9.111 | −31 |

## IV. The Four Prototropic Tautomers of Purine

Let us start with the basic problem of the four fundamental tautomers of purine. As stated before, although the molecule of purine is generally represented in the form N(9)H, in which a hydrogen atom as attached to N-9, tautomeric forms may naturally be considered, in which the proton is attached to the other nitrogens of the molecule, yielding the tautomers N(7)H, N(3)H, and N(1)H. In fact, although the majority of biological purines exist essentially as derivatives of the N(9)H tautomer, the crystalline form of purine itself, which consists of long chains of these molecules linked together by single hydrogen bonds, involves the N(7)H tautomer[88] and the same is true about the crystal structure of some more complex derivatives of purine (*vide infra*). Moreover, unambiguous methyl derivatives of the *four* tautomeric forms (with the methyl replacing the nitrogen-bound hydrogen) have been prepared—the derivatives of the N(9)H and N(7)H tautomers by Bendich *et al.* in 1954[89] and those of the N(1)H and N(3)H tautomers more recently by Townsend and Robins, in 1962 and 1966.[90-92] A larger variety of more complex derivatives of these various tautomers has also been prepared recently in the laboratory of F. Bergmann, in Jerusalem.

The four tautomeric forms have been subjected to quantum mechanical calculations by the two above-mentioned procedures.[93] Before presenting the results one important general remark must be made. It concerns the problem of molecular geometries.

The results of calculations depend to some extent, as it is well known, on the geometries adopted. Now, these, if known at all, are generally only for the most common tautomers so that the geometries of the more rare forms have to be assumed. This is a general problem that had to be dealt with in this study and the way in which it was generally solved may be illustrated using purine as an example. For this molecule, the only directly available geometry is that of the N(7)H tautomer.[88] That of the N(9)H may be deduced relatively easily. Because of the particular importance of this most common tautomer, the calculations have been carried out, in fact, for two geometries: (1) one inferred from that of the N(7)H tautomer by inverting the distances of the 7–8 and 9–10 bonds and the 5–7 and 4–9 bonds and of the corresponding angles, which we shall designate as the inversed Watson's geometry [concerning this procedure, see Sletten *et al.*[94]] and (2) the one established by Spencer[95] on the basis of a general study of purine derivatives (Spencer's geometry). (For

---

[88] D. G. Watson, R. M. Sweet, and R. Marsh, *Acta Crystallog.* **19**, 573 (1965).
[89] A. Bendich, P. J. Russell and J. J. Fox, *J. Amer. Chem. Soc.* **76**, 6073 (1954).
[90] L. B. Townsend and R. K. Robins, *J. Org. Chem.* **27**, 990 (1962).
[91] L. B. Townsend and R. K. Robins, *J. Amer. Chem. Soc.* **84**, 3008 (1962).
[92] L. B. Townsend and R. K. Robins, *J. Heterocyc. Chem.* **3**, 241 (1966).
[93] B. Pullman, H. Berthod, D. Bergmann, F. Bergmann, Z. Neiman, and H. Weiler-Feilchenfeld, *C.R. Acad. Sci.* **267**, 1461 (1968).
[94] J. Sletten, E. Sletten, and L. H. Jensen, *Acta Crystallog.* **B24**, 1692 (1968).
[95] M. Spencer, *Acta Crystallog.* **12**, 59 (1959).

other similar "statistical" evaluations see Ringertz [96] and Donohue.[97]) The geometries of the N(3)H and N(1)H tautomers are more difficult to ascertain. The guiding line in this respect was the observation that because of the possibility of writing only one Kekulé-type structure for these tautomers, the "resonance" of their mobile electrons must be more restricted than that of the N(9)H and N(7)H tautomers and therefore the bond localization should be more pronounced in them. It is obvious, however, that the geometries assumed for these particular tautomers are subject to caution. The geometries adopted in the calculations are represented in Fig. 4.

The principal results of the computations for these fundamental forms are summarized in Tables IV and V and lead to the following main conclusions.

### 1. Total Molecular Energy (Table IV)

The first fundamental question concerns, naturally, the relative stabilities of the different tautomers. In this respect, it is the CNDO results which *a priori* are particularly significant. They indicate that both from the point of view of the electronic energy and the total energy (electronic energy plus nuclear repulsion), *the N(9)H and the N(7)H tautomers should be more stable than the N(3)H or N(1)H ones.* Following the values of the total energies this excess of stability should be of the order of 40 kcal/mole. The two most stable tautomers are predicted to be of comparable energy.

### 2. Highest Occupied (HOMO) and Lowest Empty (LEMO) Molecular Orbitals (Table IV)

The energies of these two essential orbitals give information, respectively, about the electron-donating properties or ionization potentials and the electron-accepting properties or electron affinities of the compounds. It can be seen that both methods predict that the four tautomers should have remarkably and somewhat surprisingly similar ionization potentials. As may be judged from the known ionization potential of the N(9)H tautomer, whose experimental value is 9.7 eV,[98] the PPP-DBP method somewhat underestimates and

[96] H. Ringertz, Ph.D. Thesis, Karolinska Institutet, Stockholm, 1969.
[97] J. Donohue, *Arch. Biochem. Biophys.* **128**, 591 (1968).
[98] Ch. Lifschitz, E. D. Bergmann, and B. Pullman, *Tetrahedron Lett.* **46**, 4583 (1967).

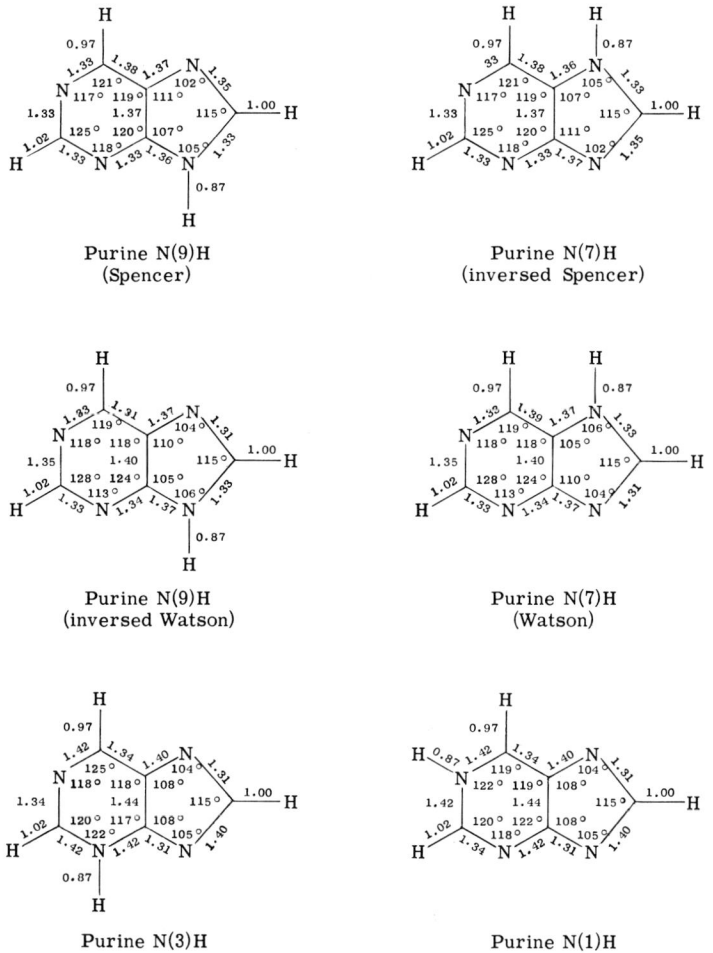

FIG. 4. The geometries adopted for purine tautomers.

the CNDO method overestimates the values of the ionization potentials.

Concerning the electron affinities of the tautomers, both methods predict that they should be greater for the N(1)H and N(3)H tautomers than for the N(9)H and N(7)H ones. These properties are also probably overestimated in the CNDO procedure. No experimental information is available on the electron affinities of purines.

TABLE IV

SOME ENERGY CHARACTERISTICS OF PURINE TAUTOMERS

| Purine tautomer | SCF–CI method | | | | | | Ring currents (with benzene = 1) | | CNDO method | | |
|---|---|---|---|---|---|---|---|---|---|---|---|
| | HOMO (eV) | LEMO (eV) | $S_1$ eV | $S_1$ m$\mu$ | $\theta^a$ (deg) | $\lambda_{max}^b$ (m$\mu$) | Hexagonal ring | Pentagonal ring | HOMO (eV) | LEMO (eV) | Energy$^c$ (kcal/mole) |
| N(1)H | −9.0 | 0.5 | 4.50 | 276 | 103 | 275 | 0.32 | 0.26 | −11.0 | 1.9 | 38 |
| N(3)H | −9.2 | 0.3 | 4.45 | 279 | 120 | 277 | 0.43 | 0.35 | −11.3 | 1.7 | 33 |
| N(7)H (Watson) | −9.1 | 1.1 | 4.73 | 262 | 38 | 266.5 | 1.03 | 0.65 | −11.9 | 3.0 | 6 |
| N(7)H (inversed Spencer) | −9.1 | 1.0 | 4.93 | 251 | 42 | | | | −11.6 | 3.0 | 0 |
| N(9)H (inversed Watson) | −9.0 | 1.2 | 4.81 | 258 | 59 | 264 | 1.00 | 0.66 | −11.5 | 3.2 | 7 |
| N(9)H (Spencer) | −8.9 | 1.1 | 4.93 | 251 | 54 | | | | −11.3 | 3.2 | 0 |

$^a$ Counted counterclockwise from the line C-4–C-5.
$^b$ In the corresponding methylated derivatives. (For some more complex derivatives see Chen and Clark.[99])
$^c$ With respect to the most stable tautomer considered as energy zero.

### 3. The Longest Wavelength of Absorption ($\lambda_{max}$) (Table IV)

In this respect it is the PPPCI results on the $\pi$-electronic system which *a priori* should be considered. They enable one to predict that the *N(1)H and N(3)H tautomers should be bathochromic by about 10 m$\mu$ with respect to the N(7)H and N(9)H ones*. In fact, the comparison of the theoretical evaluation of the energy of the first singlet transition ($S_1$) with the experimental values of $\lambda_{max}$ in the methylated derivatives of the tautomers indicates an extremely satisfactory agreement both in the ordering of all four tautomers and in the absolute values. For a more detailed study of the spectrum of the most common N(9)H tautomer see Pullman and Pullman.[63, 74]

Table IV also indicates the angle ($\theta$) of the polarization directions of this longest $\pi$–$\pi$* absorption band in the four tautomers, measured counterclockwise with respect to the C-4–C-5 axis. The angles are of the order of 38°–54° for the N(7)H and N(9)H tautomers and of the order of 103°–120° for the N(1)H and N(3)H ones (Fig. 5). Very

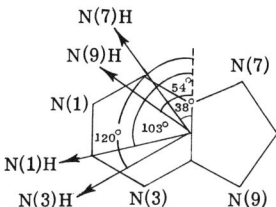

FIG. 5. Computed polarization directions of the longest $\pi$–$\pi$* absorption band in the four purine tautomers.

recently this direction has been determined experimentally through the examination of the polarized spectra of single crystals of purine.[99] The experimental data refer thus to the N(7)H tautomer. The value found (48°) is in quite satisfactory agreement with the calculated one.

### 4. Dipole Moments (Table V)

The values of the total dipole moments predicted by the two procedures are similar. Both procedures indicate that the tautomers may be divided, from that viewpoint, into two groups: *tautomers N(1)H and N(7)H which should possess relatively high dipole moments*,

[99] H. H. Chen and L. B. Clark, *J. Chem. Phys.* **51**, 1862 (1969).

## TABLE V
### Dipole Moments (Debye Units) in Purine Tautomers

| Purine tautomer | PPP–DBP method | | | | CNDO method | | | | |
|---|---|---|---|---|---|---|---|---|---|
| | $\mu_\sigma$ | $\mu_\pi$ | $\mu_{total}$ | ∢ [a] (deg) | $\mu_\sigma$ | $\mu_\pi$ | $\mu_{sp}$ [b] | $\mu_{total}$ | ∢ [a] (deg) |
| N(1)H | 1.44 | 4.21 | 5.38 | 241 | 2.11 | 5.98 | 3.02 | 6.75 | 241 |
| N(3)H | 0.29 | 3.24 | 3.37 | 320 | 1.90 | 4.72 | 1.68 | 4.19 | 320 |
| N(7)H (Watson) | 1.82 | 3.79 | 5.50 | 153 | 1.20 | 3.94 | 3.38 | 6.08 | 150 |
| N(7)H (inversed Spencer) | 1.91 | 3.98 | 5.84 | 151 | 1.41 | 4.23 | 3.44 | 6.22 | 151 |
| N(9)H (inversed Watson) | 0.44 | 2.86 | 3.27 | 48 | 1.58 | 3.29 | 2.00 | 3.75 | 49 |
| N(9)H (Spencer) | 0.64 | 3.05 | 3.68 | 46 | 1.43 | 3.60 | 2.02 | 4.19 | 45 |

[a] Angle of the dipole with the C-4–C-5 axis (counted counterclockwise).
[b] Hybridization moment.

of the order of 5–7 Debyes and tautomers $N(9)H$ and $N(3)H$ which should possess a significantly smaller dipole moment of the order of 3–4 Debyes. From the previously known[100] value of 4.3 D for the dipole moment of 9-methylpurine, it may be assumed that the CNDO results may perhaps be somewhat more satisfactory on the absolute scale. Table V indicates also the different components which make up the total moments and the angle of the dipole with the C-4–C-5 bond. The constancy of this angle in the two procedures is worth stressing. Figure 6 gives a pictorial representation of these results.

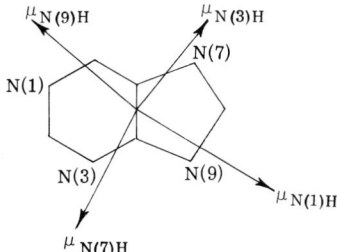

Fig. 6. Relative magnitudes and the directions of the dipole moments in the four purine tautomers (PPPCI method).

It may be observed that the dipole moments point in directions approximately opposite to the N–H axis.

Recently, dipole moments for simple derivatives of the four tautomers of purine have been measured.[93, 101] Three series of derivatives have been investigated: the $N$-methyl-8-phenylpurines, the $N$-methyl-6-thiomethylpurines, and the N-methyl-6-thiomethyl-8-phenylpurines. The results are summarized in Table VI. It can be seen that these measurements confirm the theoretical predictions very satisfactorily. It must, however, be mentioned that Dyer et al.[102] did not find any appreciable difference in the dipole moments of the 9-benzyl and 7-benzyl derivatives of 6-chloropurine (4.91 and 5.03 D, respectively). Moreover, Weiler-Feilchenfeld and Neiman[101]

---

[100] H. DeVoe and I. Tinoco, Jr., *J. Mol. Biol.* **4**, 500 (1962).
[101] H. Weiler-Feilchenfeld and Z. Neiman, in "Quantum Aspects of Heterocyclic Compounds in Chemistry and Biochemistry" (E. D. Bergmann and B. Pullman, eds.), p. 308. Israel Acad. Sci. Humanities, Jerusalem (distributed by Academic Press, New York), 1970.
[102] E. Dyer, R. E. Farris, Jr., C. E. Minnier, and M. Tokizawa, *J. Org. Chem.* **34**, 973 (1969).

TABLE VI  DIPOLE MOMENTS OF PURINES (DEBYE UNITS)

| Tautomer | N(1)H | N(3)H | N(7)H | N(9)H |
|---|---|---|---|---|
| Theory | 5.4–6.8 | 3.4–4.2 | 5.5–6.1 | 3.3(3.7)–3.8(4.2) |
| Experimental | 6.92 | 4.37 | — | 4.52 |
|  | — | 5.24 | 5.61 | 3.01 |
|  | Unstable | 4.85 | ? | 4.05 |
|  | 6.52 | | | |

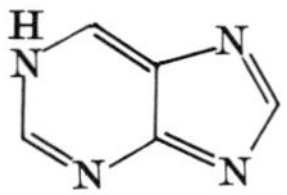

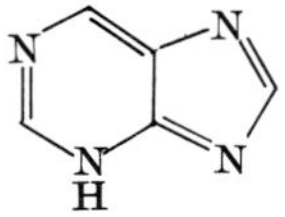

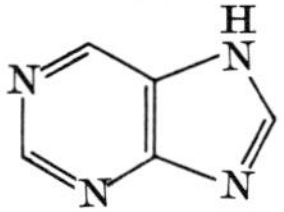

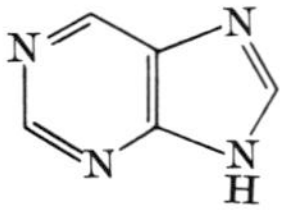

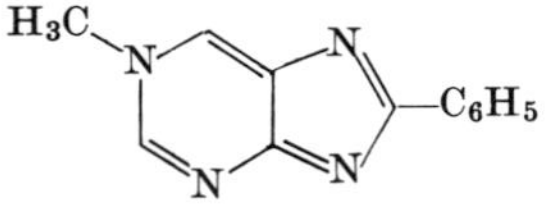

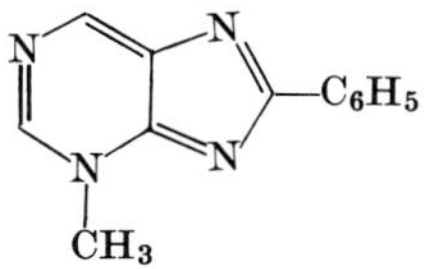

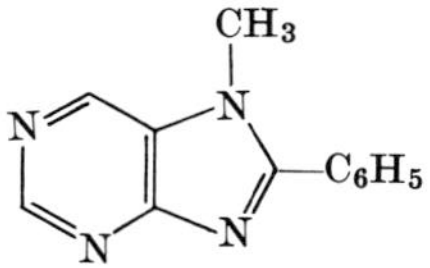

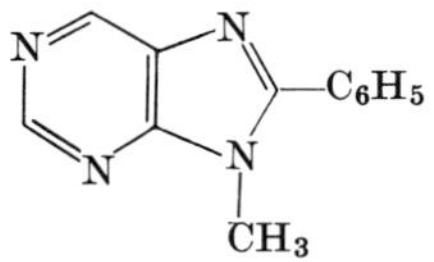

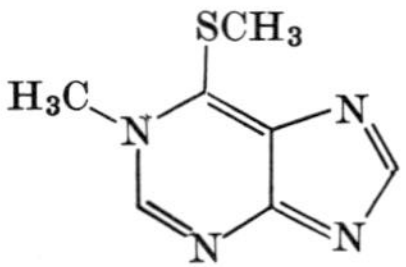

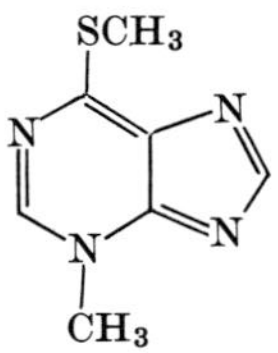

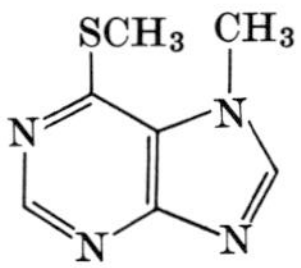

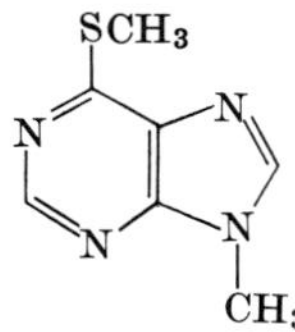

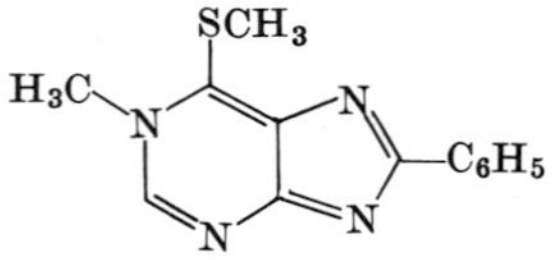

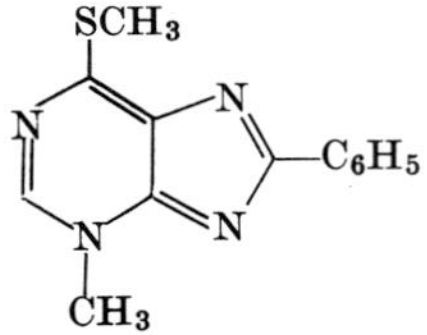

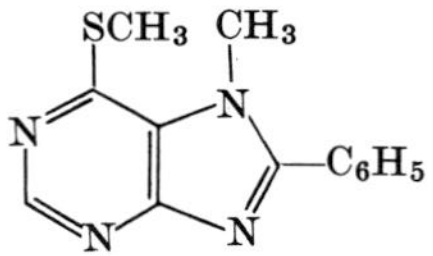

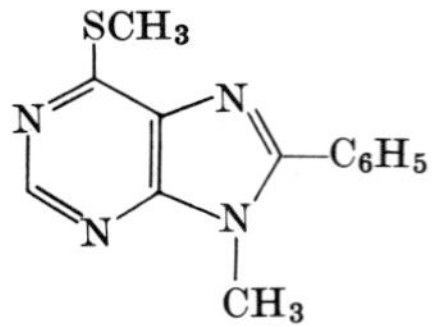

have not only studied the absolute values of the moments, but have also tried to determine their directions and succeeded in doing so for the N(9)H tautomer. The procedure employed involves the determination of the moments of 6-chloropurine (5.34 D) and 2,6-dichloropurine (5.62 D). Independently these moments have been calculated, by vector addition (with the group moment of Cl taken as 1.6 D). Assuming then that the chloropurines exist in the common N(9)H form and taking for the moment of purine the theoretical value of 4.2 D, the best fit between the calculated and the measured moments of the mono- and dichloropurines (calculated values equal 5.33 and 5.58 D, respectively) was obtained when the direction of the moment of purine was taken as 52° (counterclockwise with respect to the C-4–C-5 axis), a result in very close agreement with the theoretical value of 45°–49°.

The variation of the values and directions of the dipole moments in the four purine tautomers results, of course, from the appreciable differences in their charge distributions. Figure 7 indicates this

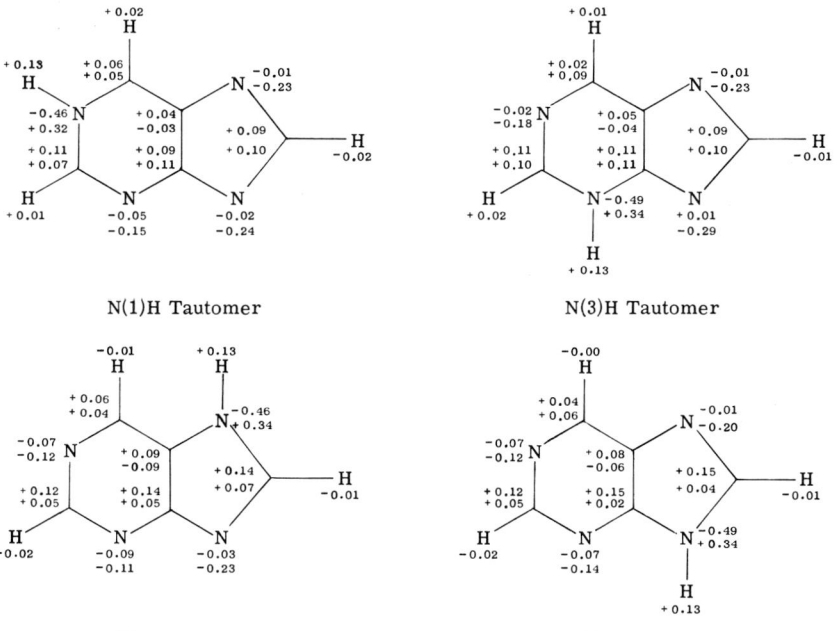

Fig. 7. Net electronic charges in purine tautomers evaluated by the CNDO/2 method. Upper numbers: $\sigma$ charges; lower numbers: $\pi$ charges.

distribution, both for the $\sigma$ and the $\pi$ electrons, as obtained by the CNDO method. The distribution obtained by the PPP-DRBP procedure is, generally speaking, analogous. For the same reason only one diagram corresponding to results obtained with the inversed Watson geometry is indicated for the N(9)H tautomer.

## 5. Ring Currents

Due to the recent development of the utilization of nuclear magnetic resonance for the study of molecular associations (in particular, stacking) involving the purine and pyrimidine bases (e.g., Ts'o et al.[103-106]), the knowledge of the *ring currents* in these compounds is of particular importance for the interpretation of the observed data and the estimation of the favored conformations. These ring currents have been evaluated[107] in the PPP approximation for a number of biological purines and, in particular, for the four tautomers of purine itself. They are listed in Table IV. It can be seen that the currents are higher, in both rings, for the tautomeric forms in which the proton is at the imidazole ring [the N(9)H and the N(7)H forms] than for those in which the proton is at the pyrimidine ring [the N(3)H and the N(1)H forms]. In all these forms, nevertheless, the current is greater in the hexagonal than in the pentagonal ring.

Very recently, an even more directly useful presentation of the results has been produced by a direct evaluation of the intermolecular nuclear shielding values for protons of purines.[108] Figure 8 represents such values (in ppm) due to the ring current in the N(9)H and N(1)H tautomers of purine in a plane 3.4 Å distant from the molecular surface of the purine (which is the mean intermolecular distance in complexes involving purines with other conjugated rings). The diagrams for the N(7)H and N(3)H tautomers are very similar to the ones for N(9)H and N(1)H.

Diamagnetic anisotropies have also been evaluated[109] for the different tautomeric forms of purine. They, too, fall into two groups:

[103] P. O. P. Ts'o, *in* "Molecular Associations in Biology" (B. Pullman, ed.), p. 39. Academic Press, New York, 1968.
[104] P. O. P. Ts'o, M. P. Schweizer, and D. P. Hollis, *Ann. N. Y. Acad. Sci.*, **158**, 256 (1969).
[105] R. S. Sarma, V. Ross, and N. O. Kaplan, *Biochemistry* **7**, 3052 (1968).
[106] R. S. Sarma, P. Dannies, and N. O. Kaplan, *Biochemistry* **7**, 4359 (1968).
[107] C. Giessner-Prettre and B. Pullman, *C. R. Acad. Sci.* **268**, 1115 (1969).
[108] C. Giessner-Prettre and B. Pullman, *J. Theoret. Biol.* **27**, 87 (1970).
[109] C. Giessner-Prettre and B. Pullman, *C. R. Acad. Sci.* **266**, 933 (1968).

the N(9)H and the N(7)H tautomers having large predicted anisotropies of the order of 1.4, the anisotropy of benzene, and the N(1)H and N(3)H tautomers having a much smaller predicted anisotropy of the order of 0.5, the anisotropy of benzene. The experimental value known only for the N(9)H tautomer[110] is in complete agreement with the theoretical one.

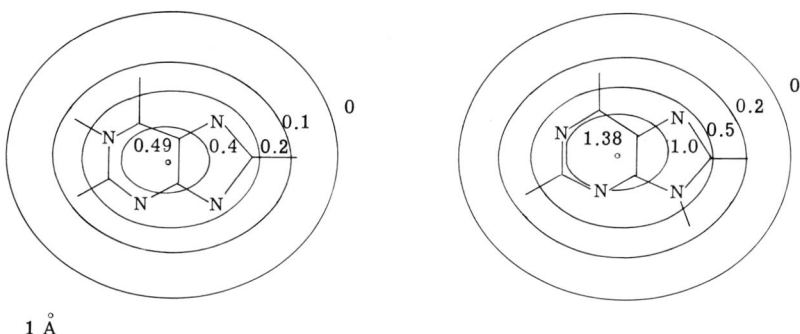

1 Å

FIG. 8. The intermolecular shielding values ($\Delta\delta$) due to the ring current in the N(1)H and N(9)H tautomers of purine (in a plane 3.4 Å distant from the molecular surface).

## V. Amine–Imine Tautomerism in Adenines

Adenine is generally represented in the amino and the N(9)H form (**12**), a conclusion based on a large number of physicochemical studies by a variety of methods. Its tautomerization to the imine form (**13**) has nevertheless been postulated to be of possible significance in spontaneous mutagenesis, as the imine form is no longer able to couple with thymine, but may, on the other hand, form a hydrogen-bonded pair with cytosine. As can be seen (Table VII) calculations carried out for these two forms by the CNDO/2 method indicate that the amine form is, in fact, predicted to be about 27 kcal/mole more stable than the imine one [the imine form (**13**) being designated in Table VII as the N(1)H–N(9)H one, a notation which specifies the

---

[110] G. Fourche, A. Pacault, P. Bothorel, and J. Hoarau, *C. R. Acad. Sci.* **262**, 1813 (1966).

TABLE VII

Molecular Energies in Adenine Tautomers (CNDO/2 Calculation) with Respect to the Most Stable Tautomer

| Tautomer | Energy (kcal/mole) |
|---|---|
| Amine[a] | |
| N(9)H | 0 |
| N(7)H | 2 |
| N(3)H | 30 |
| N(1)H | 36 |
| Imine | |
| N(1)H–N(9)H | 27 |
| N(1)H–N(7)H | 26 |
| N(3)H–N(9)H | 39 |
| N(3)H–N(7)H | 33 |

[a] Geometry following Spencer.[95]

positions of the two ring protons]. A very similar result is obtained if the N(7)H tautomer of the amine form of adenine (**14**) and the corresponding N(1)H–N(7)H tautomer of the imine form (**15**) are

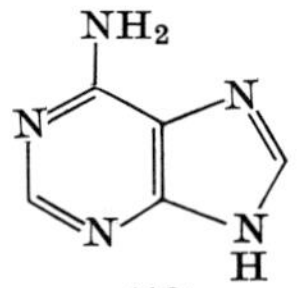

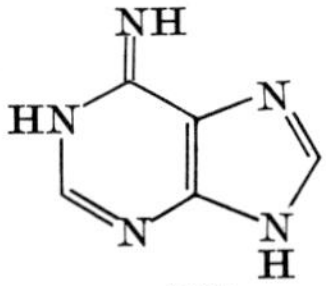

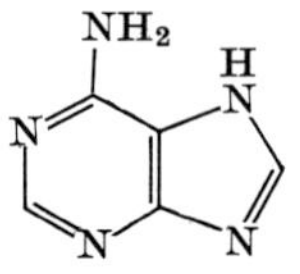

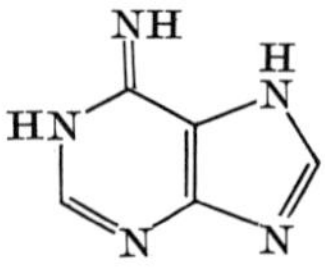

considered. A recent study of the basicity of analogs methylated at positions which prevent tautomerization confirmed that the amino tautomer of adenosine predominates largely over the imine tautomer in aqueous solution.[111]

The problem may nevertheless be raised whether the greater overall stability of the amine form is a general property of all possible pairs of amine–imine tautomers of adenine or whether a pair may be found in which the imine form would be the most stable one. Thus, it may, in fact, be observed that following a very approximate, but nevertheless significant way of reasoning, frequently used with success in resonance theory,[112] the sum of the bond energies being practically nearly the same in a pair of amine–imine tautomers, the greater stability of the amine form of the N(9)H and N(7)H tautomers may probably be attributed in a large extent to the great $\pi$-electron delocalization energy of this form due to the presence of a Kekulé-type resonance in the pyrimidine ring. Such resonance will, however, no longer be present in the amine forms of the two remaining possible tautomers, the N(1)H and the N(3)H ones (16) and (17). In 16 and 17, the $\pi$-electrons delocalization energies may therefore be of a comparable order of magnitude in the amine and the corresponding imine forms, in which case no *a priori* guess may then be easily made as to the preeminence of one or the other form. The imine forms corresponding to the N(1)H adenine (16), are 13 and 15, those corresponding to the N(3)H adenine (17) are 18 and 19. It may be worthwhile stressing that in the two amine forms no hydrogen is present at N-7 or N-9 of the imidazole ring. This hydrogen has moved over to N-1 or N-3. On the other hand, a hydrogen is attached to N-9 or

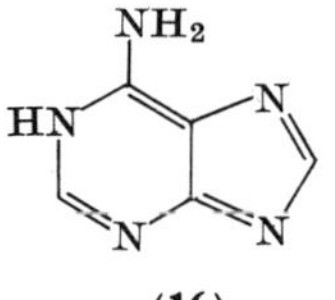

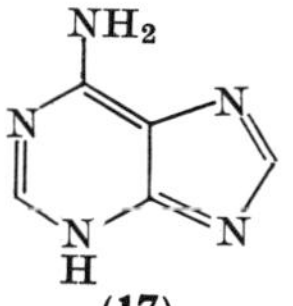

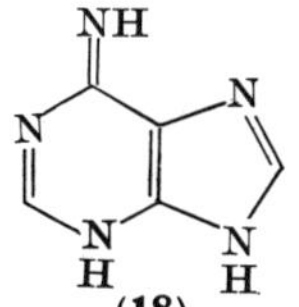

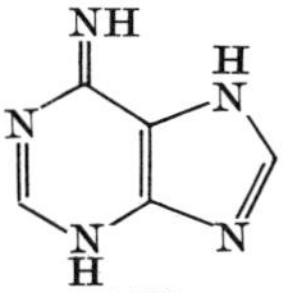

[111] R. V. Wolfenden, *J. Mol. Biol.* **40**, 307 (1969).
[112] See, e.g., B. Pullman and A. Pullman, "Les Théories Electroniques de la Chimie Organique." Masson, Paris, 1952.

to N-7 in the imine forms. Table VII indicates the answer—a positive one—to our question. Thus, it shows that while the amine form of the N(3)H tautomer of adenine (**17**) is predicted to be more stable than the associated imine forms (**18** or **19**) the *reverse is true* for the two corresponding forms of the N(1)H tautomer. The imine forms of this tautomer (**13** or **15**) are predicted to be more stable than the associated amine form (**16**). The situation may be traced back to the

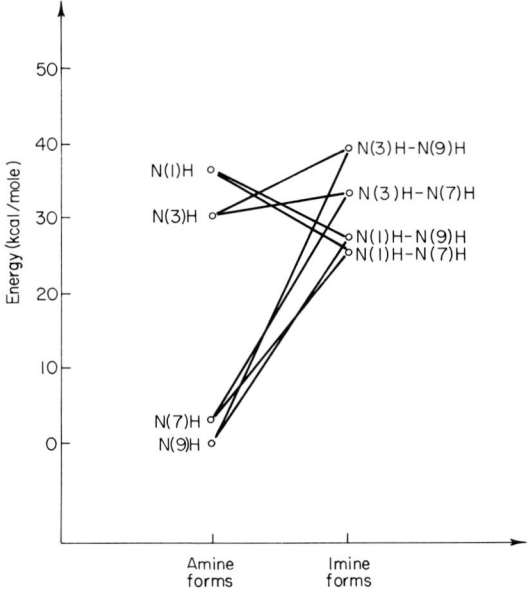

Fig. 9. Relative stabilities of the associated amine and imine forms of adenine tautomers.

fact that although the amine form of the N(1)H tautomer is *the least stable* of the amine forms of the four tautomers, its imine forms are *the most stable* of the imine forms, sufficiently stable, in fact, to lie below the corresponding amine form. The overall relationship between the different tautomeric forms is illustrated schematically in Fig. 9.

Now, it is particularly remarkable that this delicate theoretical result is confirmed by experiment. Thus, chemical evidence indicates that although 3-methyladenine exists in the amine form (20),[113, 114]

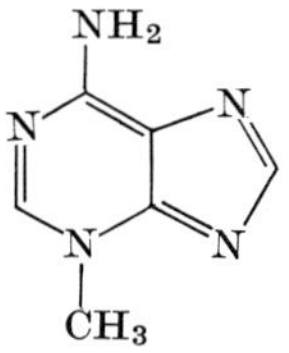

(20)

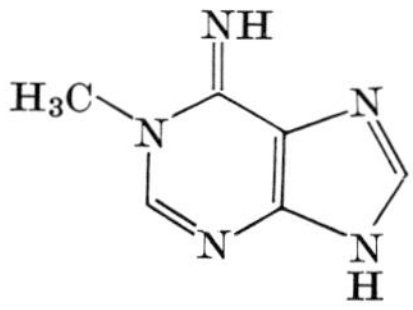

(21)

derived from 17, 1-methyladenine exists, on the contrary, in the imine form (21),[115] derived from 13. In fact, our calculations do not preclude the existence of 21 as a derivative of 15 rather than 13.

Examples of a tautomeric shift toward the imine form are relatively scarce among conjugated heterocycles. In our laboratory such calculations have been carried out, always by the CNDO/2 method, for two other cases, which seem worthwhile mentioning.

(1) The apparent, very substantial shift of the amine–imine equilibrium toward the imine form in 5,6-dihydrocytosine (22) with respect to cytosine

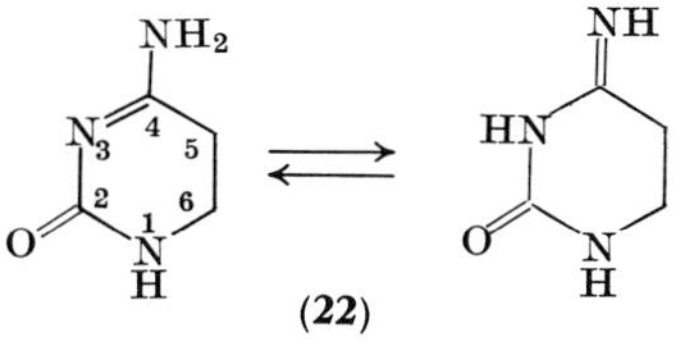

(22)

(23)[116-118] to the point that 1-methyl-5,6-dihydrocytosine while existing

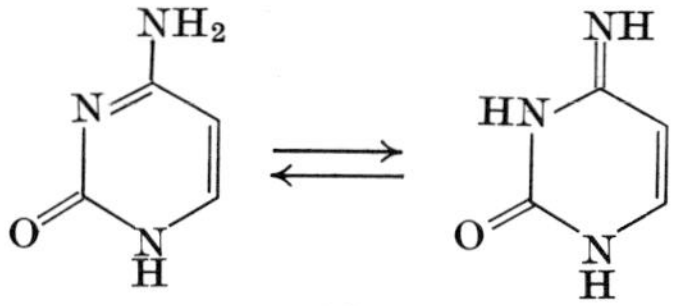

(23)

essentially in the amine form in water, exists essentially in the imine form in chloroform. The calculations indicate that while in cytosine the amine form is

[113] J. W. Jones and R. K. Robins, *J. Amer. Chem. Soc.* **84**, 1914 (1962).
[114] B. C. Pal and C. A. Horton, *J. Chem. Soc.* 400 (1964).
[115] P. Brookes and P. D. Lawley, *J. Chem. Soc.* 539 (1960).
[116] D. M. Brown, *Pure Appl. Chem.* **18**, 187 (1969).
[117] D. M. Brown and M. J. E. Hewlins, *J. Chem. Soc.* 2050 (1968).
[118] B. Singer and H. Fraenkel-Conrad, *Progr. Nucl. Acid Res. Mol. Biol.* **9**, 1 (1969).

more stable than the imine one by about 15 kcal/mole, the two forms are of comparable stability (within 1 kcal/mole) in dihydrocytosine.

(2) The existence in the imine form of compound **24**, whereas 4-amino-

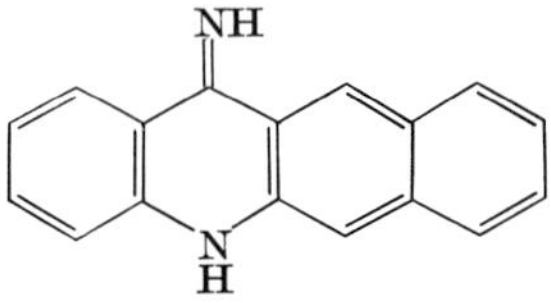

(24)

pyridine (**25**), 4-aminoquinoline (**26**), and 9-aminoacridine (**27**) all exist in the

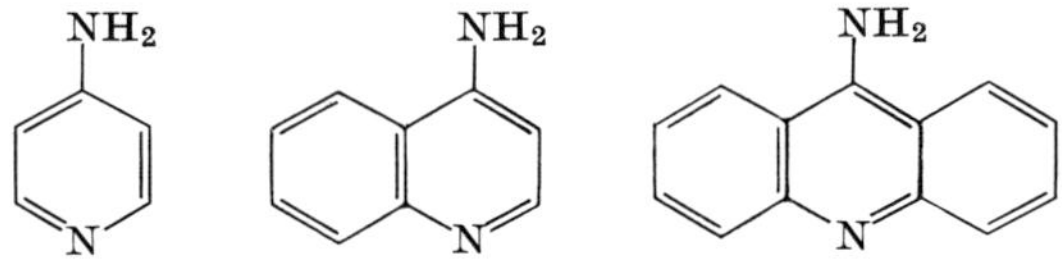

(25)  (26)  (27)

amine forms.[119] The phenomenon is accounted for in Fig. 10 which indicates the difference in energy between the imine form and the amine form of the preceding compounds as a function of the number of the fused rings. The calculations have been performed for **25–27** and they indicate a steady decrease of this difference with the increase in the number of rings. The extrapolation of the calculations to **24** correctly predicts that in this molecule the imine form should be the most stable. It should be pointed out that quantitatively the actual observed energy differences are considerably smaller than those calculated.

A number of physicochemical properties which have been investigated for a series of simple derivatives of the different tautomeric forms of adenine, prominent among which are dipole moments and ultraviolet absorption spectra, throw additional light on the problem. Table VIII indicates the results of a theoretical evaluation of the dipole moments for the different tautomeric forms (**12–19**) of adenine as well as the experimental values of the moments in methyl and benzyl derivatives of these forms with the exception of the derivatives of the N(1)H form for which, unfortunately, no experimental data could be obtained because of lack of sufficiently soluble 1-substituted derivatives.

From the comparison of the theoretical and experimental moments it is obvious that the N(9)H and the N(7)H derivatives are present

[119] S. F. Mason, *J. Chem. Soc.* 1281 (1959).

in the amine forms. For the N(3)H derivatives it can be observed that although the moment of the 6-dimethylamino compound (in which no imino tautomerization is possible) corresponds closely to the moment calculated for the amine form of N(3)H adenine, the moments of 3-methyl- and 3-benzyladenines are somewhat higher than predicted by theory thus possibly indicating a small contribution

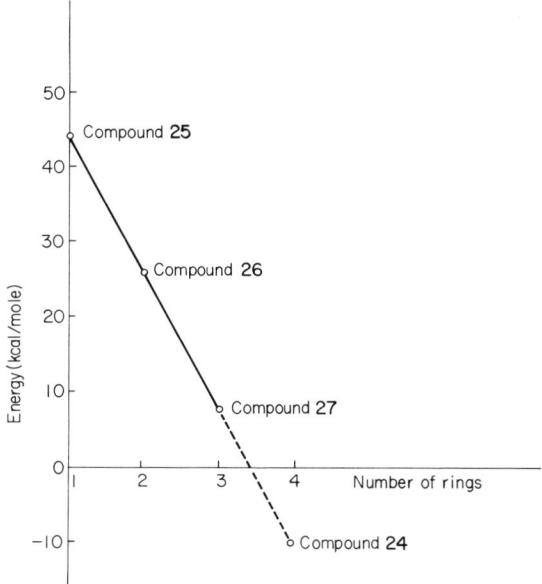

Fig. 10. Energy differences between the imine and amine forms in the compounds **24–27**.

of the imine forms which would most probably be the N(3)H–N(9)H tautomers. Figure 11 represents the distribution of the σ- and π-electronic charges, as obtained in the CNDO method for the most important tautomeric forms of adenines.

Indications about ultraviolet absorption spectra of these compounds reproduced in Table IX are less clear-cut. The predicted wavelengths for the amine forms are increasing in the order N(9)H < N(7)H < N(3)H. Although this sequence of absorptions is confirmed by experiment, the quantitative agreement is poor especially for the N(3)H tautomer. As for the absorptions of 1-methyl- and 1-benzyladenine, they have been considered as corresponding to those of the

## TABLE VIII
### Dipole Moments of Adenine Tautomers

| Tautomer | PPP–DBP method | | CNDO/2 method | | Experiment[99,120] |
|---|---|---|---|---|---|
| | $\mu$ (D) | $\theta$ (deg) | $\mu$ (D) | $\theta$ (deg) | |
| Amine | | | | | |
| N(9)H | 2.4 | 77 | 3.0 | 66 | 3.25 D in 9-methyl adenine[a] |
| N(7)H | 7.5 | 161 | 7.4 | 156 | 2.75 D in 9-benzyl adenine<br>8.10 D in 7-benzyladenine |
| N(3)H | 3.3 | −81 | 4.2 | −74 | 4.99 D in 3-methyladenine<br>4.92 D in 3-benzyladenine |
| N(1)H | 7.2 | −130 | 8.9 | −128 | 4.20 D in 3-methyl-6-dimethylaminoadenine |
| Imine | | | | | |
| N(1)H–N(9)H | 6.6 | −24 | 5.9 | −32 | |
| N(1)H–N(7)H | 2.8 | −130 | 3.8 | −138 | |
| N(3)H–N(9)H | 10.8 | 4 | 10.6 | 3 | |
| N(3)H–N(7)H | 3.5 | 20 | 3.2 | 23 | |

[a] 3.0 D following DeVoe and Tinoco.[100]

TABLE IX

TRANSITION ENERGIES IN ADENINE TAUTOMERS (PPP–DBP METHOD)

| Tautomer | $S_1$ theoretical | | | $\lambda_{max}$ experimental | |
|---|---|---|---|---|---|
| | eV | m$\mu$ | $\theta$ (deg) | m$\mu$ | Compound measured |
| Amine | | | | | |
| N(9)H | 4.82 | 257 | 19 | ⎧ 261 <br> ⎩ 262 | 9-Methyladenine[22] <br> 9-Benzyladenine[120] |
| N(7)H | 4.77 | 260 | 8 | ⎧ 272 <br> ⎩ 273 | 7-Methyladenine[121,123] <br> 7-Benzyladenine[120] |
| N(3)H | 4.17 | 297 | 65 | ⎧ 274 <br> ⎩ 274 | 3-Methyladenine[122–124] <br> 3-Benzyladenine[120] |
| N(1)H | 4.47 | 277 | 39 | 277 | 3-Isoadenosine[125] |
| | | | | — | — |
| Imine | | | | | |
| N(1)H–N(9)H | 4.32 | 287 | −80 | ⎧ 270 <br> ⎩ 267 | 1-Methyladenine[126] <br> 1-Benzyladenine[120] |
| N(1)H–N(7)H | 4.50 | 276 | −31 | — | — |
| N(3)H–N(9)H | 4.61 | 269 | −37 | — | — |
| N(3)H–N(7)H | 4.70 | 264 | 28 | — | — |

*imine* forms. The quantitative agreement is again only approximate. It is not unlikely, of course, that the alcoholic solutions of the compounds studied contain, in fact, mixtures of tautomers, the equilibrium being moreover different from the one expected in the nonpolar solvents in which dipole moments are measured.[120]

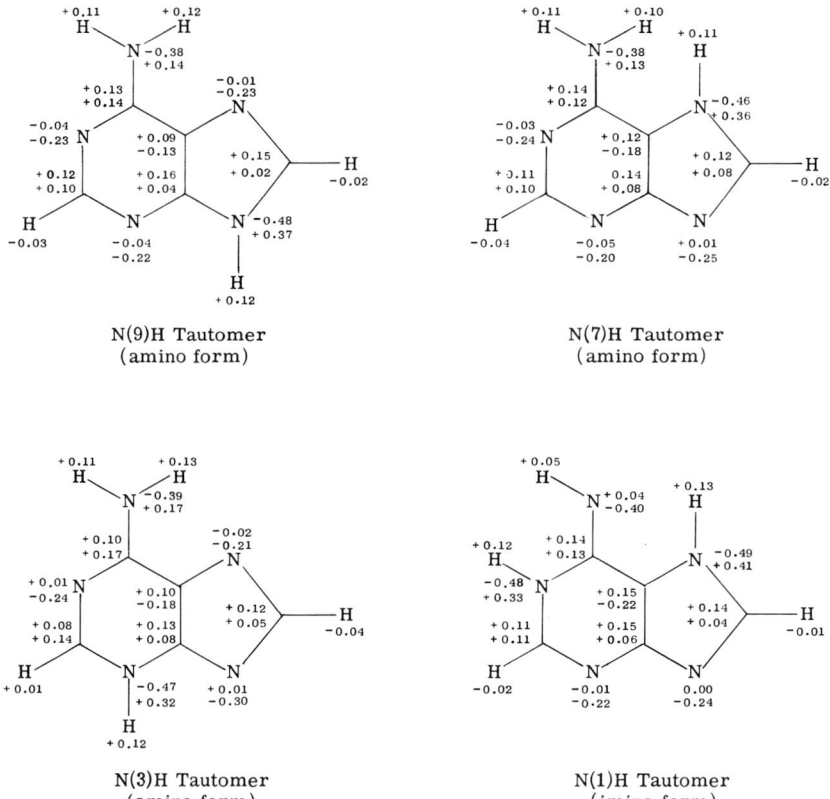

Fig. 11. Net electronic charges in adenine tautomers evaluated by the CNDO/2 method. Upper numbers: $\sigma$ charges; lower numbers: $\pi$ charges.

[120] H. Weiler-Feilchenfeld and Z. Neiman, *in* "Quantum Aspects of Heterocyclic Compounds in Chemistry and Biochemistry (E. D. Bergmann and B. Pullman, eds.), p. 308. Israel Acad. Sci. Humanities, Jerusalem (distributed by Academic Press, New York), 1970.

It may be added that the spectrum of the principal N(9)H form of adenine has been studied in more detail by a number of authors (e.g., Pullman and Pullman et al.[63, 74, 121–130]).

Concerning other outstanding electronic properties of the adenine tautomers, some of the theoretically evaluated ones are listed in Table X. They are self-explanatory. It may just indicate that the

TABLE X

MISCELLANEOUS ELECTRONIC PROPERTIES OF ADENINE TAUTOMERS

| Tautomer | SCF–CI method | | | | CNDO/2 method | |
|---|---|---|---|---|---|---|
| | HOMO (eV) | LEMO (eV) | Ring current | | HOMO (eV) | LEMO (eV) |
| | | | Hexagonal | Pentagonal | | |
| Amine | | | | | | |
| N(9)H | −7.9 | 1.5 | 0.90 | 0.66 | −10.1 | 3.8 |
| N(7)H | −8.3 | 1.4 | — | — | −10.7 | 2.9 |
| N(3)H | −8.1 | 0.7 | 0.38 | 0.43 | −10.7 | 2.4 |
| N(1)H | −8.4 | 0.9 | 0.30 | 0.33 | −10.7 | 2.7 |
| Imine | | | | | | |
| N(1)H–N(9)H | −8.2 | 0.9 | 0.38 | 0.65 | −9.2 | 3.0 |
| N(1)H–N(7)H | −8.4 | 1.0 | — | — | −9.5 | 3.1 |
| N(3)H–N(9)H | −8.3 | 1.1 | 0.20 | 0.60 | −9.4 | 3.4 |
| N(3)H–N(7)H | −8.4 | 0.9 | — | — | −9.5 | 3.3 |

[121] R. N. Prasad, and R. K. Robins, *J. Amer. Chem. Soc.* **79**, 6401 (1957).
[122] G. B. Elion, in "Chemistry and Biology of Purines," p. 39. Ciba Found. Symp. Churchill, London, 1957.
[123] J. W. Jones and R. K. Robins, *J. Amer. Chem. Soc.* **84**, 1914 (1962).
[124] B. C. Pal and C. A. Horton, *J. Chem. Soc.* 400 (1964).
[125] N. J. Leonard and R. A. Laursen, *Biochemistry* **4**, 354 (1965).
[126] P. Brookes and P. D. Lawley, *J. Chem. Soc.* 539 (1960).
[127] H. Berthod, C. Giessner-Prettre, and A. Pullman, *Int. J. Quant. Chem.* **1**, 123 (1969).
[128] M. Tanaka and S. Nagakura, *Theoret. Chim. Acta* **6**, 320 (1966).
[129] J. S. Kwiatkowski, *Theoret. Chim. Acta* **10**, 47 (1968).
[130] I. Fischer-Hjalmars and J. Nag-Chandhuri, *Acta Chem. Scand.* **23**, 2963 (1969).

ionization potential of adenine [probably the N(9)H tautomer] has been measured[98] and found to be equal to 8.91 eV, a value intermediate between that indicated by the two methods used.

Among the other possible isomeric aminopurines, some attention has been devoted to the amine–imine tautomerism in 2-aminopurine

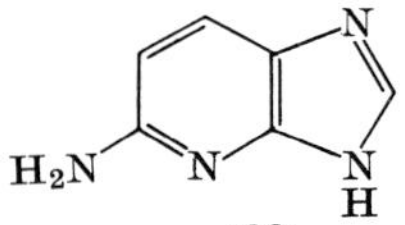

(28), because of the known mutagenic effect of this compound. Both Hückel-type and SCF calculations indicate, however, that 2-aminopurine should have a smaller tendency to exist in the rare imine form than adenine,[50, 131] a situation which makes the involvement of such a form in mutagenesis by the compound improbable. Following Freese[132] its high mutagenic activity should be attributed to its frequent mistake pairing by means of a single hydrogen bond with cytosine.

## VI. Lactam–Lactim Tautomerism in Hydroxypurines

For this particular problem the theory drags somewhat behind experimental evidence, at least insofar as the question of the relative stability of the two forms is concerned. The experimental evidence coming mainly from infrared spectroscopy unambiguously shows[12] that the three isomeric 2-, 6-, and 8-hydroxypurines (29–31) all exist essentially, both in the solid state and in solution, in the oxo form, as they all present the characteristic C=O stretching vibration (near 1670 cm$^{-1}$ in the 2- and 6-hydroxypurines and near 1740 cm$^{-1}$ in the 8-hydroxy isomer) and show no band which could be attributed

[131] V. I. Danilov, Yu A. Kruglyak, V. A. Kupriyevich, and O. V. Shramko, *Biophysics (USSR)* **12**, 840 (1967).
[132] E. Freese, *J. Mol. Biol.* **1**, 87 (1959).

to an O–H group. The same situation prevails also in guanine (**32**)[10, 133–135] in which the C=O frequency is at 1665 cm$^{-1}$.

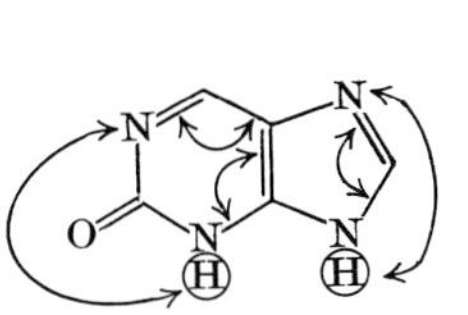

(29) 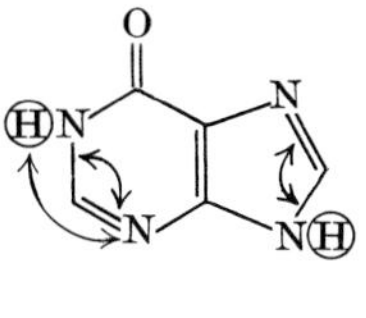    (30)

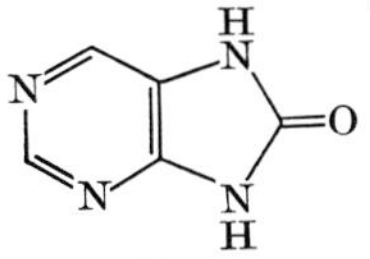

(31)

Attempts to account for this situation by the CNDO procedure surprisingly encounter difficulties. Fujita, Imamura, and Nagata[136] while having apparently no trouble in finding that the usual amino and lactam forms are the most stable ones in adenine, thymine, cytosine, and uracil, obtain the surprising result that the lactim form of guanine (**33**) should be about 12 kcal/mole more stable than its

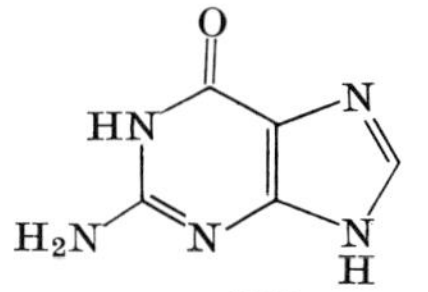

(32) 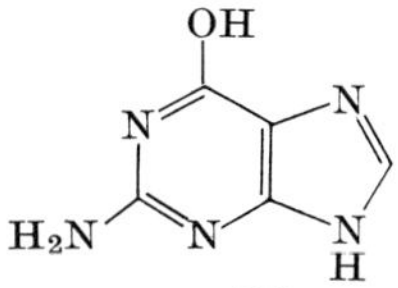    (33)

lactam form, a result which they consider as an indication that the predominant form of guanine in the gaseous state could be the lactim form. While this could be so, another possibility is that the result of the Japanese authors is perhaps an artifact of their calculations due to the selection of a particular molecular geometry for this rare form. In the absence of any direct experimental indication in this field, these authors have simply used for the geometry of the rare form the same interatomic distances as those of the usual lactam form only

[133] R. Shapiro, *Progr. Nucl. Acid Res. Mol. Biol.* **8**, 73 (1968).
[134] H. T. Miles and J. Frazier, *Biochim. Biophys. Acta* **79**, 216 (1964).
[135] J. S. Kwiatkowski, *Acta Phys. Pol.* **34**, 365 (1968).
[136] H. Fujita, A. Imamura, and Ch. Nagata, *Bull. Chem. Soc. Japan* **42**, 1467 (1969).

adopting for the C–O distance 1.47 Å and for the O–H distance 0.96 Å. Unpublished calculations from our laboratory on the same problem indicate that the *relative* stabilities of the lactam and lactim forms of guanine depend very strongly on the interatomic distances adopted for the C–O and still more for the O–H bonds in the lactim form. The situation is represented in Table XI indicating the greater (−) or smaller (+) stability of the lactim form of guanine with respect to the lactam form as a function of the lengths of the C–O and O–H bonds in the lactim form.

Unfortunately information about the C–O and O–H bond lengths in heterocyclic phenols is very scarce and does not permit a decision.

TABLE XI

RELATIVE STABILITY (kcal/mole) OF THE LACTIM FORM
WITH RESPECT TO THE LACTAM FORM OF GUANINE

| C–O bond length \ O–H bond length | 0.95 Å | 0.90 Å | 0.85 Å |
|---|---|---|---|
| 1.44 Å | −17 | +2 | +32 |
| 1.36 Å | −25 | −6 | +25 |

Whatever it be, this very strong dependence of the position of the lactam–lactim equilibrium upon such a delicate structural feature as the length of the O–H bond in the lactim form makes calculations on this problem meaningless in the absence of more detailed crystallographic information. For the same reason it may be envisaged that the equilibrium may undergo displacements relatively easily, following the conditions, and may be therefore, as a mean, less fixed than the amino–imine one studied in the case of adenines. Such seems to be the conclusion of Wolfenden's comparative study[111] of the tautomeric equilibria in inosine and adenosine, through the determination of the basicity of suitably methylated analogs.

If the study of the relative stability of the lactam and lactim forms of hydroxypurines thus escapes a rigorous quantitative treatment, the situation is more agreeable with respect to some other electronic properties of the two forms, which are much less sensitive to the details of the structure and, in particular, to the distance of the

O–H bond. Among such properties a prominent one is the ultraviolet absorption spectrum and the theory may therefore be used for the examination of some of the spectroscopic shifts which accompany the lactam–lactim tautomerization. Much caution must, however, be exercised in this respect. Thus, in a recent paper Kwiatkowski[135, 137] performed Pariser–Parr–Pople-type calculations on the electronic structure of hydroxypurines, essentially to interpret their ultraviolet spectra. In these calculations he assumed that these compounds exist predominantly in their lactim form, and the results of his calculations, at least for 6- and 8-hydroxypurine, did not seem to contradict this assumption. It is only in the case of the 2-hydroxy isomer that a particularly striking disagreement between theory and experiment led him to admit that this last compound may exist in the lactam form. Calculations carried out for this form gave, in fact, a more satisfactory agreement with experiment.[138] As we have seen, unambiguous infrared spectroscopy evidence clearly shows that all three isomers exist essentially in the lactam form. This shows that ultraviolet absorption may provide only very uncertain evidence about the lactam–lactim tautomerism in hydroxypurines and related compounds.

The reasons for this are the simplicity of the spectra and their frequent similarity in the lactam and lactim forms (see, e.g., Mason et al.[12, 133, 139]).

This precaution stated, a simultaneous theoretical investigation of the essential spectroscopic features of the lactam and lactim forms of the three fundamental monohydroxypurines yields some interesting results. These are summarized in Table XII, together with the corresponding experimental data, which in the case of lactim forms refer to the methoxy derivatives. As in all these compounds there is besides the lactam–lactim tautomerism, the possibility of an oscillation of the "imidazole proton" between the different N atoms of the ring system, Table XII indicates the exact tautomeric form or forms to which the calculations refer. In case of the lactam forms these are the most probable tautomers of such forms obtained in the study described in the next section. For the lactim forms they are the *a priori* most probable ones.

[137] J. S. Kwiatkowski, *Theoret. Chim. Acta* **10**, 47 (1968).
[138] J. S. Kwiatkowski, *Theoret. Chim. Acta* **11**, 167 (1968).
[139] J. H. Lister, *Advan. Heterocyc. Chem.* **6**, 1 (1966).

The examination of Table XII shows that the theory correctly predicts: (1) the continuous hypsochromic shift observed in the series of the lactam forms when passing from 2-hydroxy- to 6-hydroxy- and then to 8-hydroxypurine; (2) the similar hypsochromic shift in the series of the lactim forms, when passing from 2-hydroxy- to 6-hydroxypurine; (3) the hypsochromic shift when passing from the lactam to the lactim form in 2-hydroxy- and 6-hydroxypurines. A

TABLE XII

Spectroscopic Properties of Monooxopurines (PPP–CI Method)

| Tautomer | $S_1$ theoretical | | $\lambda_{max}$ experimental[12] | |
|---|---|---|---|---|
| | eV | m$\mu$ | eV | m$\mu$ |
| 2-Hydroxypurine | | | | |
| Lactam, N(3)H–N(7)H | 3.66 | 339 | 3.94 | 315 |
| Lactim, N(7)H | 4.54 | 273 ⎱ | 4.38 | 283 |
| N(9)H | 4.66 | 266 ⎰ | | |
| 6-Hydroxypurine | | | | |
| Lactam, N(1)H–N(7)H | 4.48 | 277 | 4.43 | 280 |
| Lactim, N(7)H | 4.86 | 255 ⎱ | 4.92 | 252 |
| N(9)H | 5.00 | 248 ⎰ | | |
| 8-Hydroxypurine | | | | |
| Lactam, N(7)H–N(9)H | 4.70 | 264 | 4.48 | 277 |
| Lactim, N(7)H | 4.89 | 254 | — | — |
| N(9)H | 4.81 | 258 | — | — |

similar hypsochromic shift is predicted based on the lactam–lactim tautomerization in 8-hydroxypurine. This homogeneous situation should not, however, lead to any generalization about the respective absorption maxima of the longest wavelength in pairs of other hydroxypurines or similar compounds. Thus, a small (unaccounted theoretically) bathochromic shift is observed when passing from guanine (lactam) to 6-methoxy-2-aminopurine.[133] Bathochromic shifts are also observed upon enolization of a number of lactam hydroxypteridines.[140]

[140] S. F. Mason, in "Chemistry and Biology of Pteridines," Ciba Found. Symp., p. 174. Churchill, London, 1954.

Another interesting electronic property which may be studied comparatively for the lactam and lactim forms of hydroxypurines is their dipole moment. A difficulty occurs, however, due to the fact that dipole moments are, on the one hand, extremely sensitive to the position of the "imidazole proton" and, on the other, they are also very sensitive, in the lactim form, to the C–O–H angle. Any hindrance to the free rotation of the O–H bond may therefore have a very strong influence on the value of the moment. For these reasons we shall limit ourselves here to a few general qualitative remarks.

If we take as reference the dipole moments calculated for the predicted most stable tautomers of the lactam forms of the hydroxypurines (as discussed in Section VII) and which are generally the N(7)H tautomers, we come out with a rather high dipole moment ($\approx 9$ D) for 2-hydroxypurine and relatively low dipole moments for 6-hydroxypurine (2.6 D) and 8-hydroxypurine (1.4 D) (the numbers quoted come from CNDO calculations). Under these conditions the lactimization should correspond to a decrease in the dipole moment in 2-hydroxypurine (3–7 D) and to its increase in 6-hydroxypurine (3.6–5.7 D) and 8-hydroxypurine (3.2–4.6 D). Unfortunately no experimental data are available in this field. Calculations are also available for the directions of the moments (as well as for the transition moments and the HOMO's and LEMO's of all these compounds).

## VII. The Fine Structure of Hydroxypurines

Whatever be the difficulties in dealing satisfactorily with the problem of the lactam–lactim tautomerism in hydroxypurines, the predominance of the lactam tautomer granted, there remains the problem of the detailed structure of the most probable lactam form for each isomer. The problem is essentially that of the site of location of the "imidazole proton." From that point of view forms **34–38** have to be considered for 2-hydroxypurine, forms **39–42** for 6-hydroxypurine (hypoxanthine), and forms **43–45** for 8-hydroxypurine. There are, in addition, some betaine tautomeric forms but these are probably of low stability and will not be considered further. Before describing the results of theoretical calculations, it may be useful to indicate that from the experimental point of view we may, in this respect, turn again for significant evidence to infrared spectroscopy

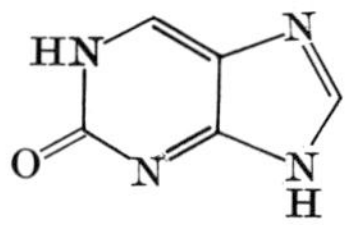

(34) 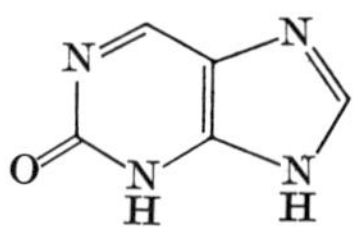  (35)

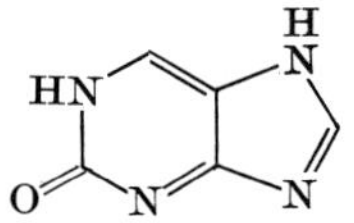

(36) 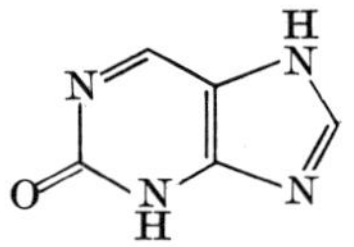  (37)

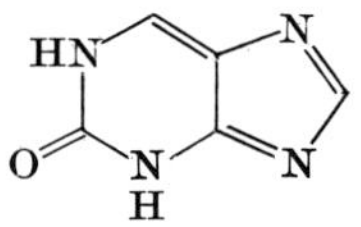

(38) 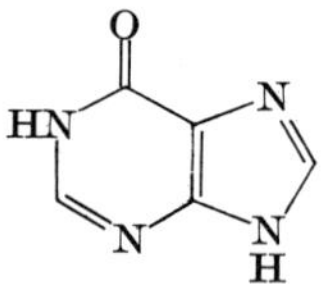  (39)

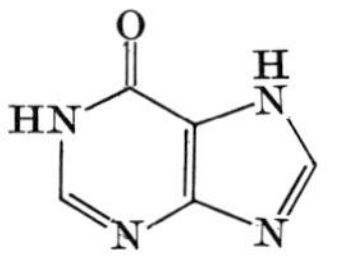

(40) 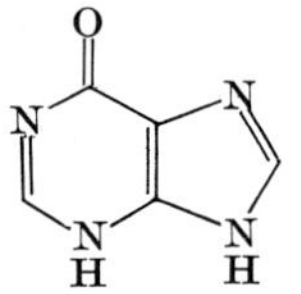  (41)

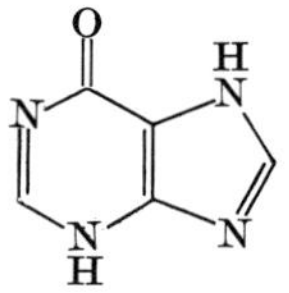

(42) 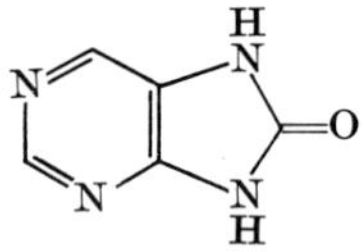  (43)

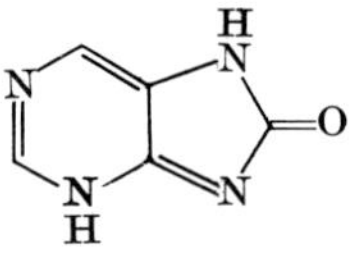

(44) 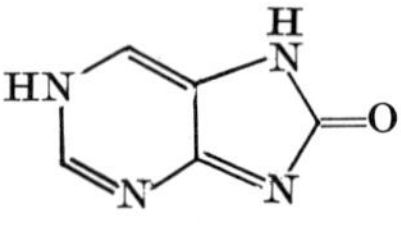  (45)

TABLE XIII

ELECTRONIC CHARACTERISTICS OF HYDROXYPURINES (LACTAM FORMS)

| Compound | Tautomer | $\mu_{(D)}$ D | SCF-CI method ||||||| CNDO method ||||
|---|---|---|---|---|---|---|---|---|---|---|---|---|---|
| | | | $\theta^a$ (deg) | HOMO (eV) | LEMO (eV) | $S_1{}^b$ (m$\mu$) Theoret. | $S_1{}^b$ (m$\mu$) Exptl. | $S_2{}^b$ (m$\mu$) Theoret. | $S_2{}^b$ (m$\mu$) Exptl. | $\mu$ (D) | $\theta^a$ (deg) | HOMO (eV) | LEMO (eV) | Energy$^c$ (kcal/mole) |
| 2-Hydroxypurine | | | | | | | | | | | | | | |
| 34 | N(1)H–N(9)H | 4.4 | 113 | −8.3 | 0.3 | 317 | — | 239 | — | 5.0 | 119 | −10.1 | 1.9 | 39 |
| 35 | N(3)H–N(9)H | 8.0 | 62 | −8.2 | 0.6 | 303 | — | 234 | — | 8.8 | 61 | −10.1 | 2.4 | 43 |
| 36 | N(1)H–N(7)H | 9.1 | 153 | −8.3 | 0.1 | 327 | — | 237 | — | 10.9 | 15 | −10.2 | 1.6 | 41 |
| 37 | N(3)H–N(7)H | 8.4 | 122 | −8.1 | 0.3 | 239 | 315$^{22}$ | 230 | 238$^{22}$ | 9.1 | 119 | −10.1 | 2.0 | 38 |
| 38 | N(1)H–N(3)H | 2.0 | −163 | −9.3 | 0.0 | 292 | — | 252 | — | 2.3 | −140 | −11.2 | 1.5 | 42 |
| 6-Hydroxypurine$^d$ (hypoxanthine) | | | | | | | | | | | | | | |
| 39 | N(1)H–N(9)H | 5.6 | −16 | −8.0 | 1.0 | 294 | — | 240 | — | 5.9 | −16 | −9.7 | 2.9 | 39 |
| 40 | N(1)H–N(7)H | 2.6 | −157 | −8.2 | 1.2 | 277 | 280$^{141}$ | 245 | 249$^{141}$ | 2.6 | −156 | −10.1 | 3.0 | 38 |
| 41 | N(3)H–N(9)H | 10.5 | 11 | −8.1 | 1.2 | 273 | — | 224 | — | 11.8 | 11 | −10.1 | 3.2 | 53 |
| 42 | N(3)H–N(7)H | 3.6 | 41 | −8.2 | 1.1 | 267 | — | 238 | — | 4.9 | 36 | −10.2 | 3.1 | 46 |
| 8-Hydroxypurine | | | | | | | | | | | | | | |
| 43 | N(7)H–N(9)H | 2.3 | −118 | −8.7 | 0.8 | 264 | 277$^{22}$ | 227 | 235$^{22}$ | 1.4 | −129 | −11.0 | 3.2 | 0 |
| 44 | N(3)H–N(7)H | 4.8 | −93 | −8.2 | 0.1 | 306 | — | 263 | — | 5.7 | −94 | −10.1 | 1.7 | 41 |
| 45 | N(1)H–N(7)H | 9.4 | −128 | −8.1 | 0.2 | 288 | — | 281 | — | 11.2 | −127 | −10.0 | 1.8 | 45 |

$^a$ Counterclockwise with respect to the C-4–C-5 axis.
$^b$ The experimental absorption maxima are listed along the lines corresponding to the most stable tautomer of each hydroxypurine. In the case of 2- and 6-hydroxypurines for which there exist two tautomers of very close total energies they may represent the absorption of a mixture of such tautomers.
$^c$ With respect to the most stable tautomer considered as zero energy.
$^d$ Geometry following Spencer.[95]

141 L. B. Clark, G. G. Perchel, and I. Tinoco, *J. Phys. Chem.* **69**, 3615 (1965).

which permits through the study of the N–H stretching vibrations to distinguish between o-quinoid ($\nu_{\text{NH}}$ near 3350 cm$^{-1}$) and p-quinonoid ($\nu_{\text{NH}}$ near 3450 cm$^{-1}$) forms.[12] The available data somewhat favor form **37** for 2-hydroxypurine, form **40** for 6-hydroxypurine, and form **43** for 8-hydroxypurine.

The results of the calculations on a number of essential electronic properties of oxypurines are presented in Table XIII and are compared, as far as possible, with the available experimental data. They indicate the following.

(1) In complete agreement with the deductions from infrared data, the calculations estimate the tautomers **37**, **40**, and **43** as the most stable forms of 2-hydroxypurine, 6-hydroxypurine, and 8-hydroxypurine, respectively, followed very closely, however, by tautomers **34** and **39** for the first two isomers. Their existence as mixtures of tautomeric forms in comparable proportions seems therefore highly probable. On a relative scale the most stable of the three isomers should be the 8-hydroxy one (certainly because of the high content of its π-electronic delocalization), which should, in fact, be appreciably more stable than the 2- or 6-isomers, which are predicted to be of comparable stability.

In connection with the forthcoming discussion on the structure of the N(3)H tautomer of guanine (Section VIII), it may be interesting to point out that the N(3)H tautomers of hypoxanthine, of the general structure **46** (with the second proton at N-7 or N-9) are predicted to be about 8–14 kcal/mole less stable than the corresponding N(1)H tautomers, the N(3)H–N(7)H tautomer being more stable by about 6 kcal/mole than the N(3)H–N(9)H one. The structure of the 3-methylhypoxanthine is, in fact, presented by Bergmann et al.[142] as **47**.

(46)      (47)

An interesting difference predicted between the N(3)H–N(7)H and N(3)H–N(9)H tautomers concerns their dipole moments, the moment

[142] F. Bergmann, G. Levin, A. Kalmus, and H. Kwietny-Govrin, *J. Org. Chem.* **26**, 1504 (1961).

predicted for the former being of the moderate value of 4.9 D, whereas the one predicted for the latter having the very large value of 11.8 D [*vide infra* the case of the N(3) tautomer of guanine].

(2) In all three isomers the most stable tautomeric form involves one proton at N-7, the second one being at N-3 in 2-hydroxypurine, at N-1 in 6-hydroxypurine, and at N-9 in 8-hydroxypurine. The preferential attachment of a proton at N-7 of these isomers seems thus a general feature of their structure. Although this situation agrees with the probable preeminence of the N(7)H tautomers in solution, it should not be considered as prejudging the nature of the tautomer present in the crystal of these substances, a problem which will be discussed in detail in Section XI.

(3) There is a *very considerable* variation in the dipole moments among the different isomers and among the different tautomeric forms of each isomer. The possible utility of dipole moment measurements for distinguishing between different tautomeric forms, in particular, between the N(7)H and N(9)H tautomer, frequently difficult to distinguish by other physicochemical techniques, is therefore worthwhile stressing. Unfortunately, because of the difficulty of dissolving these substances in nonpolar solvents, no information whatsoever is available in this field. On the other hand, for the discussion on the crystal structure of these and other purines (Section XI) it may also be useful to point out that no general rule seems to exist about the relative values of the dipole moments with respect to the intrinsic stabilities of the different tautomers. Thus, the most stable tautomers are predicted to have a relatively low dipole moment in 6- and 8-hydroxypurines (1–3 D) and a relatively high one ($\approx 9$ D) in 2-hydroxypurine.

These differences in dipole moments result, of course, from differences in the distribution of electronic charges. Figure 12 represents these distributions, as evaluated by the CNDO/2 method for the most important tautomers of hydroxypurines.

(4) Concerning the ultraviolet absorption spectra of the three tautomers, theory and experiment agree in considering 2-hydroxypurine as absorbing toward the longest wavelength, 8-hydroxypurine as absorbing toward the shortest wavelength, and 6-hydroxypurine as having an intermediate absorption, provided that the 280 m$\mu$ shoulder (solvent $H_2O$, pH 7) observed in the absorption spectrum of hypoxanthine be recognized as a separate transition. (In fact,

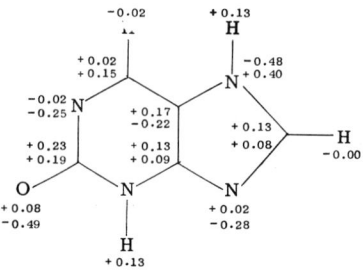

N(3)H-N(7)H Tautomer
of 2-hydroxypurine

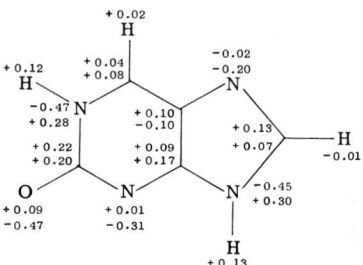

N(1)H-N(9)H Tautomer
of 2-hydroxypurine

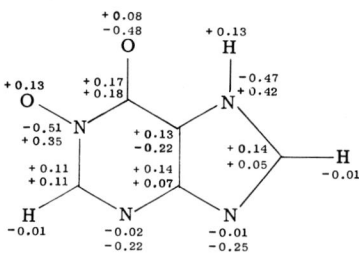

N(1)H-N(7)H Tautomer
of 6-hydroxypurine

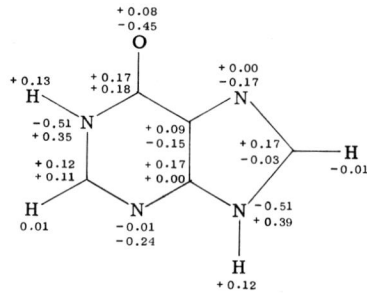

N(1)H-N(9)H Tautomer
of 6-hydroxypurine

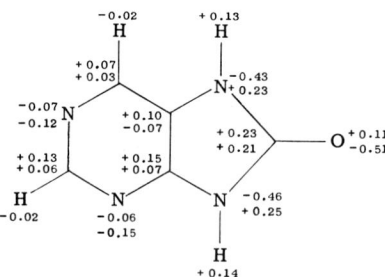

N(7)H-N(9)H Tautomer
of 8-hydroxypurine

Fig. 12. Net electronic charges in the most important tautomers of hydroxypurines evaluated by the CNDO/2 method. Upper numbers: $\sigma$ charges; lower numbers: $\pi$ charges.

following Kleinwächter et al.,[143] in dioxane solvent there are even a few shoulders in that region, the one at the longest wavelength lying at 291 mμ.) The calculations, which predict the transition of longest wavelength to lie between 277 and 294 mμ for the most probable tautomeric forms of hypoxanthine, greatly favor this viewpoint.

The agreement between theory and experiment extends also to the second absorption band in these compounds; this, following experimental evidence lies at the longest wavelength in 6-hydroxypurine, at the shortest wavelength in 8-hydroxypurine, and at intermediate wavelength in 2-hydroxypurine. The calculations reproduce this ordering and numerically agree within 10 mμ with the observed transitions.

Altogether, it appears evident that calculations based on the lactam forms of hydroxypurines are susceptible of agreement with an extensive amount of experimental data concerning these compounds and offer therefore a complementary support for the representation of these molecules in such forms.

Among the polyhydroxypurines the attention of the theoreticians, at least insofar as refined methods of calculation are concerned, has been centered essentially on xanthine and, in particular, on the problem of the N(7)H–N(9)H tautomerism in this compound

(48)            (49)

(48 ⇌ 49). The reason for this situation was the early recognition that, in distinction to the majority of purine compounds which exist preferentially as derivatives of the N(9)H form, the xanthines exist essentially as derivatives of its N(7)H form (48) which appears thus as probably its predominant tautomer.[143–145] Calculations by the CNDO method confirm this assumption entirely by indicating that

---

[143] V. Kleinwächter, J. Drobnik, and L. Augenstein, *Photochem. Photobiol.* **6**, 133 (1967).

[144] B. J. Sukhorukov and V. I. Poltev, *Biophysics* **9**, 152 (1964).

[145] H. Berthod, C. Giessner-Prettre, and A. Pullman, *C. R. Acad. Sci.* **262**, 2657 (1966).

TABLE XIV
ELECTRONIC PROPERTIES OF XANTHINE TAUTOMERS

| Tautomer | PPP–DBP method | | | | | | | CNDO method | | | |
|---|---|---|---|---|---|---|---|---|---|---|---|
| | $\mu$ (D) | $\theta$ (deg) | HOMO (eV) | LEMO (eV) | $S_1$ theoretical | | Ring current | | $\mu$ (D) | $\theta$ (deg) | HOMO (eV) | LEMO (eV) |
| | | | | | eV | m$\mu$ | Hexagonal | Pentagonal | | | | |
| N(7)H | 3.9 | 112 | −8.8 | 0.7 | 4.74 | 262 | — | — | 4.0 | 99 | −10.8 | 3.0 |
| N(9)H | 7.2 | 27 | −8.7 | 1.2 | 5.00 | 248 | 0.08 | 0.67 | 7.9 | 24 | −10.6 | 3.6 |

the N(7)H form of xanthine should be about 7 kcal/mole more stable than its N(9)H form.[76]

The interest in this problem was increased recently by the determination of a number of electronic properties of these two forms and by the confirmation given to theoretical predictions in this field. As seen in Table XIV, the most striking among these concerns the dipole moments and the ultraviolet spectra of the tautomers. Thus the theory predicts that the more stable N(7)H tautomer should have the lower dipole moment (by 3 D), a situation contrary to that observed in the purine series (Section IV) and that the N(7)H tautomer should be bathochromic with respect to the N(9)H one.

TABLE XV

Dipole Moments (D) and Longest Wavelength of Absorption (m$\mu$) in Xanthine Tautomers

| Derivative of 8-decylthioxanthine | Structure | $\mu$ (dioxane) | $\lambda_{max}$ (ethanol) |
|---|---|---|---|
| 3,9-Dimethyl- | 50 | 8.20 | 279, 260 |
| 3,7-Dimethyl- | 51 | 4.90 | 292 |
| 1,3,9-Trimethyl- | 53 | 6.54 | 279, 260 |
| 1,3,7-Trimethyl- | 52 | 4.18 | 292 |
| 1,3-Dimethyl- | 54 | 5.08 | 292 |

In order to make possible the determination of the dipole moments of the xanthine derivatives, generally insoluble in nonpolar solvents, it has been necessary to introduce a "solubilizing" substituent at C-8, the decylthio-group, in the reasonable assumption that the introduction of the same group at the same position in all compounds will not affect the *sequence* of the physical properties of the various substances. Furthermore, as customary, the tautomers have been "fixed" by N-methylation.

The results obtained [99, 146] are indicated in Table XV. It is easy to see that in very satisfactory agreement with the theoretical predictions there is indeed a pronounced difference in the dipole moments

[146] B. Pullman, E. D. Bergmann, H. Weiler-Feilchenfeld, and Z. Neiman, *Int. J. Quant. Chem.* **35**, 103 (1969).

between the 7- and 9-methyl derivatives, those of the 9-derivatives being the greater. The usefulness of the technique for the determination of unknown structures is illustrated for the case of 1,3-dimethyl-8-decylthioxanthine which can exist both in the N(7)H and the N(9)H forms. The value of its moment, 5.08 D, indicates obviously that it is the N(7)H form (54) which exists at least predominantly.

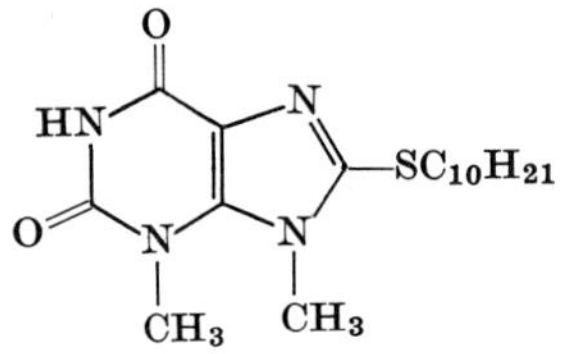

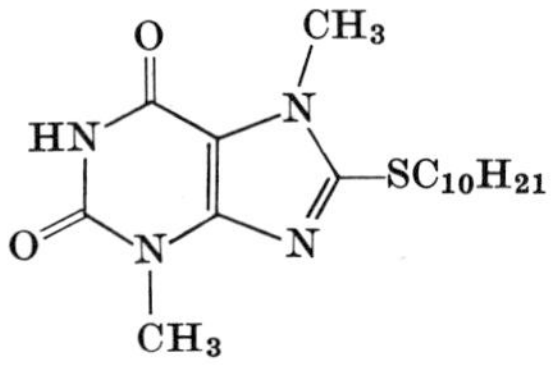

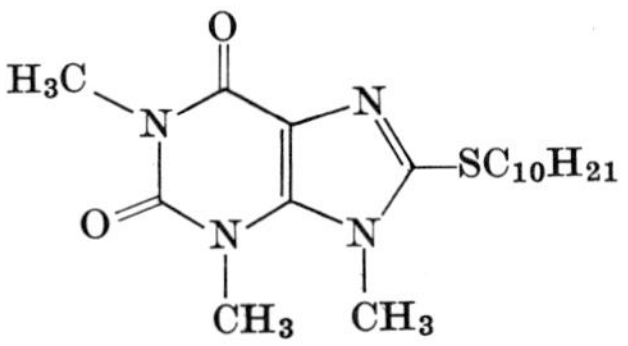

Very recently[99, 146] this type of study has been extended to the determination of the angle of the moment of xanthine with the C-4–C-5 axis, following the technique described in Section IV in relation to a similar determination in the case of purines. The predicted angle (counted counterclockwise) is, following the CNDO calculation, 99° for the N(7)H form. By comparing the moments of caffeine, which is a 1,3,7-trimethylxanthine (55) (3.70 D), 2-thiocaffeine (56) (4.76 D), 6-thiocaffeine (57) (3.76 D), and 2,6-dithiocaffeine (58) (4.62 D) and taking into account the fact that the moment of the

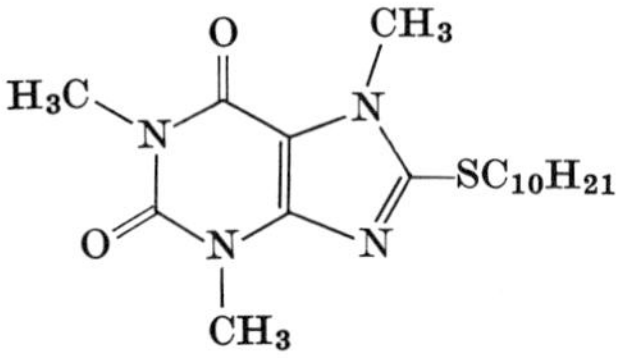

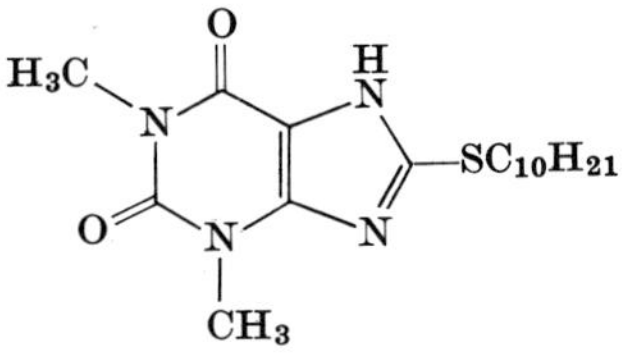

C=S bond is by 1.1 D higher than that of the C=O bond, one obtains a best fit if one assumes a value of 96° for the above-defined angle. (The moments calculated with this angle are: **55** = 3.70 D, **56** = 4.72 D, **57** = 3.74 D, and **58** = 4.64 D.)

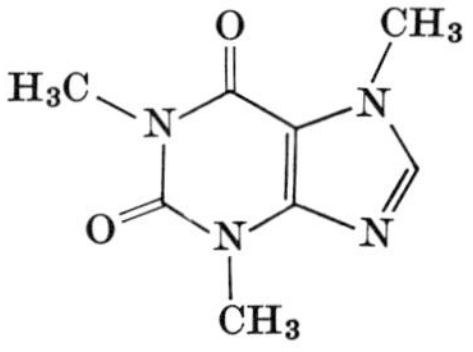

(55) 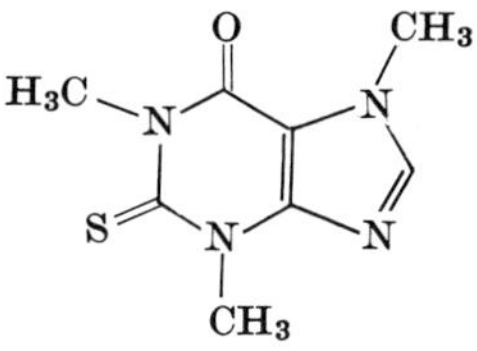  (56)
(57) 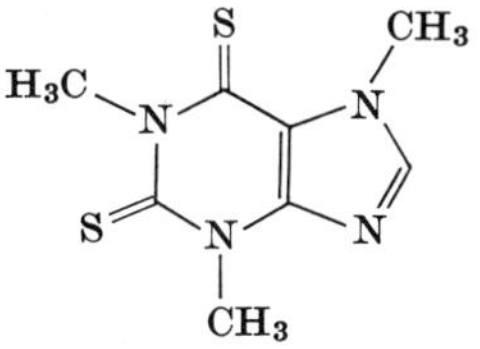  (58) 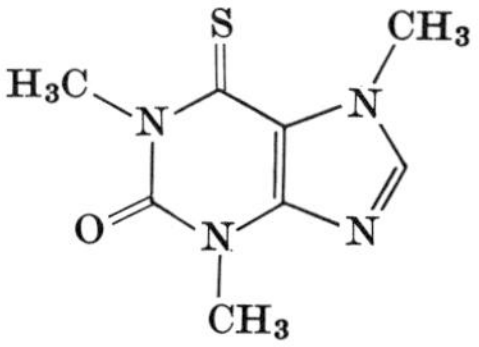

The experimental results also confirm the predictions concerning the ultraviolet absorption. The 7-methyl derivatives are bathochromic with respect to the 9-methyl ones and, in fact, the energy difference between the two wavelengths is exactly 0.3 eV as predicted. It may, however, be pointed out that a much smaller difference, although in the correct direction, has been reported by Pfleiderer and Nübel[147] between the two simpler derivatives 7-methylxanthine and 9-methylxanthine.

Concerning the other properties considered in Table XIV the ionization potential of xanthine has been found[98] equal to 9.30 eV, a value intermediate, as usual, between those predicted by the PPP and the CNDO methods. The potential predicted for the two tautomeric forms being practically the same, the experimental value does not enable to fix their identity.

The electronic properties of a number of other tautomers of xanthine (lactim forms) have been calculated in the PPP method by Kunii and Kuroda,[148] but no discussion of the results has been attempted.

[147] W. Pfleiderer and G. Nübel, *Ann. Chem.* **647**, 155 (1961).
[148] T. L. Kunii and H. Kuroda, *Rep. Computer Centre, Univ. of Tokyo* **1**, 227 (1968).

## VIII. Guanine, With a Special Discussion of Its N(3)H Tautomer

Because of the presence of a carbonyl and an amino group and the possibilities for the "imidazole" proton to oscillate between nitrogens N-7, N-9, and N-3, the molecule of guanine offers numerous and complex possibilities of tautomerization. Among those two have received special attention.

(1) The N(7)H ⇆ N(9)H tautomerization. The two tautomers (**59**) and (**60**) are predicted to have within 1 kcal/mole identical stabilities,

(**59**)    (**60**)

with perhaps a slight advantage for the N(7)H form. The prediction is in good agreement with the (very scarce, to be true) available data coming especially from the study of the ultraviolet spectra of these substances,[133, 149] which suggest the presence of the two tautomers in solution with a slight predominance of the N(7)H one.

As can be seen from Table XVIA, B the theory also predicts that the two tautomers should be characterized by very different dipole moments, that of the N(9)H tautomer being double that of the N(7)H one. Unfortunately, because of difficulties in solubilizing these substances (even when N-methylated) in nonpolar solvents, no experimental data are available. On the other hand, one notes an apparent disagreement between theory and experiment concerning the relative position of $\lambda_{max}$ in the two tautomers. It must, however, be remembered that the experimental data concern N-methylated derivatives in solution. A more recent determination of $\lambda_{max}$ in guanine in the vapor phase places its $\lambda_{max}$ at 293 m$\mu$[141] in good agreement with the calculations for the N(9)H form.

A much more curious case is that of the N(3)H tautomer (**61**) of guanine that we have studied (at least to begin with) in which the imidazole proton is shifted to position 3, a proton remaining attached to position 1. It is thus an N(1)H–N(3)H tautomer. Such a structure

---

[149] W. Pfleiderer, *Ann. Chem.* **647**, 167 (1967).

### TABLE XVIA
ELECTRONIC PROPERTIES OF GUANINE TAUTOMERS (PPP-DBP METHOD)

| | | | | | | | $S_1$ theoret. | | $\lambda_{max}$ exp. | Ring current | |
| --- | --- | --- | --- | --- | --- | --- | --- | --- | --- | --- | --- |
| Structure | Tautomer | $\mu$ (D) | $\theta$ (deg) | HOMO (eV) | LEMO (eV) | eV | m$\mu$ | eV | m$\mu$ | Hexagonal | Pentagonal |
| 59 | N(7)H | 3.4 | −119 | −7.7 | 1.5 | 4.43 | 280 | 4.38 | 283[149] | 0.38 | 0.64 |
| 60 | N(9)H | 7.2 | −30 | −7.6 | 1.5 | 4.27 | 290 | 4.52 | 270[149] | 0.30 | 0.63 |
| 61 | N(3)H−N(1)H | 13.6 | −60 | −7.4 | 0.8 | 2.41 | 514 | — | — | 0.23 | 0.70 |
| 63 | N(3)H−N(7)H | 4.3 | −7 | −7.7 | 1.4 | 4.59 | 270 | 4.70 | 264[0] | — | — |
| 64 | N(3)H−N(9)H | 12.4 | −4 | −7.6 | 1.6 | 4.49 | 276 | — | — | — | — |

### TABLE XVIB
ELECTRONIC PROPERTIES OF GUANINE TAUTOMERS (CNDO METHOD)

| Structure | Tautomer | $\mu$ (D) | $\theta$ (deg) | HOMO (eV) | LEMO (eV) | Molecular energy[a] (kcal/mole) |
| --- | --- | --- | --- | --- | --- | --- |
| 59 | N(7)H | 3.4 | −114 | −9.5 | 3.6 | 0 |
| 60 | N(9)H | 7.5 | −31 | −9.1 | 3.8 | 1 |
| 61 | N(3)H−N(1)H | 15.0 | −59 | −9.2 | 1.7 | 51 |
| 63 | N(3)H−N(7)H | 5.5 | 13 | −9.8 | 3.7 | 6 |
| 64 | N(3)H−N(9)H | 13.6 | 3 | −9.7 | 4.1 | 14 |

[a] With respect to the most stable tautomer considered as zero energy.

was attributed by Townsend and Robins[150, 151] to 3-methylguanine. The peculiarity of this case lies in the fact that it is impossible to write for such a form a usual covalent formula which would involve double bonds only between distant atoms. The simplest covalent structure which can be written for such a form would be of the type **61a** or **b** and would involve $\pi$ bonds between distant atoms. Other possibilities are ionic structures of the type **61c** or **d**, a situation which Townsend and Robins describe by representing 3-methylguanine as **62**.

The calculations (Table XV) indicate that this betaine tautomer is predicted to be about 50 kcal/mole less stable than the usual tautomers of guanine. From this point of view its existence would therefore be surprising. Moreover, if existing it should possess two other striking features; its predicted dipole moment should be of the order of 15 D and its spectrum should exhibit an enormous bathochromic shift with respect to the spectra of the usual guanines. No information is available about the dipole moment of this molecule but its absorption spectrum ($\lambda_{max} = 264$ m$\mu$ at pH 1; 273 m$\mu$ at pH 11) does not exhibit any extraordinary feature.

Under these circumstances we are very much tempted to conclude that the structural formula suggested by Townsend and Robins for 3-methylguanine is erroneous and that probably this molecule corresponds to a more classical type of tautomer. This could be, for example, the N(3)H–N(7)H one (**63**). As can be seen in Table XVI the properties of such a tautomer would be more reasonable and from the spectral point of view more in agreement with the observed properties of the substance. The distinction between **63** or the N(3)H–N(9)H tautomer could easily be made on the basis of dipole moment determination. The N(3)H–N(7)H seems to be the more probable of the two.

On the other hand, this could eventually also be the imine tautomer (**65**). Calculations carried out for this tautomer indicate, however, that it should be about 4 kcal/mole less stable than **63**. The dipole moment predicted for **65** is of the order of 4 D.

Figure 13 presents the distribution of electronic charges, as evaluated by the CNDO/2 method in the most significant guanine tautomers.

PPP calculations have been performed[148] for a number of other tautomeric forms of guanine (imine and lactim). No discussion of the calculated results has, however, been given.

[150] L. B. Townsend and R. K. Robins, *J. Org. Chem.* **27**, 990 (1962).
[151] L. B. Townsend and R. K. Robins, *J. Amer. Chem. Soc.* **84**, 3008 (1962).

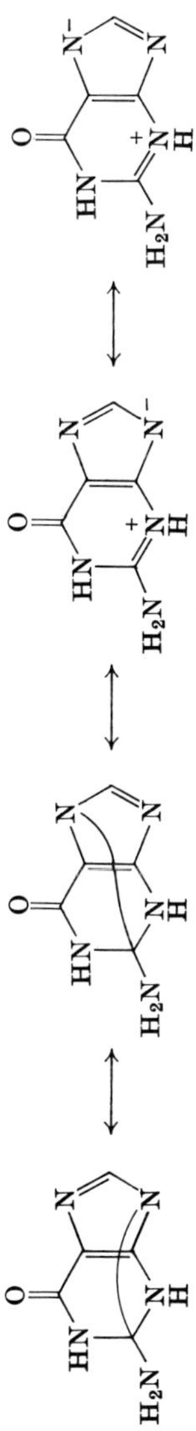

Sec. VIII.] ELECTRONIC ASPECTS OF PURINE TAUTOMERISM 141

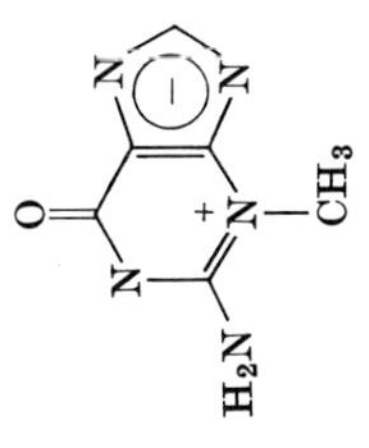

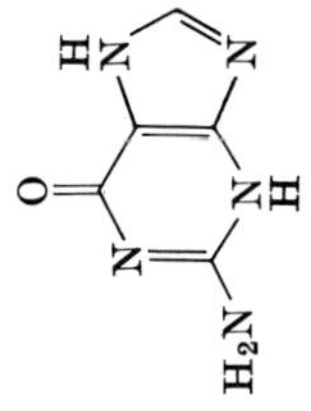

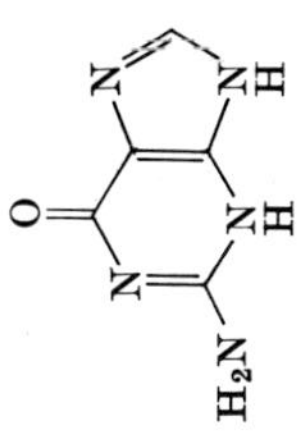

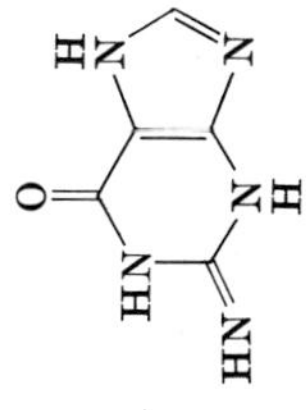

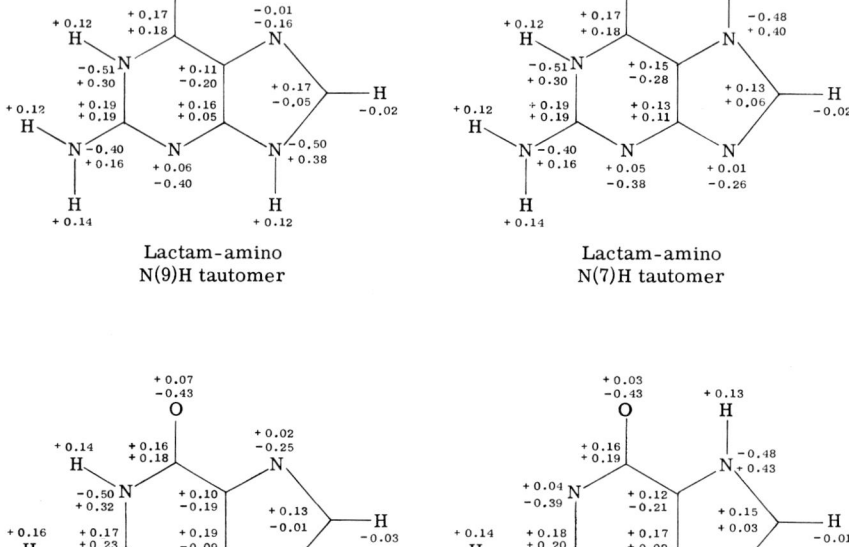

Fig. 13. Net electronic charges in guanine tautomers evaluated by the CNDO/2 method. Upper numbers: σ charges; lower numbers: π charges.

## IX. Tautomerism in 8-Azapurines and the Problem of the N(8)H Tautomer

8-Azapurines are important compounds, if only because of the involvement of some of them in cancer chemotherapy. From the viewpoint of this review they present, moreover, the interesting possibility of another tautomeric form with the proton fixed at N-8 of the pentagonal ring. Very recent experimental isolation of the N-methylated derivatives of the three tautomeric forms, N(7)H,

N(8)H, and N(9)H in a large series of 8-azapurines,[152–157] made a theoretical study of their properties particularly advantageous. Such a study[158] has been carried out for the 8-aza derivatives of purine (**66**), adenine (**67**), guanine (**68**), xanthine (**69**), and hypoxanthine (**70**). The principal results are summed up in Table XVII.

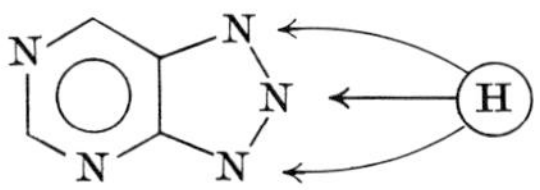

(**66**)

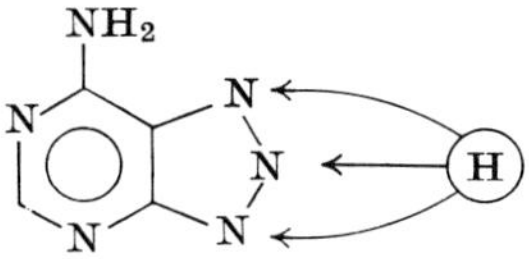

(**67**)

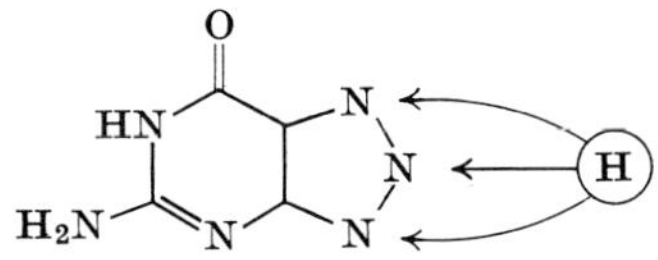

(**68**)

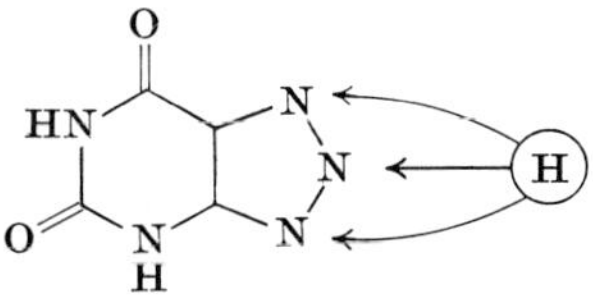

(**69**)

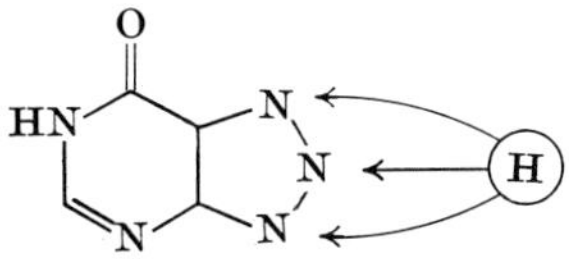

(**70**)

The calculations indicate, in the first place, that in all azapurines studied the relative stabilities of the three tautomers decrease in the order N(9)H > N(7)H > N(8)H and this even in the case when the N(7)H tautomer is more stable than the N(9)H tautomer in the corresponding purine. The N(8)H tautomers always appear as fundamentally the least stable, about 20–30 kcal/mole less stable than the two other tautomers. Although the problem of the crystal structure of the purines and, in particular, of the occurrence in the crystal of the N(7)H or N(9)H tautomers will be discussed in a later section, it may be useful to note here that while the presence of the

[152] G. Nübel and W. Pfleiderer, *Chem. Ber.* **98**, 1060 (1965).
[153] A. Albert, *J. Chem. Soc.* 427 (1966).
[154] J. W. Bunting and D. D. Perrin, *J. Chem. Soc.* 433 (1966).
[155] A. Albert and K. Tratt, *J. Chem. Soc.* 344 (1968).
[156] A. Albert, *J. Chem. Soc.* 2076 (1968).
[157] A. Albert, *J. Chem. Soc.* 152 (1969).
[158] B. Pullman and H. Berthod, *C. R. Acad. Sci.* **268**, 2958 (1969).

TABLE XVII

Some Electronic Properties of Prototropic Tautomers of 8-Azapurine

| Compound | Tautomeric form | PPP–DBP method | | | | | | CNDO method | | | | |
|---|---|---|---|---|---|---|---|---|---|---|---|---|
| | | HOMO (eV) | LEMO (eV) | $S_1$ theoret. (m$\mu$) | $\lambda_{max}$ exp.[a] (m$\mu$) | $\mu$ (D) | $\theta$ (deg) | HOMO (eV) | LEMO (eV) | Molecular energy CNDO method[b] (kcal) | $\mu$ (D) | $\theta$ (deg) |
| 8-Azapurine | N(7)H | −9.3 | 0.4 | 270 | 276[155] | 4.6 | −177 | −11.9 | 2.1 | 2 | 4.3 | 176 |
| | N(8)H | −9.0 | 0.1 | 265 | 270[156] | 2.4 | 134 | −11.2 | 1.2 | 34 | 2.8 | 120 |
| | N(9)H | −9.5 | 0.5 | 256 | 264[153,154] | 2 | −6 | −11.9 | 2.4 | 0 | 2.0 | 6 |
| 8-Azadenine | N(7)H | −8.4 | 0.7 | 282 | 285[155] | 6.5 | −177 | −10.6 | 2.5 | 6 | 6.2 | 178 |
| | N(8)H | −8.2 | 0.5 | 285 | 291[156] | 4.4 | 155 | −10.2 | 1.8 | 32 | 4.4 | 145 |
| | N(9)H | −8.3 | 0.9 | 264 | 277[157] | 0.5 | −108 | −10.8 | 3.0 | 0 | 0.1 | 8 |
| 8-Azaxanthine | N(7)H | −9.0 | 0.2 | 276 | 274[152] | 1.7 | 144 | −11.1 | 1.9 | 2 | 1.4 | 120 |
| | N(8)H | −9.3 | 0.4 | 256 | 272[152] | 3.5 | 55 | −11.7 | 2.0 | 23 | 3.6 | 50 |
| | N(9)H | −9.1 | 0.9 | 250 | 255[152] | 6.1 | 5 | −11.3 | 3.3 | 0 | 6.8 | 4 |
| 8-Azahypoxanthine | N(7)H | −8.5 | 0.7 | 281 | 264[155] | 4.0 | −127 | −10.4 | 2.3 | 6 | 3.7 | −124 |
| | N(8)H | −8.6 | 0.5 | 263 | 267[156] | 1.9 | −57 | −10.7 | 2.2 | 27 | 1.8 | −67 |
| | N(9)H | −8.5 | 0.4 | 294 | 254[156] | 6.3 | −43 | −10.6 | 2.4 | 0 | 6.5 | −41 |
| 8-Azaguanine | N(7)H | −7.9 | 0.9 | 290 | — | 5.6 | −108 | −9.7 | 2.8 | 9 | 5.6 | −107 |
| | N(8)H | −8.1 | 0.9 | 270 | — | 3.9 | −67 | −9.9 | 2.6 | 30 | 4.2 | −74 |
| | N(9)H | −8.0 | 1 | 288 | — | 8.6 | −50 | −9.9 | 3.4 | 0 | 8.7 | −51 |

[a] In the corresponding methyl derivatives.
[b] With respect to the most stable tautomer of each 8-azapurine, considered as zero energy.

N(9)H tautomer has been established in the crystal of 8-azaguanine monohydrate,[38, 42] the existence of the N(8)H tautomer has been advocated[159] for the crystal of 8-azaxanthine monohydrate. This situation, if confirmed,[160] could then be attributed, following the general discussion of Section XI, to the influence of crystal forces, which would then be particularly interesting to compute.

Among the physicochemical properties established experimentally and which may thus be compared with the theoretical computations, one finds in first place the ultraviolet absorption spectra. An inspection of the theoretical data indicates that no general rule seems to exist about the relative positions of the $\lambda_{max}$ of the three tautomeric forms N(7)H, N(8)H, and N(9)H of an azapurine. Thus, the theoretical order of decreasing bathochromic shift is N(7)H > N(8)H > N(9)H for 8-azapurine and 8-azaxanthine and N(8)H > N(7)H > N(9)H for 8-azadenine. It is remarkable to observe that these diversified predictions are confirmed by the order of the effects observed in the corresponding N-methylated derivatives of these compounds. The only exception to the agreement between theory and experiment is presented by the tautomers of 8-azahypoxanthine, in which case the disagreement is complete. Following the discussion concerning the longest wavelength of absorption in hypoxanthine (Section VII), it does not seem impossible that this could be due to the fact that the transition around 260 m$\mu$ is not the transition of longest wavelength in these molecules and that another band (shoulder?) could perhaps occur in the gaseous phase toward 280 m$\mu$.

No experimental data are available about the dipole moments or ionization potentials or electron affinities of these different forms. It may just be said that the N(8)H tautomers studied are not expected to have any particularly outstanding property in comparison with the N(7)H or N(9)H ones. In particular, they are predicted to have moderate dipole moments of the order of 4–5 D.

## X. The Thione–Thiol Tautomerism in Mercaptopurines and the Fine Structure of Thiopurines

Because of the difficulties encountered in the study of the lactam–lactim tautomerism in hydroxypurines, difficulties due to the delicate

[159] W. Nowacki and H. Bürki, *Z. Kristallogr.* **106**, 339 (1955).
[160] It is confirmed: H. C. Mez and J. Donohue, *Z. Kristallogr.* **130**, 376 (1969).

dependence of the phenomenon on geometrical factors such as the length of the C–O and O–H bonds, no attempt has been made so far to investigate quantum mechanically the problem of the thione–thiol tautomerism of mercaptopurines. The available experimental evidence is in favor of the existence of these and related compounds preferentially in the thione form, both in solution and in the solid state,[22, 161–163] and the calculations which have thus far been carried out have been devoted essentially to the determination of the electronic properties of these forms, although some electronic properties of the thiol forms, less dependent on the detailed molecular geometry, have also been investigated.

Altogether, the results are less abundant than for the hydroxypurines, in particular, concerning the utilization of the CNDO method. In fact, this method has been applied essentially to the evaluation of the relative stabilities of the N(7)H and N(9)H tautomer of 6-mercaptopurine (**71**) and (**72**), because of the interest of this particular datum in connection with the crystal structure of the

(**71**)   (**72**)

substance (*vide infra*, Section XI). The result is that the N(7)H tautomer (**71**) is about 3.5 kcal/mole more stable than the N(9)H tautomer (**72**). Other electronic properties have only been investigated by the self-consistent field semiempirical Pople–Pariser–Parr method. The three isomeric 2-, 6-, and 8-mercaptopurines have been analyzed, but less completely than the corresponding hydroxypurines, the tautomeric forms studied being **71** and **72** for 6-mercaptopurine, **73** and **74** for 2-mercaptopurine, and **75** for 8-mercaptopurine, as well as the corresponding thiol forms **76–81**.

[161] A. Albert and G. B. Barlin, *in* "Current Trends in Heterocyclic Chemistry" (A. Albert *et al.*, eds.), p. 51. Butterworths, London, 1958.
[162] E. Sletten, J. Sletten, and L. H. Jensen, *Acta Crystallog.* **B25**, 1330 (1969).
[163] G. M. Brown, *Acta Crystallog.* **B25**, 1338 (1969).

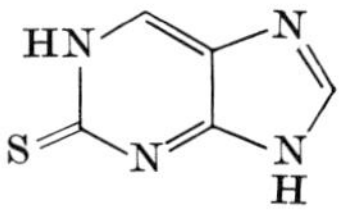

### Sec. X.] ELECTRONIC ASPECTS OF PURINE TAUTOMERISM 147

(73), 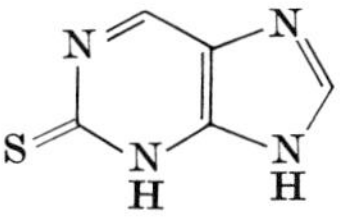 (74), 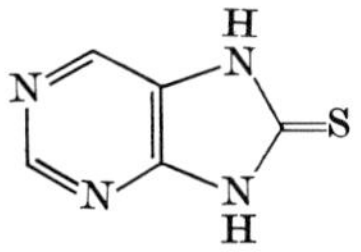 (75), 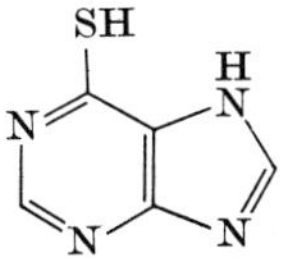 (76), 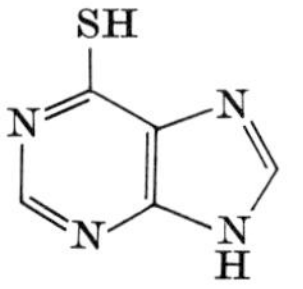 (77), 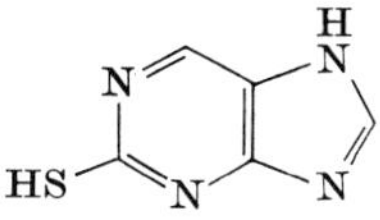 (78), 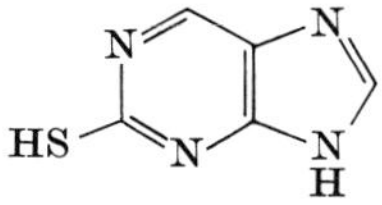 (79), 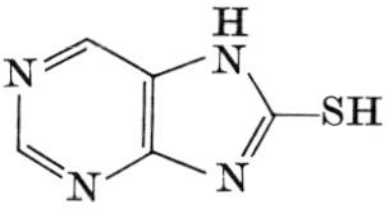 (80), 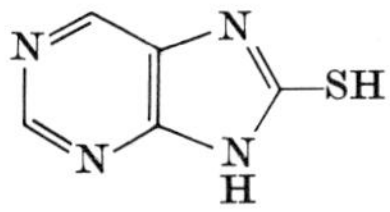 (81)

The results, summed up in Table XVIII, lead to the following main conclusions.

(1) Concerning the spectroscopic properties—the only ones for which experimental data are also available—it may be observed that although the numerical agreement between the calculated and observed values of the longest wavelength of absorption ($S_1$) is not as satisfactory as it was in the case of hydroxypurines, the theory correctly predicts the relative bathochromic order: 2-mercaptopurine > 6-mercaptopurine > 8-mercaptopurine of the thione forms, and the order 2-mercaptopurine > 6-mercaptopurine of the thiol forms. It predicts a slight hypsochromic shift when passing from 6- to 8-mercaptopurine in the thiol forms, while the experimental data seem to indicate a stationary absorption.

When the calculations for the mercaptopurines are compared with those of the hydroxypurines (Tables XII and XIII), they correctly reproduce the general bathochromic shift of the former with respect to the latter.

(2) Concerning the dipole moments, about which no experimental indications exist so far for the mercaptopurines, it may be observed by comparison with the results of Table XIII, that generally speaking the moments of the mercaptopurines in their thione forms should be somewhat greater than those of the corresponding hydroxypurines

## TABLE XVIII
### Electronic Properties of Mercaptopurines (SCF–IC Method)

| Compound | Structure | Tautomer | $S_1$ (m$\mu$) Theoretical | $S_1$ (m$\mu$) Experimental | $\mu$ (D) | $\theta^a$ | HOMO (eV) | LEMO (eV) |
|---|---|---|---|---|---|---|---|---|
| **Thione forms** | | | | | | | | |
| 2-Mercaptopurine | 73 | N(1)H–N(9)H | 380 } 384 | 348 | 5.4 | 115 | −7.5 | 0.1 |
| | 74 | N(3)H–N(9)H | | | 8.8 | 68 | −7.3 | 0.2 |
| 6-Mercaptopurine | 71 | N(1)H–N(7)H | 349 } 347 | 325 | 0.4 | −116 | −7.6 | 0.2 |
| | 72 | N(1)H–N(9)H | | | 6.8 | −8 | −7.3 | 0.7 |
| 8-Mercaptopurine | 75 | N(7)H–N(9)H | 327 | 310 | 3.8 | −107 | −7.7 | 0.4 |
| **Thiol forms** | | | | | | | | |
| 2-Mercaptopurine | 78 | N(7)H | 280 } 272 | 305$^b$ | — | — | −7.6 | 1.2 |
| | 79 | N(9)H | | | — | — | −7.6 | 1.4 |
| 6-Mercaptopurine | 76 | N(7)H | 272 } 272 | 290$^b$ | — | — | −7.9 | 1.2 |
| | 77 | N(9)H | | | — | — | −7.7 | 1.4 |
| 8-Mercaptopurine | 80 | N(7)H | 261 } 261 | 290$^b$ | — | — | −8.2 | 1.3 |
| | 81 | N(9)H | | | — | — | −8.0 | 1.4 |

$^a$ Counted counterclockwise with respect to the C-4–C-5 axis.
$^b$ In methylthiopurines.

in their lactam forms, with the exception of the N(1)H–N(7)H tautomer of 6-mercaptopurine (**71**), which should have a lower moment than the corresponding tautomer of hypoxanthine (**40**). (For 6-mercaptopurine the theoretical dipole moments are available following also the CNDO/2 calculations. They are equal to 0.18 D for **71** and 7.47 D for **72** in close agreement with the PPP results.)

A general check of the overall correctness of the predictions in this field is given by the comparison of the theoretical and experimental results in the series caffeine (**82**), 2-thiocaffeine (**83**), 6-thiocaffeine (**84**), and 2,6-dithiocaffeine (**85**) for which both are known. The theoretical moments are 4.3, 5.7, 4.2, and 5.6 D, respectively, and the experimental ones 3.7, 4.8, 3.8, and 4.6 D, respectively.[164] One notes the strong modification of the moment produced by the replacement of the O atom linked to C-2 by an S atom and the practical absence of any change on a similar replacement of the O atom linked to C-6.

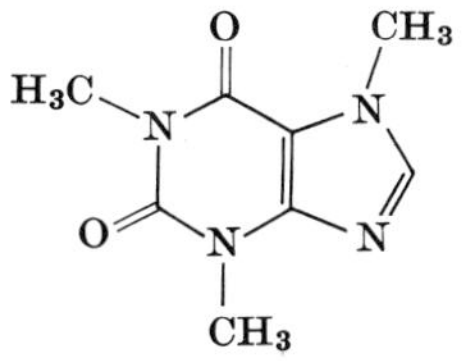

(**82**) 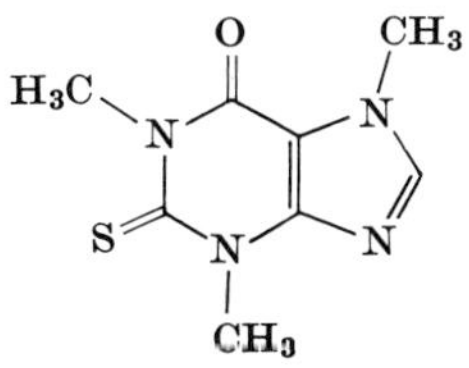 (**83**)

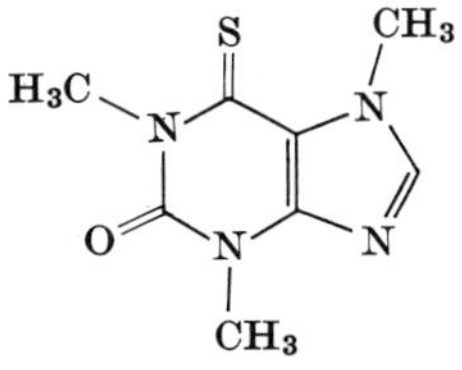

(**84**) 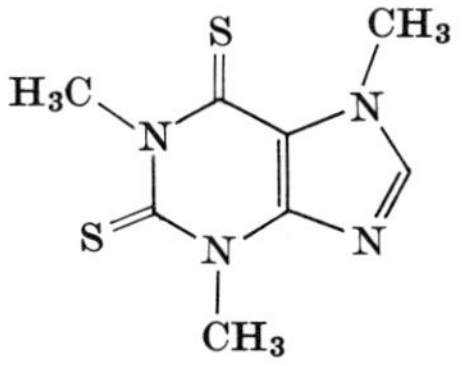 (**85**)

Concerning the moments of the thiol forms, their values depend to an appreciable extent on the conformation of the S–H bond (see Section VI for the similar case of the lactim forms of hydroxypurines).

[164] H. Weiler-Feilchenfeld, *J. Chem. Soc. B* 596 (1970).

## XI. The N(7)H ⇌ N(9)H Tautomerism and the Crystal Structure of Purines

The problem of the N(9)H ⇌ N(7) tautomerism of purines has recently gained increased interest following the already mentioned discovery by Watson et al.[34] that the crystal of purine contains the N(7)H form of the molecule. It consists of long chains of this tautomer, linked together by single hydrogen bonds, in the way illustrated in **86** for a dimer.

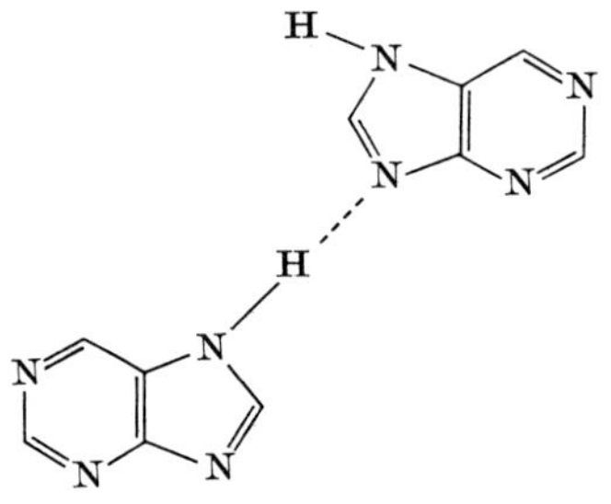

(86) 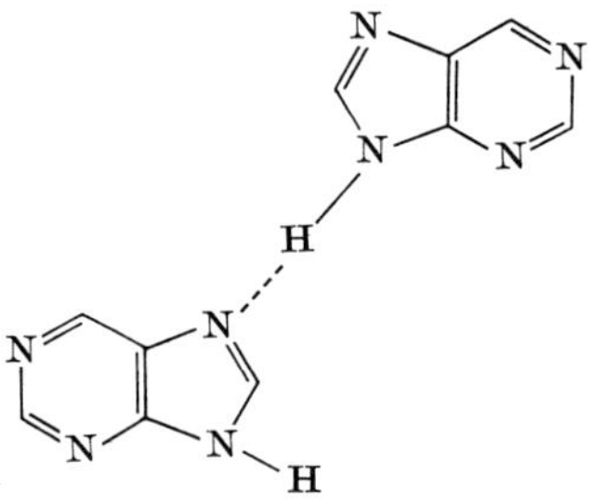     (87)

*A priori*, this situation could be ascribed to two principal factors: (a) the greater intrinsic stability of the form N(7)H; (b) the higher value of the crystal packing forces in the case of the N(7)H form, a complex problem which may perhaps be simplified by comparing the intermolecular interaction energies in the dimer (**86**) with those in the hypothetical dimer (**87**), which differs from **86** only in the shift of the proton from N-7 to N-9 of the bases involved.

Concerning the relative stabilities of the two forms, we have already seen (Section IV) that the two tautomers are probably of comparable intrinsic stability. Without forgetting the difficulties of drawing conclusions about relative stabilities from studies of chemical reactivity, it may nevertheless be added that this situation is to some extent corroborated by chemical evidence indicating, for example, the formation in almost equal amounts of 7- and 9-acetylpurine on treatment of the base under appropriate conditions with acetic anhydride, the 7 position being slightly favored.[165] The explanation

[165] G. S. Reddy, L. Mandell, and J. H. Goldstein, *J. Chem. Soc.* 1414 (1963).

for the presence of the N(7)H tautomer in the crystal of purine must therefore be looked for essentially in factor (b).

This situation leads us to say a few words about the general problem of base–base interactions, a very wide problem extending beyond that of the crystal structure of purines into those of base–base (including purine–pyrimidine) interactions in solution and upon cocrystallization, the structure of di-, oligo-, and polynucleotides, and culminating with the structure of the nucleic acids themselves. It is generally customary to distinguish in such interactions two principal types—*in-plane interactions* related to hydrogen bonding and *vertical interactions* between stacked bases.

In recent years a great amount of theoretical work has been carried out in a number of laboratories, in order to establish the nature of the forces responsible for such associations and to evaluate them.[63, 74, 166-169] We cannot, of course, enter here into the details of the procedures and shall therefore only summarize briefly the principal lines of approach.

The theoretical studies referred to above made it evident that the most important forces contributing to the type of interactions that we are considering are the intermolecular forces known under the general designation of van der Waals–London forces.

These forces are usually evaluated and have been so initially for the interactions between the nucleic acid bases[166] in the "dipole" approximation which considers these interactions ($E_D$) as the sum of three principal contributions

$$E_D = E_{\mu\mu} + E_{\mu\alpha} + E_L$$

where $E_{\mu\mu}$ is the electrostatic dipole–dipole, $E_{\mu\alpha}$ the induction or polarization dipole-induced dipole, and $E_L$ the London or dispersion interaction energies. These are defined respectively as follows:

---

[166] H. De Voe and I. Tinoco, *J. Mol. Biol.* **4**, 518 (1962).
[167] D. F. Bradley, S. Lifson, and B. Honig, *in* "Electronic Aspects of Biochemistry" (B. Pullman, ed.), p. 77. Academic Press, New York, 1964.
[168] P. Claverie, *in* "Molecular Associations in Biology" (B. Pullman, ed.), p. 115. Academic Press, New York, 1968.
[169] R. Rein, N. S. Goel, N. Fukuda, M. Pollak, and P. Claverie, *Ann. N. Y. Acad. Sci.* **153**, 805 (1969).

$$E_{\mu\mu} = \frac{1}{R_{AB}^3}\left[\vec{\mu}_A \vec{\mu}_B - \frac{3}{R_{AB}^2}(\vec{\mu}_A \vec{R}_{AB})(\vec{\mu}_B \vec{R}_{AB})\right]$$

$$E_{\mu\alpha} = -\frac{1}{2}\frac{1}{R_{AB}^6}\left\{\alpha_A\left[3\left(\frac{\vec{\mu}_B \vec{R}_{AB}}{R_{AB}}\right)+1\right] + \alpha_B\left[3\left(\frac{\vec{\mu}_A \vec{R}_{AB}}{R_{AB}}\right)+1\right]\right\}$$

$$E_L = -\frac{3}{2}\frac{I_A I_B}{I_A + I_B}\frac{\alpha_A \alpha_B}{R_{AB}^6}$$

where $\vec{\mu}_A$ and $\vec{\mu}_B$ are the respective dipole moments of molecules A and B (with $R_{AB}$ the distance between the points of location of these dipoles), $\alpha_A$ and $\alpha_B$ their polarizabilities, and $I_A$ and $I_B$ their ionization potentials.

It may, however, be observed [63, 167] that because of the shortage of the intermolecular distances with respect to the molecular dimensions, this "dipole" approximation may be rather inaccurate in the particular case of the interactions between the purine and pyrimidine bases and that it is preferable to treat the problem in the "monopole" approximation, i.e., by considering all the negative and positive charges in the system as interacting in a simple Coulombic fashion. In this "monopole" approximation the total energy ($E_M$) may then be considered as the sum of three main contributions

$$E_M = E_{\rho\rho} + E_{\rho\alpha} + E_L$$

where $E_{\rho\rho}$ is the monopole–monopole, $E_{\rho\alpha}$ the monopole-induced dipole, and $E_L$ the dispersion interaction energies. The first two are defined by:

$$E_{\rho\rho} = \sum_{i \in A} \sum_{j \in B} \frac{\rho_i \rho_j}{R_{ij}}$$

$$E_{\rho\alpha} = -\tfrac{1}{2}[\vec{\mu}_{A\to B} \vec{E}_{A\to B} + \vec{\mu}_{B\to A} \vec{E}_{B\to A}]$$

$$= -\tfrac{1}{2}[\alpha_B(\vec{E}_{A\to B})^2 + \alpha_A(\vec{E}_{B\to A})^2]$$

since

$$\vec{\mu}_{A\to B} = \alpha_B \vec{E}_{A\to B} \quad \text{and} \quad \vec{\mu}_{B\to A} = \alpha_A \vec{E}_{B\to A}$$

with

$$\vec{E}_{A\to B} = \sum_{i \in A} \frac{\rho_1}{(R_{iB})^3} \vec{R}_{iB} \quad \text{and} \quad \vec{E}_{B\to A} = \sum_{j \in B} \frac{\rho_j}{(R_{jA})^3} \vec{R}_{jA}$$

where index $i$ designates the atoms of molecule A and $j$ those of molecule B, the $\rho$'s are the net charges of the corresponding atoms, $\vec{\mu}_{A\to B}$ represents the dipole moment induced in B by the distribution of charges in A, and $\vec{E}_{A\to B}$ the field induced in B by the charges in A at the point of location of the induced dipole, $(\vec{R}_{iB})$ designates the vector from atom $i$ in A to this point, etc.

This transformation from the dipole to the monopole approximation in the calculations of the intermolecular forces has very significant consequences [74] and represents an essential step in the improvement of the calculations. More recently [168, 169] further refinements have still been introduced in the calculations, prominent among which are the replacement of molecular polarizability by bond polarizabilities and the introduction of a supplementary short-range repulsion term generally in the form of a semiempirical function of the type proposed by Kitaygorodskii [170] and used more extensively by Favini and Simonetta.[171] Details of these refinements should be looked for in the original papers.

The calculations of such intermolecular forces have been applied to the study of the interaction energies in dimers **86** and **87**.[172] The results of the calculations carried out in the different aforementioned approximations are indicated in Table XIX. It is believed that they are listed in the order of increasing refinement.

The results of Table XIX indicate that, in fact, the intermolecular interaction energies are significantly greater in the dimer **86** than in the dimer **87**. It may certainly be extrapolated that the same situation would prevail in higher polymers of the two types. It is also interesting to underline that the major part of the bonding energy comes from the electrostatic component.

On the other hand, in the crystal, the purines, besides being hydrogen-bonded, are also packed on top of one another (stacked).

[170] A. I. Kitaygorodskii, *Tetrahedron* **14**, 230 (1961).
[171] G. Favini and M. Simonetta, *Theoret. Chim. Acta* **1**, 294 (1963).
[172] B. Pullman, H. Berthod, and J. Caillet, *Theoret. Chim. Acta* **10**, 43 (1968).

## TABLE XIX
### INTERACTION ENERGIES IN PURINE DIMERS[a]

| Compounds | Approximation | Components | | | | Total |
|---|---|---|---|---|---|---|
| | | Electrostatic | Inductive | Dispersion | Repulsion | |
| Dimer (86) (observed) | Dipole-induced dipole | −1.12 | −0.30 | −0.39 | — | −1.81 |
| | Monopole-induced dipole | −6.84 | −0.73 | −0.39 | — | −7.96 |
| | Monopole-bond polarizabilities | −6.84 | −1.36 | −1.32 | — | −9.52 |
| | Monopole-bond polarizabilities +repulsion | −6.84 | −1.36 | −1.32 | +2.42 | −7.10 |
| Dimer (87) (hypothetical) | Dipole-induced dipole | −0.21 | −0.12 | −0.37 | — | −0.69 |
| | Monopole-induced dipole | −5.68 | −0.43 | −0.37 | — | −6.48 |
| | Monopole-bond polarizabilities | −5.68 | −1.03 | −1.34 | — | −8.05 |
| | Monopole-bond polarizabilities +repulsion | −5.68 | −1.03 | −1.34 | +4.37 | −3.68 |

[a] Data given in kcal/mole.

The general theory of the phenomenon [74, 166] indicates that the forces responsible for stacking are basically the same as those operating in hydrogen bonding with a shift of the dominant contribution toward the dispersion (London) component. Explicit calculations of the stacking energies for a pair of N(7)H tautomers and a pair of N(9)H ones indicate that this part of the total interaction energy is also greater in the former than in the latter pair; values equal, respectively, to 6.8 and 5.6 kcal/mole for the most favorable arrangements.

It therefore appears plausible to admit that the main reason for the occurrence of the N(7)H tautomer in the crystal of purine resides, at least to a large part, in the greater interaction energy that may be obtained with this tautomeric form.

The problem may and has been considered also in relation to the crystal structure of other purines, although in somewhat less detail.[173] Recent findings indicate that although the crystals of guanine, hypoxanthine, and 8-azaguanine contain the N(9)H tautomer of the bases,[174, 175] the crystal of 6-mercaptopurine monohydrate is made of the N(7)H form.[176, 177]

From the results quoted in different sections of this paper it may be observed that although the N(9)H tautomer is predicted to be more stable than the N(7)H tautomer in 8-azaguanine, the two forms, N(7)H and N(9)H, are predicted to be of comparable intrinsic stabilities in hypoxanthine and in guanine (with a consistent although very slight advantage for the N(7)H form), and the N(7)H form is predicted to be more stable than the N(9)H one in 6-mercaptopurine. Obviously no general straightforward correlations seem to exist between the intrinsic stabilities of the isolated tautomeric forms and their presence in the crystals of the substances, a conclusion which could, of course, have been anticipated from the previous example of purine. The problem has therefore to be examined more profoundly from the viewpoint of the interaction energies. Explicit calculations have been performed for adenine and guanine[173] leading to the prediction that it is the N(9)H form which should be present in the crystals of these compounds. Work is in progress in our laboratory on this subject in connection with other compounds.

[173] B. Pullman, H. Berthod, and J. Langlet, *C. R. Acad. Sci.* **266**, 1063 (1968).
[174] Ch. E. Bugg, U. F. Thewalt and R. E. Marsh, *Biochem. Biophys. Res. Commun.* **33**, 436 (1968).
[175] J. Sletten, E. Sletten, and L. H. Jensen, *Acta Crystallog.* **B24**, 1692 (1968).
[176] E. Sletten, J. Sletten, and L. H. Jensen, *Acta Crystallog.* **B25**, 1330 (1969).
[177] G. M. Brown, *Acta Crystallog.* **B25**, 1338 (1969).

It may perhaps be useful to remember that the dipole moments of the tautomers cannot and should not be considered as indicative of the relative values of such interactions in the first place, because appropriate calculations must be carried out in this case (as we have seen) in the "monopole" approximation and, second, because even in the "dipole" approximation, the mutual orientation of the dipoles of the interacting molecules is important. A glance at the data on the dipole moments of the different compounds mentioned here indicates that, in fact, there is no relation between the value of this moment in the different tautomers and the presence of such tautomers in the crystal. Thus, the dipole moments are predicted to be greater for the N(7)H form than for the N(9)H one in purine and adenine, but greater in the N(9)H form than in the N(7)H one in guanine, hypoxanthine, xanthine, 8-azaguanine, 8-azaxanthine, and 6-mercaptopurine. Also, no general relationship seems to exist between the relative values of the dipole moments and the stabilities of the different tautomers.

## XII. Conclusion

This paper has presented a discussion of a large number of tautomeric forms for the fundamental purines. It is obviously a rich field of study in which the conjunction of theory and experiment seems particularly fruitful.

Although the problem is basically a physicochemical one, the biological importance of purines is so overwhelming that it seems inevitable to reconsider at least some aspects of that importance in this conclusion, in the light of the preceding discussion.

The essential possible interference of purine tautomerism in biochemistry and, in particular, in biochemical evolution concerns, as mentioned in the introduction to this chapter, its possible role in mutagenesis. This could occur through the mispairing of the bases when present in rare tautomeric forms.

Two obviously essential questions which can then be raised in relation to such a viewpoint concern: (1) the relative tendency or capacity of the different nucleic acid bases to exist in a rare form, this tendency possibly determining the relative frequency of their involvement in spontaneous mutations and (2) the effect that the formation of a miscoupled pair of bases may have on some of their fundamental electronic, and thus biological, properties.

The first of these questions has, of course, to be considered, in a broader way, for the probability of tautomerism in both the purines and the pyrimidines. In this respect, the general theoretical prediction, made by us on the basis of simpler calculations as early as 1962[178] and confirmed later by other authors (e.g., Danilov[179]) and by the present refined calculations, is that it is the guanine and cytosine constituents of the nucleic acids which should have the greatest tendency to exist in their respective (lactim and imine) rare forms. These are therefore the bases which should have the greatest probability of being involved in spontaneous mutations insofar as tautomerization may be considered as a cause of such mutations. The transformation G–C → A–T should thus be more frequent than the reverse one.

At this point, it may be interesting to indicate that the fact that the G–C pairs constitute the unstable part of the genome and that they mutate spontaneously more frequently than the A–T pairs has been reported in a number of publications.[180–182] One cannot and should not, however, conclude from this situation that miscouplings through tautomerization of bases *are* the principal cause of spontaneous mutations, because those are definitely known to be due to a large series of causes, the relative importance of which is difficult to ascertain. The concordance between the theoretical predictions and the experimental facts is, however, worthwhile stressing, as an eventual discordance would certainly have been by some.

The second problem, namely, the different effects that the presence of rare tautomers and the subsequent miscouplings may introduce into the physicochemical properties of DNA, is, of course, a very broad one. As a particularly interesting aspect we may consider its influence on the stability of the nucleic acid.[183] Thus, Table XX indicates the van der Waals–London interaction energies calculated, following the

[178] B. Pullman and A. Pullman, *Biochim. Biophys. Acta* **64**, 403 (1962).
[179] V. I. Danilov, *Biophysics* (*URSS*) **12**, 621 (1967).
[180] E. Freese, *Proc. Int. Congr. Biochem., Moscow*, Vol. I, p. 204. Pergamon Press, London, 1961.
[181] E. Freese, *in* "Molecular Genetics" (J. H. Taylor, ed.), Part I, p. 207. Academic Press, New York, 1963.
[182] J. W. Drake, *Proc. Nat. Acad. Sci. U.S.* **55**, 738 (1966).
[183] B. Pullman and J. Caillet, *C.R. Acad. Sci.* **264**, 1900 (1967).

general procedure described in Section X, for the "natural" hydrogen-bonded G–C and A–T pairs and for the "miscoupled" pairs involving a base in its rare tautomeric form, indicated by an asterisk (*).[183]

Table XX confirms, as indicated in Section I of this review, that among the "natural" complementary pairs, the G–C pair has a much greater interaction energy than the A–T pair.* It shows also that

TABLE XX

Interaction Energies in the Watson–Crick and Miscoupled Base Pairs[a]

| Pair  | $E_{\rho\rho}$ | $E_{\rho\alpha}$ | $E_L$  | $E_{total}$ |
|-------|----------------|------------------|--------|-------------|
| A–T   | −4.61          | −0.27            | −0.77  | −5.65       |
| G–C   | −15.91         | −2.02            | −1.25  | −19.18      |
| A*–C  | −16.78         | −1.67            | −0.99  | −19.44      |
| C*–A  | −7.85          | −0.49            | −0.90  | −9.24       |
| G*–T  | −6.83          | −0.05            | −1.11  | −7.99       |
| T*–G  | −14.86         | −1.60            | −0.99  | −17.45      |

[a] Data in kcal/mole.

mispairings introduce very appreciable variations in the interaction energies and thus in the stabilities of the associations. Moreover, it may be predicted that miscoupled pairs containing the rare forms of guanine or cytosine should be appreciably less stable than those containing the rare forms of adenine or thymine. The formation of the first ones (which, as we have seen previously, is more probable than that of the second ones) may therefore introduce local elements of instability into the nucleic acids.

Modifications in the stabilities of the pairs are not, of course, the only result of the presence of rare tautomeric forms. There are also deep modifications in the other properties of the bases, such as the distribution of electronic charges.[185] Such transformations must have

---

* The calculations refer to interaction in vacuum. For their relation to measurements in solution see Pullman.[184]

[184] B. Pullman, *Proc. Seattle Symp. Quantum Biol.* 1969, Springer-Verlag, in press.

[185] B. Pullman, *Isr. J. Chem.* **1**, 412 (1963).

a profound influence on the chemical properties of the bases and, by way of consequence, on their biological properties as well. Very little is known about such phenomena.

The investigation of the chemical and physicochemical properties of purine tautomers, as well as their possible biological role, is still in a primitive form. Further developments are desirable and may have far reaching consequences for our understanding of the mechanism of life processes.

### Acknowledgment

The authors wish to acknowledge useful discussions with Professors E. D. Bergmann, F. Bergmann, and Dr. H. Weiler-Feilchenfeld and to thank them for communicating data prior to publication. They wish also to acknowledge the collaboration of Drs. H. Berthod, C. Giessner-Prettre, and J. Caillet in preparing the different calculations described in this paper.

# 1,6,6a$S^{IV}$-Trithiapentalenes and Related Structures

NOËL LOZAC'H

*Ecole Nationale Supérieure de Chimie de Caen, Caen, France*

I. Introduction and Nomenclature . . . . . . 162
II. Preparation of α-(1,2-Dithiol-3-ylidene)carbonyl Compounds . 167
   A. Formation of the Dithiole Ring . . . . . 167
   B. Reaction of Dithiole Derivatives with Reactive Methylene Groups . . . . . . . . . 172
   C. Reaction of Active Methylene Compounds with Dithiole Derivatives . . . . . . . . . 173
   D. Reaction of 1,2-Dithiole-3-thiones with Carbene Precursors 177
   E. Modification of Other α-(1,2-Dithiol-3-ylidene) Functions . 179
   F. Substitution or Modification of Substituents . . . 181
III. Preparation of 1,6,6a$S^{IV}$-Trithiapentalenes . . . . 182
   A. Formation of the Heterocyclic System . . . . 182
   B. Reaction of Dithiole Derivatives with Reactive Methylene Groups . . . . . . . . . 186
   C. Reaction of Active Methylene Compounds with Dithiole Derivatives . . . . . . . . . 187
   D. Condensation of 1,2 Dithiolo-3-thiones with Acetylenes . 191
   E. Modification of Other α-(1,2-Dithiol-3-ylidene) Functions . 192
   F. Substitution or Modification of Substituents . . . 193
IV. Other Heterocycles Derived from 1,6,6a$S^{IV}$-Trithiapentalenes . 195
   A. Thiophene Derivatives . . . . . . . 195
   B. Thiopyrone Derivatives . . . . . . . 196
   C. Positive Ions Derived from 1,6,6a$S^{IV}$-Trithiapentalenes . 197
   D. Nitrogen Compounds Analogous to 1,6,6a$S^{IV}$-Trithiapentalenes . . . . . . . . . . 199
   E. Selenium Compounds Analogous to 1,6,6a$S^{IV}$-Trithiapentalenes . . . . . . . . . . 202
V. Extended Structures . . . . . . . . 204
VI. Theoretical Studies and Physical Properties . . . . 207
   A. Valence Problems . . . . . . . 207
   B. X-Ray Diffractometry . . . . . . . 212
   C. Infrared Spectra . . . . . . . . 218
   D. Dipole Moments . . . . . . . . 222
   E. Visible and Ultraviolet Spectra . . . . . 224
   F. NMR Spectra . . . . . . . . 229

## I. Introduction and Nomenclature

The first preparation of a 1,6,6a$S^{IV}$-trithiapentalene* was described in 1925 by Arndt, Nachtwey, and Pusch,[1] who obtained a compound $C_7H_8S_3$ by reaction of 2,4,6-heptanetrione with phosphorus pentasulfide. Formula **1** was tentatively proposed. Further investigations corroborating this formula were published by Arndt in 1948.[2]

In 1953, Traverso and Sanesi,[3] reacting 4-thiopyrones with potassium hydrogen sulfide, obtained two compounds for which they proposed formulas **2a** and **2b**. The reason for this choice was that these products, treated by phosphorus pentasulfide, gave compounds similar to those already described by Arndt and his co-workers.[1,2]

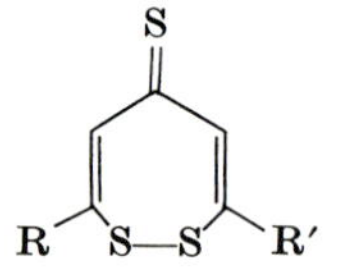

(**1**) R = R′ = $CH_3$

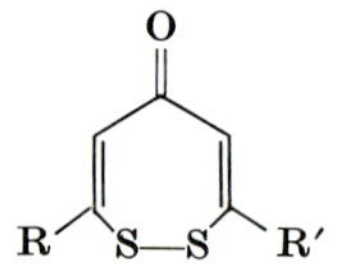

(**2a**) R = $C_6H_5$, R′ = H
(**2b**) R = $C_6H_5$, R′ = $CH_3$

In the following years, further investigations were conducted on these compounds until, in 1958, the correct formulas (**3** and **4**) were at last established almost simultaneously by taking into account the carbonyl stretching vibration of **4**[4] and an X-ray structural determination of **3** (R = R′ = $CH_3$).[5]

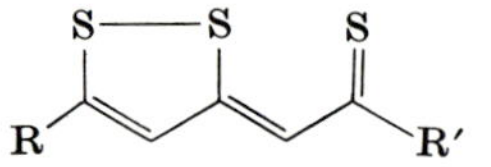

(**3**) 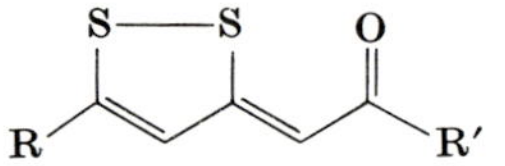          (**4**)

---

* *Chemical Abstracts* lists this compound as 1,6,6a-trithia(6a-$S^{IV}$)pentalene.

[1] F. Arndt, P. Nachtwey, and J. Pusch, *Chem. Ber.* **58**, 1633 (1925).
[2] F. Arndt, *Rev. Fac. Sci. Univ. Istanbul, Ser. A* **13**, 57 (1948).
[3] G. Traverso and M. Sanesi, *Ann. Chim. (Rome)* **43**, 795 (1953).
[4] G. Guillouzo, *Bull. Soc. Chim. Fr.* 1316 (1958).
[5] S. Bezzi, M. Mammi, and C. Garbuglio, *Nature* **182**, 247 (1958); S. Bezzi, C. Garbuglio, M. Mammi, and G. Traverso, *Gazz. Chim. Ital.* **88**, 1226 (1958); M. Mammi, R. Bardi, C. Garbuglio, and S. Bezzi, *Acta Crystallogr.* **13**, 1048 (1960).

The compounds we shall discuss in this paper have aroused particular interest since 1958, when it was proven that they exemplify a peculiar type of electron delocalization, sometimes referred to as "single bond–no bond resonance." This concept refers to a delocalization of $\sigma$ bonds rather similar to the well-known delocalization of $\pi$ bonds.

When $\pi$ bonds are delocalized in a given structure, systematic names are generally based on one given resonance structure and it is left to the reader to understand that the real structure does not have the electronic distribution implied by the systematic name. For instance, the name "cyclopentadienide anion" really applies to structure **5**, although formula **6**, with six delocalized $\pi$ electrons, gives a better picture.

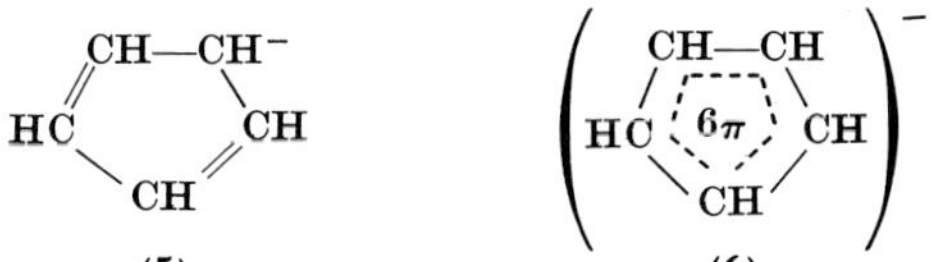

(5)    (6)

When, as in the preceding example, only $\pi$ bonds are affected by resonance, the arbitrary choice of the structure named usually affects only the ending of the name, e.g., the locants of the double bonds. The root of the name, based on $\sigma$-bonded atoms, is generally unambiguous.

Things are much more complicated when $\sigma$ electrons are delocalized. When a systematic name is needed, a resonance structure has to be chosen arbitrarily and the resulting name may be quite different from the one applying to another resonance structure.

Let us consider, for example, structures **7** and **8** which, in fact, describe a single substance.

(7)    (8)

These structures may be named as follows: **7**, 1-(1,2-benzodithiol-3-ylidene)propane-2-thione and **8**, 6-(5-methyl-1,2-dithiol-3-ylidene)-2,4-cyclohexadiene-1-thione.

These names are not precise because they contain no indication of the configuration, Z or E,[5a] of the carbon–carbon double bond which is not part of a ring. In fact, as we shall see, the three sulfur atoms lie almost on a straight line, and the name of the compound should indicate this.

That is why compounds such as **7** and **8** are generally described as containing a fused ring system, which can be written as **9** or **10**; these structures have the further advantage of showing the symmetry of the system.

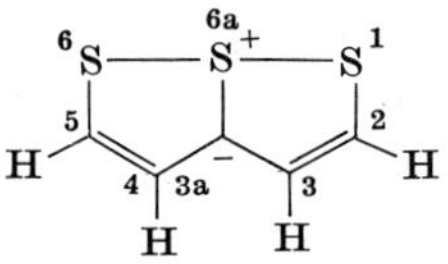

(9) 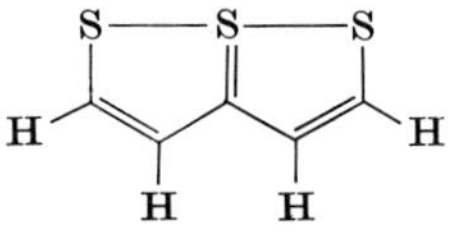    (10)

Structure **9** is not very likely because, by electron transfer it would lead to structures having a *negative* charge on carbons 2 and 5. As we shall see, this is incorrect, the charge on these atoms being clearly positive. In "The Ring Index,"[6] compound **10** is named [1,2]dithiolo-

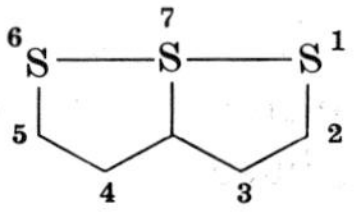

(11)

[1,5-*b*][1,2]dithiole, and numbered as in **11**. In our opinion this name is difficult to understand because the valency of the central sulfur atom is uncertain. Usually in such fusion names, the heteroatom common to two rings retains its standard valency, which is clearly not the case here.

In order to circumvent this difficulty, the simplest way is to resort to nomenclature based on pentalene and name structure **10** 1,6,6a$S^{IV}$-trithiapentalene. This name brings out the abnormal valency of the sulfur atom numbered 6a (6a$S^{IV}$) and permits us to apply the rule stating that such fusion names are related to the structure containing the greatest number of noncumulative double bonds.

[5a] Nomenclature according to J. E. Blackwood, C. L. Gladys, K. L. Loenig, A. E. Petrarca, and J. E. Rush, *J. Amer. Chem. Soc.* **90**, 509 (1968).

[6] "The Ring Index," Suppl. I (1957–1959), p. 27. American Chemical Society, Washington, D.C., 1963.

It should be understood that the choice of this name, based mainly on practical nomenclature considerations, does *not* define the real electronic structure of the molecule, which will be discussed later in Section VI.

Other names can be found in the literature for structure **10**, namely, thiothiophthene and thiathiophthene. Thiothiophthene is questionable because, in systematic replacement nomenclature, the affix "thio" must indicate the replacement of *oxygen* by sulfur and not of *carbon* by sulfur.

Thiathiophthene is a better choice, provided a locant be given to thia ($6aS^{IV}$-thiathiophthene). However, it is contrary to IUPAC nomenclature to derive a replacement name from the trivial name of a heterocyclic system (thiophthene).

Other structures such as α-(1,2-dithiol-3-ylidene)ketones (**12**) can be named according to common nomenclature because they do not possess the symmetry of the $1,6,6aS^{IV}$-trithiapentalenes. However, even with such unsymmetrical systems, pentalene names may be useful, either to stress the fact that three heteroatoms are in line or to show the similarity with trithiapentalene derivatives, for instance, in the case of selenium analogs (**13**).

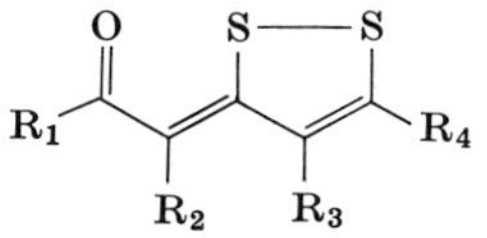

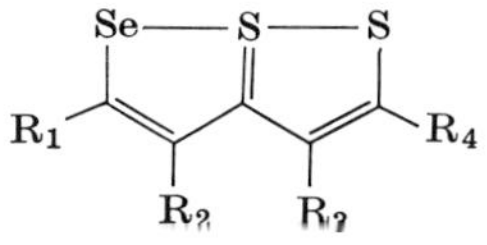

Last, it should be noted that some compounds which could be formally named as dithiolylidene ketones (**14**) are better represented by an ionic structure (**15**) or a pentalene structure (**16**).

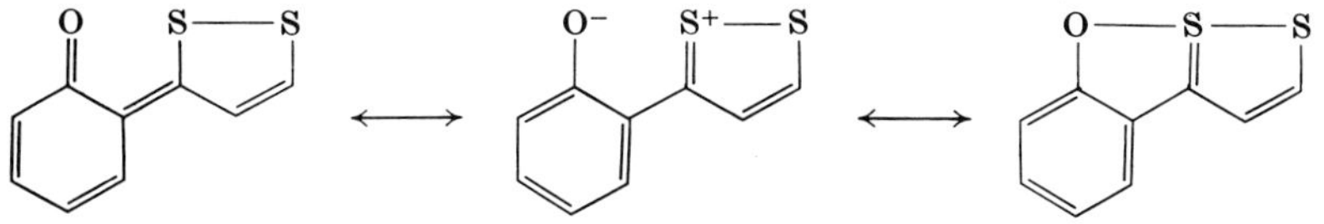

(**14**) 6-(1,2-Dithiol-3-ylidene)-2,4-cyclohexadien-1-one
(**15**) 2-(1,2-Dithiol-3-ylio)phenolate
(**16**) $8\text{-Oxa-}1,8aS^{IV}$-dithiacyclopenta[*a*]indene.

Besides dithiolylidene ketones (**12**), which can be regarded as 1-oxa-6,6a$S^{IV}$-dithiapentalenes (**17**), some other systems similar to 1,6,6a$S^{IV}$-trithiapentalenes have been described (**18–23**), and X-ray diffractometry and IR and NMR spectrometry suggest that the bonding of the three heteroatoms is somewhat similar to the bonding of the three sulfur atoms in 1,6,6a$S^{IV}$-trithiapentalenes.

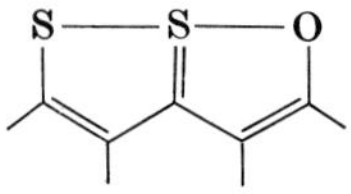

**(17)** 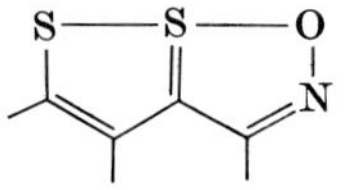 **(18)** 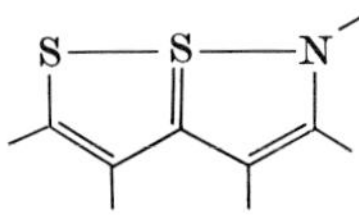 **(19)** 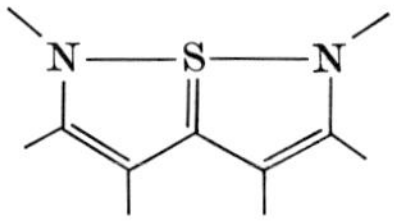 **(20)** 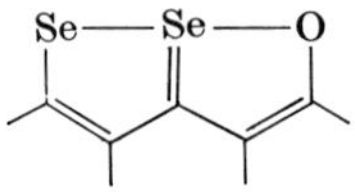 **(21)** 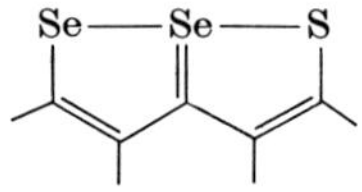 **(22)** 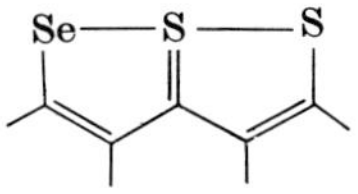 **(23)**

There are indications that anomalous bonding may also exist in extended structures **24, 25**, and **26**.

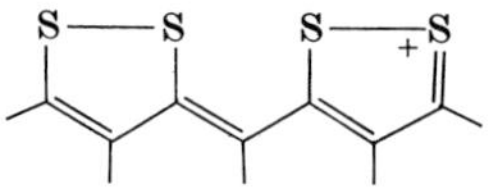

**(24)** 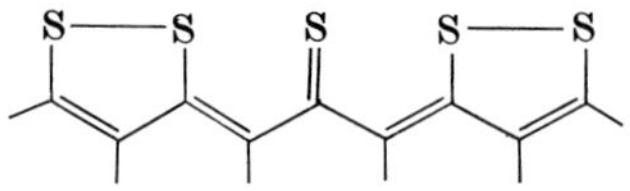 **(25)** 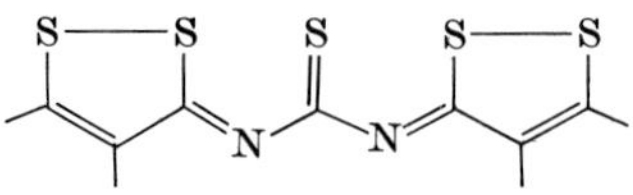 **(26)**

## II. Preparation of α-(1,2-Dithiol-3-ylidene)carbonyl Compounds

### A. Formation of the Dithiole Ring

#### 1. *From Pyran-4-thiones*

Pyran-4-thiones reacting with potassium hydrogen sulfide lead to α-(1,2-dithiol-3-ylidene) ketones and/or thiopyran-4-thiones (Eq. 1).[3, 7–9]

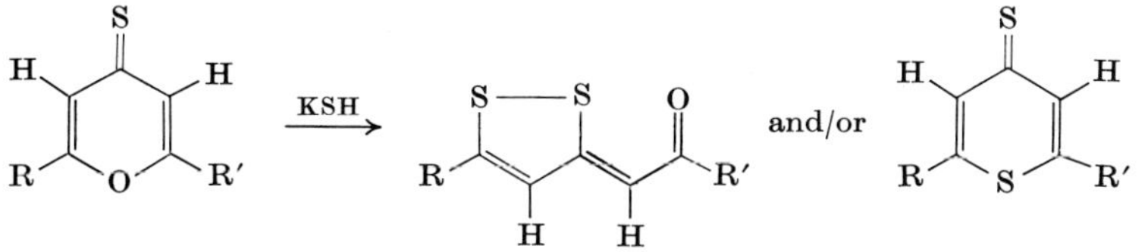

Thiopyran-4-thiones give adducts with mercury dichloride which, by reaction with aqueous sodium carbonate, give a mixture of α-(1,2-dithiol-3-ylidene)ketone and thiopyran-4-one (Eq. 2).[9, 10]

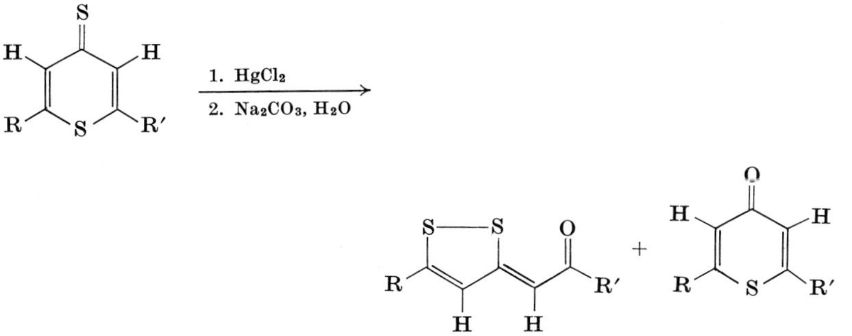

A better synthesis of α-(1,2-dithiol-3-ylidene)ketones from thiopyran-4-thiones consists in the reaction with sodium hydroxide in dimethylformamide (DMF), followed by ferricyanide oxidation (Eq. 3).[11]

[7] G. Pfister-Guillouzo and N. Lozac'h, *Bull. Soc. Chim. Fr.* 3254 (1964).
[8] G. Traverso, *Ann. Chim.* (*Rome*) **44**, 1018 (1954).
[9] G. Traverso, *Chem. Ber.* **91**, 1224 (1958).
[10] H. G. Hertz, G. Traverso, and W. Walter, *Ann. Chem.* **625**, 43 (1959).
[11] J. G. Dingwall, D. H. Reid, and J. D. Symon, *Chem. Commun.* 466 (1969).

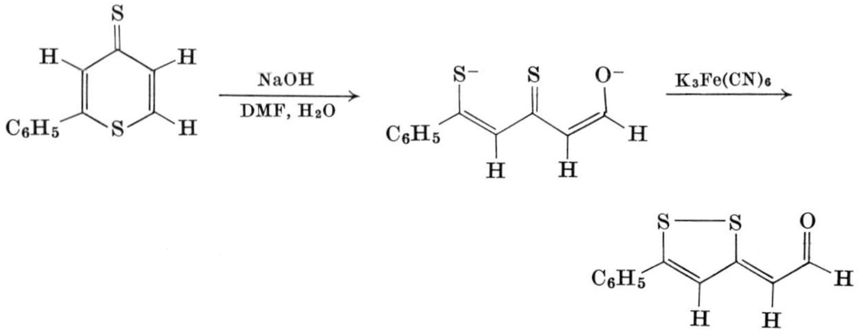

## 2. From β,δ-Diketophenols

Phosphorus pentasulfide reacts with β,δ-triketones, giving mainly 1,6,6a$S^{IV}$-trithiapentalenes (Section III, A, 1), but in some cases α-(1,2-dithiol-3-ylidene)ketones may be obtained.[12] When this reaction is applied to β,δ-diketophenols, the keto groups alone are thionated, the phenolic oxygen remaining unchanged.[12] This behavior may be taken as a proof that the oxygen is much more of the phenol than of the ketone type, owing to the aromatic resonance of the rings.

This fact may be interpreted by "mesoionic" or $S^{IV}$ formulas.[13] In this reaction γ-pyranthiones are also obtained (Scheme 1).

## 3. From α,γ-Diethylenic Carbonyl Compounds

At approximately 200°C, cinnamylidene malonic esters react with elemental sulfur and (1,2-dithiol-3-ylidene)malonic esters (**27**) are obtained (Eq. 4).[14, 15]

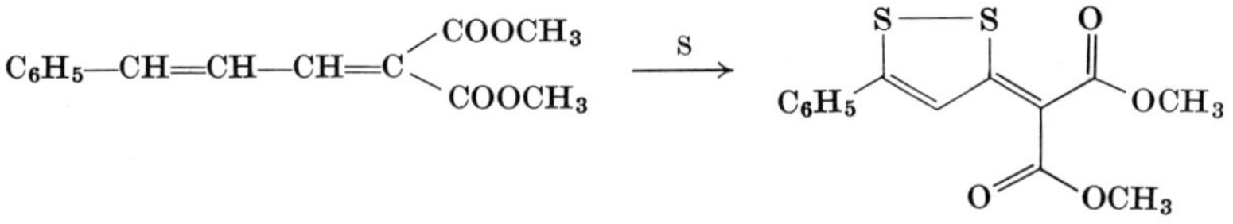

---

[12] M. Stavaux and N. Lozac'h, *Bull. Soc. Chim. Fr.* 2082 (1967).
[13] R. Pinel, Y. Mollier, and N. Lozac'h, *Bull. Soc. Chim. Fr.* 856 (1967).
[14] H. Quiniou, *Bull. Soc. Chim. Fr.* 213 (1960).
[15] H. Quiniou and N. Lozac'h, *Bull. Soc. Chim. Fr.* 1171 (1963).

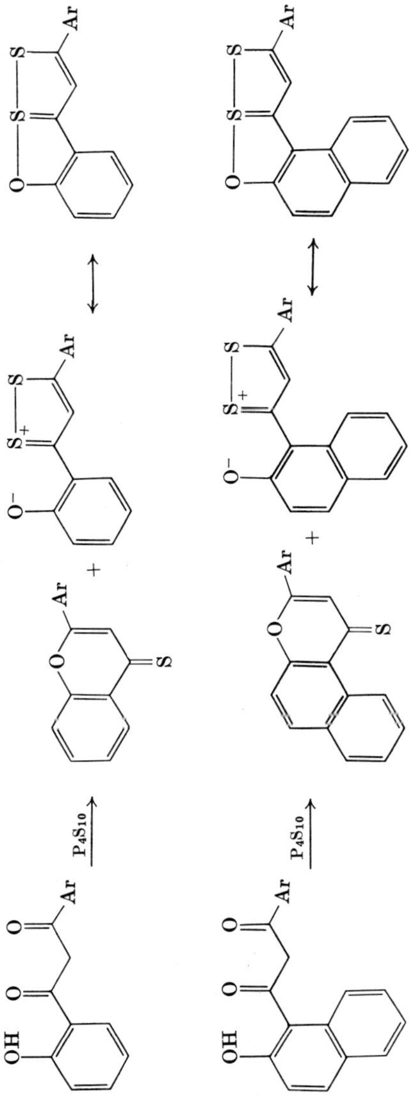

SCHEME 1

α-Cinnamylidene ketones[16–18] and β-keto esters[19] react in a similar way, but thiophene derivatives can also be obtained (Scheme 2).

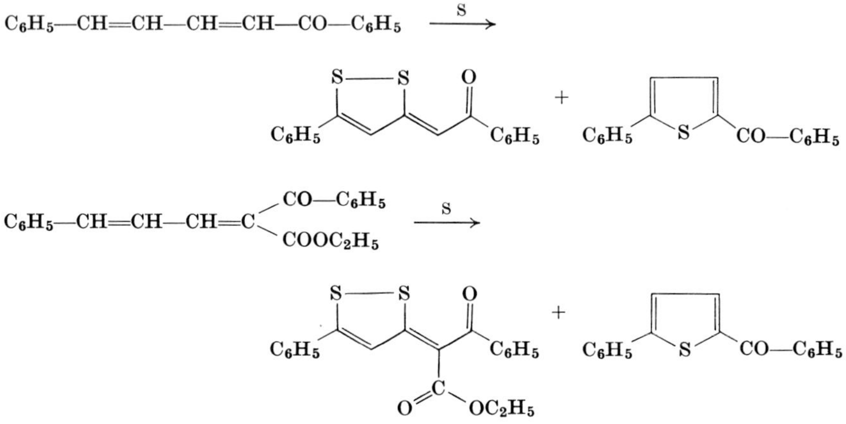

$$C_6H_5-CH=CH-CH=CH-CO-C_6H_5 \xrightarrow{S}$$

[structure: $C_6H_5$-substituted isothiazole-type ring with S—S, =CH—CO—$C_6H_5$] + [thiophene: $C_6H_5$—(S)—CO—$C_6H_5$]

$$C_6H_5-CH=CH-CH=C\begin{array}{l}CO-C_6H_5\\COOC_2H_5\end{array} \xrightarrow{S}$$

[structure with S—S ring, $C_6H_5$, CO—$C_6H_5$, and C(=O)OC$_2$H$_5$ branch] + [thiophene: $C_6H_5$—(S)—CO—$C_6H_5$]

SCHEME 2

With 2-cinnamylidenecyclohexanone, the cyclohexane ring is dehydrogenated by sulfur and the compound thus obtained does not show ketonic properties and for this reason a $S^{IV}$ formula or an ionic formula is preferred (Eq. 5).[13, 20]

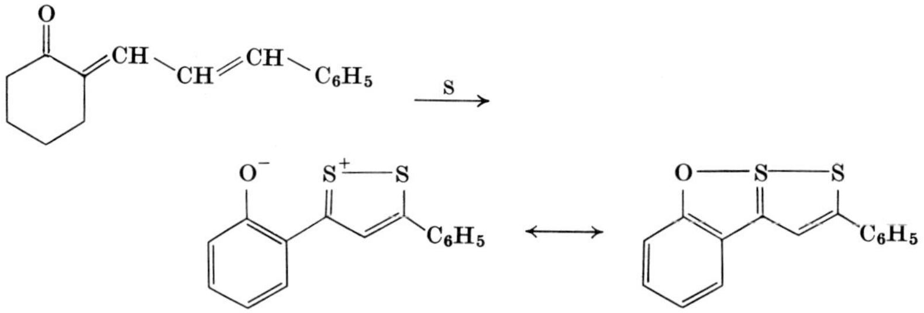

[Eq. 5: cyclohexanone with =CH—CH=CH—$C_6H_5$ side chain $\xrightarrow{S}$ benzo-fused ring with $O^-$, $S^+$—S, $C_6H_5$ ↔ benzo-fused O—S—S, $C_6H_5$] (5)

A similar reaction is observed with sulfur and 2-cinnamylidene-1-tetralone, but in this case significant quantities of 2-phenylbenzo[h]-chromene-4-thione are also obtained (Scheme 3).[21]

---

[16] G. Pfister-Guillouzo and N. Lozac'h, *Bull. Soc. Chim. Fr.* 153 (1963).
[17] G. Pfister-Guillouzo and N. Lozac'h, *Bull. Soc. Chim. Fr.* 3252 (1964).
[18] R. Pinel, Y. Mollier, and N. Lozac'h, *Bull. Soc. Chim. Fr.* 1049 (1966).
[19] Nguyen Kim Son, F. Clesse, H. Quiniou, and N. Lozac'h, *Bull. Soc. Chim. Fr.* 3466 (1966).
[20] R. Pinel, Y. Mollier, and N. Lozac'h, *C.R. Acad. Sci.* **260**, 5065 (1965).
[21] Y. Poirier and N. Lozac'h, *Bull. Soc. Chim. Fr.* 865 (1967).

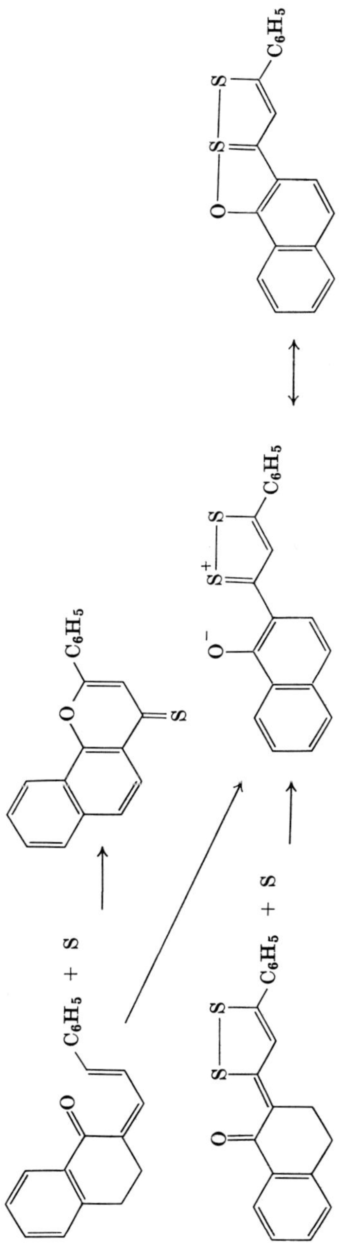

Scheme 3

## 4. Sulfuration of α,γ-Diacetylenic Carbonyl Compounds

Very good yields (ca. 90%) are obtained by reacting sodium disulfide in methanol, at pH 9, with α,γ-diacetylenic carbonyl compounds (Eq. 6).[22]

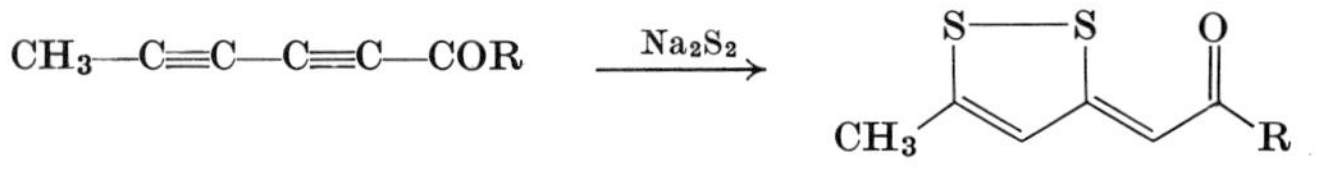

$$CH_3-C\equiv C-C\equiv C-COR \xrightarrow{Na_2S_2} \text{[dithiole product]} \quad (6)$$

R = H, CH$_3$, or OCH$_3$

## B. Reaction of Dithiole Derivatives with Reactive Methylene Groups

Carbon oxysulfide and sodium hydride react with 4,5,6,7-tetrahydro-1,2-benzodithiole-3-thione in dimethyl sulfoxide (DMSO), giving a dianion (**28**) which, by action of methyl iodide, gives a dithiolic thio ester (**29**). On standing, **28** partly isomerizes to **30**, which, by methylation gives the dithiolone **31**.[23]

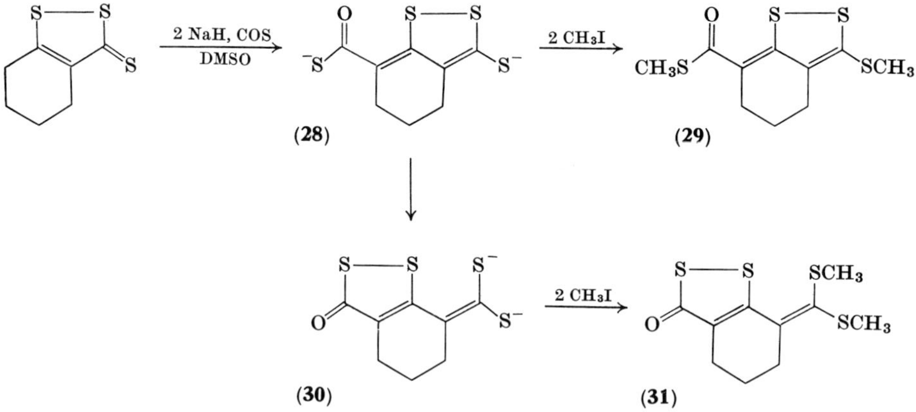

Another procedure taking advantage of the methylene reactivity of some dithiole compounds is described in Section II, E, 3.

[22] R. Bohlmann and E. Bresinsky, *Chem. Ber.* **100**, 107 (1967).
[23] J. L. Burgot and J. Vialle, *Bull. Soc. Chim. Fr.* 3333 (1969).

## C. Reaction of Active Methylene Compounds with Dithiole Derivatives

### 1. 1,2-*Dithiole*-3-*thiones* and 1,2-*Dithiol*-3-*imines*

Relatively stable 1,2-dithiole-3-thiones, such as the 5-aryl-substituted compounds, react with 1-acenaphthenone [24] or 1-tetralone [21] in the presence of sodium hydroxide, giving a 2-(1,2-dithiol-3-ylidene)-1-acenaphthenone (or tetralone) (Eq. 7).

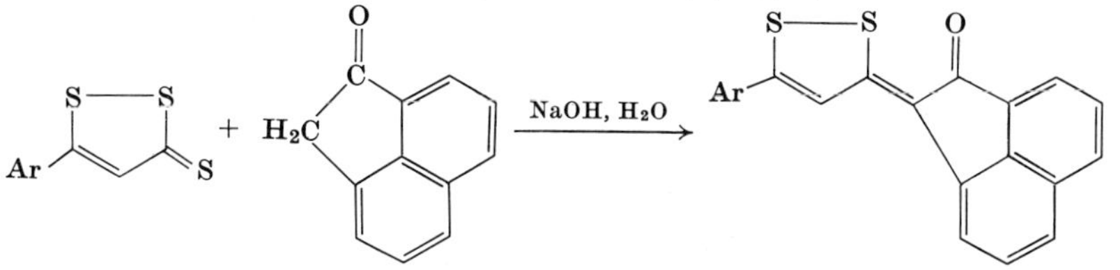

With 4-aryl-1,2-dithiole-3-thiones, a dithiopyrone is obtained [24] probably because the dithiole ring of the dithiolylidene ketone is less stabilized by a 4-aryl than by a 5-aryl group (see Section IV, B).

In propionic acid, with a small quantity of pyridine, 5-*p*-methoxyphenyl-1,2-dithiole-3-(*N*-phenylimine) reacts with acenaphthenone,[25] giving the corresponding dithiolylidene ketone, but this reaction has not been systematically investigated.

### 2. 1,2-*Dithiolium Cations without Leaving Groups*

4-Phenyl-1,2-dithiolium* cation reacts with acetone and ethyl vinyl ether in the following way (Scheme 4).[26]

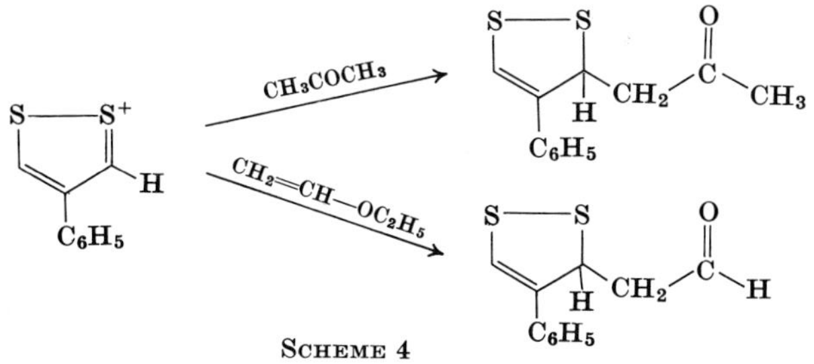

SCHEME 4

\* *Chemical Abstracts* preferred name; by I.U.P.A.C. rules the parent ion is called "dithiolylium."

[24] Y. Mollier and N. Lozac'h, *Bull. Soc. Chim. Fr.* 700 (1960).
[25] Y. Mollier and N. Lozac'h, *Bull. Soc. Chim. Fr.* 614 (1961).
[26] E. Klingsberg, *J. Org. Chem.* **31**, 3489 (1966).

The corresponding dithiolylidene ketones are obtained by subsequent oxidation with chloranil.[26]

Most commonly, however, an oxidation takes place during the condensation, probably at the expense of some dithiolium cation, and a dithiolylidene carbonyl compound is formed directly (Eq. 8).

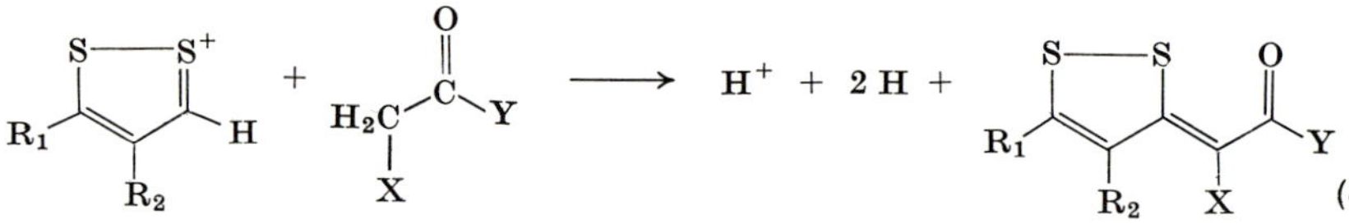

$$\underset{R_2}{\overset{S-S^+}{R_1}}\!\!\!\!\!\!\!\!\!\!\!\!\!\!\!\!\!\!\!\!\!\!\!\!\!\!\!\!\!\!\!\!\!\!\! H + H_2C(X)C(O)Y \longrightarrow H^+ + 2H + \text{dithiolylidene product} \quad (8)$$

This reaction has been made with various carbonyl compounds, e.g., cyclopentanone,[27] cyclohexanone,[20] cycloheptanone,[27] cyclooctanone,[27] 1-tetralone,[21] 2-tetralone,[28] variously substituted acetophenones,[29] phenyl benzyl ketone,[29] dibenzyl ketone,[29] dibenzoylmethane,[18,30] ethyl benzoylacetate,[18] ethyl malonate,[30] malonanilide,[30] and 2-(bisalkylthio)methylenecyclopentanones and cyclohexanones.[27]

In conjugated systems, $\gamma$-condensation can occur, as in the case of anthrone (Eq. 9).[31]

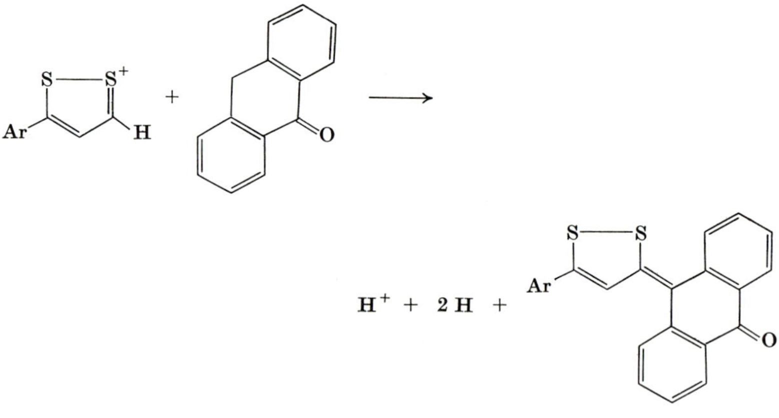

$$\text{Ar-dithiolium}^+ + \text{anthrone} \longrightarrow H^+ + 2H + \text{Ar-dithiolylidene-anthrone} \quad (9)$$

[27] O. Coulibaly and Y. Mollier, *Bull. Soc. Chim. Fr.* 3208 (1969).
[28] Y. Poirier and N. Lozac'h, *Bull. Soc. Chim. Fr.* 2090 (1967).
[29] E. Klingsberg, *J. Amer. Chem. Soc.* **85**, 3244 (1963).
[30] Y. Mollier, F. Terrier, R. Pinel, N. Lozac'h, and C. Menez, *Bull. Soc. Chim. Fr.* 2074 (1967).
[31] R. Pinel and Y. Mollier, *C.R. Acad. Sci. C* **264**, 1768 (1967).

### 3. *3-Chloro-1,2-dithiolium Cations*

Phosphorus oxychloride reacts with 1,2-dithiol-3-ones, giving 3-chloro-1,2-dithiolium ions. If this reaction is carried out in the presence of an active methylene compound, a condensation follows immediately, giving a dithiolylidene derivative. For instance, 5-phenyl-1,2-dithiol-3-one reacts with phosphorus oxychloride and benzoylacetonitrile, giving benzoyl-(5-phenyl-1,2-dithiol-3-ylidene)-acetonitrile (Eq. 10).[29]

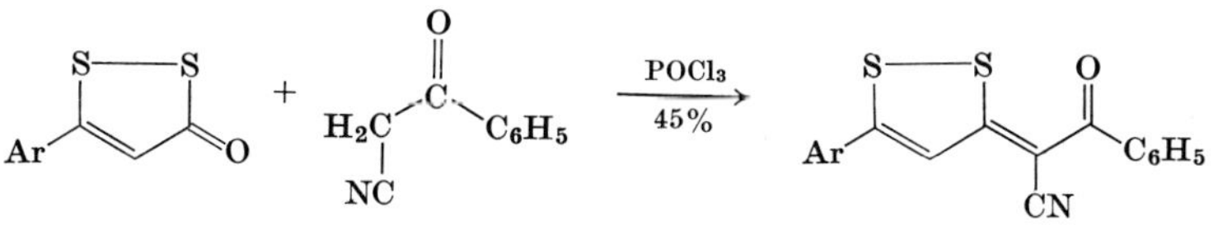

(10)

In similar conditions, hippuric acid leads to a dithiolylidene-oxazolone[32] and malononitrile gives a dithiolylidenemalononitrile which can be hydrolyzed to a monoamide.[26]

3-Chloro-1,2-dithiolium cations can be used in the form of crystalline perchlorates.[33] These compounds react with hydroxy-substituted and/or alkoxy-substituted aromatic compounds to give 3-aryl-1,2-dithiolium cations (Eq. 11).

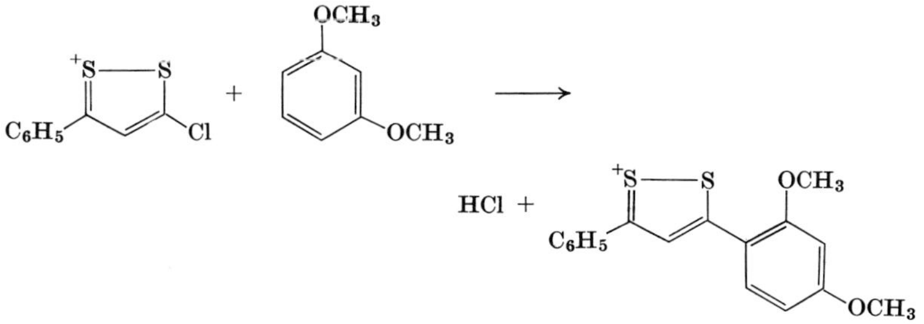

(11)

Whereas 2-naphthol is attacked at position 1, 1-naphthol is substituted in the 4-position. The phenolic cations thus obtained

---

[32] R. J. S. Beer, K. C. Brown, R. P. Carr, and R. A. Slater, *Tetrahedron Lett.* 1961 (1965).

[33] G. A. Reynolds, *J. Org. Chem.* **33**, 3352 (1968).

when treated with a tertiary amine, give neutral compounds of the dithiolyliophenolate (or dithiolylidene ketone) type (Scheme 5).[33, 34]

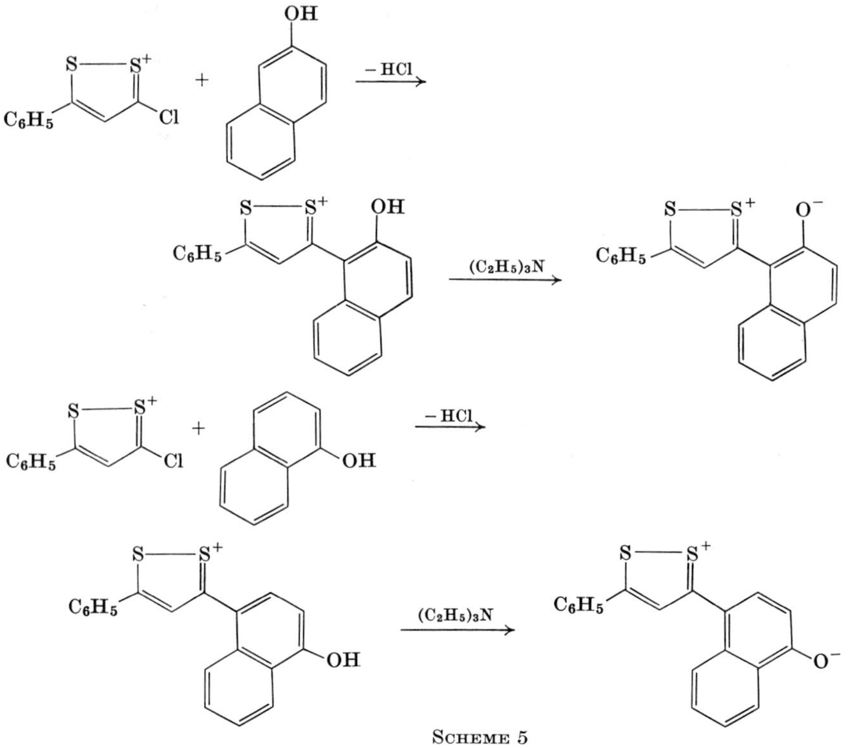

SCHEME 5

## 4. 3-Methylthio-1,2-dithiolium Cations

The reaction of these cations with active methylene compounds is often performed in acetic acid containing pyridine.[24, 25, 29, 35–37] Sometimes pyridine[35] or ethanol[38, 39] has been used as solvent (Eq. 12). In this reaction, pyridine acts as a basic catalyst, but it may also

[34] N. Lozac'h and C. Th. Pedersen, *Acta Chem. Scand.* **24**, 3189 (1970).
[35] Y. Mollier, N. Lozac'h, and F. Terrier, *Bull. Soc. Chim. Fr.* 157 (1963).
[36] Y. Mollier, F. Terrier, and N. Lozac'h, *Bull. Soc. Chim. Fr.* 1778 (1964).
[37] U. Schmidt, R. Scheuring, and A. Luttringhaus, *Ann. Chem.* **630**, 116 (1960).
[38] H. Behringer, M. Ruff, and R. Wiedenmann, *Chem. Ber.* **97**, 1732 (1964).
[39] J. Bignebat and H. Quiniou, *C.R. Acad. Sci. C* **270**, 83 (1970).

be N-methylated, regenerating the corresponding 1,2-dithiole-3-thione.[25]

$$\underset{R_2}{\overset{R_1}{\underset{\|}{\bigvee}}}\overset{S\!-\!S^+}{\underset{SCH_3}{\bigvee}} + \underset{X}{\overset{H_2C}{\underset{|}{\bigvee}}}\!\!\!\overset{COY}{\longrightarrow}$$

$$H^+ + CH_3SH + \underset{R_2\ \ X}{\overset{S\!-\!S}{\underset{R_1}{\bigvee}}}\!\!\!\overset{O}{\underset{}{\bigvee}}Y \quad (12)$$

This reaction has been applied to a great number of carbonyl compounds, e.g., cyclopentanone,[27] 1,3-cyclohexanedione,[36] 5,5-dimethyl-1,3-cyclohexanedione,[35] 1-acenaphthenone,[24, 25] 1,3-indanedione,[35] 1,3-phenalanedione,[35, 36] sodium salts of 3-aryl-3-oxopropanals,[39] 2,3-dihydrobenzo[b]thiophen-3-one,[40] sodium benzoylacetate,[41] benzoylacetonitrile,[29, 30, 38] ethyl cyanoacetate,[30, 35] cyanoacetanilide,[30] barbituric acid,[35] rhodanine,[35] N-phenylrhodanine,[37] and N-methylenebenzothiazoline.[37]

Structural proofs have been given by comparing some of the preceding condensation products with compounds obtained by reacting elemental sulfur with cinnamylidene compounds. Condensation of 5-p-methoxyphenyl-3-methylthio-1,2-dithiolium iodide with ethyl cyanoacetate [35, 42] followed by the ethanolysis of the cyano group gave the same dithiolylidene malonate as the reaction of elemental sulfur on ethyl p-methoxycinnamylidene malonate.[15]

Para condensation with anthrone,[31] as well as secondary reactions leading to γ-dithiopyrones,[40] has also been described.

3-Methylthio-1,2-dithiolium cations attack sodium phenoxide in the 2-position[13, 20] and sodium 2,6-dimethylphenoxide in the 4-position.[31]

## D. Reaction of 1,2-Dithiole-3-thiones with Carbene Precursors

These reactions are not necessarily carbene reactions, although this may well be the case with diazoketones. Their common feature

[40] D. B. Easton, D. Leaver, and D. M. McKinnon, *J. Chem. Soc. C* 642 (1968).
[41] D. Leaver and D. M. McKinnon, *Chem. Ind. (London)* 461 (1964); E. I. G. Brown, D. Leaver and D. M. McKinnon, *J. Chem. Soc. C* 1202 (1970).
[42] Y. Mollier, *Bull. Soc. Chim. Fr.* 213 (1960).

is the replacement of sulfur by an $\overset{R}{\underset{R'}{>}}C=$ group according to the overall schematic reaction:

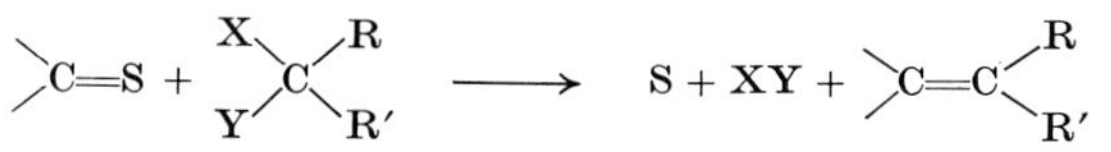

$$>C=S + \overset{X}{\underset{Y}{>}}C\overset{R}{\underset{R'}{<}} \longrightarrow S + XY + >C=C\overset{R}{\underset{R'}{<}}$$

A common intermediate could be a thiirane which loses a sulfur atom.

$$>C\underset{S}{\overset{}{\diagdown\diagup}}C\overset{R}{\underset{R'}{<}} \longrightarrow S + >C=C\overset{R}{\underset{R'}{<}}$$

## 1. Diazoketones

At 150°C, α-diazoketones react with 1,2-dithiole-3-thiones, giving α-(1,2-dithiol-3-ylidene)ketones (Scheme 6). This may be done either with cyclic or acyclic diazoketones[28]; α-diazo esters react similarly.[28]

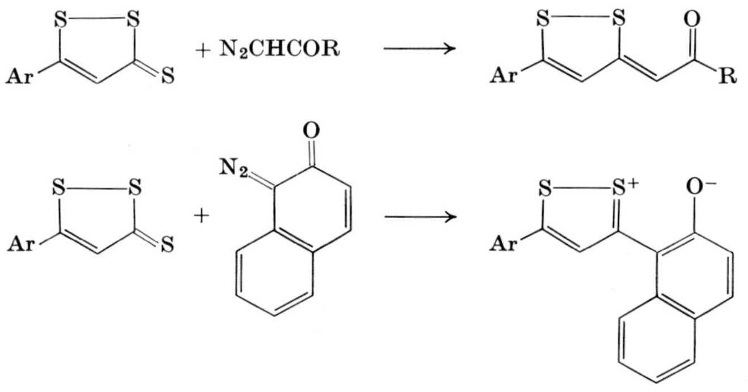

SCHEME 6

## 2. α-Halogenoketones

Bromoacetone or phenacyl bromide react with 1,2-dithiole-3-thiones giving a 3-acetonylthio- or a 3-phenacylthio-1,2-dithiolium bromide, which yields a dithiolylidene ketone, through sulfur extrusion, by moderate heating.[43]

[43] G. Caillaud and Y. Mollier, *Bull. Soc. Chim. Fr.* 2018 (1970).

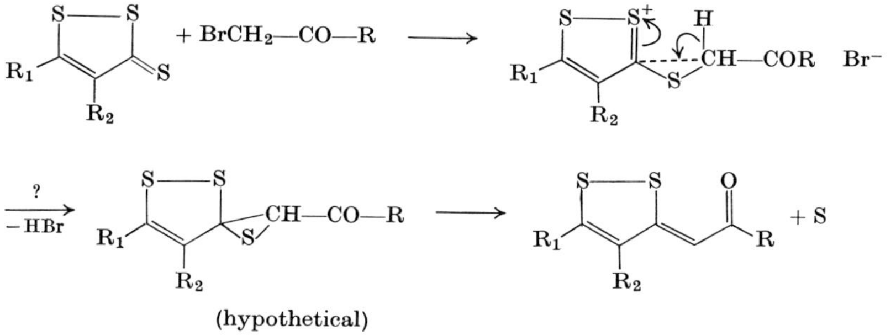

SCHEME 7

## E. MODIFICATION OF OTHER α-(1,2-DITHIOL-3-YLIDENE) FUNCTIONS

### 1. 1,6,6a$S^{IV}$-Trithiapentalenes

(5-Methyl-1,2-dithiol-3-ylidene)acetone is formed by moderate heating of 2,5-dimethyl-1,6,6a$S^{IV}$-trithiapentalene with either 70% perchloric acid or 96% sulfuric acid.[2] The structure of the dithiolylidene ketone has been proved by Raney nickel desulfurization.[44]

From 2-methyl-5-phenyl-1,6,6a$S^{IV}$-trithiapentalene, (5-phenyl-1,2-dithiol-3-ylidene)acetone only is obtained, and not the other possible isomer. The structure was proved by Raney nickel desulfurization (Eq. 13).[44]

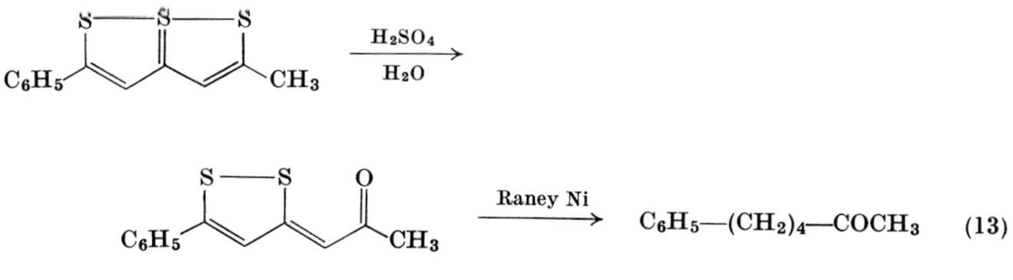

Mercuric acetate may also be used. In experimental conditions where 2,5-diphenyl-1,6,6a$S^{IV}$-trithiapentalene does not react, the 2,4-diphenyl isomer is oxidized, giving only one of the two possible isomers (Eq. 14).[29]

[44] H. Behringer, H. Reimann, and M. Ruff, Angew. Chem. **72**, 415 (1960).

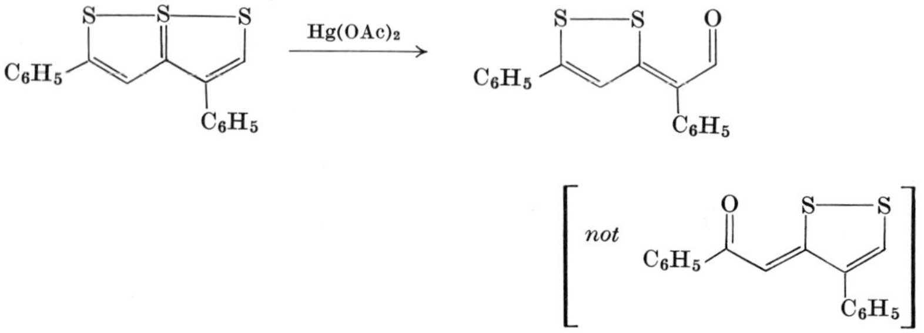

[Structural diagram of equation (14): 1,2-dithiol-3-ylidene reaction with Hg(OAc)₂] (14)

Selective hydrolysis by sulfuric acid is also observed in the formation of 2-(5-aryl-1,2-dithiol-3-ylidene)cyclanones from the corresponding trithiapentalenes, in low yield.[27]

2,5-Diaryl-3-formyl-1,6,6a$S^{IV}$-trithiapentalenes are selectively oxidized on carbon 2, and not on carbon 5 either by mercuric acetate or by peracetic acid (Eq. 15).[39] A 3-alkoxycarbonyl substituent has the same orienting effect in this oxidation.[45]

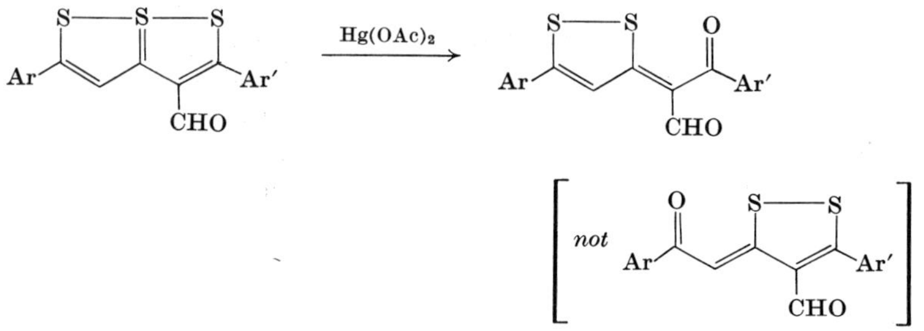

[Structural diagram of equation (15)] (15)

3-Acyl-2-methylthio-1,6,6a$S^{IV}$-trithiapentalenes, on reaction with mercuric acetate, undergo a degradation leading to a dithiolylidene ketone (Eq. 16).[46]

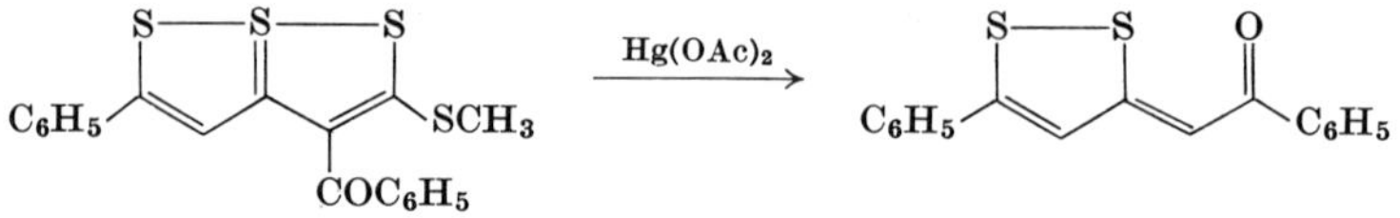

[Structural diagram of equation (16)] (16)

Dithiolylidene esters can be obtained in a similar way.[46]

---

[45] C. Trebaul and J. Teste, *Bull. Soc. Chim. Fr.* 3790 (1966).
[46] R. J. S. Beer, R. P. Carr, D. Cartwright, D. Harris, and R. A. Slater, *J. Chem. Soc. C* 2490 (1968).

## 2. 1,2-*Dithiol-3-ylidenemalononitriles*

α-(1,2-Dithiol-3-ylidene)malononitriles may be prepared by reacting 3-methylthio- or 3-chloro-1,2-dithiolium cations with malononitrile.[26, 47] They can also be obtained with 1,2-dithiole-3-thiones and tetracyanoethylene or its oxide.[48]

These dithiolylidenemalononitriles are hydrolyzed in aqueous ethanol in the presence of hydrochloric acid, leading to α-(1,2-dithiol-3-ylidene)carboxamides (Eq. 17).[48]

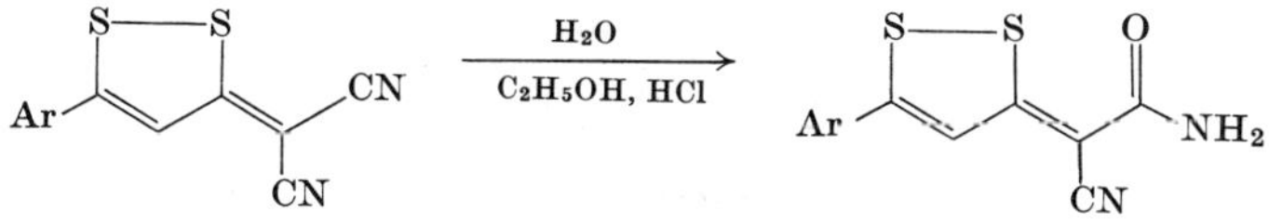

## 3. α-(1,2-*Dithiol-3-ylidene)alkyliminium Salts*

Dimethylthioformamide reacts at a methylene group in the 3- or 5-position of a 1,2-dithiolium cation. This reaction may be performed in acetic anhydride. The resulting Vilsmeier salt (**32**), treated with aqueous sodium hydroxide, gives an α-(1,2-dithiol-3-ylidene) aldehyde.[49]

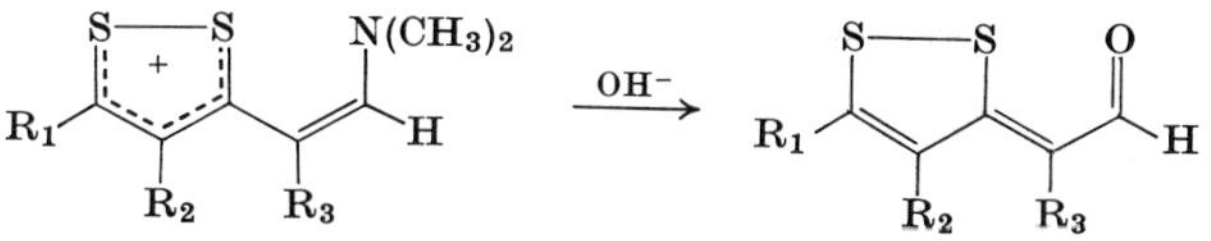

### F. Substitution or Modification of Substituents

A hydrogen atom on the carbon atom between the carbonyl and the dithiole ring can be substituted by a nitroso group by reaction with nitrous acid in acetic acid at 5°C.[50] Sulfuryl chloride can also replace this hydrogen atom with chlorine.[26] It is also replaced on Vilsmeier–Haack formylation of a (5-aryl-1,2-dithiol-3-ylidene)acetophenone.[39]

---

[47] A. Luttringhaus, E. Futterer, and H. Prinzbach, *Tetrahedron Lett.* 1209 (1963).
[48] A. Rouessac and J. Vialle, *Bull. Soc. Chim. Fr.* 2054 (1968).
[49] J. G. Dingwall, S. McKenzie, and D. H. Reid, *J. Chem. Soc. C* 2543 (1968).
[50] R. J. S. Beer, D. Cartwright, R. J. Gait, R. A. W. Johnstone, and S. D. Ward, *Chem. Commun.* 688 (1968).

In aqueous–alcoholic sodium hydroxide, α-(1,2-dithiol-3-ylidene) β-keto esters are hydrolyzed. The corresponding ketocarboxylic acid,

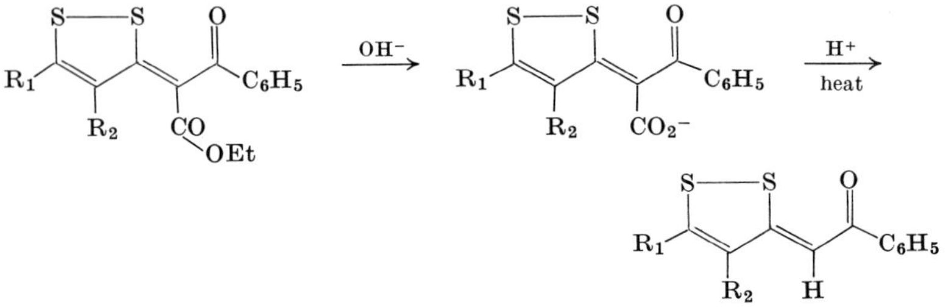

liberated by acidification, loses carbon dioxide by heating (Eq. 18).[18, 19]

Alkyl α-(1,2-dithiol-3-ylidene)malonates may be partially hydrolyzed; the resulting acid is easily decarboxylated (Eq. 19).[51]

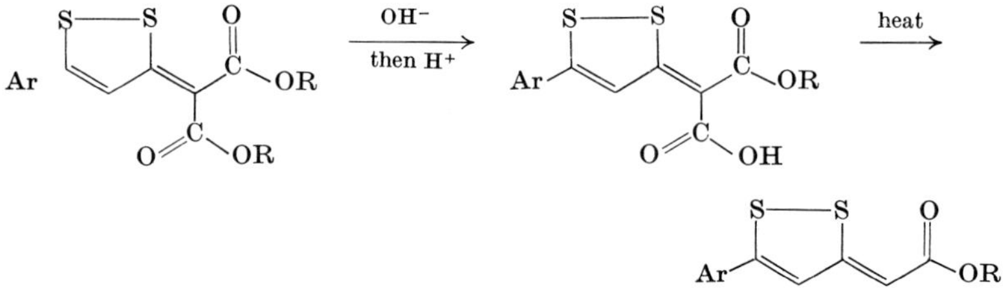

## III. Preparation of 1,6,6a$S^{IV}$-Trithiapentalenes

### A. Formation of the Heterocyclic System

#### 1. From β,δ-Triketones

Reaction of phosphorus pentasulfide with β,δ-triketones leads to 1,6,6a$S^{IV}$-trithiapentalenes.[1, 2, 8, 52] This was the first synthesis of trithiapentalenes known, but its development was delayed until practical methods for preparing β,δ-triketones were available. One of these procedures is the acylation of ketones in the presence of sodium hydride.[53] Another method begins with the diacylation of the enamine derived from a cyclanone.[38]

Several 1,6,6a$S^{IV}$-trithiapentalenes have now been obtained from triketones (Eq. 20).[2, 12, 38]

[51] G. Duguay, H. Quiniou, and N. Lozac'h, *Bull. Soc. Chim. Fr.* 2763 (1967).
[52] G. Traverso, *Ann. Chim. (Rome)* **45**, 687 (1955).
[53] M. L. Miles, T. M. Harris, and C. R. Hauser, *J. Org. Chem.* **30**, 1007 (1965).

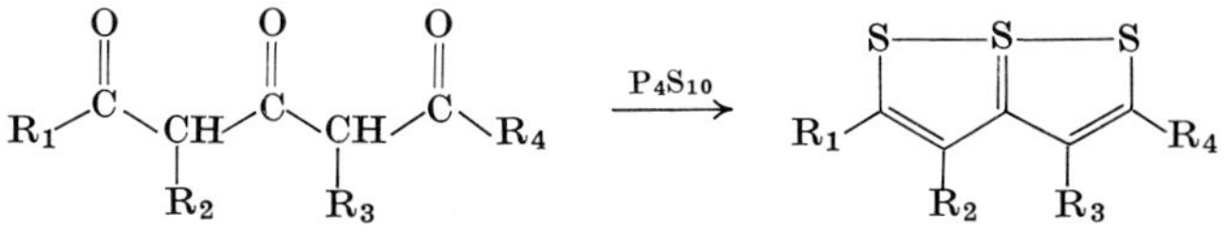

$1,6,6aS^{IV}$-TRITHIAPENTALENES

(20) [reaction scheme: triketone + P$_4$S$_{10}$ → trithiapentalene]

| R$_1$ | R$_2$ | R$_3$ | R$_4$ | Ref. |
|---|---|---|---|---|
| CH$_3$ | H | H | CH$_3$ | 1, 2 |
| C$_2$H$_5$ | H | H | C$_2$H$_5$ | 52 |
| CH$_3$ | H | H | Ar | 8, 12 |
| C$_6$H$_5$ | H, CH$_3$, or C$_6$H$_5$ | H, CH$_3$, or C$_6$H$_5$ | C$_6$H$_5$ | 8, 12, 29 |
| C$_6$H$_5$ | H | H | C$_6$H$_5$CH=CH– | 12 |
| C$_6$H$_5$CH=CH– | H | H | C$_6$H$_5$CH=CH– | 12 |
| C$_6$H$_5$ | H | H | 3-Pyridyl | 12 |
| Ar[a] | –CH$_2$CH$_2$– | | Ar | 12 |
| C$_2$H$_5$ | –CH$_2$CH$_2$CH$_2$– | | C$_2$H$_5$ | 38 |
| Ar | –CH$_2$CH$_2$CH$_2$– | | Ar | 12 |
| Ar | –CH$_2$CH(CH$_3$)–CH$_2$– | | Ar | 12 |

[a] Ar represents C$_6$H$_5$, $p$-CH$_3$C$_6$H$_4$, $p$-ClC$_6$H$_4$, or $p$-CH$_3$OC$_6$H$_4$.

## 2. From γ-Dithiopyrones

The ring of γ-dithiopyrones may be opened by aqueous sodium sulfide in dimethyl sulfoxide. By subsequent oxidation by potassium ferricyanide, trithiapentalenes are obtained in rather good yields, as in the following example (Eq. 21).[11]

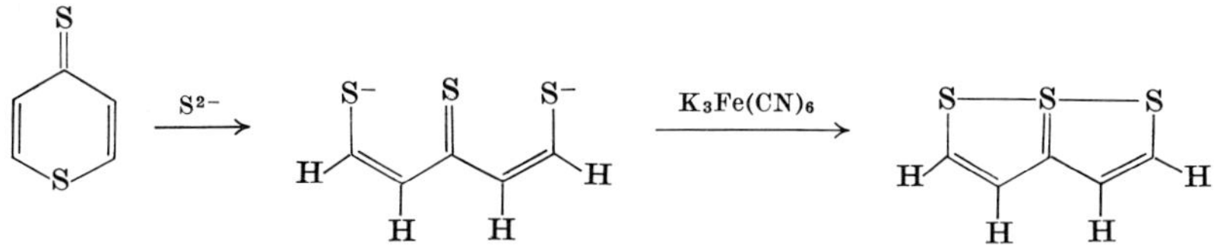

(21) [reaction scheme: γ-dithiopyrone → S$^{2-}$ → open-chain intermediate → K$_3$Fe(CN)$_6$ → trithiapentalene]

## 3. By Simultaneous Acylation and Sulfuration

In the presence of sodium acetate, thioacetic acid reacts with 2,4-pentanedione and various α-acetylenic ketones to give $1,6,6aS^{IV}$-trithiapentalenes (Scheme 8). In this reaction thioacetic acid reacts both as acylating and sulfurating agent.[54]

[54] H. Behringer and A. Grimm, *Ann. Chem.* **682**, 188 (1965).

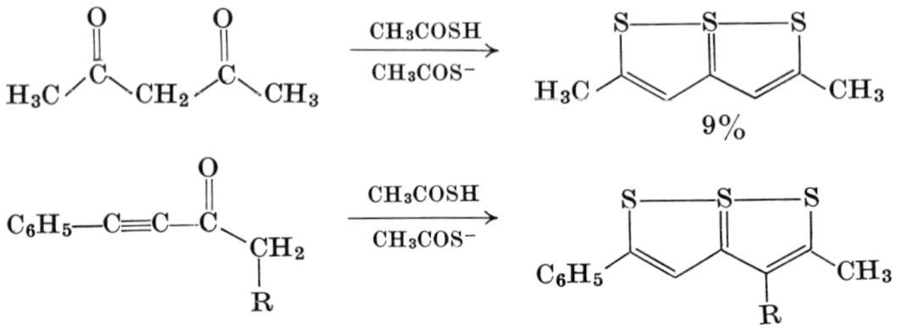

[SCHEME 8 reaction diagrams]

where R = H, the yield is 36%; R = CH$_3$, 39%; and R = C$_6$H$_5$, 81%.

SCHEME 8

When an alkyl acetylaroylacetate reacts with phosphorus pentasulfide in pyridine, a complex reaction occurs, involving sulfuration, degradation, and condensation, that leads to an alkyl 2,5-diaryl-1,6,6a$S^{IV}$-trithiapentalene-3-carboxylate in moderate yields (Eq. 22).[45]

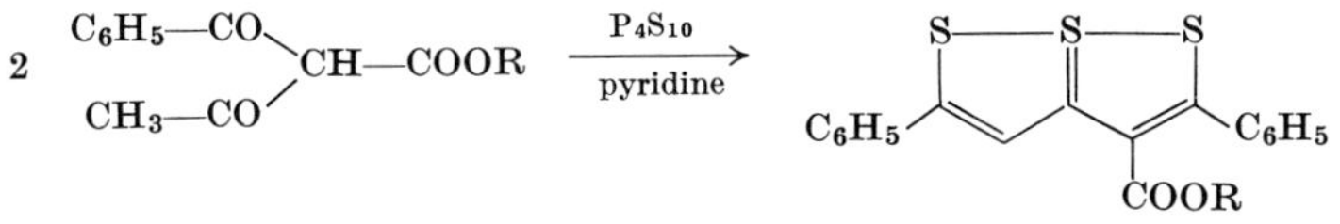

[Equation 22: reaction of acetylbenzoylacetate with P$_4$S$_{10}$ in pyridine giving 2,5-diphenyl trithiapentalene-3-carboxylate] (22)

### 4. *From α,α′-Bis(dialkylthiomethylene)ketones*

In the presence of a strong base, such as sodium *t*-amyloxide, one mole of carbon disulfide reacts with one mole of ketone R$_1$–CH$_2$–CO–CH$_2$–R$_2$, giving a dianion (**33**) which can be methylated to **34**.

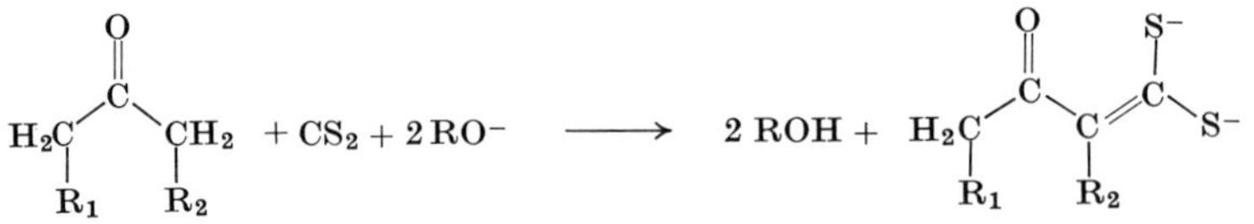

[Reaction scheme showing ketone + CS$_2$ + 2 RO$^-$ → 2 ROH + dianion (33); 33 + 2 CH$_3$I → 2 I$^-$ + (34)]

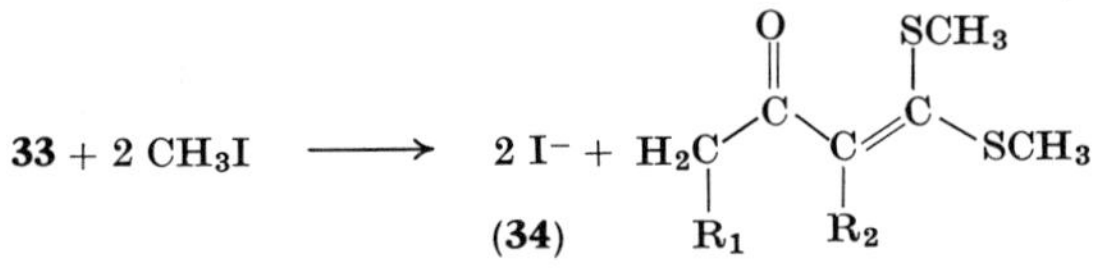

If **34** is, in turn, treated by carbon disulfide in basic medium, a dianion (**35**) is formed, which can lose a methanethiolate ion if conformation **36** is possible, giving the monoanion (**37**).[55-57]

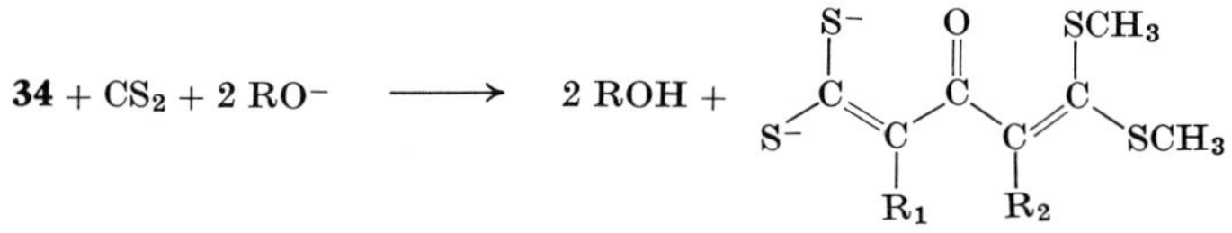

Anions (**35**) are easily methylated by methyl iodide, giving **38**. With medium-sized cyclanones ($C_5$ to $C_8$), the formation of **37** is not possible,[55] but with larger rings the thiopyrones can be obtained.[58] When the tetrakisalkylthio compound (**38**) is reacted with phosphorus pentasulfide, a bis(alkylthio)trithiapentalene is formed.[55]

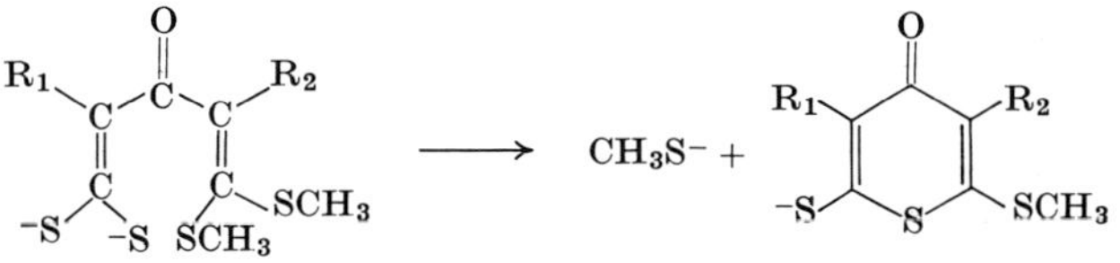

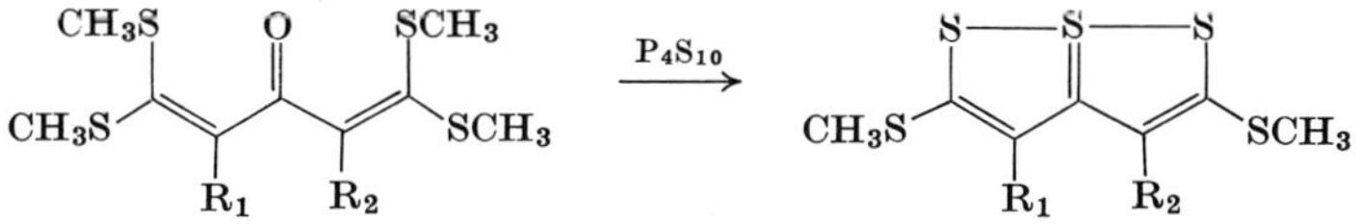

The fact that the final product is a trithiapentalene and not a dithiolethione has been proved by the following synthesis (Scheme 9).[55]

[55] A. Thuillier and J. Vialle, *Bull. Soc. Chim. Fr.* 2194 (1962).
[56] A. Thuillier and J. Vialle, *Bull. Soc. Chim. Fr.* 2182 (1962).
[57] A. Thuillier and J. Vialle, *Bull. Soc. Chim. Fr.* 2187 (1962).
[58] C. Portail and J. Vialle, *Bull. Soc. Chim. Fr.* 3790 (1968).

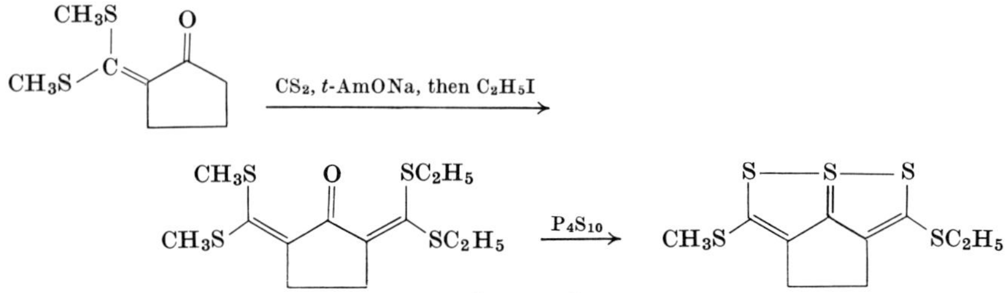

SCHEME 9

The final product contains one methylthio and one ethylthio group, and this would not have occurred if either of the two possible dithiole-thiones (**39** and **40**) had been obtained.

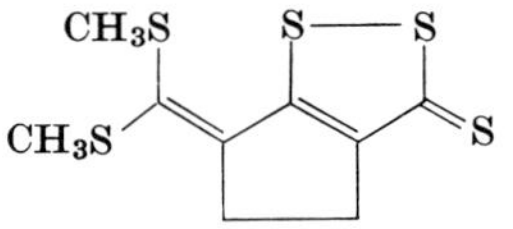

(**39**) 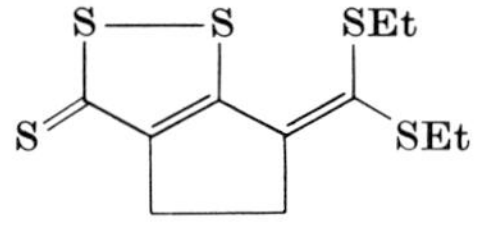   (**40**)

## B. REACTION OF DITHIOLE DERIVATIVES WITH REACTIVE METHYLENE GROUPS

5-Methyl or 5-methylene substituents in a 1,2-dithiole-3-thione are particularly reactive. In the presence of *t*-amyloxide, carbon disulfide reacts with such groups giving a dianion which, when treated by methyl iodide, forms a bis(methylthio)trithiapentalene (Scheme 10).[59, 60]

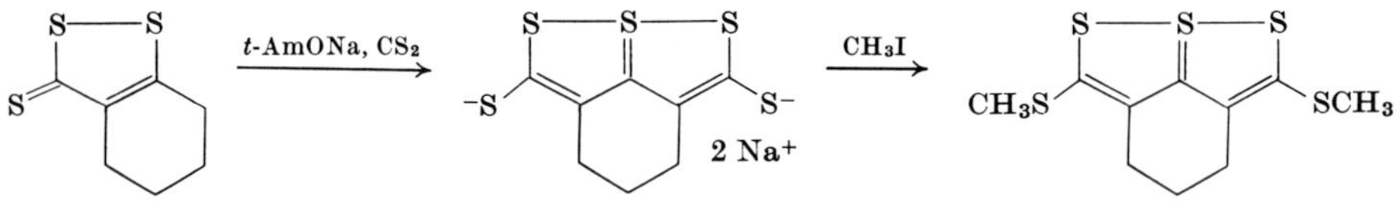

SCHEME 10

A similar reaction is observed with sodium hydride as basic agent in tetrahydrofuran (THF) (Scheme 11).[61]

[59] C. Portail and J. Vialle, *Bull. Soc. Chim. Fr.* 451 (1964).
[60] C. Portail and J. Vialle, *Bull. Soc. Chim. Fr.* 3187 (1966).
[61] R. J. S. Beer, D. Cartwright, and D. Harris, *Tetrahedron Lett.* 953 (1967).

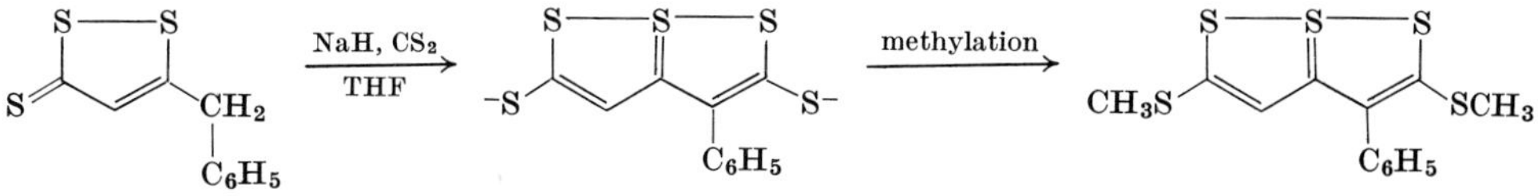

SCHEME 11

3-Methyl-1,2-dithiolium cations also react with dithio esters which serve as acylating agents (Eq. 23).[41]

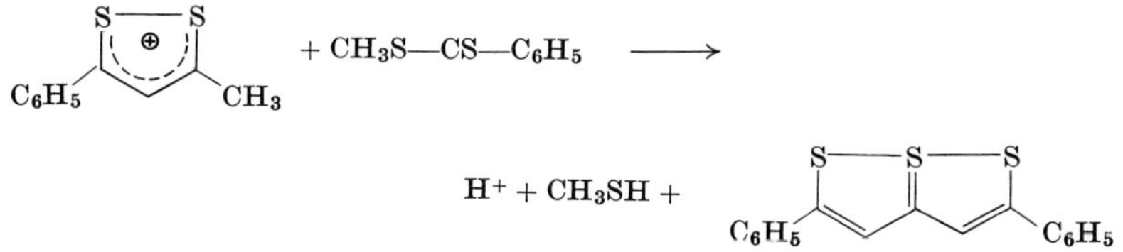

$$H^+ + CH_3SH + \quad (23)$$

An indirect method, using the reactivity of methylene groups substituted in position 3 of 1,2-dithiolium cations, is given later (Section III, E, 3).

### C. Reaction of Active Methylene Compounds with Dithiole Derivatives

#### 1. 1,2-Dithiolium Cations

3-Aryl-1,2-dithiolium salts are attacked by triethylamine in aqueous ethanol. Apparently one molecule of salt is first attacked by water, with loss of sulfur. The hypothetical 3-aryl-3-thioxopropanal (or its corresponding enol) would then react on another molecule of dithiolium salt to give a 2,5-diaryl-1,6,6a$S^{IV}$-trithiapentalene-3-carbaldehyde (Scheme 12).[62]

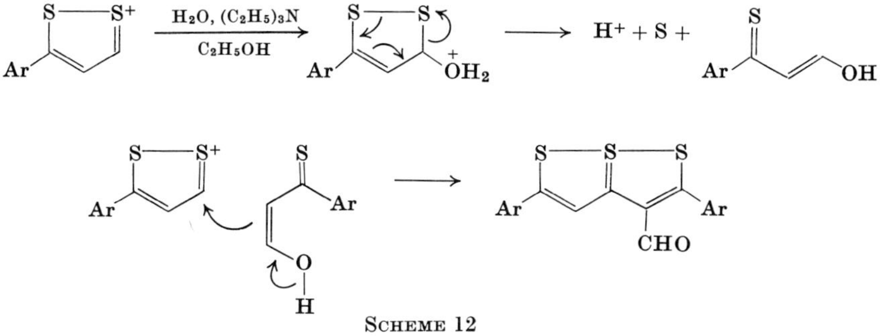

SCHEME 12

[62] J. Bignebat and H. Quiniou, *C.R. Acad. Sci. C* **267**, 180 (1968).

The structure of this aldehyde was proved by H. Quiniou through an independent synthesis—formylation of the diaryltrithiapentalene by dimethylformamide and phosphorus oxychloride (see Section III,F,1).

When a 3-aryl-1,2-dithiolium cation attacks methyl benzoyldithioacetate, two reactions may occur, probably with a common intermediate (**41**).[63] In a neutral medium, **41** undergoes only dehydrogenation, probably by the action of unreacted dithiolium cations, as it is often seen with carbonyl compounds. In this way, the same trithiapentalene **42** is obtained as with the corresponding 5-aryl-3-methylthio-1,2-dithiolium cation (see Section III,C,2).

In pyridine, however, the same reactants do not give a trithiapentalene. In this basic solvent, a nucleophilic attack of **41** prevails and a benzoyl dithiopyrone is obtained (Scheme 13).

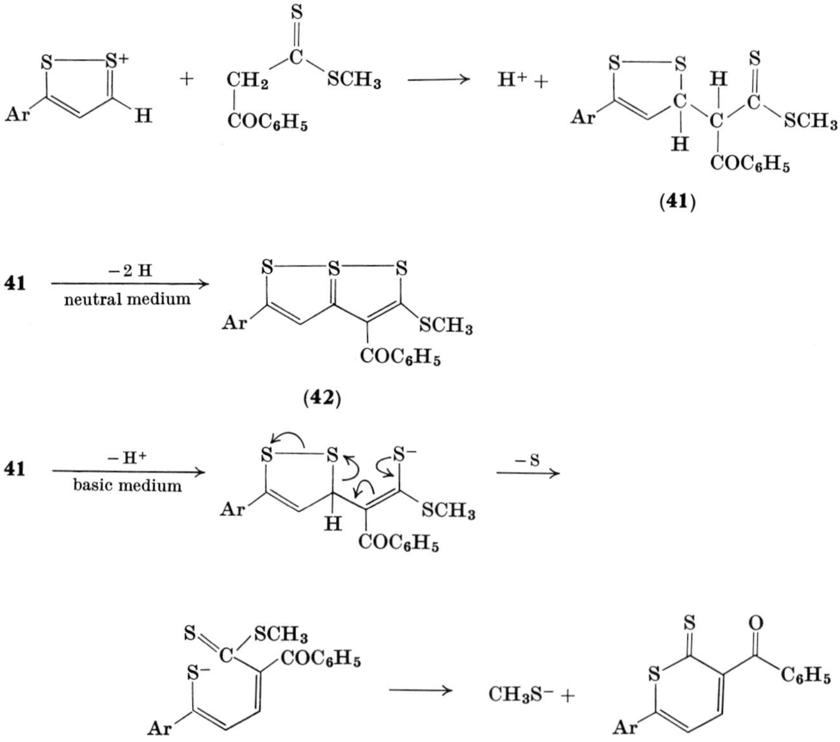

SCHEME 13

[63] F. Clesse and H. Quiniou, *C.R. Acad. Sci. C* **268**, 637 (1969).

## 2. 3-Methylthio-1,2-dithiolium Cations

By heating a 5-aryl-3-methylthio-1,2-dithiolium cation in anhydrous pyridine, a 5-aryl-1,2-dithiole-3-thione is obtained together with $N$-methylpyridinium cation.[25] However, in the presence of water, the dithiole ring may be opened, giving an acyldithioacetic ester which condenses with unreacted dithiolium cation.[32, 63a] This mechanism has been proved by reacting methyl benzoyldithioacetate with 5-phenyl-3-methylthio-1,2-dithiolium methosulfate in acetic acid containing a little pyridine (Eq. 24).[32]

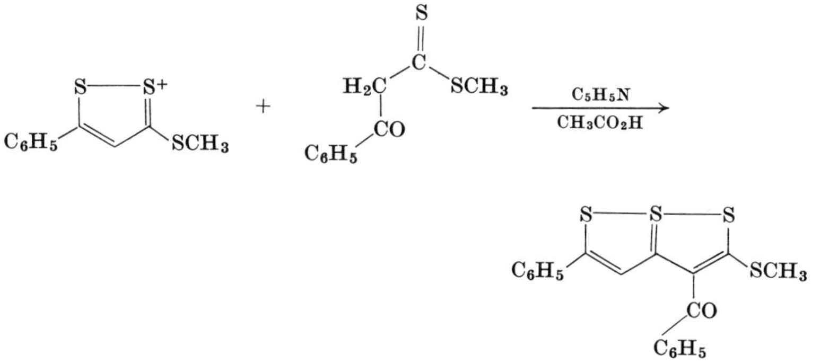

$$\text{(24)}$$

In like fashion, methyl ethoxycarbonyldithioacetate gives 3-ethoxycarbonyl-2-methylthio-5-phenyl-1,6,6a$S^{IV}$-trithiapentalene (**43**)[46, 61] and cyanothioacetamide gives 2-amino-3-cyano-5-phenyl-1,6,6a$S^{IV}$-trithiapentalene (**44**).[30, 64]

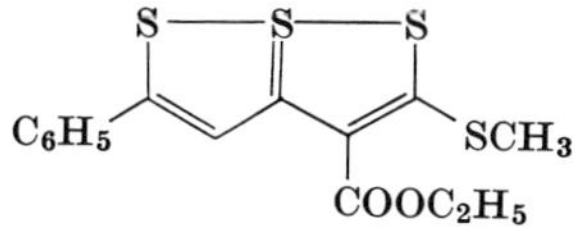

(**43**) 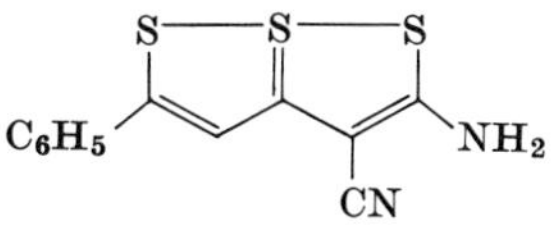   (**44**)

Another semidegradative synthesis is the reaction of 3-methylthio-1,2-benzodithiolium sulfate with malonic acid. It does not give, as initially thought,[65] a 3-[(1,2-benzodithiol-3-ylidene) methyl]-1,2-benzodithiolium ion, but in fact a 2-(1,2-benzodithiol-3-ylidene)-2$H$ benzo[$b$]thiophene-3-thione (**45**).[40]

---

[63a] C. Bouillon and J. Vialle, *Bull. Soc. Chim. Fr.* 4560 (1968).
[64] H. Behringer and R. Wiedenmann, *Tetrahedron Lett.* 3705 (1965).
[65] U. Schmidt, *Ann. Chem.* **635**, 109 (1960).

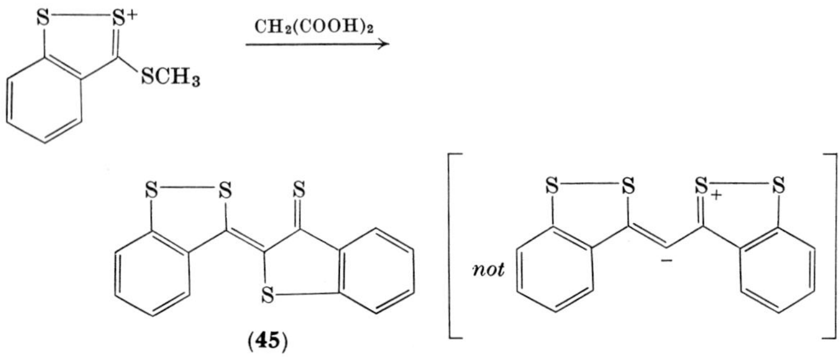

### 3. 3,5-*Bismethylthio*-1,2-*dithiolium Cations*

These compounds react just as the corresponding monomethylthio derivatives do.[46, 61] Bismethylthiotrithiapentalenes are obtained in this way (Eq. 25).

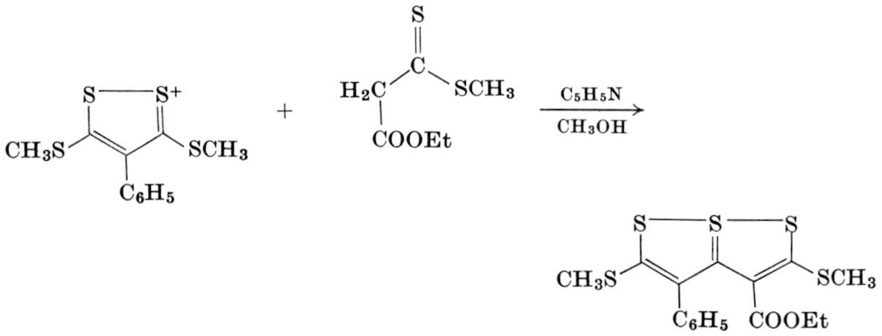

With methylbenzoyldithioacetate, the same condensation is unexpectedly accompanied by loss of the benzoyl group (Eq. 26).[46, 61]

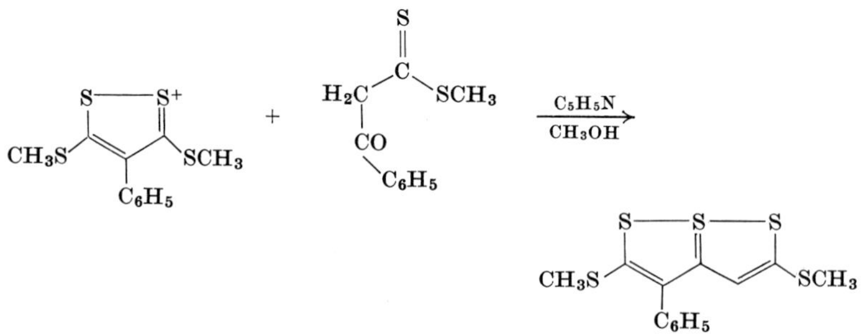

The debenzoylation is probably not complete.[46] Moreover, noticeable quantities of the bismethylthio salt are converted into 5-methylthio-4-phenyl-1,2-dithiole-3-thione,[46] probably by pyridine demethylation.

## D. Condensation of 1,2-Dithiole-3-thiones with Acetylenes

Various acetylenes give a 1,3-dipolar addition with 1,2-dithiole-3-thiones,[64, 66] leading to a 1,3a$S^{IV}$,4-trithiapentalene ("isothiathiophthene") (Eq. 27).

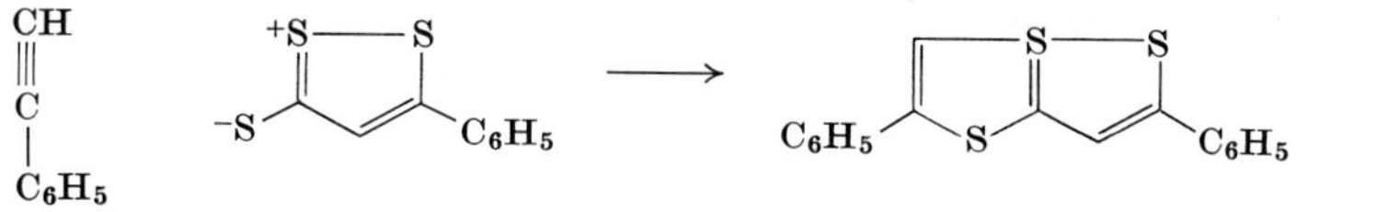

$$(27)$$

However, it has been shown[67-69] that the structure of the final product may depend on the experimental procedure. For example, if phenylacetylene reacts with 5-phenyl-1,2-dithiole-3-thione in xylene in the presence of hydrochloric acid, the 1,3a$S^{IV}$,4-trithiapentalene is obtained; but by using neutral dry xylene as solvent, a 1,6,6a$S^{IV}$-trithiapentalene is formed which can also be obtained by heating the 1,3a$S^{IV}$,4-isomer with phosphorus pentasulfide in tetrahydronaphthalene (Scheme 14).[67, 68]

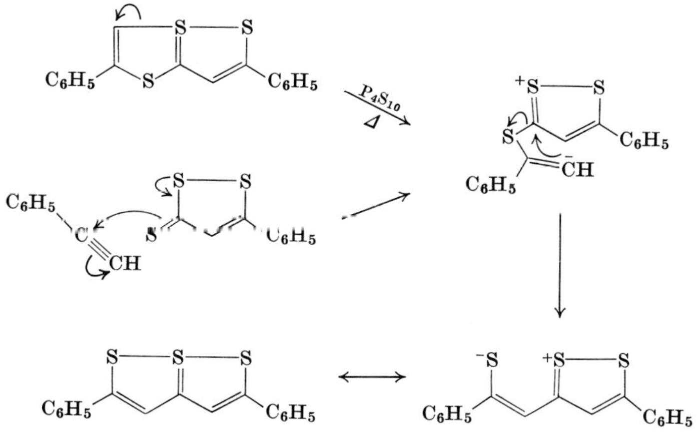

SCHEME 14

[66] D. B. J. Easton and D. Leaver, *Chem. Commun.* 585 (1965).
[67] H. Davy, M. Demuynck, D. Paquer, A. Rouessac, and J. Vialle, *Bull. Soc. Chim. Fr.* 1150 (1966).
[68] H. Davy, M. Demuynck, D. Paquer, A. Rouessac, and J. Vialle, *Bull. Soc. Chim. Fr.* 2057 (1968).
[69] H. Behringer, D. Bender, J. Falkenberg, and R. Wiedenmann, *Chem. Ber.* **101**, 1428 (1968).

### E. Modification of Other α-(1,2-Dithiol-3-ylidene) Functions

#### 1. α-(1,2-Dithiol-3-ylidene)ketones and -amides

Reaction of phosphorus pentasulfide has frequently been used for preparing trithiapentalenes from the corresponding 1,2-dithiol-3-ylideneketones. Various solvents have been used, e.g., benzene,[3, 16–19, 27] toluene,[19, 29, 45] xylene,[19, 29] pyridine,[18, 35] and diglyme.[38]

This synthesis gave the first chemical proof of the symmetry of the trithiapentalene system, the same trithiapentalene being obtained by reacting phosphorus pentasulfide with two different dithiolylidene ketones[16] (Scheme 15).

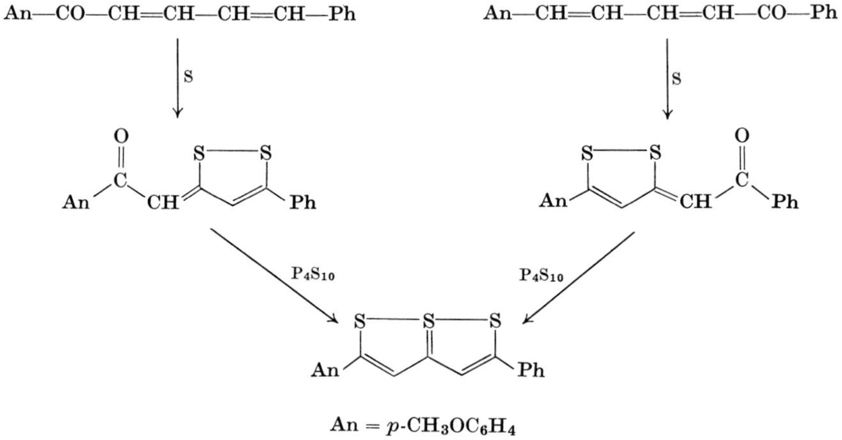

$$An = p\text{-}CH_3OC_6H_4$$

Scheme 15

Dithiolyliophenolates do not react with phosphorus pentasulfide under the same conditions and benzotrithiapentalenes cannot be obtained in this way. This sulfuration also seems difficult if a chlorine atom is substituted on the carbon between the dithiole ring and the carbonyl.[26]

α-(1,2-Dithiol-3-ylidene)amides give 2-amino-1,6,6a$S^{IV}$-trithiapentalenes in a similar way (Eq. 28).[26]

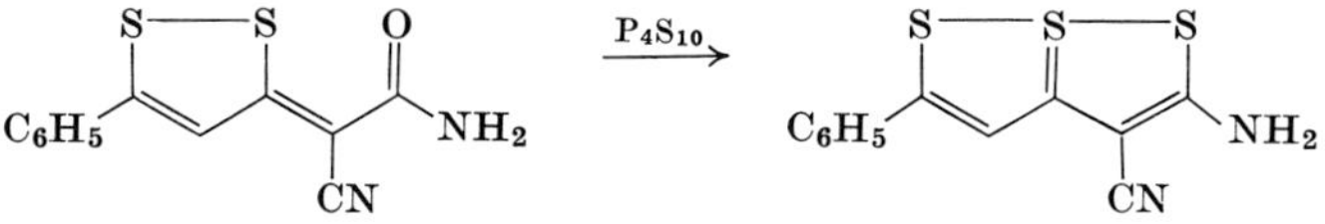

## 2. α-(1,2-Dithiol-3-ylidene)malononitriles

These nitriles, when treated with hydrogen sulfide in pyridine and triethylamine, give 2-amino-3-cyano-1,6,6a$S^{IV}$-trithiapentalenes (Eq. 29).[48]

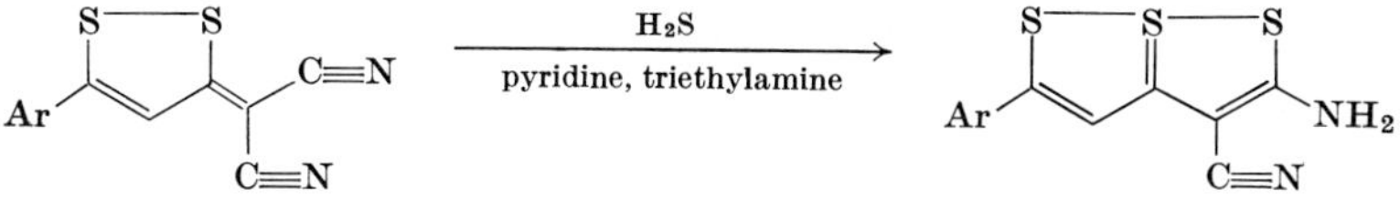

## 3. α-(1,2-Dithiol-3-ylidene)alkyliminium Salts

In boiling acetic anhydride or in the presence of phosphorus oxychloride,[70] 1,2-dithiolium cations with 3- (or 5-) R–CH$_2$ substituents react with dimethylthioformamide, giving a Vilsmeier salt. This compound reacts smoothly with sodium hydrogen sulfide to give a trithiapentalene (Scheme 16).[49, 70, 71]

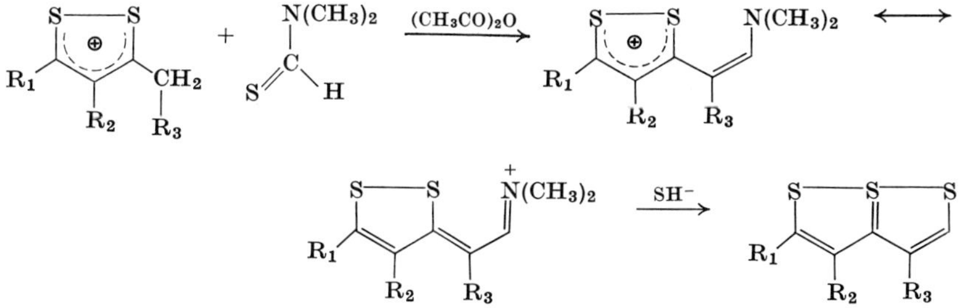

SCHEME 16

### F. SUBSTITUTION OR MODIFICATION OF SUBSTITUENTS

These reactions are directed by the fact that electrophilic reagents attack in position 3 or 4, whereas nucleophilic reagents attack in position 2 or 5.

[70] J. G. Dingwall, D. H. Reid, and K. Wade, *J. Chem. Soc. C* 913 (1969).
[71] J. G. Dingwall and D. H. Reid, *Chem. Commun.* 863 (1968).

## 1. Electrophilic Substitutions

In an inert solvent, bromine can replace a 3- or 4-hydrogen atom on the trithiapentalene system.[50, 61] Nitration and nitrosation in position 3 (or 4) can also be performed.[50] However, attempted nitration or nitrosation of 2,5-diphenyl-1,6,6a$S^{IV}$-trithiapentalene led to simultaneous oxidation of S-1 and nitrosation on C-3.[50]

Formylation according to the Vilsmeier–Haack method occurs also in position 3 (or 4) in good yield (Eq. 30).[62, 72] $N,N$-Dimethylthioformamide may advantageously replace $N,N$-dimethylformamide (Eq. 31).[70]

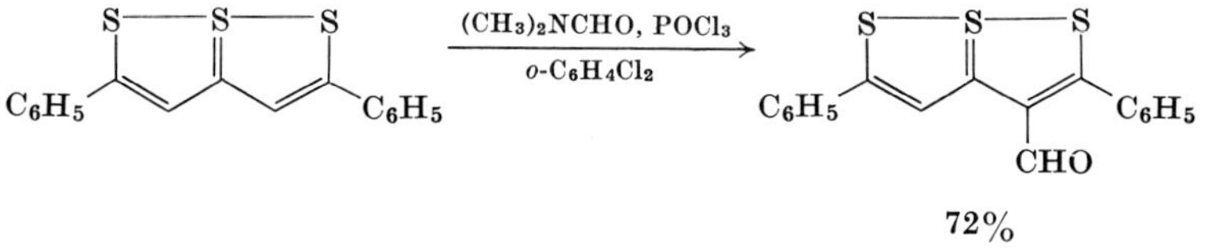

$$\text{(30)}$$

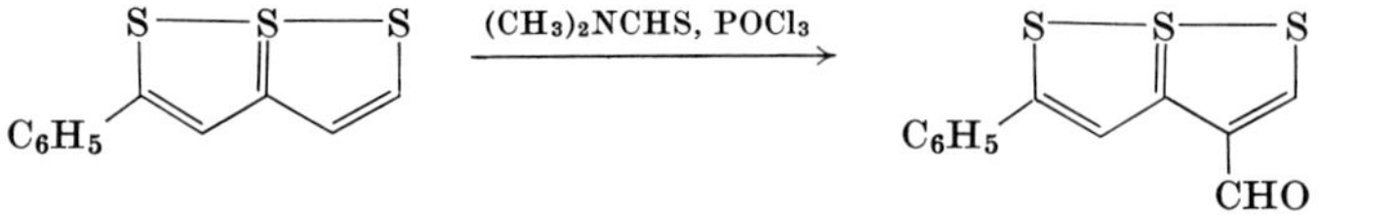

$$72\%$$

$$\text{(31)}$$

## 2. Nucleophilic Substitution

Methylthio substituents in position 2 or 5 are replaced by ethoxy groups on reaction with sodium ethoxide.[46, 50] Similarly, aliphatic amines replace methylthio substituents by alkylamino groups.[50]

## 3. Hydrolysis of Cyano or Ester Groups

A cyano group in position 3 may be hydrolyzed and decarboxylated by heating with hydrogen bromide in aqueous acetic acid.[38] Similarly, ethoxycarbonyl substituents in position 3 may be eliminated by heating with a small quantity of aqueous hydrochloric or hydrobromic acid in acetic acid.[46, 61]

## 4. Reactions of Active Methylene Groups in Side Chains

A methyl or methylene fixed in position 2 or 5 on the trithiapentalene system reacts with aromatic aldehydes in the presence of a weak base such as piperidine. Styryl-1,6,6a$S^{IV}$-trithiapentalenes are obtained in this way (Eq. 32).[73]

[72] J. Bignebat and H. Quiniou, *C.R. Acad. Sci. C* **269**, 1129 (1969).
[73] M. Stavaux and N. Lozac'h, *Bull. Soc. Chim. Fr.* 2077 (1968).

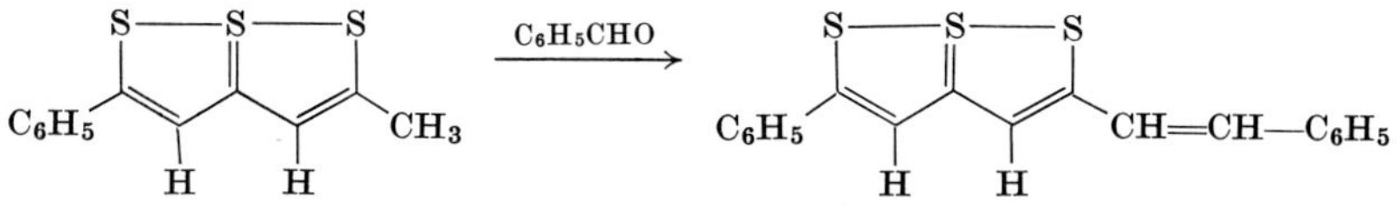

[Diagram of Eq. (32): condensation with C6H5CHO]

$$\text{(32)}$$

Condensation of carbon disulfide on the same methylene substituents is described in Section V.

## 5. Oxidation

The trithiapentalene system is not attacked as easily as free thione groups, as is shown in the potassium permanganate oxidation in the following reaction (Eq. 33).[35]

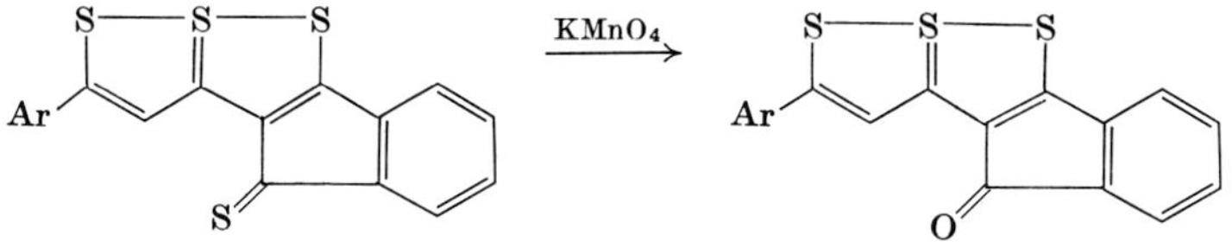

[Diagram of Eq. (33): KMnO4 oxidation]

$$\text{(33)}$$

## IV. Other Heterocycles Derived from 1,6,6a$S^{IV}$-Trithiapentalenes

### A. THIOPHENE DERIVATIVES

Methyl or methylene groups at positions 2 and 5 are activated by the partial positive charge in these positions on the trithiapentalene system. The corresponding carbanions can rearrange[2, 74, 75] (Scheme

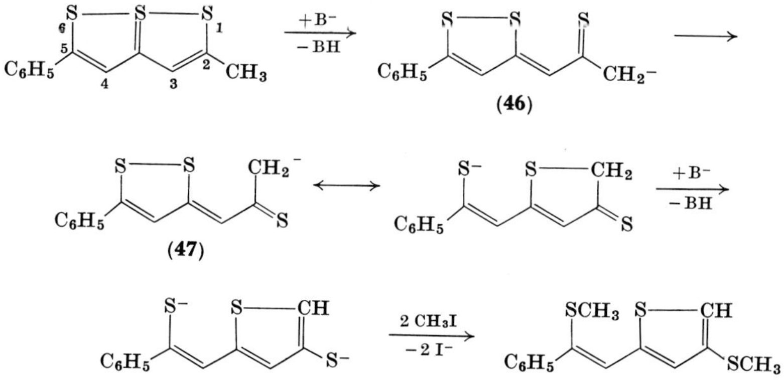

[Scheme 17 structural diagrams showing compounds (46), (47), and subsequent rearrangements with +B⁻/−BH and 2 CH₃I/−2 I⁻]

SCHEME 17

[74] M. Stavaux and N. Lozac'h, *Bull. Soc. Chim. Fr.* 3557 (1967).
[75] F. Arndt and W. Walter, *Chem. Ber.* **94**, 1757 (1961).

17). This rearrangement implies a change of conformation, from **46** to **47**, corresponding to a rotation around the bond between carbon atoms 2 and 3 in the trithiapentalene structure. If this rotation is prevented, for instance, if the carbon atoms are part of a six-membered carbon ring, the carbanion (**46**) is stable enough to condense with carbon disulfide in reasonable yield.[74] This reaction is described in the synthesis of extended structures (Section V).

## B. Thiopyrone Derivatives

In dimethylformamide, $1,6,6aS^{IV}$-trithiapentalenes having a free 2- or 5-position are attacked by hydrosulfide or sulfide ion and the trithiapentalene system rearranges into a γ-dithiopyrone.[71] The following mechanism has been suggested (Scheme 18).

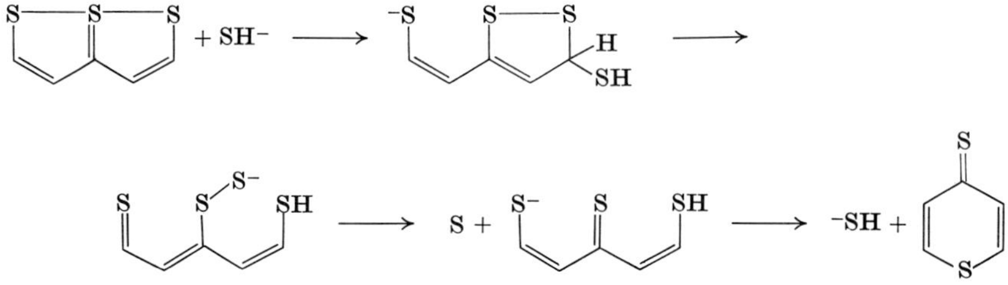

SCHEME 18

This mechanism has been preferred to the reductive scission of S–S bonds because the rearrangement can also be promoted by OH⁻.[71]

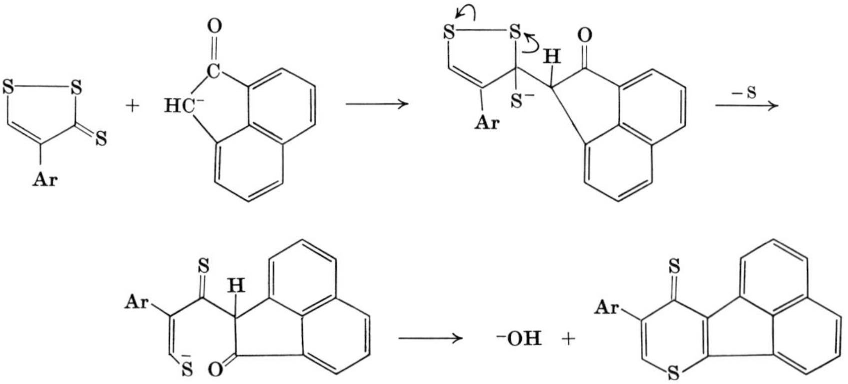

SCHEME 19

This transposition explains why the condensation in alkaline media of 4-aryl-1,2-dithiole-3-thiones with acenaphthenone gives a γ-dithiopyrone instead of a dithiolylidene ketone.[24] A possible mechanism is as in Scheme 19.

## C. Positive Ions Derived from 1,6,6a$S^{IV}$-Trithiapentalenes

Sulfur atoms in positions 1 and 6 in the trithiapentalene system may sometimes react as the thione sulfur atom of 1,2-dithiole-3-thiones. For instance, 2,4-diphenyl- and 2,3,4-triphenyl-1,6,6a$S^{IV}$-trithiapentalene add methyl iodide (Eq. 34).[76]

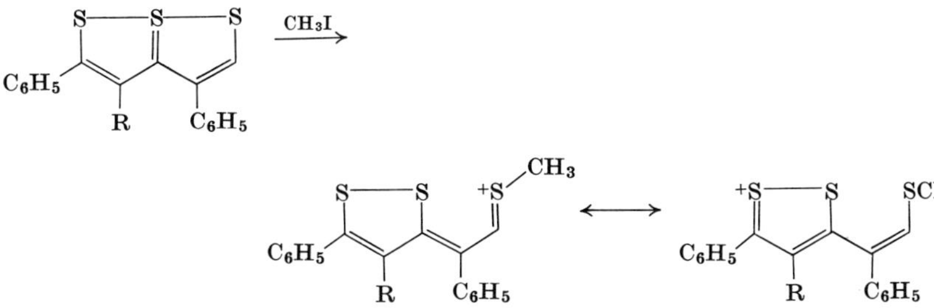

$$R = H, C_6H_5 \tag{34}$$

S-Alkylation of 2,5-dimethyl- or of 2,5-diphenyl-1,6,6a$S^{IV}$-trithiapentalene is more difficult and it is necessary to resort to triethyloxonium tetrafluoroborate.[77] A cyano substituent in the 3-position favors alkylation on sulfur atom in the 1-position (Eq. 35).[77]

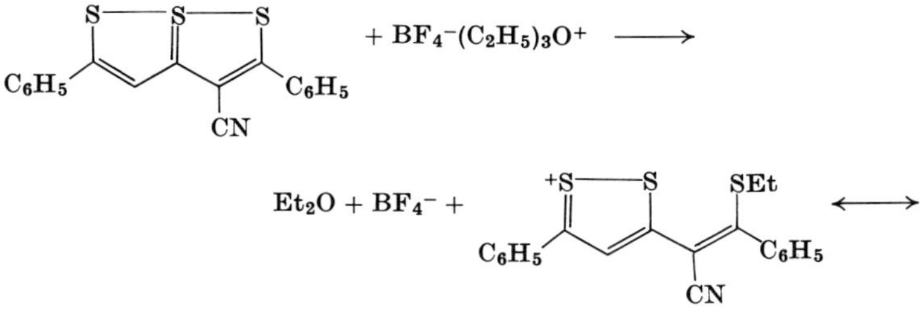

$$\tag{35}$$

[76] E. Klingsberg, *J. Org. Chem.* **33**, 2915 (1968).
[77] H. Behringer and J. Falkenberg, *Chem. Ber.* **102**, 1585 (1969).

The reactivity of these positive ions toward aromatic amines is considered in Section IV,D. They also react in acetic acid containing a small quantity of pyridine with active methylene compounds such

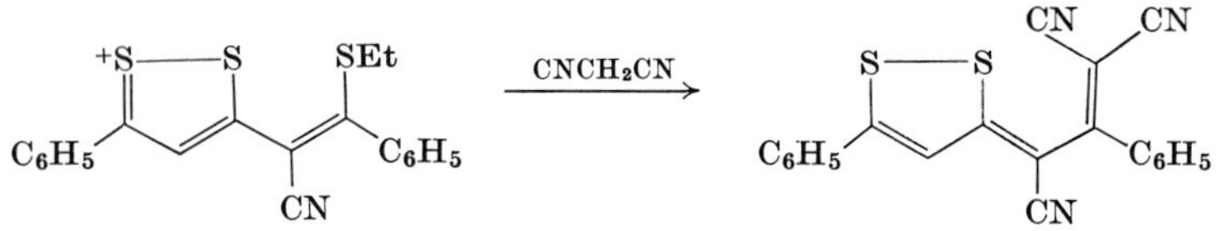

$$\text{(36)}$$

as malononitrile, ω-cyanoacetophenone, or N-methylrhodanine (Eq. 36). Treatment of the condensed ring dithiolylideneketone (**48**) with triphenylmethyl tetrafluoroborate gave a 7-acyl-1,2-benzodithiolium cation (**49**).[78]

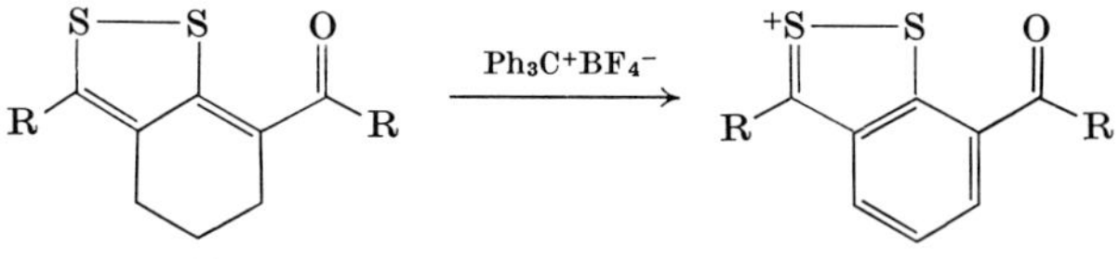

$R = C_6H_5, p\text{-}CH_3C_6H_4, p\text{-}CH_3OC_6H_4, \text{ or } SCH_3$

In a similar way, trithiapentalene (**50**) leads to the positive ion (**51**). NMR shows magnetic equivalence of the two groups R in the compounds **51** (where R is p-tolyl, p-methoxyphenyl, or methylthio).[78]

At the present time it cannot be said whether this apparent symmetry of **51** is due to resonance or to a rapidly established equilibrium.

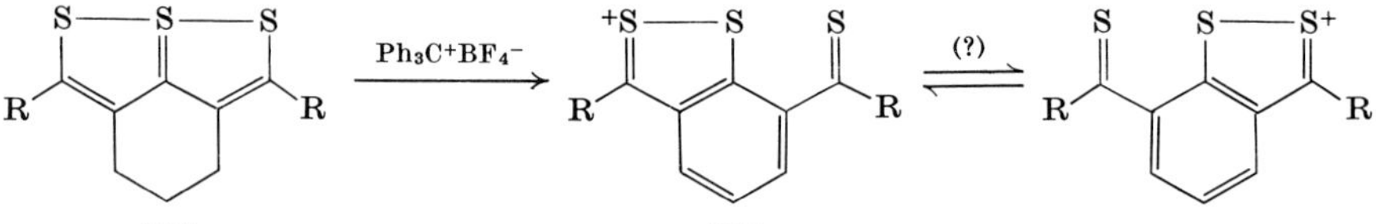

According to their structure, α-(1,2-dithiol-3-ylidene) ketones are more or less easily protonated. This protonation is often effected by perchloric acid and a crystallized perchlorate may be obtained. This has been observed with dithiolyliophenolates[20, 31, 33] and various α-(5-aryl-1,2-dithiol-3-ylidene)cyclanones.[13, 21, 27]

[78] E. I. G. Brown, D. Leaver, and T. J. Rawlings, *Chem. Commun.* 83 (1969).

## D. Nitrogen Compounds Analogous to 1,6,6a$S^{IV}$-Trithiapentalenes

Some compounds have been prepared in which one or two carbon atoms of the trithiapentalene system have been replaced by nitrogen.

5-Aryl-1,2,4-dithiazole-3-thiones react with phenylacetylene giving two compounds, namely, *N*-(4-aryl-1,3-dithiol-2-ylidene) thiobenzamide (**52**) and 2,5-diaryl-1,6,6a$S^{IV}$-trithia-3-azapentalene, which may also be considered as *N*-(5-aryl-1,2-dithiol-3-ylidene)thiobenzamide (**53**). This compound is converted into the corresponding benzamide (**54**) by mercuric acetate oxidation, the opposite reaction being realized with phosphorus pentasulfide.[79]

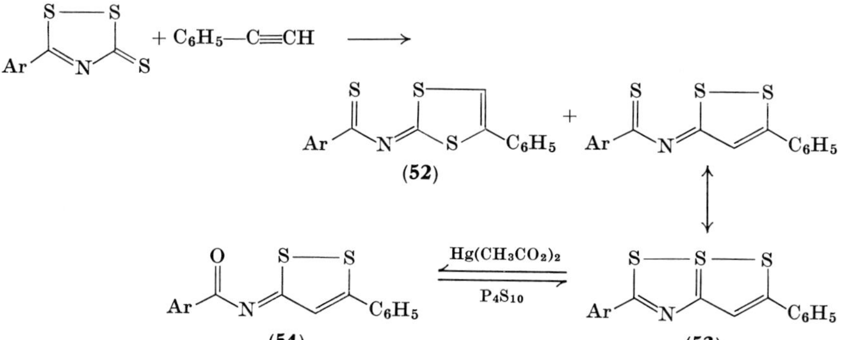

The benzamides (**54**) are also obtained by the following reactions (Scheme 20).[80]

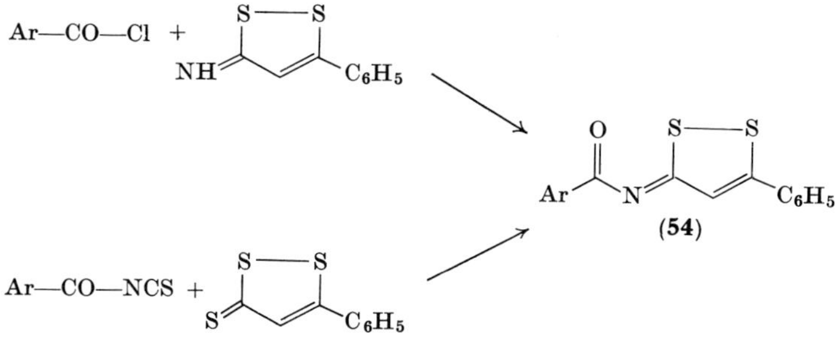

SCHEME 20

[79] G. Lang and J. Vialle, *Bull. Soc. Chim. Fr.* 2865 (1967).
[80] A. Grandin and J. Vialle, *Bull. Soc. Chim. Fr.* 1851 (1967).

2,5-Diaryl-1,6,6a$S^{IV}$-trithia-3,4-diazapentalenes (55) are obtained in good yields by reacting phosphorus pentasulfide with $N,N'$-diaroyl-$S$-methylisothiourea.[81]

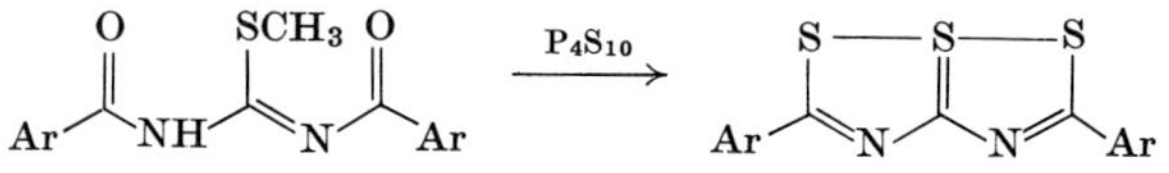

(55)

Arylisothiocyanates react in pyridine with 5-arylamino-1,2,4-dithiazol-3-imines (arylthiurets) at ordinary temperature, giving colorless trans compounds (56) whose structure is probably stabilized by hydrogen bonding. By heating, these *trans* compounds are converted into their cis isomers (57), which can be considered as 2,5-bis(arylamino)-1,6,6a$S^{IV}$-trithia-3,4-diazapentalenes (58).[82]

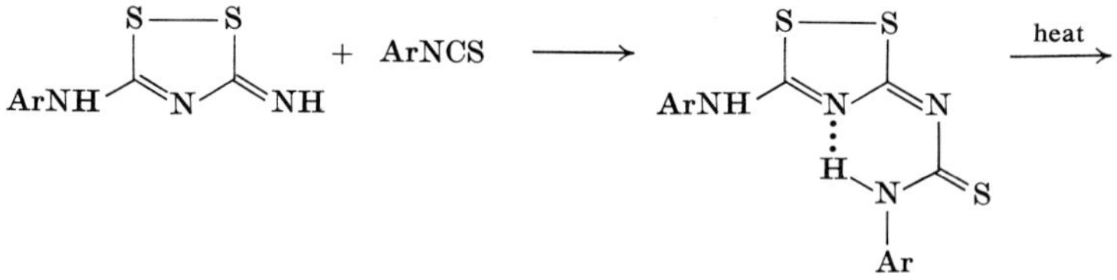

(56)

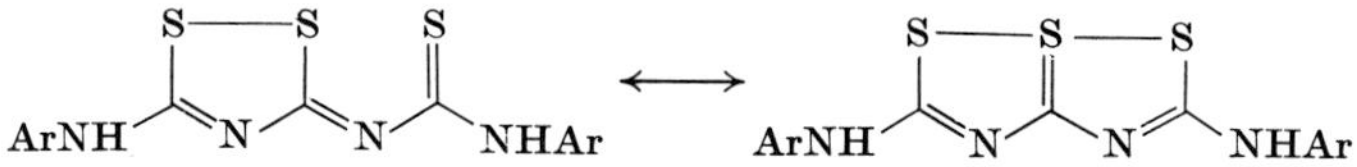

(57)    (58)

For these compounds, the bonding of the three sulfur atoms is probably quite similar to the bonding of the same atoms in 1,6,6a$S^{IV}$-trithiapentalenes.

Similarity is less evident when a sulfur atom is replaced by a nitrogen atom. The following structures have been described, generally under other names:

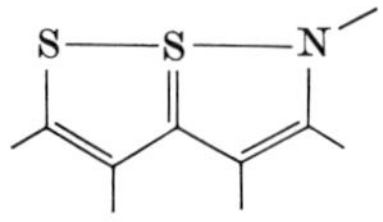

1,6a$S^{IV}$-Dithia-6-azapentalenes[76,77] 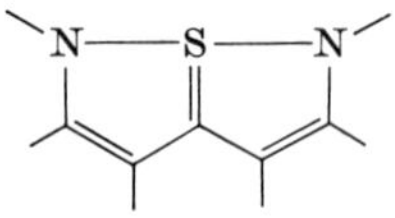    6a$S^{IV}$-Thia-1,6-diazapentalenes[83]

[81] J. L. Derocque, M. Perrier, and J. Vialle, *Bull. Soc. Chim. Fr.* 2062 (1968).
[82] H. Behringer and D. Weber, *Chem. Ber.* **97**, 2567 (1964).
[83] D. H. Reid and J. D. Symon, *Chem. Commun.* 1314 (1969).

For these systems we have adopted a bicyclic structure by analogy with trithiapentalenes and also to account for a recent NMR proof of the apparent symmetry of the last structure. It should be remembered, however, that this formulation is at present tentative and needs experimental confirmation.

Methyl iodide adducts of 2,4-diphenyl- or of 2,3,4-triphenyl-1,6,6a$S^{IV}$-trithiapentalene react with aniline giving a compound which can be considered as α-(1,2-dithiol-3-ylidene)aldimine (**59**) or perhaps also as α-(isothiazolin-5-ylidene)thioketone (**60**).[76]

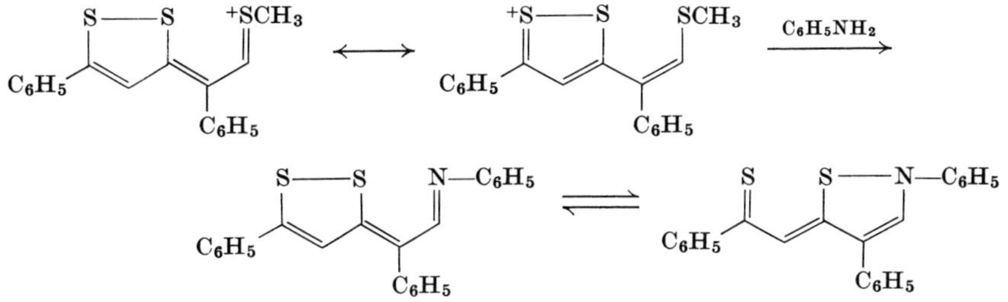

It cannot be said with certainty, at the present time, whether this system is an equilibrium, or is best described by a mesomeric structure (**61**) analogous to trithiapentalene (Section VI, A).

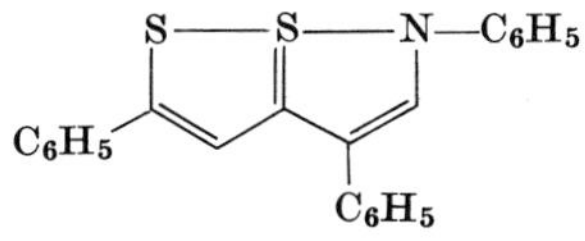

The same aldimine (**59** or **60**) is also obtained by reacting aniline with the sulfur dichloride adduct of the trithiapentalene.[76] Cations obtained with the use of triethyloxonium tetrafluoroborate react similarly (Eq. 37).[77]

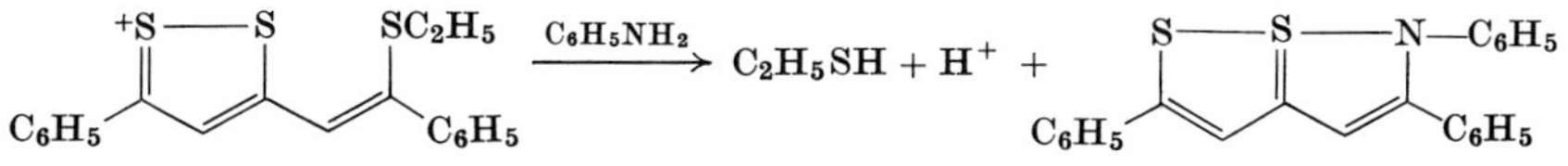

Compounds **61** may, in turn, react with methyl iodide, giving a cation (**62**).[76]

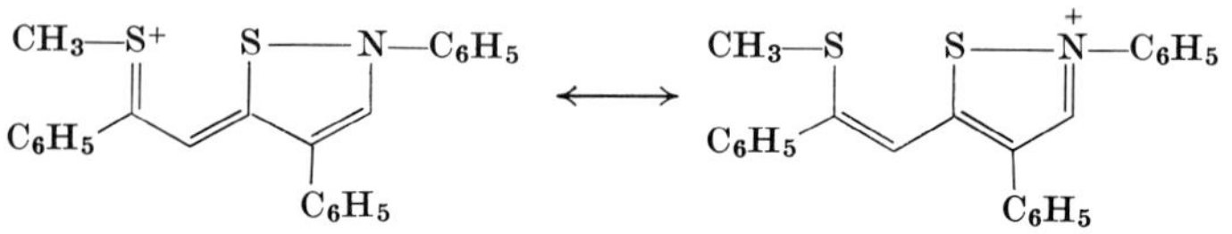

(**62**)

In another synthesis, the Vilsmeier salt (**63**) with aqueous methylamine gave the aza compound (**64**). The methyl iodide adduct of **64** was afterward reacted with methylamine, forming the diaza derivative (**65**).[83]

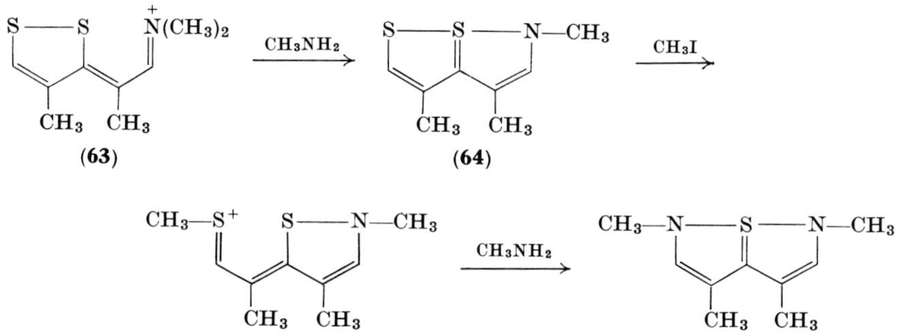

(**63**) (**64**)

(**65**)

It is interesting to note that the NMR spectrum of **65** even at $-70°$ corresponds to an apparently symmetrical structure, which is evidently in agreement with formula **65**, but could also be explained by an equilibrium between rapidly interconverting isomers.[83]

### E. Selenium Compounds Analogous to 1,6,6a$S^{IV}$-Trithiapentalenes

2,6-Dimethyl-γ-pyrone reacts with sodium selenide (or sodium hydrogen selenide) to give 2,6-dimethyl-4-seleno-γ-pyrone (**66**). This, in turn, reacts with sodium selenide to give 4,6-bis(hydroseleno)-3,5-heptadien-2-one (**67**), which is readily oxidized by air to 5-methyl-1,2-diselenol-3-ylidene acetone (**68**).[84, 85]

[84] M. Sanesi and G. Traverso, *Chem. Ber.* **93**, 1566 (1960).
[85] G. Traverso, *Ann. Chim.* (*Rome*) **47**, 3 (1957).

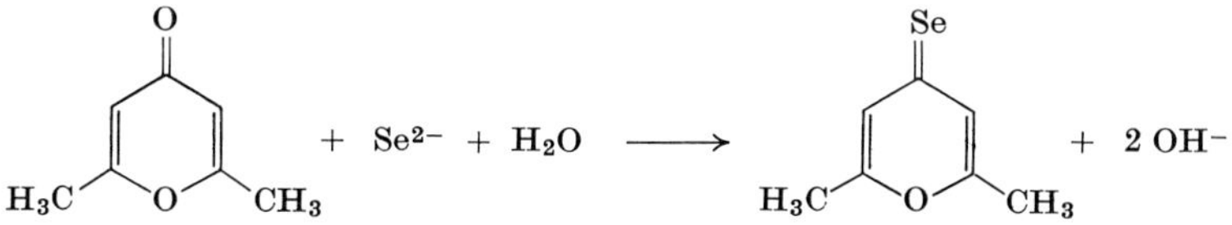

Sec. IV. E.]  1,6,6a$S^{IV}$-TRITHIAPENTALENES  203

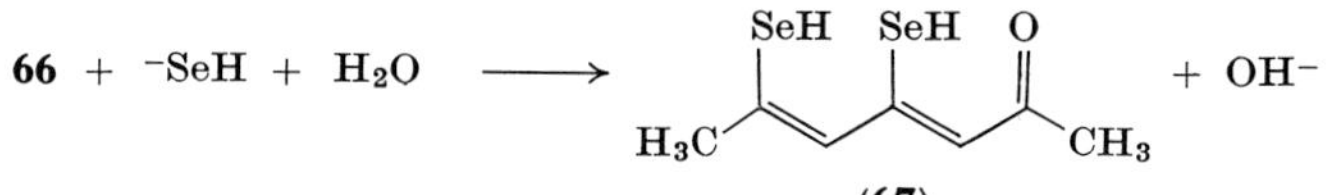

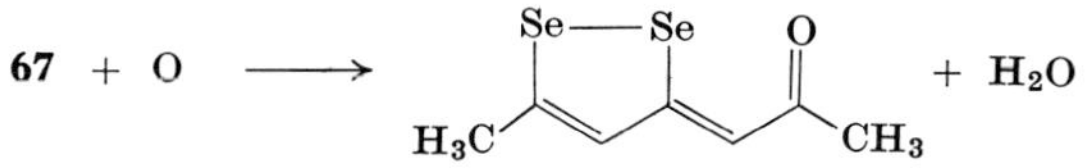

[Reaction scheme showing 4H-2,6-dimethylpyran-4-one + Se²⁻ + H₂O → selenone analog (**66**) + 2 OH⁻]

**66** + ⁻SeH + H₂O ⟶ H₃C–C(SeH)=CH–C(SeH)=CH–C(O)–CH₃ (**67**) + OH⁻

**67** + O ⟶ [Se–Se bridged compound] (**68**) + H₂O

Compound **67** or **68**, treated by phosphorus pentasulfide in boiling benzene, give 1-thia-6,6a$S^{IV}$-diselenapentalene (**69**).[85]

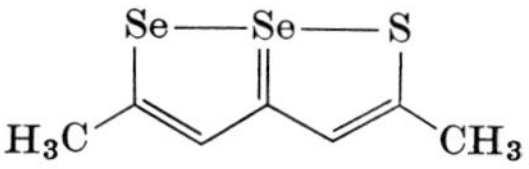

[Structure **69**: Se—Se—S ring system with two CH₃ groups]

2,4-Diphenyl-1,6a$S^{IV}$-dithia-6-selenapentalene has been prepared by two different methods: (a) reaction of phosphorus pentaselenide with 2-phenyl-(5-phenyl-1,2-dithiol-3-ylidene)ethanal[86] and (b) reaction of sodium hydrogen selenide with the Vilsmeier salt (**70**).[49]

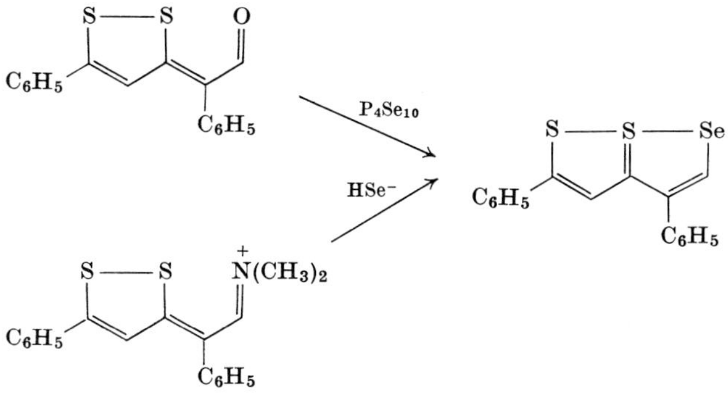

[Reaction scheme: aldehyde precursor + P₄Se₁₀ → 2,4-diphenyl-dithia-selenapentalene product; Vilsmeier salt (**70**) + HSe⁻ → same product]

[86] J. H. van den Hende and E. Klingsberg, *J. Amer. Chem. Soc.* **88**, 5045 (1968).

## V. Extended Structures

1,2-Dithiolium ions, 1,2-dithiole-3-thiones, and 1,6,6a$S^{IV}$-trithiapentalenes show structural relationships which can be further extended. Compounds of this series are of peculiar interest because of the structural problems involved concerning the nature of the chemical bond and the aromaticity of the ring systems.

When the compounds contain an odd number of sulfur atoms, they are neutral molecules, and monomeric thioketones may be considered as the first member of the series. Compounds containing an even number of sulfur atoms are positive ions. In the following formulas, only one resonance structure is given for the sake of simplicity.

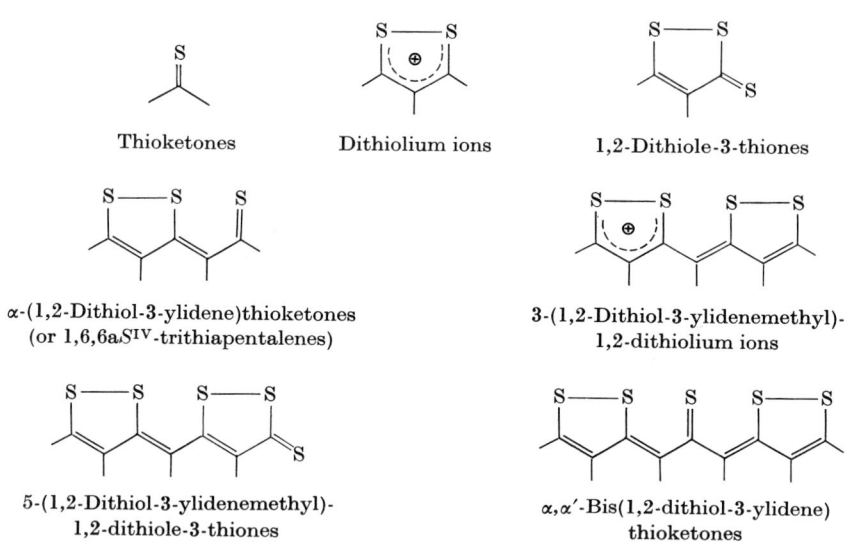

Thioketones

Dithiolium ions

1,2-Dithiole-3-thiones

α-(1,2-Dithiol-3-ylidene)thioketones (or 1,6,6a$S^{IV}$-trithiapentalenes)

3-(1,2-Dithiol-3-ylidenemethyl)-1,2-dithiolium ions

5-(1,2-Dithiol-3-ylidenemethyl)-1,2-dithiole-3-thiones

α,α'-Bis(1,2-dithiol-3-ylidene) thioketones

3-(1,2-Dithiol-3-ylidenemethyl)-1,2-dithiolium ions may be prepared by reacting a 3-methylthio-1,2-dithiolium ion with a 3-methyl-1,2-dithiolium ion, either in ethanol[87] or in acetic acid with catalytic amounts of pyridine (Eq. 38).[40]

[87] E. Klingsberg, *J. Heterocycl. Chem.* **3**, 243 (1966).

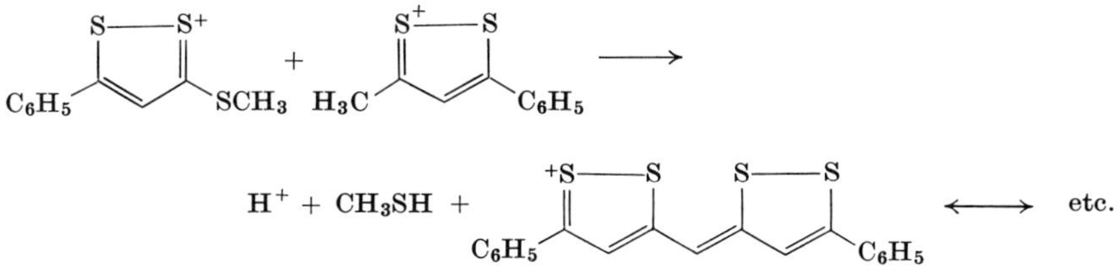

(38)

It has been noted earlier (Section IV, A) that carbon disulfide can be condensed with trithiapentalenes having a 2- or 5-methylene group, provided that the corresponding carbanion does not rearrange too easily.

For example, trithiapentalene (71) reacts with carbon disulfide in the presence of a strong base (t-AmONa; NaH) giving the dianion 72 which can be methylated to 73 or oxidized to a dithiolylidenealkyl-dithiolethione (74).[74]

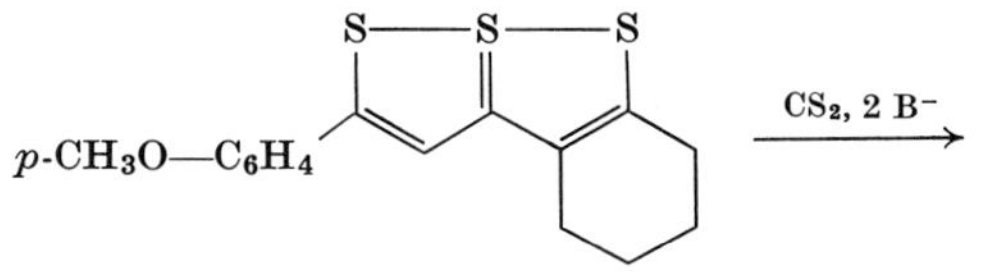

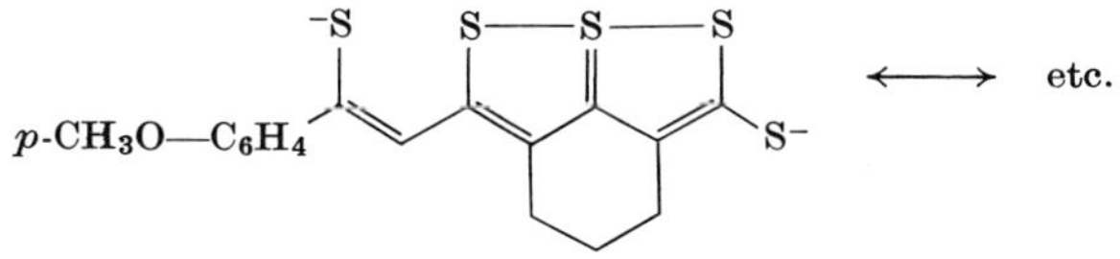

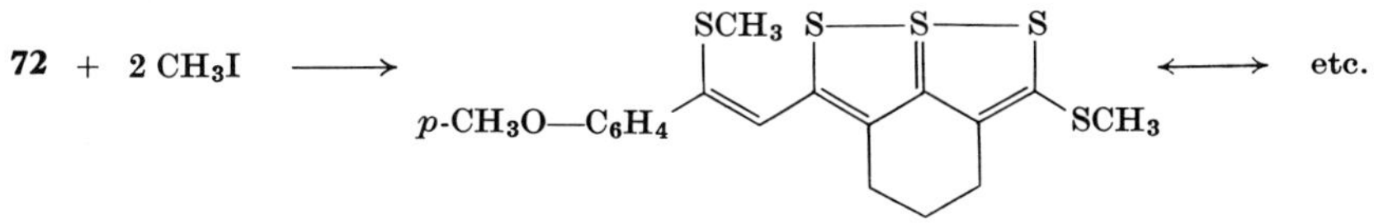

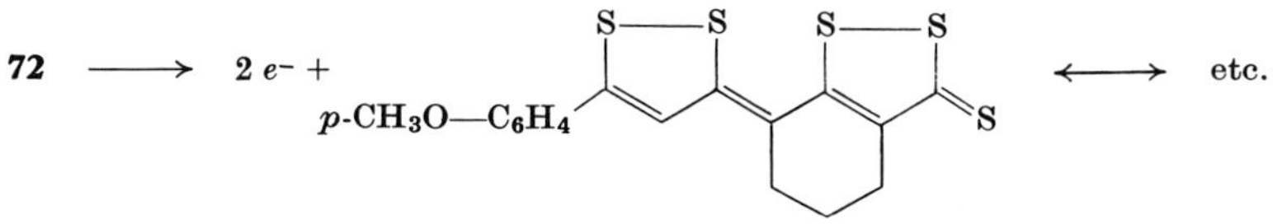

When the dianion **75** reacts with ethylene dibromide, a compound **76** is obtained whose structure has been established by X-ray crystallography.[88]

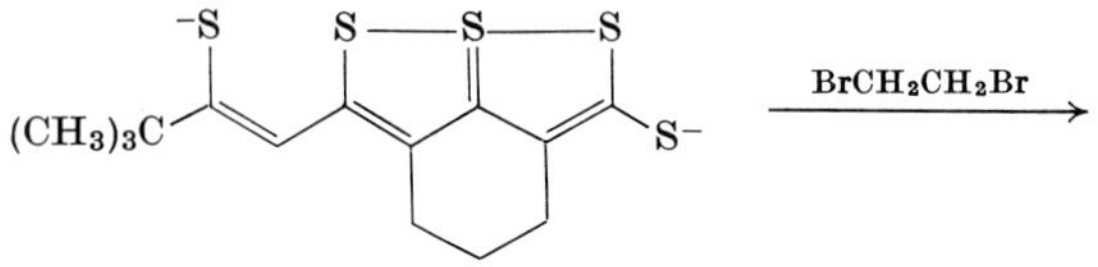

(**75**)

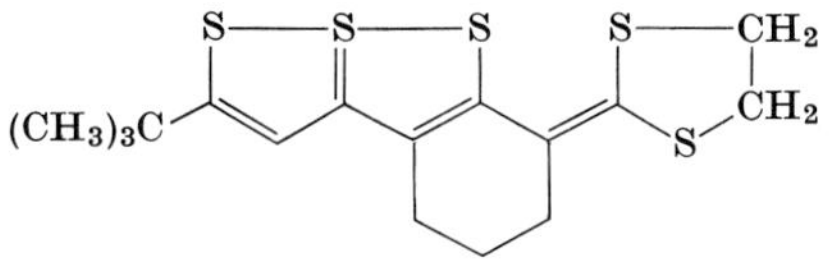

(**76**)
(other resonance structures possible)

From 5-[(1,2-dithiol-3-ylidene)alkyl]-1,2-dithiole-3-thiones (**74**) other extended structures can be obtained,[89, 90] e.g., (a) α-(1,2-dithiol-3-ylidene)-α'-(1,3-dithiol-2-ylidene)thioketones (**77**), by reaction

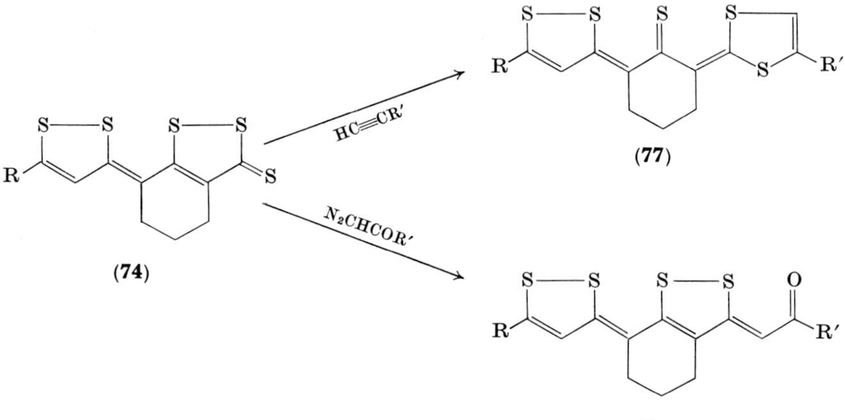

(**74**)

(**77**)

(**78**)

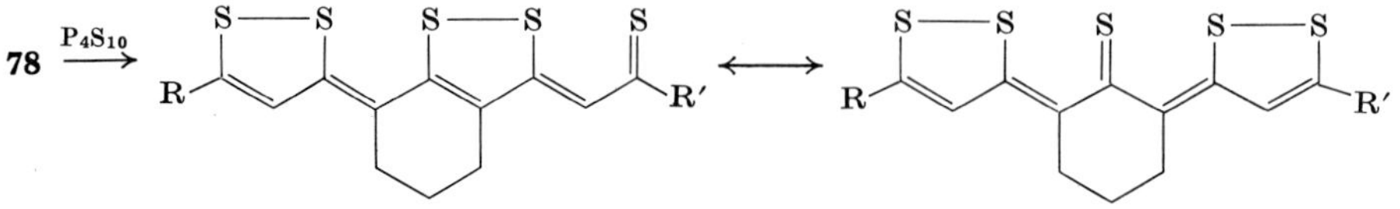

(**79**)

[88] J. Sletten, *Chem. Commun.* 688 (1969).
[89] M. Stavaux and N. Lozac'h, *Bull. Soc. Chim. Fr.* 4273 (1968).
[90] M. Stavaux and N. Lozac'h, *Bull. Soc. Chim. Fr.* 4184 (1969).

with acetylenes and (b) 5-[(1,2-dithiol-3-ylidene)alkyl]-1,2-dithiol-3-ylidene ketones (**78**) by reaction of a diazoketone with **74**. The carbonyl compound **78** with phosphorus pentasulfide forms **79**.

For the compound **79** in which $R = C_6H_5$ and $R' = (CH_3)_3C$, X-ray crystallography has shown that the five sulfur atoms are in line and that partial bonding exists between any neighboring pair of sulfur atoms.

For the compound **79** in which $R = R' = (CH_3)_3C$, the NMR spectrum shows that the structure is symmetrical notwithstanding the fact that the method of synthesis would normally lead to an unsymmetrical arrangement of sulfur atoms.

Another extended structure **80**, with five sulfur atoms able to have the same type of bonding as 1,6,6a$S^{IV}$-trithiapentalene, has been obtained through the synthesis shown in Scheme 21.[91]

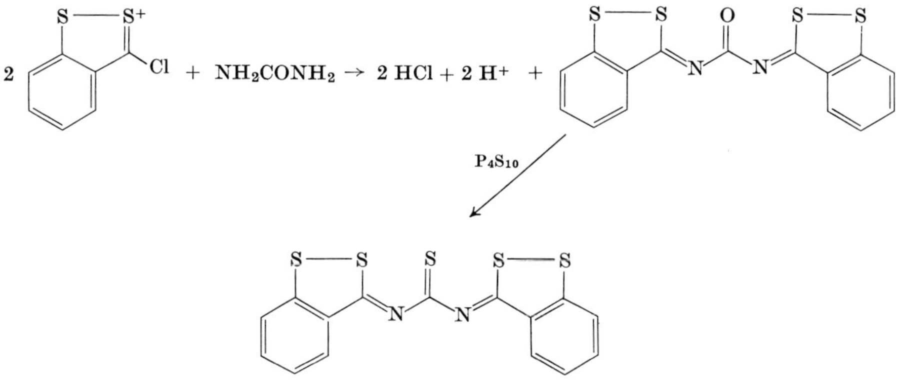

SCHEME 21

## VI. Theoretical Studies and Physical Properties

### A. VALENCE PROBLEMS

Theoretical implications of the structure of 1,6,6a$S^{IV}$-trithiapentalenes are currently open to discussion. Classical theory of $\sigma$ bonds and $\pi$ orbitals does not explain all the properties of these compounds and various theoretical explanations have been put

---

[91] E. Klingsberg, *Chem. Ind. (London)* 1813 (1968).

forward. A survey [92] recently published gives a good summary of the problems involved.

This discussion could be conducted in two ways. First, we could try to deduce structural information from experimental facts through unequivocal mathematical calculations. This would be most satisfying and certainly constitutes the ultimate goal for structural studies. Unfortunately, divergencies of opinions currently observed show that we are still far from such achievement. One important reason for these difficulties may be that quantum mechanical treatment of $d$ orbitals has not reached the precision attained for $s$ and $p$ orbitals. So, in many cases, it cannot be said whether discrepancies between experience and theory are due to inadequacy of theoretical concepts or to shortcomings of mathematical tools.

A less ambitious, but more practical, approach is to examine if and how known experimental facts may agree with a limited set of general assumptions, and this is what we shall try to do.

Let us first consider structures such as **81** or **82**. Whenever X-ray determinations have been made, it has been found that the three

(**81**)  X = O
(**82**)  X = S

(**83**)  X = O
(**84**)  X = S

atoms S, S, and X are approximately in a straight line. When X = O, the carbonyl stretching vibration is markedly affected. This can be explained only by a special effect operating with structure **81** and ineffective with structure **83**.

Moreover, structure **82** (or **84**) should explain why usual thione properties are markedly diminished, though not entirely suppressed, for these compounds.

So it appears that some action by the rest of the molecule is exerted on the atom X. The magnitude of this interaction is much greater than what can be expected by inductive or tautomeric effects, so it is reasonable to assume that some bonding exists between X and one of the sulfur atoms. G. Guillouzo [4] was first to stress this fact, pointing

[92] R. A. W. Johnstone and S. D. Ward, *Theor. Chim. Acta* **14**, 420 (1969).

out the similarity of the CO stretching vibration with the same vibration in chelated β-diketones.

At this point, we have only considered the interaction between X and S atoms which is not accounted for by the classical valence theory. What is the operating mechanism is another question. The first explanation given [5] was the existence of a "single bond–no bond" resonance between structures **82a** and **82b**.

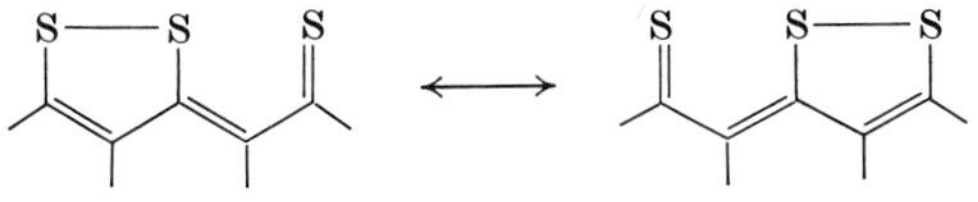

(82a) ⟷ (82b)

Before discussing the significance of this theory, we must consider also the possibility of a rapid equilibrium between valence isomers **82a** ⇌ **82b**.

This hypothesis has been sometimes put forward, but, for the time being, real proof of individual existence of such valence isomers is still lacking. Although we cannot exclude the possibility that future refinements of experimental techniques may eventually prove the existence of such isomers, it does not seem advisable to accept as a fact the existence of isomers for which no conclusive evidence has ever been produced. We shall then prefer the "single bond–no bond" resonance theory to the rapid isomerization theory. In so doing we admit that there exists, in trithiapentalenes, an array of atoms which can be represented as in **85**, in which the lines between atomic symbols mean only that some sort of bonding exists, without prejudice of the nature of this bonding.

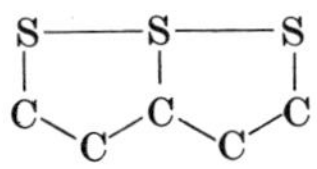

(85)

The existence of S–S–S bonding being admitted, the question arises as to the nature of this bonding. Ordinary conventions for writing chemical formulas do not work well here, so particular methods had to be developed.

Historically, the first proposal was the "single bond–no bond resonance" concept,[5] later used for calculations by Giacometti and Rigatti[93] carried out according to the Hückel MO method. In this

[93] G. Giacometti and G. Rigatti, *J. Chem. Phys.* **30**, 1633 (1959).

hypothesis, there are bonds between the central sulfur atom of **82** and both other sulfur atoms.

If we now consider the $\sigma$ and $\pi$ bondings, formula **82** for both resonance structures has twelve or ten $\pi$ electrons according to whether or not we admit that every sulfur atom retains only two unshared electrons on the M level. So the "no bond" resonance does not show the existence of a naphthalene-like $10\pi$-electron system. If dotted lines are used to represent "half bonds," the system con-

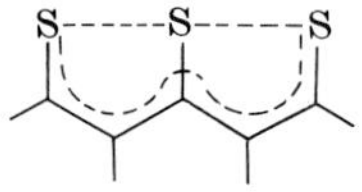

(86)

sidered can be represented by **86**. It is rather difficult to translate formula **86** in terms of $\sigma$ and $\pi$ electrons mainly because one does not know how to deal with the delocalized S–S bonding electrons.

It has been suggested that $\pi$-orbital overlap could explain the bonding in **86**,[94] but the strength of S–S–S bonds would then be rather small. The sulfoniumylid structure **87** complies with common valence theory and the octet rule for sulfur atoms.

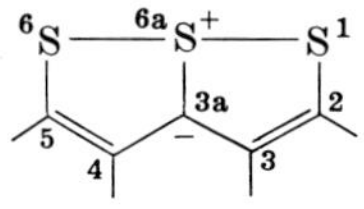

(87)

In **87**, every carbon or sulfur atom is surrounded by eight valence electrons, two of which are unshared in the case of sulfur atoms. The total number of $\pi$ electrons is ten—four from the double bonds, two from the negative carbon atom, and two from each of the external sulfur atoms.

Thus, **87** gives a satisfactory picture for a ten $\pi$-electron naphthalene-like system, but the localization of charges is unsatisfactory because abundant chemical evidence shows that carbon atoms numbered 2 and 5 are more positive than carbon atoms 3 and 4, and according to **87**, mesomeric displacement of the negative charge on atom 3a would lead to the opposite conclusion.

In order to overcome this difficulty, one may admit that instead of opposite electrical charges on atoms 3a and 6a, there exists a double

---

[94] E. M. Shustorovich, *Zh. Obshch. Khim.* **29**, 2424 (1959).

bond between these atoms. This introduces the concept of a quadricovalent ($S^{IV}$) sulfur atom and requires the use of $d$ orbitals to explain the bonding of this atom.

Maeda[95-97] has shown that $\sigma$ bonding between sulfur atoms could result from hybrid $pd$ orbitals of the central sulfur atom. In an interesting paper, Gleiter and Hoffmann[98] have studied the bonding of the three sulfur atoms by four electrons. They have shown that, if hybridized $pd$ orbitals are used for the central sulfur atom, there is a very flat energy minimum when the central sulfur atom is approximately equidistant from the external sulfur atoms. Supposing these external atoms fixed, calculated energy curves allow a displacement of the central atom of approximately $\pm 0.2$ Å from the symmetrical position. The interesting point is that this theory explains why in some trithiapentalene reactions, interatomic distances in the S–S–S system do not give any indication concerning the relative stability of the bonds.

Johnstone and Ward[92] have made SCF-MO calculations corroborated by ultraviolet spectra. Like Maeda, they admit the intervention of the $3d$ orbitals of the central sulfur atom, for which they consider a $p^2d$ hybridization.

There is now fairly good agreement between theoretical and experimental studies, and the best picture of the electronic distribution appears to be the following: Each sulfur atom retains two unshared $3s$ electrons. The terminal sulfur atoms contribute two electrons to $\sigma$ bonds and two to $\pi$ bonds, and the central sulfur atom contributes three electrons to $\sigma$ bonds and one to $\pi$ bonds. Together with the five $\pi$ electrons from the carbon atoms, the $\pi$ electrons from the sulfur atoms form a ten-electron $\pi$ system which can be adequately represented by structures **88** or **89**, omitting unshared electrons.

**(88)**   **(89)**

[95] K. Maeda, *Bull. Chem. Soc. Japan* **33**, 1466 (1960).
[96] K. Maeda, *Bull. Chem. Soc. Japan* **34**, 785 (1961).
[97] K. Maeda, *Bull. Chem. Soc. Japan* **34**, 1166 (1961).
[98] R. Gleiter and R. Hoffmann, *Tetrahedron* **24**, 5899 (1968).

This discussion should not lead one to overestimate the similarity between **88** or **89** and the naphthalene nucleus which also has ten $\pi$ electrons. A much greater similarity with naphthalene is exhibited by thiophthene (**90** or **91**).

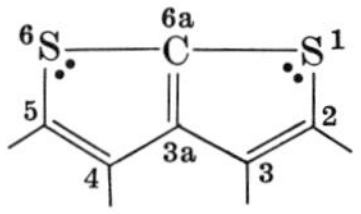

(**90**) 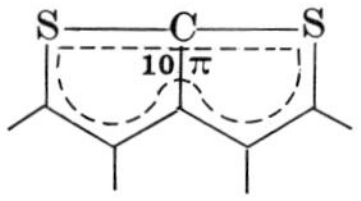    (**91**)

The difference lies in the 6a atom—a carbon atom in **90**, a sulfur in **88**. The octet rule, which is obeyed by the 6a carbon atom in **90** is broken by the 6a sulfur atom in **88**, which has a strong absorption in the visible spectrum, while there is no such absorption with naphthalene or thiophthene (**90**).

For other compounds, such as 1,2,4,6-tetraphenylthiabenzene (**92**), similar sulfur bonding has been considered, as well as sulfoniumylid structures (**93**).[99] These thiabenzenes, like trithiapentalenes, are highly colored, but they are much less stable than the latter class.

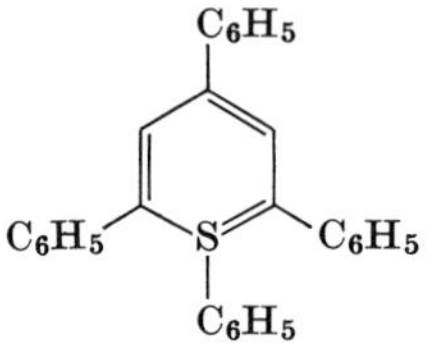

(**92**) 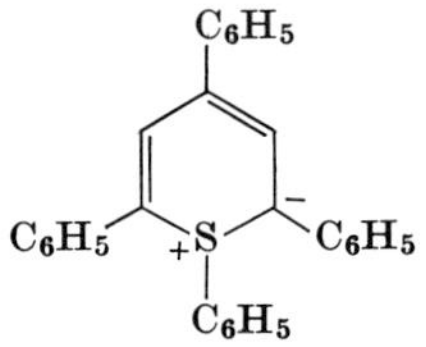    (**93**)

## B. X-Ray Diffractometry

Interatomic distance is one of the best criteria we have for deciding whether two given atoms are bound, but it is not completely unambiguous in cases of abnormal bonding such as those we shall discuss here.

We shall consider first only the sulfur atoms. If two remote atoms are progressively brought nearer, when shall we say that close contact vanishes and loose bonding appears? Some information can already be found in the literature concerning cases where an S–S bond is

[99] C. C. Price and S. Oae, "Sulfur Bonding," p. 160. Ronald Press, New York, 1962.

particularly long or where S–S nonbonding contact is particularly close.

For the van der Waals radius of sulfur, Pauling[100] has given the value 1.85 Å, which is commonly accepted without discussion, so that the generally accepted contact distance between sulfur atoms is 3.70 Å.

However, smaller values have been given for this van der Waals radius in bis-2-iodoethyltrisulfide, 1.60[101] and 1.65 Å[102]; in $S_8$ rhombic sulfur, 1.65[103] and 1.69 Å.[104]

Bondi[105] indicates that, for single-bonded sulfur, X-ray diffraction points to 1.83 Å, whereas various physical properties suggest a radius of 1.80 Å. The smaller values found for crystalline sulfur and polysulfides may be indicative of some double-bond character of the S–S bonds.

According to these various data, it seems reasonable to take 1.7 Å as a lower limit of the van der Waals radius for sulfur, and consequently to consider the possibility of bonding phenomena between two sulfur atoms only when their distance is noticeably less than 3.4 Å.

Let us now consider the bond lengths between two sulfur atoms. The following values have been found: for the $S_8$ stable molecule, 2.12[103] and 2.037 Å[104,106]; for bis-2-iodoethyltrisulfide, 2.05 Å[101]; and for 4-methyl-1,2-dithiole-3-thione, 2.04 Å.[107]

The latter case, however, differs from the two preceding in two major points: (a) in the dithiole ring, the dihedral angle C–S–S–C is very small, if not zero, whereas it is near 90° in noncyclic disulfides[108]; (b) in the dithiole ring, there is an appreciable aromatic character, although the importance of the ring current remains subject to speculation.

At any rate, it then seems possible to conclude that some special bonding effects exist when the S–S distance observed lies between

---

[100] L. Pauling, "The Nature of the Chemical Bond," 3rd ed, p. 260. Cornell Univ. Press, Ithaca, New York, 1960.
[101] J. Donohue, *J. Amer. Chem. Soc.* **72**, 2701 (1950).
[102] B. S. Sharma and J. Donohue, *Acta Crystallogr.* **16**, 891 (1963).
[103] B. E. Warren and J. T. Burwell, *J. Chem. Phys.* **3**, 6 (1935).
[104] S. C. Abrahams, *Acta Crystallogr.* **8**, 661 (1955).
[105] A. Bondi, *J. Phys. Chem.* **68**, 441 (1964).
[106] J. Donohue, *in* "Organic Sulfur Compounds" (N. Kharasch, ed.), p. 1. Pergamon, New York, 1961.
[107] W. L. Kehl and G. A. Jeffrey, *Acta Crystallogr.* **11**, 813 (1958).
[108] A. Hordvik, *Acta Chem. Scand.* **20**, 1885 (1966).

2.2 and 3.4 Å, and the experimental results given here must be interpreted accordingly.

X-Ray diffractometry has played an important role in the structural study of trithiapentalene since the first results determined by Bezzi, Mammi, and Garbuglio[5] which laid the foundation of the "no bond" resonance theory concerning these compounds.

### 1. 1,6,6a$S^{IV}$-Trithiapentalenes

Results concerning these compounds are given in Table I.

TABLE I

S–S DISTANCES IN 1,6,6a$S^{IV}$-TRITHIAPENTALENES

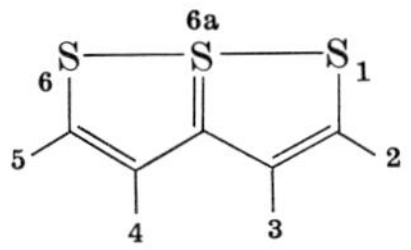

| Substituents | S-1–S-6a (Å) | S-6–S-6a (Å) | Ref. |
|---|---|---|---|
| 2,5-Dimethyl | 2.36 | 2.36 | 5 |
|  | 2.358 | 2.358 | 109 |
| 2-Methyl-4-phenyl | 2.475 ± 0.002 | 2.237 ± 0.002 | 110 |
| 2,5-Diphenyl | 2.355 ± 0.003 | 2.297 ± 0.003 | 111 |
| 2,4-Diphenyl | 2.510 ± 0.008 | 2.216 ± 0.008 | 112 |
|  | 2.499 ± 0.003 | 2.218 ± 0.003 | 113 |
| 3,4-Diphenyl | 2.434 ± 0.004 | 2.232 ± 0.004 | 114 |
| 3-Benzoyl-5-$p$-bromophenyl-2-methylthio | 2.47–2.57 ± 0.007 | 2.18 ± 0.007 | 115 |

[109] F. Leung and S. C. Nyburg, *Chem. Commun.* 137 (1969).
[110] A. Hordvik and K. Julshamn, *Acta Chem. Scand.* **23**, 3611 (1969).
[111] A. Hordvik, *Acta Chem. Scand.* **22**, 2397 (1968).
[112] A. Hordvik, E. Sletten, and J. Sletten, *Acta Chem. Scand.* **20**, 2001 (1966).
[113] A. Hordvik, E. Sletten, and J. Sletten, *Acta Chem. Scand.* **23**, 1852 (1969).
[114] P. L. Johnson and I. C. Paul, *Chem. Commun.* 1014 (1969).
[115] S. M. Johnson, M. G. Newton, I. C. Paul, R. J. S. Beer, and D. Cartwright, *Chem. Commun.* 1170 (1967).

From the limited data available, we may assume that the two S–S distances are normally equal for isolated molecules of symmetrically substituted 1,6,6a$S^{IV}$-trithiapentalenes, but these S–S bonds involving $d$ orbitals seem very sensitive to extramolecular influences in the crystal lattice[111] and to intramolecular perturbations, such as unsymmetrical substitution. Unequal spacing of the sulfur atoms may perhaps also result from steric interactions between symmetrically placed substituents.[114]

## 2. *Isosteres of* 1,6,6a$S^{IV}$-*Trithiapentalenes*

Particularly in the case of α-dithiolylidene carbonyl compounds, the classical formulas are generally preferred, so we shall give also the corresponding name. However, for sake of consistency, we shall use also the pentalene names throughout.

a. 2,5-*Dimethyl*-1-*oxa*-6,6a$S^{IV}$-*dithiapentalene or* 5-*methyl*-1,2-*dithiol*-3-*ylidene acetone*.[116]

$$a = 2.12 \text{ Å}$$
$$b = 2.41 \text{ Å}$$

The S–S bond a appears to be almost "normal," but there is also some shortening of the O–S distance b by comparison with the van der Waals contact distance. We have already accepted 1.70 Å as the van der Waals radius of sulfur. According to Bondi,[105] a mean value acceptable for oxygen seems to be 1.50 Å. The S–O "contact distance" would then be 3.20 Å, which is much larger than the measured b distance, even if we make substantial allowance for the uncertainty in the determination of van der Waals radii. For instance, even if we accept errors of 0.2 Å for both radii, we find a minimum S–O contact distance of 2.80 Å, still considerably larger than the measured value.

It should also be noted that the fact that O–S–S atoms are nearly aligned is evidence by itself, and from all these considerations we may infer that a bonding of moderate strength exists between the S and O atoms, a conclusion corroborated by much infrared evidence.

---

[116] M. Mammi, R. Bardi, G. Traverso, and S. Bezzi, *Nature* **192**, 1282 (1961).

b. 3,5-*Diphenyl-1-oxa-6,6aS$^{IV}$-dithiapentalene* or 2-(*5-phenyl-1,2-dithiol-3-ylidene*)-*2-phenylethanal*.[117]

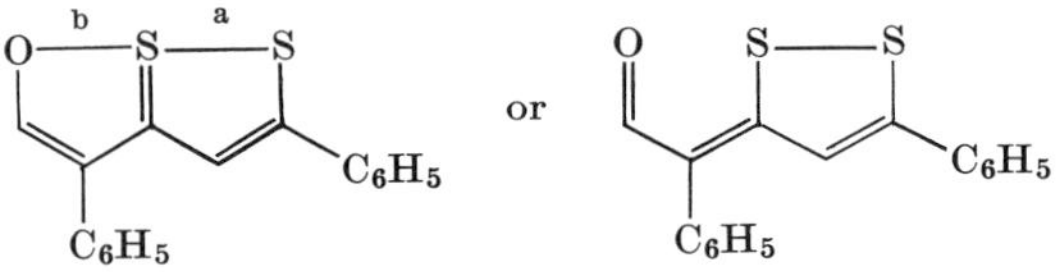

$a = 2.106 \pm 0.003$ Å
$b = 2.382 \pm 0.006$ Å

The S–O distance is relatively long, denoting a moderate interaction, although the three atoms O–S–S are nearly aligned, the O–S–S angle being $174.4° \pm 0.2°$.

c. 3-*Benzoyl-1-oxa-6,6aS$^{IV}$-dithia-2-azapentalene* or α-(*5-phenyl-1,2-dithiol-3-ylidene*)-α-*nitrosoacetophenone*.[118]

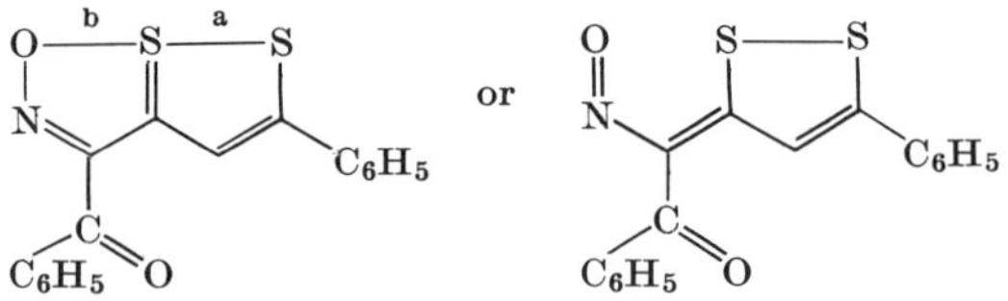

$a = 2.178 \pm 0.002$ Å
$b = 2.034 \pm 0.005$ Å

This structure was based upon the IR spectrum, which shows a "normal" carbonyl stretching band at 1640 cm$^{-1}$,[50] and it was confirmed by an X-ray study.[118] The interesting fact is that the nitroso oxygen atom prevails over the carbonyl for close contact with sulfur. A nitro group is less attracted than the nitroso group by sulfur.[119, 120]

d. 2,4-*Diphenyl-1,6aS$^{IV}$-dithia-6-selenapentalene*.[80]

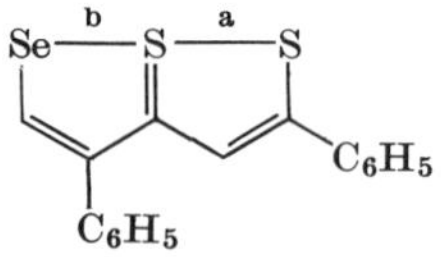

$a = 2.492 \pm 0.003$ Å
$b = 2.333 \pm 0.003$ Å

[117] A. Hordvik, E. Sletten, and J. Sletten, *Acta Chem. Scand.* **23**, 1377 (1969).
[118] P. L. Johnson and I. C. Paul, *J. Amer. Chem. Soc.* **91**, 781 (1969).
[119] R. J. S. Beer and R. J. Gait, *Chem. Commun.* 328 (1970).
[120] K. I. G. Reid and I. C. Paul, *Chem. Commun.* 329 (1970).

Let us remember that for the corresponding trithiapentalene already discussed,[112] we had the following distances: a = 2.510 and b = 2.216 Å. It is remarkable that the analogy with this trithiapentalene system is such that the Se–S distance b is *shorter* than the S–S distance a. According to Bondi,[105] the van der Waals radius for selenium is approximately 1.90 Å, but the data are scarce. So it seems safer to make a correction for sulfur data based on the comparison of $H_2S$ and $H_2Se$,[105] which shows, for the latter, an increment of 0.1 Å for the van der Waals radius of the central atom. The contact distance evaluated as 3.4 Å for S–S should be approximately 3.5 Å for S–Se. At any rate, the values found are good evidence for abnormal S–S–Se bonding.

3. *Extended Structures*

a. 3-*Phenyl-5-(5-phenyl-1,2-dithiol-3-ylidenemethyl)-1,2-dithiolium cation.*[121]

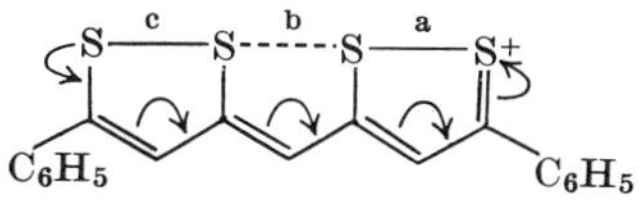

Structural study has shown that the dithiole rings are coplanar and that the four sulfur atoms are nearly aligned, the distances a and c being nearly equal. The two dithiole rings must therefore play the same role and the positive charge should be spread over both, as indicated by the arrows on the formula.

Assuming that a = c = 2.05 Å, b should then be between 3.00 and 3.10 Å. This is indicative of some bonding between the two central sulfur atoms, which is also suggested by the fact that the four sulfur atoms are in line.

b. 3,5-*Bisacetamido-1,2-dithiolium cation.*[122]

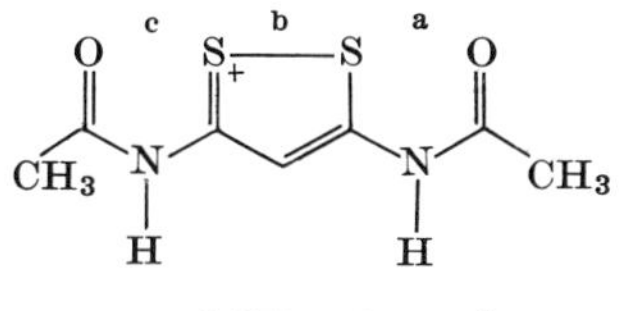

$$a = 2.571 \pm 0.010 \text{ Å}$$
$$b = 2.080 \pm 0.005 \text{ Å}$$
$$c = 2.515 \pm 0.011 \text{ Å}$$

[121] A. Hordvik, *Acta Chem. Scand.* **19**, 1253 (1965).
[122] A. Hordvik and H. M. Kjøge, *Acta Chem. Scand.* **20**, 1923 (1966).

Other resonance formulas can be written with the positive charge on the nitrogen atoms, giving some double bond character to the affected C–N bonds. Notwithstanding the close O–S contacts and the fact that the four O–S–S–O atoms are in a linear arrangement, the S–S bond has a length practically normal for 1,2-dithiolium cations.

c. *2-(5-t-Butyl-1,2-dithiol-3-ylidene)-6-(ethylenedithiomethylene)-cyclohexanethione.*[88]

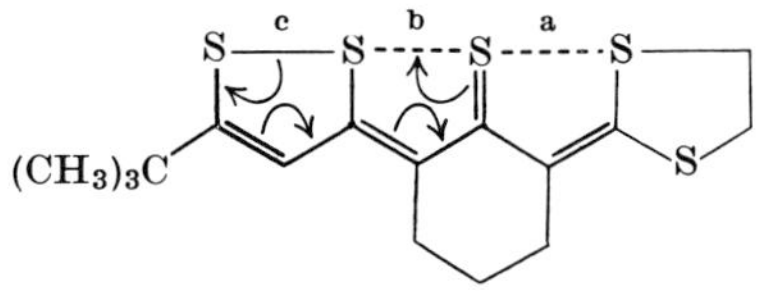

$$a = 2.966 \pm 0.001 \text{ Å}$$
$$b = 2.208 \pm 0.001 \text{ Å}$$
$$c = 2.482 \pm 0.001 \text{ Å}$$

The structure given and named is, of course, only one resonance structure. Bonds b and c are rather short and apparently constitute a normal trithiapentalene system. Bond a is much longer, and similar to the central bond b of the first cited cationic structure.

d. *2-(5-t-Butyl-1,2-dithiol-3-ylidene)-6-(5-phenyl-1,2-dithiol-3-ylidene)cyclohexanethione.*[123]

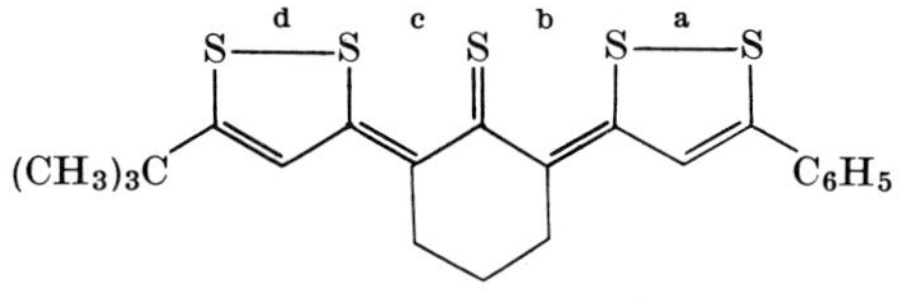

$$a = 2.14 \pm 0.02 \text{ Å}$$
$$b = 2.62 \pm 0.02 \text{ Å}$$
$$c = 2.55 \pm 0.02 \text{ Å}$$
$$d = 2.16 \pm 0.02 \text{ Å}$$

The distances measured show clearly that here is a "no bond" resonance system extended to five sulfur atoms.

## C. INFRARED SPECTRA

Infrared spectroscopy has led to the determination of the correct structure for α-(1,2-dithiol-3-ylidene)ketones independently from the

[123] J. Sletten, *Acta Chem. Scand.* **24**, 1464 (1970).

X-ray crystallographic data, the infrared spectra of these compounds showing some similarity with those of chelated β-diketones.[4]

The prominent feature is the absence of the usual carbonyl frequencies in the 1620–1720 cm$^{-1}$ range, together with one or more strong bands in the 1500–1610 cm$^{-1}$ range. The location of the carbonyl frequency is difficult to obtain because the modes are somewhat mixed, as it is for pyran-4-ones.[124] The carbonyl character of absorption bands has been deduced from solvent effects, as shown in Table II for some (5-aryl-1,2-dithiol-3-ylidene)acetophenones (**94**). Two bands are more-or-less affected by solvent effects in the 1525–1560 cm$^{-1}$ range.

TABLE II

INFRARED SPECTRA OF (5-ARYL-1,2-DITHIOL-3-YLIDENE)ACETOPHENONES IN 1500–1610 cm$^{-1}$ RANGE[18,125]

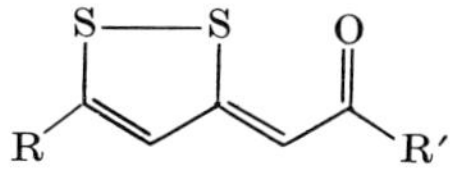

(**94**)

| R = R′ = C$_6$H$_5$ | | R = C$_6$H$_5$<br>R′ = p-ClC$_6$H$_4$ | | R = C$_6$H$_5$<br>R′ = p-CH$_3$OC$_6$H$_4$ | | R = p-CH$_3$OC$_6$H$_4$<br>R′ = C$_6$H$_5$ | |
|---|---|---|---|---|---|---|---|
| KBr | CH$_2$Cl$_2$ | KBr | CH$_2$Cl$_2$ | KBr | CH$_2$Cl$_2$ | KBr | CH$_2$Cl$_2$ |
| 1597 | 1597 | 1591 | 1591 | 1602 | 1604 | 1600 | 1602 |
| 1586 | 1589 | 1585 | 1585 | 1585 | 1587 | 1589 | 1589 |
| 1544 | 1558 | 1550 | 1553 | 1551 | 1555 | 1542 | 1550 |
| 1526 | 1536 | 1531 | 1532 | 1541 | 1537 | 1528 | 1529 |

5-Methyl-1,2-dithiol-3-ylideneacetone shows a strong band at 1578 cm$^{-1}$ (KBr) which is displaced around 1590–1596 cm$^{-1}$ in various solvents, such as methylene chloride or tetrahydrofuran.[18] This band has a clear carbonyl character if we accept that solvent effects are a proof of this character. However, it has been suggested that C–C vibrations may also contribute to this band.[18]

[124] A. R. Katritzky and R. A. Jones, *Spectrochim. Acta* **17**, 64 (1961).
[125] Y. Mollier, Private communication.

These attributions of carbonyl frequencies based on solvent effects have been challenged, in the case of 5-methyl-1,2-dithiol-3-ylideneacetone (**96**) on the basis of some results concerning a nonketonic compound **98**.[22]

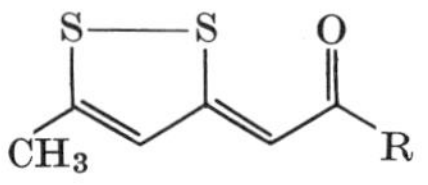

(**95**) R = H
(**96**) R = CH₃
(**97**) R = OCH₃

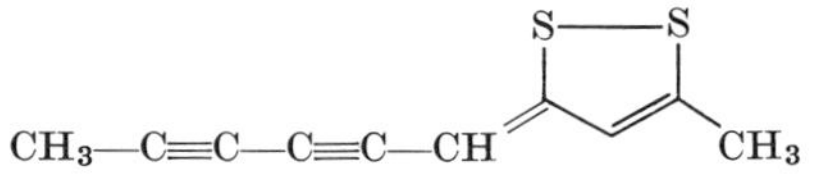

(**98**)

The authors notice that the band shown near 1600 cm$^{-1}$ by **96** also exists for **98**. From this fact they draw the conclusion that the 1600 cm$^{-1}$ band of **96** does not come from the stretching vibration of the carbonyl, to which they attribute a band at 1460 cm$^{-1}$ for **95** and **96** and a band at 1540 cm$^{-1}$ for **97**.[22]

Owing to this disagreement, more conclusive methods were needed, and $^{18}$O isotopic studies were conducted[126] for the following tabulated compounds. Only one band showed a noticeable isotopic effect, although much smaller than the one found with simpler ketones, and these results give another proof of O–S interaction.

| Dithiolylidene ketone | Isotopically affected IR band (cm$^{-1}$) |
| --- | --- |
| ω-(5-*t*-Butyl-1,2-dithiol-3-ylidene)-ω-cyanoacetophenone | 1554 |
| ω-(5-*p*-Tolyl-1,2-dithiol-3-ylidene)-ω-cyanoacetophenone | 1549 |
| α-(4-Phenyl-1,2-dithiol-3-ylidene)acetone | 1574 |

More data are needed to get a general idea of the carbonyl vibrations, but it appears that these preliminary findings are in complete agreement with solvent effect data, attributing to the carbonyl a band near 1575 cm$^{-1}$ for 1,2-dithiol-3-ylideneacetones and one near 1550 cm$^{-1}$ for 1,2-dithiol-3-ylideneacetophenones.

[126] D. Festal and Y. Mollier, *Tetrahedron Lett.* 1259 (1970).

The difference between *cis*- and *trans*-C=O (relative to the sulfur atoms) is well exemplified in derivatives of α-(1,2-dithiol-3-ylidene)-β-diketones given in Table III. In Table III, C=O assignments have been made mainly on the basis of solvent effects.

TABLE III

CARBONYL STRETCHING VIBRATIONS (cm$^{-1}$) OF α-(1,2-DITHIOL-3-YLIDENE)-β-DIKETONES

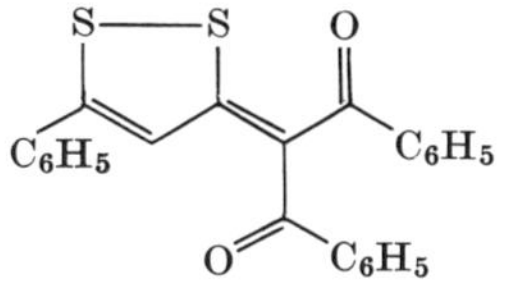

(99)

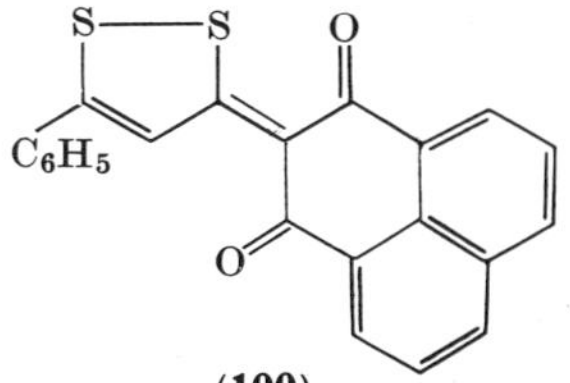

(100)

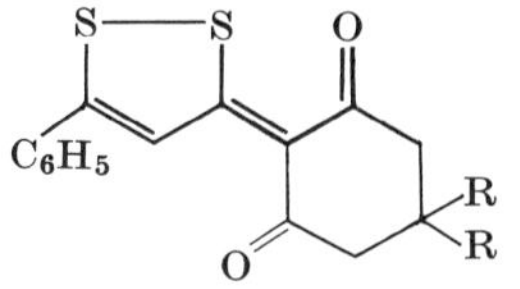

(101) R = H
(102) R = CH$_3$

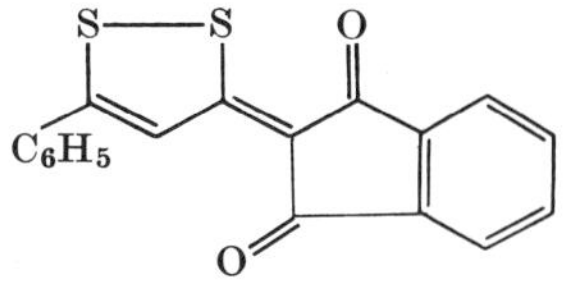

(103)

| α-(1,2-Dithiol-3-ylidene)-β-diketones | *Cis*-carbonyl (cm$^{-1}$) | | *Trans*-carbonyl (cm$^{-1}$) | | Ref. |
|---|---|---|---|---|---|
| | (KBr) | (CHCl$_3$) | (KBr) | (CHCl$_3$) | |
| 99 | 1535 | 1542 | 1624 | 1638 | 18, 30 |
| 100 | 1536 | 1542 | 1632 | 1637 | 36 |
| 101 | 1547 | 1554 | 1624 | 1643 | 36 |
| 102 | 1546 | 1558 | 1638 | 1646 | 36 |
| 103 | 1637 | 1644 | 1680 | 1693 | 36 |

It is clear that, from the point of view of conjugation, the *cis*- and *trans*-carbonyls exhibit the same effect, and the huge difference in frequency of the two carbonyls indicates that there is a special effect between one sulfur atom and the neighboring carbonyl. This effect may be called "no bond" resonance, or partial or abnormal bonding;

the fact remains that there is something here which needs further explanation.

It appears also that the *cis*-carbonyl frequency is less affected when the carbonyl is part of a five-membered ring. Structural strains enlarge then the distance between the sulfur and *cis*-oxygen atoms.

If we compare, for instance, **100** and **103** (KBr), we find for **100** that $\nu_{trans} - \nu_{cis} = 96$ cm$^{-1}$ and for **103** $\nu_{trans} - \nu_{cis} = 43$ cm$^{-1}$.

The *trans*- and *cis*-carbonyls are affected in the same way by bond angles, but in the case of **103** we note that the *cis*-carbonyl is more similar to the *trans*-carbonyl than in **100**.

## D. Dipole Moments

Various dipole moments have been measured and some of them are given in Table IV. These measurements may be useful for testing

### TABLE IV
#### Dipole Moments at 20°C

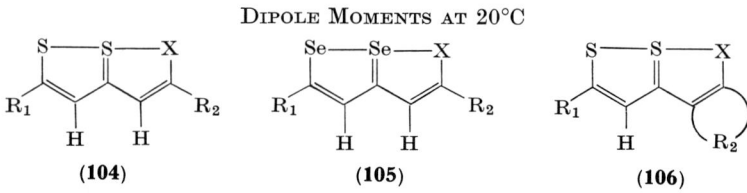

(104)  (105)  (106)

| Type of compound | R$_1$ | R$_2$ | Dipole moments | | | |
|---|---|---|---|---|---|---|
| | | | X=O | Ref. | X=S | Ref. |
| 104 | H | H | 3.78 | 127 | 3.01 | 127 |
| 104 | CH$_3$ | CH$_3$ | 4.00 | 84 | 3.52 | 84 |
| 104 | C$_6$H$_5$ | C$_6$H$_5$ | 4.25 | 13 | 3.56 | 13 |
| 105 | CH$_3$ | CH$_3$ | 3.31 | 127 | 2.76 | 127$^a$ |
| 106 | C$_6$H$_5$ | CH$_2$CH$_2$CH$_2$ | 4.69 | 13 | 3.74 | 13 |
| 106 | C$_6$H$_5$ | CH$_2$CH$_2$CH$_2$CH$_2$ | 4.18 | 13 | 4.01 | 13 |
| 106 | C$_6$H$_5$ | CH=CHCH=CH | 4.40 | 13 | 3.55 | 13 |
| 106 | C$_6$H$_5$ | CH=C(CH$_3$)CH=CH | 4.40 | 13 | | |
| 106 | C$_6$H$_4$OCH$_3$-*p* | CH=CHCH=CH | 4.96 | 13 | | |

$^a$ Value given in Sanesi *et al.*[127] for another structure (S–Se–S) which is not correct. The preparation used gives the structure indicated here, according to a previous paper.[85]

[127] M. Sanesi, G. Traverso, and M. Lazzarone, *Ann. Chim.* (*Rome*) **53**, 548 (1963).

the charge distributions of the compounds under consideration. For example, it can be seen that compounds **106** (X = O) have dipole moments too low for a ionic formula such as **107**.

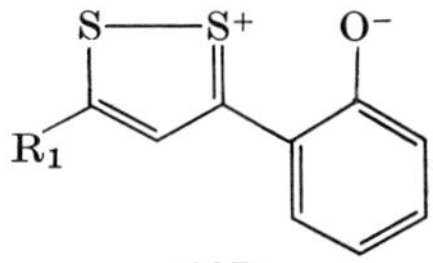

(**107**)

IR spectra show that the C–O band has a very low ketonic character. These results tend to show the significance of 1-oxa-6,6a$S^{IV}$-dithiapentalene resonance structures. However, compounds containing one oxygen atom have generally larger dipole moments than the compouuds containing only sulfur.

This can be seen also in compounds **106**. The oxygen derivative (X = O) with a fused benzene ring has a larger moment than the corresponding tetrahydro derivative (4.40 D vs. 3.97). The benzene resonance energy favors the ionic form, but to a limited extent only; in fact, the experimental value (4.40 D) is much lower than the value calculated for an entirely ionic structure (13 D).

Another interesting clue is given by "para" compounds **108** and **109**.

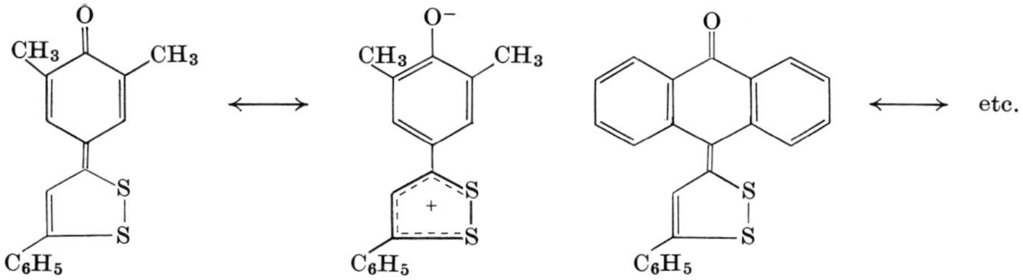

(**108**)　　$\mu = 7.28$ D　　　　　　　(**109**)　　$\mu = 4.51$ D

Compound **108** is clearly more polar than the corresponding "ortho condensation" products **110**.

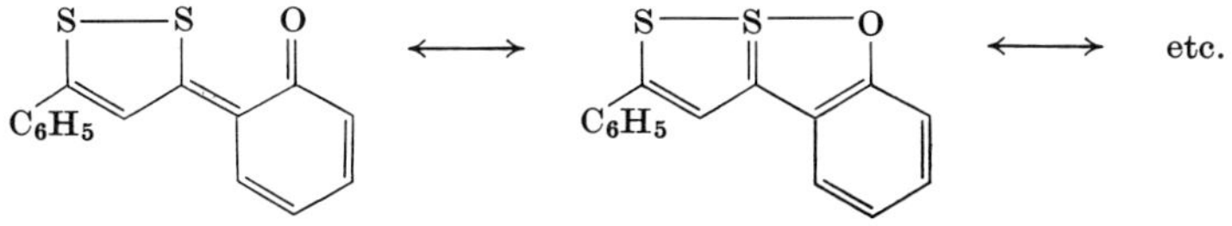

(**110**)

This difference is probably due to the fact that no O–S bonding can exist in **108**, whereas it can for **110**. Another point is that **108** is more polar than **109**, probably because a zwitterion formula is more favored in **108** than in **109**, the energy liberated by the transformation cyclohexadienone–phenol being larger than the energy liberated by the transformation anthrone–anthranol.

## E. Visible and Ultraviolet Spectra

Visible and ultraviolet spectra of 1,6,6a$S^{IV}$-trithiapentalenes have been measured for numerous compounds and have often been taken as a structural proof for this system. A SCF-MO calculation based on structure **89** has given a satisfactory explanation for the visible band, between 440 and 520 m$\mu$, assigned to a $\pi \to \pi$ transition.[92] Ionization potentials are also correctly predicted by this calculation.[92]

Electronic spectra of 1,6,6a$S^{IV}$-trithiapentalenes are characterized by strong absorption bands near 500 m$\mu$ in the visible and near 260 m$\mu$ in the ultraviolet. The 500 m$\mu$ band is responsible for the orange to red color of trithiapentalenes. The parent compound and its alkyl derivatives may show shoulders of the 260 m$\mu$ band near 300 and 230 m$\mu$. Aryl derivatives exhibit a shift of the 500 m$\mu$ band toward 470 m$\mu$ and the spreading of the 260 m$\mu$ band.[38, 49, 92]

These results are summarized in Table V.

UV spectra show some similarity when comparing 1,6,6a$S^{IV}$-trithiapentalenes with their oxygen and selenium isosteres such as compounds **111**.[49, 80]

$$\text{S———S———X}$$
$$C_6H_5 \qquad \qquad C_6H_5$$

(**111**)

| X  | UV Spectra in cyclohexane $\lambda$ m$\mu$ ($\epsilon$ $10^{-3}$) | | | | |
|----|----------|----------|----------|----------|---------|
| O  | 211 (20) | 235 (21) | 275 (13) | 296 (13) | 452 (14) |
| S  | 210 (21) | 252 (38) | 265 (35) | 322 (12) | 497 (9)  |
| Se | 210 (26) | 245 (29) | 277 (41) | 325 (16) | 528 (8)  |

## TABLE V
## UV AND VISIBLE SPECTRA OF 1,6,6a$S^{IV}$-TRITHIAPENTALENES

| $R_1$ | $R_2$ | $R_3$ | $R_4$ | $\lambda\, m\mu\, (\epsilon \cdot 10^{-3})^a$ | | | | Solvent$^b$ | Ref. |
|---|---|---|---|---|---|---|---|---|---|
| H | H | H | H | 230 (16) | 254 (49) | — | — | Cycl. | 49 |
| H | H | H | CH$_3$ | 235 (15) | 262 (48) | — | — | Cycl. | 49 |
| CH$_3$ | H | H | CH$_3$ | — | 261 (54) | — | — | | 61 |
| H | H | H | C$_6$H$_5$ | 209 (17) | 249 (41) | 270 (36) | — | Cycl. | 49 |
| H | H | C$_6$H$_5$ | H | 209 (7) | 230 (8) | 259 (17) | — | Cycl. | 49 |
| H | C$_6$H$_5$ | H | C$_6$H$_5$ | — | 244 (39) | — | 332 (11) | EtOH | 76 |
| H | H | H | C$_6$H$_5$ | 210 (21) | 252 (38) | 265 (35) | 322 (12) | Cycl. | 49 |
| C$_6$H$_5$ | H | H | C$_6$H$_5$ | — | 252 (48) | — | 305 (23) | Diox. | 38 |
| C$_6$H$_5$ | H | H | C$_6$H$_4$CH$_3$-$p$ | — | 256 (49) | — | 313 (24) | Diox. | 38 |
| | | | | — | 260 (30) | — | 313 (19) | CHCl$_3$ | 38 |
| C$_6$H$_4$CH$_3$-$p$ | H | H | C$_6$H$_4$CH$_3$-$p$ | — | 257 (45) | — | 321 (25) | Diox. | 38 |
| C$_6$H$_5$ | H | H | C$_6$H$_4$Cl-$p$ | — | 256 (45) | — | 310 (26) | Diox. | 38 |
| C$_6$H$_4$Cl-$p$ | H | H | C$_6$H$_4$Cl-$p$ | — | 259 (46) | — | 316 (27) | Diox. | 38 |
| C$_6$H$_4$CH$_3$-$p$ | H | H | C$_6$H$_4$Cl-$p$ | — | 258 (47) | — | 318 (26) | Diox. | 38 |
| | | | | — | 260 (28) | — | 319 (18) | CHCl$_3$ | 38 |
| H | CH$_3$ | H | C$_6$H$_5$ | 210 (21) | 252 (38) | 265 (35) | 322 (12) | Cycl. | 49 |
| H | C$_6$H$_5$ | H | CH$_3$ | — | 252 (44) | — | — | C$_2$H$_5$OH | 26 |
| CH$_3$ | H | H | C$_6$H$_5$ | — | 241 (50) | 268 (50) | 330 (10) | | 54 |

| | | | | 472 (5) | Cycl. | 49 |
| | | | | 470 (5) | Cycl. | 49 |
| | | | | 474 (7) | | 61 |
| | | | | — | Cycl. | 49 |
| | | | | — | Cycl. | 49 |
| | | | | 480 (18) | EtOH | 76 |
| | | | | 500 (10) | Cycl. | 49 |
| | | | | 497 (9) | Diox. | 38 |
| | | | | 509 (14) | Diox. | 38 |
| | | | | 511 (16) | CHCl$_3$ | 38 |
| | | | | 508 (13) | Diox. | 38 |
| | | | | 512 (15) | Diox. | 38 |
| | | | | 509 (16) | Diox. | 38 |
| | | | | 512 (15) | Diox. | 38 |
| | | | | 513 (16) | CHCl$_3$ | 38 |
| | | | | 510 (12) | Cycl. | 49 |
| | | | | 497 (9) | C$_2$H$_5$OH | 26 |
| | | | | 484 (6) | | 54 |
| | | | | 490 (10) | | |

*continued*

TABLE V—continued

| $R_1$ | $R_2$ | $R_3$ | $R_4$ | $\lambda\ m\mu\ (\epsilon\cdot 10^{-3})^a$ | | | | | | Solvent$^b$ | Ref. |
|---|---|---|---|---|---|---|---|---|---|---|---|
| $C_6H_5$ | H | $CH_3$ | $CH_3$ | — | 246 (71) | 278 (73) | 332 (23) | — | 490 (25) | | 54 |
| $C_6H_5$ | H | $C_6H_5$ | $CH_3$ | — | 258 (54) | 279 (55) | 327 (20) | — | 490 (25) | | 54 |
| $C_2H_5$ | $CH_2CH_2CH_2$ | | $C_2H_5$ | 235 (21) | 261 (54) | — | — | — | 481 (8) | Cycl. | 38 |
| $C_6H_5$ | H | $CH=CH-CH=CH$ | | — | 258 (20) | 270 (42) | 331 (11) | 367 (6) | 515 (13) | $CHCl_3$ | 38 |
| H | H | H | $COOC_2H_5$ | 232 (18) | 257 (31) | — | — | — | 501 (5) | Cycl. | 49 |
| $C_6H_4CH_3$-$p$ | H | CN | $C_6H_5$ | — | 259 (46) | 307 (22) | 346 (13) | — | 489 (15) | Diox. | 38 |
| $CH_3$ | H | Br | $CH_3$ | — | 263 (50) | — | — | — | 481 (7) | | 61 |
| $C_6H_5$ | H | Br | $C_6H_5$ | — | 254 (52) | — | 293 (23) | — | 510 (13) | | 61 |
| $C_6H_5$ | H | Br | $CH_3$ | — | 251 (48) | 273 (32) | 334 (11) | — | 504 (11) | | 50 |
| $C_6H_5$ | H | H | $OC_2H_5$ | — | 258 (30) | — | 285 (19) | — | 478 (10) | | 50 |
| $C_6H_5$ | H | $COC_6H_5$ | $OC_2H_5$ | — | 252 (46) | 287 (22) | 318 (14) | — | 484 (14) | $CH_2Cl_2$ | 46 |
| $SCH_3$ | H | H | $SCH_3$ | — | 251 (47) | — | 321 (12) | 354 (6) | 500 (17) | | 61 |
| $C_6H_5$ | H | H | $SCH_3$ | — | 261 (49) | — | 313 (22) | 354 (6) | 514 (15) | | 61 |
| $SCH_3$ | H | $C_6H_5$ | $SCH_3$ | — | 252 (46) | — | 323 (17) | 356 (7) | 507 (17) | $CH_2Cl_2$ | 46 |
| $C_6H_5$ | H | $COC_6H_5$ | $SCH_3$ | — | 257 (68) | — | 314 (2) | 348 (7) | 498 (15) | $CH_2Cl_2$ | 46 |
| $C_6H_5$ | H | $COOC_2H_5$ | $SCH_3$ | — | 260 (49) | — | 315 (24) | 357 (5) | 492 (14) | | 61 |
| $SCH_3$ | $C_6H_5$ | $COOC_2H_5$ | $SCH_3$ | — | 244 (41) | 258 (46) | 323 (18) | 362 (5) | 505 (16) | $CH_2Cl_2$ | 46 |
| $C_6H_5$ | H | CN | $NH_2$ | — | 255 (15) | 301 (26) | — | 360 (5) | 465 (12) | Diox. | 64 |
| | | | | — | 252 (39) | 297 (20) | — | — | 465 (8) | $C_2H_5OH$ | 26 |
| $C_6H_5$ | H | $NO_2$ | $SCH_3$ | — | 254 (34) | — | 322 (18) | — | 480 (11) | | 50 |

$^a$ $\epsilon$ are rounded to the nearest thousand. $\epsilon \cdot 10^{-3}$ is given in parentheses.
$^b$ Cycl., cyclohexane; diox., dioxane.

## TABLE VI

UV and Visible Spectra of Various α-(1,2-Dithiol-3-ylidene)carbonyl Compounds

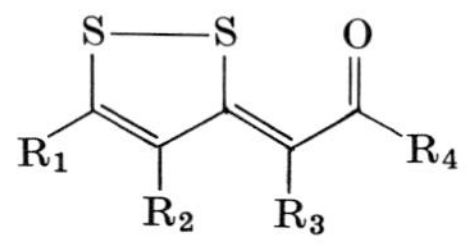

| $R_1$ | $R_2$ | $R_3$ | $R_4$ | $\lambda\,m\mu\,(\epsilon\cdot 10^{-3})$ | | | | Solvent | Ref. |
|---|---|---|---|---|---|---|---|---|---|
| H      | $C_6H_5$ | H      | H      | 221 (20) | 243 (11) | 286 (3)  | —        | 415 (10) | 436 (9)  | Cycl. | 49 |
| $C_6H_5$ | H      | H      | H      | 215 (15) | 232 (18) | 293 (13) | —        | —        | 432 (11) | Cycl. | 49 |
| H      | $C_6H_5$ | H      | $CH_3$ | —        | —        | 275 (4)  | —        | 410 (13) | 427 (13) | EtOH | 26 |
| $C_6H_5$ | H      | $CH_3$ | H      | 218 (16) | 236 (17) | 295 (12) | —        | —        | 449 (10) | Cycl. | 49 |
| $C_6H_5$ | H      | $C_6H_5$ | H    | 211 (20) | 235 (21) | 275 (13) | 296 (13) | —        | 452 (14) | Cycl. | 49 |
| $C_6H_5$ | H      | CN     | $NH_2$ | —        | —        | —        | 303 (13) | —        | 425 (15) | EtOH | 26 |

## TABLE VII

UV AND VISIBLE SPECTRA OF VARIOUS $o$-(1,2-DITHIOL-3-YLIDENE)NAPHTHALENONES AND PARA ANALOGS

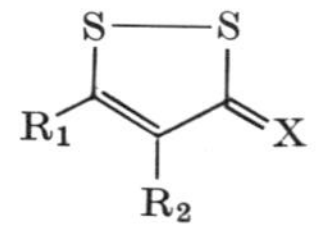

| $R_1$ | $R_2$ | X | | | $\lambda\ m\mu\ (\epsilon \cdot 10^{-3})$ | | | Solvent | Ref. |
|---|---|---|---|---|---|---|---|---|---|
| $C_6H_5$ | H | 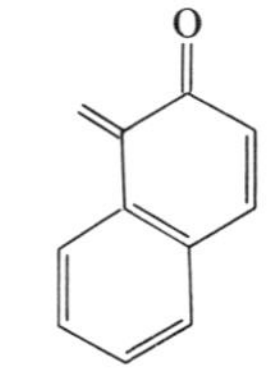 (2-naphthalenone) | — | — | — | 324 (11) | 496 (12) | $CH_3CN$ | 34 |
| | | | — | 292 (9) | — | 362 (12) | 479 (11) | $CH_3CN, HClO_4$ | 34 |
| $C_6H_4OCH_3$-$p$ | H | | — | — | — | 346 (10) | 499 (12) | $CH_3CN$ | 34 |
| | | | — | 275 (8) | — | 419 (12) | 468 (11) | $CH_3CN, HClO_4$ | 34 |
| $C_6H_4OCH_3$-$p$ | H | 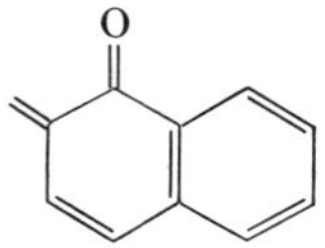 (1-naphthalenone, 2-ylidene) | — | — | 300 (11) | 354 (12) | 525 (12) | $CH_3CN$ | 34 |
| | | | — | 303 (11) | 368 (12) | 420 (12) | 467 (12) | $CH_3CN, HClO_4$ | 34 |
| $C_6H_5$ | H | 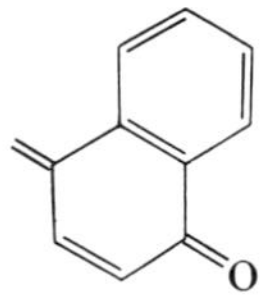 (1-naphthalenone, 4-ylidene) | 243 (20) | — | 305 (15) | 380 (3) | 517 (27) | $CH_3CN$ | 33 |
| | | | 228 (30) | 245 (11) | — | 355 (13) | 490 (15) | $CH_3CN, HClO_4$ | 33 |
| $C_6H_4OCH_3$-$p$ | H | | — | — | 349 (12) | 383 (8) | 532 (15) | $CH_3CN$ | 33 |
| | | | — | 270 (10) | 298 (12) | 416 (14) | 483 (14) | $CH_3CN, HClO_4$ | 33 |

With replacement of oxygen by sulfur, and then by selenium, visible absorption is weakened and shifted to longer wavelengths. On the other hand, in the near-ultraviolet region the selenium compound is intermediate between the oxygen and the sulfur isosteres, both with respect to wavelength and intensity of absorption.

UV spectra of α-(1,2-dithiol-3-ylidene)aldehydes or ketones [1-oxa-6,6a$S^{IV}$-dithiapentalenes (**111**), X = O] have been less systematically studied than those of **111** (X = S). Some of these data are given in Table VI.

Compared with 1,6,6a$S^{IV}$-trithiapentalenes, α-(1,2-dithiol-3-ylidene) compounds absorb more weakly in the 250 m$\mu$ region and more strongly in the visible, but at lower wavelength. The presence of condensed rings shifts the absorption toward the longer wavelengths. Paraquinonoid compounds absorb at longer wavelengths than the corresponding orthoquinonoid derivatives[33, 34] (Table VII).

Protonation by perchloric acid shifts the 500 m$\mu$ band to shorter wavelengths.

Electronic spectra of other isosteres or isomers have been studied, and sometimes used as a proof of the structural similarity of these compounds: 1,6a$S^{IV}$-dithia-6-azapentalenes,[76] 1,6,6a$S^{IV}$-trithia-3,4-diazapentalenes,[82] 1,3a$S^{IV}$,4-trithiapentalenes,[64, 69] and 3,3a$S^{IV}$,6-trithia-1-azapentalenes.[69]

## F. NMR Spectra

Typical chemical shifts and coupling constants for protons in 1-oxa-6,6a$S^{IV}$-dithiapentalenes (**112**) [α-(1,2-dithiol-3-ylidene)carbonyl compounds] and for 1,6,6a$S^{IV}$-trithiapentalenes (**113**) are given in Table VIII. These values are given for protons not subject to the influence of neighboring substituents with strong magnetic effects. They are compared with the values found in furan and thiophene for the corresponding protons; the numbering of the rings happens to be the same, starting from the heteroatom.

The fact that the H$_2$ ("aldehydic") proton of **112** is more deshielded than the corresponding proton in **113** is in favor of the dithiolylidene-aldehyde formula for **112**, whereas the chemical shifts in **113** favor the bicyclic formula with a relatively strong ring current.[13, 49]

TABLE VIII

CHEMICAL SHIFTS ($\delta$) AND COUPLING CONSTANTS (Hz) IN 1-OXA-6,6a$S^{IV}$-DITHIAPENTALENES (112) AND 1,6,6a$S^{IV}$-TRITHIAPENTALENES (113)

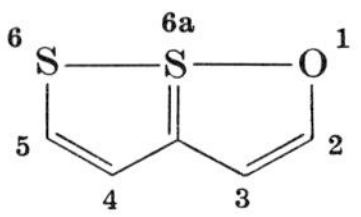

(112)

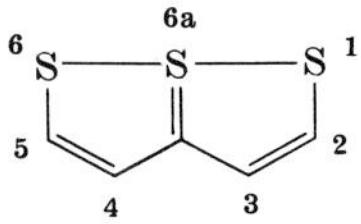

(113)

| Structure | $H_2$ (ppm) | $H_3$ (ppm) | $H_4$ (ppm) | $H_5$ (ppm) | $J_{H_2H_3}$ (Hz) | $J_{H_4H_5}$ (Hz) |
|---|---|---|---|---|---|---|
| 112 | 9.2–9.4 | 6.6–6.7 | 6.8–7.1 | 7.8–8.0 | 1.6 | 5.0 |
| (a) | 7.40 | 6.30 | 7.04 | 7.19 | 1.8 | 4.7 |
| 113 | 9.1–9.2 | 7.5–8.0 | 7.5–8.0 | 9.1–9.2 | 6.3–6.4 | 6.3–6.4 |
| (b) | 7.19 | 7.04 | 7.04 | 7.19 | 4.7 | 4.7 |

(a) corresponding values for furan ($H_2$; $H_3$) and thiophene ($H_4$; $H_5$).
(b) corresponding values for two thiophene rings ($H_2$ to $H_5$).[128]

More detailed results for compounds **112** are given in Table IX and for compounds **113** in Table X.

An interesting feature of the NMR spectra of compounds **113** is that they show the magnetic symmetry of symmetrically substituted structures. We have already pointed out (Section VI,A) that this may be the result of a rapid exchange between valence isomers, but it seems more and more accepted that this is a real case of equivalence corresponding to a symmetrical pattern of bonding.

Similar results have been obtained with 6a$S^{IV}$-thia-1,6-diazapentalenes[83] (Section IV,D) and extended structures[90] (Section V).

NMR spectra of protonated forms of **112** have been obtained and have proved useful for structure determinations.[34]

[128] L. M. Jackson and S. Sternhell, "Applications of N.M.R. Spectroscopy in Organic Chemistry," 2nd ed., chemical shifts, p. 209; coupling constants, pp. 306–307. Pergamon, Oxford, 1969.

## TABLE IX
### NMR CHEMICAL SHIFTS IN 1-OXA-6,6a$S^{IV}$-DITHIAPENTALENES[a]

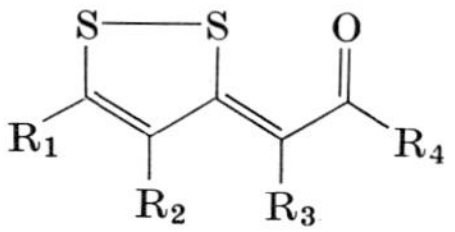

(112)

| $R_1$ | $R_2$ | $R_3$ | $R_4$ | $\delta_1$ | $\delta_2$ | $\delta_3$ | $\delta_4$ | $J_{12}$ | $J_{34}$ | Solvent | Ref. |
|---|---|---|---|---|---|---|---|---|---|---|---|
| $CH_3$ | H | H | H | 2.47 | 6.90 | 6.62 | 9.27 | — | — | $CCl_4$ | 22 |
| $CH_3$ | H | H | $CH_3$ | 2.43 | 6.82 | 6.62 | 2.22 | — | — | $CCl_4$ | 22 |
| H | H | H | $C_6H_5$ | 2.45 | 6.86 | 6.70 | 2.23 | 1.0 | 0 | $CDCl_3$ | 7 |
| H | $C_6H_5$ | H | H | 8.05 | 7.50 | 7.58 | — | 5.0 | — | $CDCl_3$ | 7 |
| $C_6H_5$ | H | H | H | 7.88 | — | 6.72 | 9.39 | — | 1.6 | $CDCl_3$ | 49 |
| H | $C_6H_5$ | H | $C_6H_5$ | — | 7.49 | 6.82 | 9.41 | — | 1.6 | $CDCl_3$ | 49 |
| $C_6H_5$ | H | H | $C_6H_5$ | 7.84 | — | 7.38 | — | — | — | $CDCl_3$ | 49 |
| $C_6H_5$ | H | H | $C_6H_5$ | — | 7.48 | 7.45 | — | — | — | $CDCl_3$ | 7 |

*continued*

TABLE IX—continued

| $R_1$ | $R_2$ | $R_3$ | $R_4$ | $\delta_1$ | $\delta_2$ | $\delta_3$ | $\delta_4$ | $J_{12}$ | $J_{34}$ | Solvent | Ref. |
|---|---|---|---|---|---|---|---|---|---|---|---|
| $C_6H_5$ | H | H | $C_6H_4OCH_3$-$p$ | — | 7.48 | 7.45 | — | — | — | $CDCl_3$ | 7 |
| $C_6H_4OCH_3$-$p$ | H | H | $C_6H_5$ | — | 7.45 | 7.45 | — | — | — | $CDCl_3$ | 7 |
| $C_6H_5$ | H | $C_6H_5$ | H | — | 7.61 | — | 9.38 | — | — | $CDCl_3$ | 49 |
| H | $C_6H_5$ | H | $CH_3$ | 7.77 | — | 6.67 | 2.24 | — | — | $CDCl_3$ | 49 |
| $CH_3$ | H | H | $C_6H_5$ | 2.51 | 7.10 | 7.56 | — | 1.0 | — | $CDCl_3$ | 7 |
| $C_6H_5$ | H | $CH_3$ | H | — | 7.49 | 2.25 | 9.23 | — | — | $CDCl_3$ | 49 |
| H | $C_6H_5$ | $C_6H_5$ | $CH_2C_6H_5$ | 7.77 | — | — | 3.65 | — | — | $CDCl_3$ or $CCl_4$ | 26 |
| H | $C_6H_5$ | $C_6H_5$ | $CHClC_6H_5$ | 7.85 | — | — | 5.58 | — | — | $CDCl_3$ or $CCl_4$ | 26 |
| $C_6H_5$ | H | $CH_2CH_2CH_2CH_2$ | | — | 7.01 | — | — | — | — | $CDCl_3$ | 13 |
| $C_6H_5$ | H | $CH_2CH_2CH_2CH_2$ | | — | 7.33 | — | — | — | — | $CDCl_3$ | 13 |
| $C_6H_5$ | H | CH=CHCH=CH | | — | 8.35 | — | — | — | — | $CDCl_3$ | 13 |
| $C_6H_4OCH_3$-$p$ | H | CH=CHCH=CH | | — | 8.35 | — | — | — | — | $CDCl_3$ | 13 |
| $C_6H_5$ | H | CH=C($CH_3$)CH=CH | | — | 8.30 | — | — | — | — | $CDCl_3$ | 13 |
| $C_6H_5$ | H | $COOC_2H_5$ | $C_6H_5$ | — | 8.68 | — | — | — | — | $CDCl_3$ | 45 |
| 2-Thienyl | H | $COOC_2H_5$ | 2-Thienyl | — | 8.27 | — | — | — | — | $CDCl_3$ | 45 |
| $CH_3$ | H | H | $OCH_3$ | 2.33 | 6.54 | 5.99 | 3.75 | — | — | $CCl_4$ | 22 |
| Ar | H | $COOC_2H_5$ | $OC_2H_5$ | — | 8.32 | — | — | — | — | $CDCl_3$ | 30 |
| Ar | H | CN | $OC_2H_5$ | — | 7.52 | — | — | — | — | $CDCl_3$ | 30 |

[a] $\alpha$-(1,2-Dithiol-3-ylidene)carbonyl compounds. Protons directly linked to the nucleus or included in the following substituents: —$CH_3$, —$CH_2(Ar)$, —$CHCl(Ar)$, —$OCH_3$.

TABLE X

NMR CHEMICAL SHIFTS IN 1,6,6a$S^{IV}$-TRITHIAPENTALENES[a]

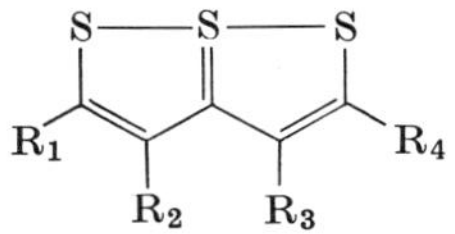

(113)

| $R_1$ | $R_2$ | $R_3$ | $R_4$ | $\delta_1$ | $\delta_2$ | $\delta_3$ | $\delta_4$ | $J_{12}$ | $J_{34}$ | Solvent | Ref. |
|---|---|---|---|---|---|---|---|---|---|---|---|
| H | H | H | H | 9.18 | 7.96 | 7.96 | 9.18 | 6.3 | 6.3 | CDCl$_3$ | 49 |
| H | H | H | CH$_3$ | 9.13 | 7.77 | 7.72 | 2.67 | 6.3 | — | CDCl$_3$ | 49 |
| CH$_3$ | H | H | CH$_3$ | 2.60 | 7.53 | 7.53 | 2.60 | 1.0 | 1.0 | CDCl$_3$ | 7 |
| | | | | 2.64 | 7.56 | 7.56 | 2.64 | 1.0 | 1.0 | CDCl$_3$ | 49 |
| H | H | H | C$_6$H$_5$ | 9.18 | 7.94 | 8.26 | — | 6.4 | — | CDCl$_3$ | 49 |
| H | H | C$_6$H$_5$ | H | 9.14 | 7.79 | — | 8.84 | 6.4 | — | CDCl$_3$ | 49 |
| H | C$_6$H$_5$ | H | C$_6$H$_5$ | 8.81 | — | 8.10 | — | — | — | CDCl$_3$ | 49 |
| C$_6$H$_5$ | H | H | C$_6$H$_5$ | — | 8.21 | 8.21 | — | — | — | CDCl$_3$ | 7 |
| | | | | — | 8.22 | 8.22 | — | — | — | CDCl$_3$ | 49 |
| C$_6$H$_5$ | H | H | C$_6$H$_4$OCH$_3$-? | — | 8.21 | 8.21 | — | — | — | CDCl$_3$ | 7 |

*continued*

TABLE X—continued

| $R_1$ | $R_2$ | $R_3$ | $R_4$ | $\delta_1$ | $\delta_2$ | $\delta_3$ | $\delta_4$ | $J_{12}$ | $J_{34}$ | Solvent | Ref. |
|---|---|---|---|---|---|---|---|---|---|---|---|
| H | $CH_3$ | H | $C_6H_5$ | 8.60 | 2.44 | 8.12 | — | — | — | $CDCl_3$ | 49 |
| H | $C_6H_5$ | H | $CH_3$ | 8.82 | — | 7.58 | 2.57 | — | — | $CDCl_3$ | 49 |
| $CH_3$ | H | H | $C_6H_5$ | 2.62 | 7.73 | 8.08 | — | 1.0 | — | $CDCl_3$ | 7 |
| $C_6H_5$ | H | $CH_2CH_2CH_2$ | | — | 7.77 | — | — | — | — | — | 13 |
| $C_6H_5$ | H | $CH_2CH_2CH_2CH_2$ | | — | 7.98 | — | — | — | — | — | 13 |
| $C_6H_5$ | H | $CH=CHCH=CH$ | | — | 8.72 | — | — | — | — | — | 13 |
| H | CHO | H | $C_6H_5$ | 9.98 | 10.11 | 9.92 | — | — | — | $CDCl_3$ | 49 |
| $C_6H_5$ | H | CHO | $C_6H_5$ | — | 9.87 | 10.10 | — | — | — | — | 62 |
| $C_6H_4CH_3$-$p$ | H | CHO | $C_6H_4CH_3$-$p$ | — | 9.83 | 10.00 | — | — | — | — | 62 |
| $C_6H_5$ | H | CHO | $C_6H_4OCH_3$-$p$ | — | 9.84 | 10.01 | — | — | — | — | 72 |
| H | H | H | $COOC_2H_5$ | 9.26 | 8.07 | 8.55 | — | 6.2 | — | $CDCl_3$ | 49 |
| $COOC_2H_5$ | H | H | $COOC_2H_5$ | — | 8.72 | 8.72 | — | — | — | $CDCl_3$ | 49 |
| $C_6H_5$ | H | $COOC_2H_5$ | $C_6H_5$ | — | 8.49 | — | — | — | — | $CDCl_3$ | 45 |
| 2-Thienyl | H | $COOC_2H_5$ | 2-Thienyl | — | 8.16 | — | — | — | — | $CDCl_3$ | 45 |
| $CH_3$ | H | Br | $CH_3$ | 2.63 | 7.94 | — | 2.72 | — | — | $CDCl_3$ | 50 |
| $C_6H_5$ | H | Br | $CH_3$ | — | 8.49 | — | 2.73 | — | — | $CDCl_3$ | 50 |
| $C_6H_5$ | H | Br | $C_6H_5$ | — | 8.74 | — | — | — | — | $CDCl_3$ | 50 |
| $SCH_3$ | $(CH_2)_n$ $n=2,3,4,5$ | | $SCH_3$ | 2.61 | — | — | 2.61 | — | — | $CDCl_3$ | 60 |
| $C_6H_5$ | H | H | $SCH_3$ | — | 7.72 | 7.68 | 2.62 | — | — | $CDCl_3$ | 46, 61 |
| $SCH_3$ | H | $C_6H_5$ | $SCH_3$ | 2.49 | 6.98 | — | 2.49 | — | — | $CDCl_3$ | 46 |
| $C_6H_5$ | H | $COOC_2H_5$ | $SCH_3$ | — | 8.30 | — | 2.54 | — | — | $CDCl_3$ | 46, 61 |

[a] Protons directly linked to the nucleus or included in the following substituents: –$CH_3$, –$CHO$, –$SCH_3$.

# Electrophilic Substitutions of Five-Membered Rings

GIANLORENZO MARINO

*Istituto di Chimica Organica, Università di Perugia, Perugia, Italy*

|     |                                                                 |     |
| --- | --------------------------------------------------------------- | --- |
| I.  | Introduction                                                    | 235 |
|     | A. Scope of the Review                                          | 235 |
|     | B. Note on the "Aromaticity" of Furan, Thiophene, and Pyrrole   | 237 |
| II. | Kinetic Studies and Mechanism                                   | 243 |
|     | A. General Aspects                                              | 243 |
|     | B. Hydrogen Exchange                                            | 244 |
|     | C. Halogenation                                                 | 246 |
|     | D. Nitration                                                    | 254 |
|     | E. Acylation and Related Reactions                              | 256 |
|     | F. Other Reactions                                              | 260 |
| III.| Relative Reactivity of Rings and Ring Positions                 | 263 |
|     | A. Monocyclic Systems with One Heteroatom                       | 263 |
|     | B. Monocyclic Systems with Two or More Heteroatoms              | 280 |
|     | C. Polycyclic Systems                                           | 284 |
| IV. | Effects of Substituents                                         | 293 |
|     | A. Directing Effects                                            | 293 |
|     | B. Hammett-Type Correlations                                    | 298 |

## I. Introduction

### A. Scope of the Review

The present chapter attempts to provide a sufficiently complete summary of all the quantitative and semiquantitative information available concerning electrophilic substitution in five-membered heteroaromatic rings.

The review is organized in the following manner. After a preliminary introduction to the "aromaticity" of the five-membered rings, the discussion (Section II) is centered mainly on the substitution process itself, and the mechanism of the various electrophilic substitutions are considered separately. Afterward attention is shifted from the electrophile to the heteroaromatic substrate (Section III) and the relative reactivity of different rings and ring positions is widely examined. Finally, the problem of the transmission of electronic effects of substituents across these rings and the application of Hammett-type correlations is discussed (Section IV). The importance of these reactions in synthetic chemistry is not treated at all.

Systems considered are the parent heterocycles with one heteroatom (furan, thiophene, selenophene, and pyrrole), their monoaza and polyaza derivatives, and the monobenzo derivatives of the above systems. Dibenzo derivatives such as dibenzofuran and carbazole are not taken into account, since they have no position in the five-membered ring susceptible to electrophilic attack. Likewise, no consideration is given to more complicated systems, formed by the fusion of two heterocyclic rings such as thienothiophenes or pyrrolopyridines, for which, in any case, no quantitative work is available.

General surveys on the chemistry of furan,[1] thiophene,[2] pyrrole,[3] benzothiophene,[4] and selenophene[4a] have been published in recent years; reference will be made to them, whenever necessary. In connection with the present compilation, the reviews of de la Mare and Ridd,[5] Ridd,[6] Eisch,[7] Katritzky and Johnson,[8] and Jones,[9] and the volume by Norman and Taylor[10] are also of special interest.

The literature has been covered to the end of 1969. Important work published in leading Journals until June 1970 and some unpublished work (mainly from the author's laboratory) has also been included.

[1] C. H. Eugster and D. P. Bosshard, *Advan. Heterocycl. Chem.* **7**, 377 (1966).
[2] S. Gronowitz, *Advan. Heterocycl. Chem.* **1**, 1 (1963).
[3] K. Schofield, "Heteroaromatic Nitrogen Compounds." Butterworths, London, 1967.
[4] B. Iddon and R. M. Scrowston, *Advan. Heterocycl. Chem.* **11**, 178 (1970).
[4a] N. N. Magdesieva, *Advan. Heterocycl. Chem.* **12**, 1 (1970).
[5] P. B. D. de la Mare and J. Ridd, "Aromatic Substitution." Butterworths, London, 1959.
[6] J. Ridd, *in* "Physical Methods in Heterocyclic Chemistry" (A. R. Katritzky, ed.), Vol. 1, p. 109. Academic Press, New York, 1963.
[7] J. J. Eisch, *Advan. Heterocycl. Chem.* **7**, 1 (1966).
[8] A. R. Katritzky and C. D. Johnson, *Angew. Chem., Int. Ed. Engl.* **6**, 608 (1967).
[9] R. A. Jones, *Advan. Heterocycl. Chem.* **11**, 383 (1970).
[10] R. O. C. Norman and R. Taylor, "Electrophilic Substitution in Benzenoid Compounds." Elsevier, Amsterdam, 1965.

## B. Note on the "Aromaticity" of Furan, Thiophene, and Pyrrole

Furan, thiophene, and pyrrole, the three heterocycles "parent" to all the others of this class, have a pentagonal planar structure. Bond lengths and angles are quantities of relevance to our discussion. The values reported below have been chosen because they are more comparable having been determined by the same authors[11-13] using the same procedure. In the valence bond description, these molecules

are considered as resonance hybrids of the following contributing structures (1–5):

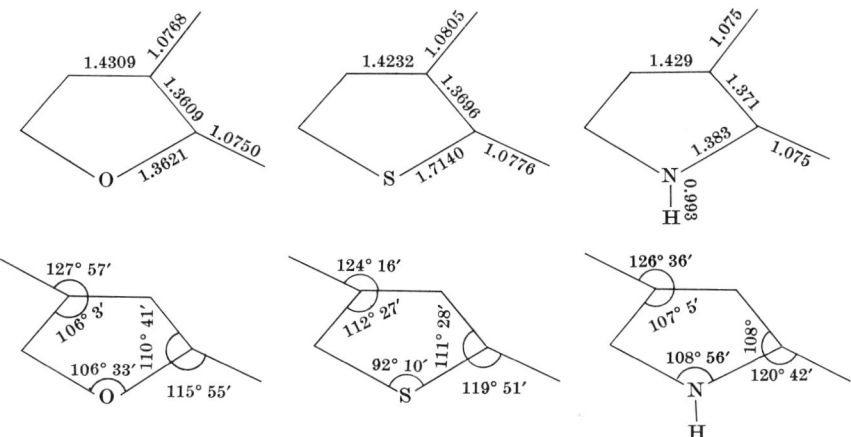

For thiophene, besides the limiting structures analogous to 1–5, structures such as 6–8 are possible owing to the capacity of the sulfur

---

[11] B. Bak, D. Christensen, W. B. Dixon, L. Hansen-Nygaard, R. Anderson, and M. Schottlander, *J. Mol. Spectrosc.* **9**, 124 (1962).

[12] B. Bak, D. Christensen, L. Hansen-Nygaard, and J. Restrup-Andersen, *J. Mol. Spectrosc.* **7**, 58 (1961).

[13] B. Bak, D. Christensen, L. Hansen-Nygaard, and J. Restrup-Andersen, *J. Chem. Phys.* **24**, 720 (1956).

atom to use $d$ orbitals. For furan, in view of its "dienic" character, structures such as **9** should also be considered.

(6)    (7)    (8)    (9)

Because of their high reactivity toward electrophilic reagents, Gilman[14] described them as "superaromatic" molecules. This term, used afterward in many textbooks of organic chemistry, is misleading since all these nuclei are actually *less* aromatic than benzene.

Different criteria may be used for comparing the "aromaticities" of these rings with each other and with benzene. A classic example is that based on the resonance energies; values for resonance energies for these rings from various sources have been assembled in Table I. The values differ widely according to the procedure used for the calculations; however, if the comparison is restricted to values similarly obtained, two points can be established beyond any doubt: (*i*) all three rings are less aromatic than benzene (resonance energy, 36 kcal/mole) and (*ii*) furan is the least aromatic of the series.

TABLE I

RESONANCE ENERGIES AND "AROMATICITIES" OF FURAN, PYRROLE, AND THIOPHENE

| Compound | Resonance energy[a] | | | Aromaticity (relative to benzene)[b] | | |
|---|---|---|---|---|---|---|
| | I | II | III | IV | V | VI |
| Furan | 23 | 22 | 16 | 46 | 52 | 67 |
| Pyrrole | 31 | 24.5 | 21 | 59 | 67 | — |
| Thiophene | 31 | 28 | 29 | 75 | 90 | 76 |

[a] Data in kcal/mole. I, L. Pauling, "The Nature of the Chemical Bond," 3rd ed., pp. 196–197. Cornell Univ. Press, New York, 1960. II, G. W. Wheland, "Resonance in Organic Chemistry," p. 99. Wiley, New York, 1955. III. F. Klages, *Chem. Ber.* **82**, 358 (1949).

[b] Based on "ring current" measurements (benzene = 100). IV, J. A. Elvidge, *Chem. Commun.* 160 (1965). V, D. W. Davies, *Chem. Commun.* 258 (1965). VI, H. A. P. De Jough and H. Wynberg, *Tetrahedron* **21**, 515 (1965).

[14] H. Gilman and E. B. Towne, *Rec. Trav. Chim. Pays-Bas* **51**, 1054 (1932).

Another criterion is that based on the ratio of the C-2–C-3 and C-3–C-4 lengths (the more aromatic the ring, the closer the ratio approaches 1). These ratios (based on Bak's values) are: thiophene, 0.964; pyrrole, 0.959; and furan, 0.950.

Applying the criterion of the induced "ring current,"[15] Elvidge[16] obtained the following comparative percentages of aromaticity: benzene, 100; thiophene, 75; pyrrole, 59; and furan, 46. However, other authors,[17-19] using the same method, arrived at somewhat different results.

To sum up, although each criterion has limits of validity, nevertheless all the criteria agree in establishing the following order of decreasing *ground state aromaticities*: benzene > thiophene > pyrrole > furan.

Because of conjugation, the $\pi$ electrons are displaced from the heteroatom *toward* the carbon atoms. On the other hand, the heteroatoms O, S, and N, being more electronegative than C, withdraw electrons *from* the ring by an inductive mechanism. Wherein lies the balance between these two opposite effects? What will be the electron distribution in the ground states of these molecules? Will it be similar to that of chloro- and bromobenzene or to that of anisole and aniline? Most textbooks of organic and heterocyclic chemistry incline toward the latter. The main argument favoring the latter is based on the chemical behavior of these molecules, which approaches more that of anisole and aniline than that of the halobenzenes.

In my opinion this is not a good argument. The great reactivity toward electrophiles is not, of necessity, an indication of a great ground-state electron density at the carbons of the ring (and, vice versa, the calculated electron densities *are not* a good reactivity index for electrophilic substitution). This statement is supported by, among others, the following observations.

The electron-attracting inductive effect of oxygen is much greater than that of sulfur; on the other hand, the conjugative effect is much smaller in furan than in thiophene, as shown by the larger degree of "bond fixation" observed in the former. Nevertheless, furan is much

---

[15] J. A. Elvidge and L. M. Jackman, *J. Chem. Soc.* 859 (1961).
[16] J. A. Elvidge, *Chem. Commun.* 160 (1965).
[17] H. A. P. de Jongh and H. Wynberg, *Tetrahedron* **21**, 515 (1965).
[18] D. W. Davies, *Chem. Commun.* 258 (1965).
[19] R. J. Abraham, R. C. Sheppard, W. A. Thomas, and S. Turner, *Chem. Commun.* 43 (1965); R. J. Abraham and W. A. Thomas, *J. Chem. Soc. B* 127 (1966).

more reactive than thiophene in all the electrophilic substitutions (see Section III). Furan and thiophene rings are also more susceptible than benzene to *nucleophilic* attack. Thus, halogeno derivatives of furan react about ten times as fast as the corresponding benzene derivatives in the reaction with methoxide ions.[20] Also, in the substitutions activated by nitro groups, furan and thiophene rings are much more reactive than the benzene ring. The relative rates of piperidino-debromination[21, 22] are 1-bromo-4-nitrobenzene, 1; 2-bromo-5-nitrothiophene, $4.7 \times 10^2$; 2-bromo-5-nitrofuran, $8.9 \times 10^4$.

The great reactivity toward electrophiles is due to the low localization energies, and is possibly also a consequence of the fact that the Wheland intermediates have the same number of covalent bonds as the starting molecules, as Dewar first pointed out.[23]

We consider now other approaches to the problem, namely, MO calculations, dipole moment measurements, and $pK_a$'s of carboxylic acids.

## 1. *MO Calculations*

Many calculations of $\pi$-electron densities are available in the literature, but the data are far from satisfactory. According to the method employed and the values of the parameters selected, different results are obtained.

The situation for thiophene is typical; practically every possible combination has been obtained in the order of electron densities at $\alpha$-C, $\beta$-C, and benzene (b) positions. (These calculations do not generally take into account the contributions of the $\sigma$ electrons.)

$\alpha = \beta = b$   Longuet-Higgins[24]; De Heer[25]

$b < \alpha < \beta$   Solony *et al.*[26]; McKreevoy[27]

---

[20] D. G. Manly and E. D. Amstutz, *J. Org. Chem.* **22**, 133 (1957).
[21] R. Motoyama, S. Nishimura, Y. Murakami, K. Hari, and E. Imoto, *Nippon Kagaku Zasshi* **78**, 950 (1957); *Chem. Abstr.* **54**, 14224 (1960).
[22] D. Spinelli, G. Guanti, and C. Dell'Erba, *Boll. Sci. Fac. Chim. Ind. Bologna* **25**, 71 (1967); *Chem. Abstr.* **68**, 2389 (1968); C. Dell'Erba, A. Salvemini, and D. Spinelli, *Ann. Chim. Rome* **52**, 1156 (1962).
[23] M. J. S. Dewar, "The Electronic Theory of Organic Chemistry," p. 188. Oxford, Clarendon Press, London, 1949.
[24] H. C. Longuet-Higgins, *Trans. Faraday Soc.* **45**, 173 (1949).
[25] J. de Heer, *J. Amer. Chem. Soc.* **76**, 4802 (1954).
[26] N. Solony, F. W. Birss, and J. B. Greenshields, *Can. J. Chem.* **43**, 1569 (1965).
[27] M. M. McKreevoy, *J. Amer. Chem. Soc.* **80**, 5543 (1958).

$b < \beta < \alpha$   Berthier and Pullman[28]; Pilar and Morris[29]
Fabian et al.[30]; Mangini and Zauli[31]

$\beta < b < \alpha$   Chiorboli and Manaresi[32]

$\alpha < b < \beta$   K. Kikuchi[33]

The situation is similar for furan and pyrrole. It is the opinion of the author that much time must pass before a satisfactory answer can be obtained from this approach.

## 2. *Dipole Moments*

The dipole moment of pyrrole $(1.80)^{34}$ is greater than that of tetrahydropyrrole (1.57 D); this seems to suggest that the $\pi$ moment is larger that the $\sigma$ moment and that the positive charge of the total dipole is on the nitrogen atom.

The dipole moments of furan (0.72 D) and thiophene (0.53 D) are smaller than those of the corresponding saturated heterocycles (tetrahydrofuran, 1.68 D; tetrahydrothiophene, 1.87 D.) Many authors[35-38] seem to believe that, also in these cases, the direction of the dipole is from the heteroatom (positive pole) to the ring (negative pole); others, however, are of a different opinion.[12, 32, 39, 40]

The values of the dipole moments of 2- and 3-substituted thiophens and furans,[34, 40a, 40b] however, seem to indicate that the latter authors are in the right.

[28] G. Berthier and B. Pullman, *C. R. Acad. Sci.* **231**, 744 (1950).
[29] F. L. Pilar and J. R. Morris, *J. Chem. Phys.* **34**, 389 (1961).
[30] J. Fabian, A. Mehlhorn, and R. Zahradnik, *J. Phys. Chem.* **72**, 3975 (1968).
[31] A. Mangini and B. Zauli, *J. Chem. Soc.* 2210 (1960).
[32] P. Chiorboli and P. Manaresi, *Gazz. Chim. Ital.* **84**, 248 (1954).
[33] K. Kikuchi,, *Sci. Rep Tohoku Univ.*, Ser. 1 **40**, 133 (1956); *Chem. Abstr.* **52**, 8727 (1958).
[34] A. M. McClellan, "Tables of Experimental Dipole Moments." Freeman, San Francisco, 1963.
[35] R. M. Acheson, "An Introduction to the Chemistry of Heterocyclic Compounds," 2nd ed., p. 143. Wiley (Interscience), New York, 1967.
[36] M. H. Palmer, "The Structure and Reactions of Heterocyclic Compounds," p. 255. Arnold, London, 1967.
[37] L. A. Paquette, "Principles of Modern Heterocyclic Chemistry," p. 104. Benjamin, New York, 1968.
[38] A. Albert, "Heterocyclic Chemistry," 2nd ed. Athlone Press, London, 1968.
[39] W. Adam and A. Grimison, *Theor. Chim. Acta* **7**, 342 (1967).
[40] F. Momicchioli and G. Del Re, *J. Chem. Soc. B* **674** (1969).
[40a] H. Lumbroso and C. Carpanelli, *Bull. Soc. Chim. France*, 3198 (1964).
[40b] P. Pigenet, J. P. Morizur, Y. Pascal, and H. Lumbroso, *Bull. Soc. Chim. France*, 361 (1969).

## 3. $pK_a$'s of Carboxylic Acids

In Table II the p$K$'s of the 2- and 3-carboxylic acids of thiophene, furan, and pyrrole are reported. While pyrrole carboxylic acids, like the alkoxy- and amino-substituted benzoic acids, are weaker, thiophene- and furancarboxylic acids, like the chloro-and bromobenzoic acids, are stronger than unsubstituted benzoic acid. This behavior is confirmed by other side-chain reactivity data.[41]

TABLE II

IONIZATION CONSTANTS OF THE CARBOXYLIC ACIDS[a]

| Acid | $pK_a$ | Ref.[b] |
|---|---|---|
| Benzoic | 4.21 | 1 |
| 2-Furoic | 3.16 | 2 |
| 3-Furoic | 3.95 | 3 |
| 2-Thiophene carboxylic | 3.53 | 2 |
| 3-Thiophene carboxylic | 4.10 | 4 |
| 2-Pyrrole carboxylic | 4.45 | 2 |
| 3-Pyrrole carboxylic | 5.07 | 5 |

[a] In water, at 25°C.
[b] KEY TO REFERENCES:
1 G. Kortum, W. Vogel, and K. Andrussow, "Dissociation Constants of Organic Acids in Aqueous Solutions," p. 352. Butterworth, London, 1961.
2 P. O. Lumme, *Suom. Kemistilehti B* **33**, 87 (1960); *Chem. Abstr.* **55**, 5104 (1961).
3 W. E. Catlin, *Iowa State Coll. J. Sci.* **10**, 65 (1935); *Chem. Abstr.* **30**, 935 (1936).
4 J. M. Loven, *Z. Phys. Chem.* **19**, 456 (1896).
5 H. Rapoport and C. D. Willson, *J. Org. Chem.* **26**, 1102 (1961).

To sum up, for pyrrole the situation seems to be clearer in the sense that the conjugative effect overcomes the inductive, leading to a charge distribution similar to that of aniline. This hypothesis is confirmed by the fact that pyrrole itself is a weak acid and that the basic center is not the nitrogen atom but the $\alpha$-carbon.[42] For furan

[41] H. H. Jaffé and H. L. Jones, *Advan. Heterocycl. Chem.* **3**, 238 (1963).
[42] R. J. Abraham, E. Bullock, and S. S. Mitra, *Can. J. Chem.* **37**, 1859 (1969).

and thiophene, the available data seem in favor of the opposite situation (the inductive effect prevailing).

## II. Kinetic Studies and Mechanism

### A. General Aspects

To a first approximation, electrophilic substitution in five-membered heteroaromatic substrates proceeds through a mechanism which does not differ substantially from that usually accepted for homocyclic aromatic substrates.

The most widely accepted mechanism for electrophilic aromatic substitution involves a change from $sp^2$ to $sp^3$ hybridization of the carbon under attack, with formation of a species (the Wheland or $\sigma$ complex) which is a real intermediate, i.e., a minimum in the energy-reaction coordinate diagram. In most of cases the rate-determining step is the formation of the $\sigma$ intermediate; in other cases, depending on the structure of the substrate, the nature of the electrophile, and the reaction conditions, the decomposition of such an intermediate is kinetically significant. In such cases a positive primary kinetic isotope effect and a base catalysis are expected (as Melander[43] first pointed out).

Prior to, and perhaps also after, the formation of the $\sigma$ complex, a $\pi$ complex (with the aromatic ring behaving as electron donor) can form, although it has not been proved that its formation is a *necessary* step in the reaction path.[44] It has recently been suggested that the formation of the $\pi$ complex could in some cases become rate-determining, when the electrophile is a very powerful one.[45] This hypothesis, although questioned both from the experimental and theoretical point of view,[46-48] is a possibility and could be applicable also when the aromatic substrate is a powerful nucleophile (as is the case with many five-membered heterocyclic rings).

A number of factors often contributes to making the substitution in heteroaromatic substrates more complicated.

[43] L. Melander, *Ark. Kemi* **2**, 211 (1950).
[44] E. Berliner, *Prog. Phys. Org. Chem.* **2**, 253 (1964).
[45] G. A. Olah, S. J. Kuhn, and S. H. Flood, *J. Amer. Chem. Soc.* **83**, 4571 (1961).
[46] W. S. Tolgyesi, *Can. J. Chem.* **43**, 3279 (1965).
[47] R. Nakane and A. Natsubori, *J. Amer. Chem. Soc.* **88**, 3011 (1966).
[48] S. Y. Caille and R. J. P. Corriu, *Chem. Commun.* 1251 (1967).

(1) When a basic center is present in the molecule (as in most nitrogen-containing heterocycles) and the substitution is achieved in strongly acidic media, the substrate is almost completely protonated. Even in more weakly acidic media (as in acetic acid), hydrogen bonding to the nitrogen atom can alter the orientation pattern and the relative rates.

(2) When a secondary nitrogen atom is present (as in pyrrole and imidazole) reaction can occur through the dissociated anion, with formation of either the $N$- or the $C$-substituted product. The $N$-substituted isomer, may, in turn, rearrange to the $C$-substituted one.

(3) Often the heteroatom is the most nucleophilic center in the molecule, for instance, in imidazole, In such cases the electrophile, e.g., a halogen molecule or the catalyst (a Lewis acid for Friedel–Crafts type reactions) may interact with the heteroatom with formation of complexes of various kinds ($n$-donor, charge-transfer, etc.).

(4) Since heterocyclic systems exhibit a greater or lesser degree of "bond fixation," addition (either of the 1,2- or 1,4-type) can accompany substitution. Alternatively, the addition products may be converted into the final substitution products through an elimination process.

(5) The substitution reaction may occur on a prototropic "non-aromatic" form. Both types of tautomerism, i.e., that prevailing in the parent heterocycles (pyrrole–pyrrolenine, indole–indolenine) and that typical of the hydroxy and amino derivatives, must be considered.

Whereas some of these complications are also encountered in substitutions of six-membered rings,[7, 8, 49–51] others are specifically characteristic of the reactions of the five-membered rings.

In the following sections, the mechanism of specific electrophilic substitutions will be discussed on the basis of the available kinetic or quantitative work.

## B. Hydrogen Exchange

Reactions in which one isotope of hydrogen on an aromatic nucleus is replaced by another are called by the general term—hydrogen exchange. According to the particular isotope involved as displaced

---

[49] R. D. Brown, in "Current Trends in Heterocyclic Chemistry" (A. Albert, ed.), p. 13. Butterworths, London, 1958.
[50] K. Schofield, Quart. Rev. 4, 382 (1950).
[51] J. H. Ridd, Z. Chem. 8, 201 (1968).

group and as electrophile, there are, in theory, six different reactions: protodedeuteriation (H/D), protodetritiation (H/T), deuteriation (D/H), tritiation (T/H), deuteriodetritiation (D/T), and tritiodedeuteriation (T/D). Hydrogen exchange reactions have several peculiar characteristics which make them particularly suitable for studies of the mechanism of electrophilic substitution and of reactivity–structure correlations.[52]

A number of kinetic studies on hydrogen exchange of derivatives of furan,[53] thiophene,[53-58] pyrrole,[53] selenophene,[59] imidazole,[60] oxazole,[60] thiazole,[60] and indole[61] have been reported. The kinetic picture is the same as for benzene derivatives; reactions follow pseudo first-order kinetics and the $k_1$ values increase markedly as the acidity of the medium increases.

Butler and Hendry[56] have found that there is a linear correlation between log $k$ and $-H_0$ for protodetritiation at the 2- and 3- positions of thiophene and the linearity extends over two $H_0$ units. The slopes are very close to those reported for hydrogen exchange in benzene derivatives.[62] These authors report the existence of a general acid catalysis in the protodetritiation of 5-methoxy-2-tritiothiophene and conclude that the reaction follows an A-$S_E2$ mechanism. Such a mechanism is also confirmed by the values of the activation entropies, which fit onto a plot of log $k$ against $\Delta S^{\ddagger}$ given by Matesich[63] for reactions following this mechanism.

[52] See Norman and Taylor,[10] p. 202.
[53] K. Schwetlick, K. Unverferth, and R. Mayer, *Z. Chem.* **7**, 58 (1967).
[54] A. I. Shatenshtein, A. G. Kamrad, I. O. Shapiro, Yu. I. Ranneva, and E. N. Zvyagintseva, *Dokl. Akad. Nauk SSSR* **168**, 364 (1966); *Chem. Abstr.* **65**, 8695g (1966).
[55] E. N. Zvyagintseva, T. A. Yakushina, and A. I. Shatenshtein, *Zh. Obshch. Khim.* **38**, 1993 (1968); *Chem. Abstr.* **70** 19292 (1969).
[56] A. R. Butler and J. B. Hendry, *J. Chem. Soc. B* 852 (1970).
[57] K. Halvarson and L. Melander, *Ark. Kemi* **8**, 29 (1956).
[58] B. Östman and S. Olsson, *Ark. Kemi* **15**, 275 (1960).
[59] A. I. Shatenshtein, N. N. Magdesieva, Yu. I. Ranneva, I. O. Shapiro, and A. I. Serebryanskaya, *Teor. Eksp. Khim.* **3**, 343 (1967); *Chem. Abstr.* **68**, 58799 (1968).
[60] H. A. Staab, M. T. Wu, A. Mannschreck, and G. Schwalbach, *Tetrahedron Lett.* 854 (1964).
[61] B. C. Challis and F. A. Long, *J. Amer. Chem. Soc.* **85**, 2524 (1963).
[62] V. Gold and D. P. N. Satchell, *J. Chem. Soc.* 3609, 3619, 3622 (1955).
[63] M. A. Matesich, *J. Org. Chem.* **32**, 1258 (1967).

Östman and Olsson[58] have determined the kinetic isotope effect in the isotopic exchange of thiophene, by comparing the rates of protodedeuteration and protodetritiation at both positions of the ring. The values obtained for the $k_D/k_T$ ratios are 1.94 for the $\alpha$ position and 1.70 for the $\beta$ position. Thus the ratio $k_D/k_T$ seems to increase with increasing reactivity, a tendency already observed in the hydrogen exchange of benzene and toluene.[64]

Most papers on this reaction deal with relations between structure and reactivity, and the results will be discussed more extensively in Sections III and IV.

Hydrogen exchange reactions of heteroaromatics[54, 59, 65] carried out in strongly alkaline media, such as potassium amide/liquid ammonia, alcoholic solutions of alkoxides, or solutions of potassium $t$-butoxide in dimethyl sulfoxide, proceed through an entirely different mechanism (sometimes called "protophilic") involving a carbanion-type intermediate[66, 67]; they are not electrophilic substitutions as such and will not be treated in this review.

## C. Halogenation

The main electrophilic reagents concerned in halogenation reactions are the neutral molecules $X_2$ (or $XY$), positively charged species, such as $X^+$ or $XOH_2^+$, and the hypohalous acids,[68] $HOX$. Since different mechanisms involving different electrophilic species can operate in the same system, a subdivision of this section based on the nature of the halogen (chlorination, bromination, iodination) has been preferred to that based on the type of electrophilic species.

### 1. *Chlorination*

Thiophene reacts with chlorine to give a mixture of substitution

---

[64] L. Melander and S. Olsson, *Acta Chem. Scand.* **10**, 879 (1956).
[65] I. O. Shapiro, L. I. Belen'kii, I. A. Romanskii, F. M. Stoyanovich, Ya. L. Gol'dfarb, and A. I. Shatenshtein, *Zh. Obshch. Khim.* **38**, 1998 (1968); *Chem. Abstr.* **70**, 19289 (1969).
[66] A. I. Shatenshtein, *Advan. Phys. Org. Chem.* **1**, 155 (1963).
[67] A. Streitwieser, D. E. Van Sickle, and L. Reif, *J. Amer. Chem. Soc.* **82**, 1513 (1960).
[68] See de la Mare and Ridd,[5] p. 116–129.

and addition products.[69–71] However, under more controlled conditions, i.e., in dilute acetic acid solution and in the dark, the amount of addition products is reduced to a minimum and practically only substitution products are formed.[72] Under these conditions, the reaction follows second-order kinetics, first-order with respect to thiophene and chlorine, a behavior analogous to that observed in benzene derivatives.[73, 74] The relative rate of thiophene to benzene was established[72] as $1.3 \times 10^7$. Substitution occurs mainly at the $\alpha$ position; only 1% $\beta$ isomer was detected by gas chromatography.[75]

The chlorination of benzothiophene (main product—3-chlorobenzothiophene) also follows second-order kinetics.[76, 77] A change in value of the rate constant has been observed, as the initial concentration of the reactants was varied over an 8-fold range. This variation, however small, has been ascribed to the formation of a weak complex (of the $\pi$ type) between benzothiophene and chlorine.[76]

2,3-Dimethylbenzothiophene when treated with chlorine in acetic acid and in the dark gives 2-chloromethyl-3-methylbenzothiophene nearly quantitatively.[76] Values of the $k_2$ constants decrease considerably when the initial concentration of the reactants increases. The presence of oxygen does not affect the $k$ values, a fact which should exclude interference from free radical chlorination.

The kinetic pattern could be accounted for by the formation of a stable intermediate of type **10**.

**(10)**

[69] W. Steinkopf, "Die Chemie des Thiophens," p. 35. Th. Steinkopf, Dresden, 1941.
[70] H. L. Coonradt and H. D. Hartough, *J. Amer. Chem. Soc.* **70**, 1158 (1948).
[71] H. L. Coonradt, H. D. Hartough, and G. C. Johnson, *J. Amer. Chem. Soc.* **70**, 2564 (1948).
[72] G. Marino, *Tetrahedron* **21**, 843 (1965).
[73] K. J. P. Orton and A. E. Bradfield, *J. Chem. Soc.* 986 (1927); A. E. Bradfield and B. Jones, *ibid.* 1006, 3073 (1928).
[74] H. C. Brown and L. M. Stock, *J. Amer. Chem. Soc.* **79**, 5175 (1957).
[75] S. Clementi, P. Linda, and G. Marino, *J. Chem. Soc. B* 1153 (1970).
[76] E. Baciocchi and L. Mandolini, *J. Chem. Soc. B* 397 (1968).
[77] S. Clementi, P. Linda, and G. Marino, *J. Chem. Soc. B* 79 (1971).

This intermediate can then rearrange to the final product (2-chloromethyl-3-methylbenzothiophene) by a mechanism similar to that postulated for the side-chain chlorination of polymethylbenzenes.[78,79] Other types of intermediate complex for this reaction have been also considered,[76] but the kinetic data do not distinguish between them.

Furan reacts with chlorine giving a mixture of polysubstitution and addition products.[80] However, the introduction of an electron-withdrawing group, such as methoxycarbonyl, makes the reaction yield only one product, and 5-chloro-2-carbomethoxyfuran is the sole product of a second-order reaction.[81]

## 2. Bromination

In contrast with chlorination, bromination of thiophene always gives substitution products exclusively, and no addition products have been isolated under a variety of experimental conditions.[82] Lauer[83] first studied the kinetics of the reaction of thiophene with molecular bromine; however, the reported value for the rate relative to benzene ($2 \times 10^4$) is not reliable, because of the uncorrected value of $k$ for bromination of benzene (for a discussion of this point, see Marino[72]). Later, the rate of bromination of thiophene relative to benzene in acetic acid was determined as $1.9 \times 10^9$, by comparing the times necessary to achieve 10% reaction in the bromination of thiophene and mesitylene, under the same conditions.[72]

In the presence of a large excess of bromide ions, the bromination of thiophene follows second-order kinetics,[84] first-order in each reagent. The effects of the water content of the solvent, the ionic strength of the solution, the bromide ion concentration, and substituents in the aromatic substrate[85] are the same in the bromination of

[78] E. Baciocchi and G. Illuminati, *Tetrahedon Lett.* 63 (1962).
[79] E. Baciocchi, A. Ciana, G. Illuminati, and C. Pasini, *J. Amer. Chem. Soc.* **87**, 3593 (1965).
[80] A. P. Dunlop and F. N. Peters, "The Furans," Reinhold, New York, 1953.
[81] P. Linda and G. Marino, unpublished observations.
[82] S.-O. Lawesson, *Ark. Kemi* **11**, 373 (1957).
[83] K. Lauer, *Ber.* **69**, 2618 (1936).
[84] P. Linda and G. Marino, *J. Chem. Soc. B* 392 (1968).
[85] G. Marino, *Rend. Accad. Naz. Lincei* **38**, 700 (1965); *Chem. Abstr.* **64**, 1917 (1966).

thiophene and benzene,[86] showing that the reaction proceeds in both aromatic systems by a substantially similar mechanism.

These findings have been confirmed by the more recent work of Butler and Hendry,[87] who studied the kinetics of bromination of thiophene in 85% aqueous acetic acid. Another argument supporting a mechanism similar to that of bromination of benzene derivatives is based on the observed values for the Arrhenius parameters; activation enthalpies and entropies for bromination of thiophene and mesitylene are very similar.[87] These authors report a small kinetic isotope effect for the bromination of thiophene ($k_H/k_D = 1.35$) but, on the basis of other arguments, they conclude that it is a secondary one, as in the bromination of benzene derivatives,[88] and that the rate-determining step is the attack on the thiophene ring by molecular bromine to give the Wheland intermediate. The difference in the size of the isotope effects for bromination from one side, and nitration and H/D exchange from the other, may reflect—according to these authors—the different amount of hyperconjugation (between the aromatic $\pi$ system and the C–H bond undergoing substitution) occurring in these reactions.[89]

Kinetic studies on the bromination of selenophene[90] and benzothiophene[77] have also been reported.

Furan and pyrrole are decomposed under bromination conditions. However, the introduction of an electron-withdrawing substituent, such as the methoxycarbonyl group, stabilizes the ring; bromination of the 2-methoxycarbonyl derivatives of furan and pyrrole follows regular kinetics, and the relative rates for these rings with respect to thiophene have been determined[84] (see Section III, A, 1).

Bromination of thiophene derivatives by $N$-bromosuccinimide (NBS) in 50:50 (v/v) chloroform–acetic acid proceeds by an electrophilic rather than a radical mechanism.[91] Many experimental facts favor such a hypothesis: (*i*) only ring-substituted products are formed, with exclusion of addition or side-chain substitution; (*ii*) the reaction is very selective, only $\alpha$-substituted products are formed; and (*iii*) the

---

[86] P. W. Robertson, *J. Chem. Soc.* 1267 (1954) and previous papers of the series.
[87] A. R. Butler and J. B. Hendry, *J. Chem. Soc. B* 170 (1970).
[88] E. Berliner and K. E. Schueller, *Chem. Ind. (London)* 1444 (1960).
[89] A. Streitwieser, R. H. Jagow, R. C. Fahey, and S. Suzuki, *J. Amer. Chem. Soc.* **80**, 2326 (1958).
[90] P. Linda and G. Marino, *J. Chem. Soc. B* 43 (1970).
[91] R. M. Kellogg, A. P. Schaap, E. T. Harper, and H. Wynberg, *J. Org. Chem.* **33**, 2902 (1968).

reaction is catalyzed by acetic acid. Bromination by NBS may be a very convenient procedure when the substrate or the product is sensitive to strong acids, since the hydrobromic acid generated during reaction is immediately consumed.[91]

The bromination of imidazole derivatives is more complicated. Imidazole, treated with an equimolecular amount of bromine in chloroform, yields, after removal of solvent and boiling the residue with water, 2,4,5-tribromoimidazole hydrobromide, ammonium bromide, and a very small amount of 4,5-dibromoimidazole.[92]

Linda[93] was able to isolate an orange complex in the bromination of imidazole, which proved to be a dicoordinate complex of unipositive bromine. The bromination of imidazole should proceed according to Scheme 1.

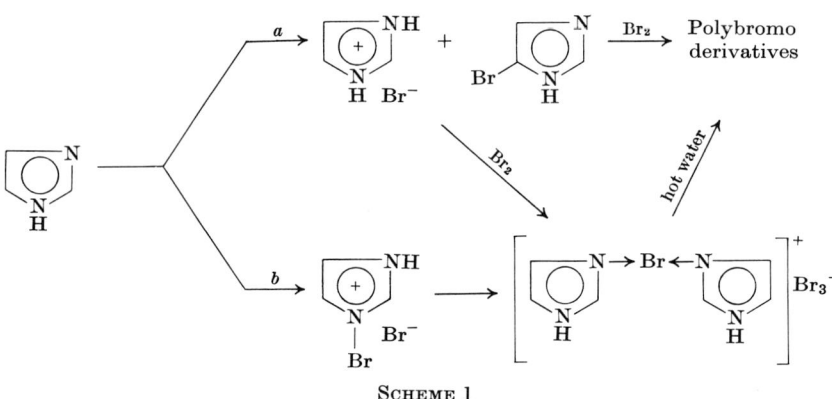

SCHEME 1

Path *a* is the preferred one. Polysubstitution (also when bromine is used in deficient amount) is due to the fact that unsubstituted imidazole is more basic than bromoimidazole and is therefore preferentially converted into the inactive protonated form.

Path *b*, on the contrary, is preferred in the bromination of benzimidazole, since the reactive positions 4 and 5 are blocked by the benzene ring. Under similar conditions, *N*-methylbenzimidazole, which has no reactive positions, gives a 1:1 complex of the *n*-donor type.[93]

[92] J. E. Balaban and F. L. Pyman, *J. Chem. Soc.* 947 (1922).
[93] P. Linda, *Tetrahedron* **25**, 3297 (1969).

### 3. Iodination

A number of kinetic studies on the iodination of reactive five-membered heterocycles in buffered aqueous solutions have been reported.

Doak and Corwin[94] have studied the kinetics of the iodination of some trisubstituted derivatives of pyrrole and $N$-methylpyrrole. Both free iodine and hypoiodous acid are supposed to be the iodinating agents in this reaction. The similar reactivity exhibited by pyrrole and $N$-methylpyrrole contradicts the hypothesis previously proposed,[95] that the great reactivity of pyrroles in iodination may be due to a reaction either of the dissociated anion or the pyrrolenine tautomer.

Many research workers have studied the kinetics and the mechanism of iodination of imidazole.[96-100] The amphoteric character of the imidazole molecule and the existence of tautomeric equilibria make the reaction more complicated. In aqueous solution, at about pH 7, the rate equation consists of two terms, one suggesting reaction of the imidazole anion with $I^+$ or $IOH_2^+$, and the other, involving dependence on the square of the imidazole concentration, indicating self-catalysis by imidazole.[96, 97] According to Ridd, the uncatalyzed term could also represent an internal rearrangement of an $N$-iodo to a $C$-iodo derivative, and the catalyzed reaction could be due to assisted proton loss in this transition state.

Further information about the position of attack and the nature of the rate-determining step was obtained by Grimson and Ridd[98] who studied the kinetics of iodination of some deuterio derivatives of imidazole. The kinetic data on the iodination of 2,4,5-trideuterio-imidazole enabled them to calculate the isotope effects for both the uncatalyzed ($k_H/k_D = 4.36$) and the self-catalyzed ($k_H/k_D = 4.47$) reactions. A comparison of the istope effects observed in the iodination

---

[94] K. W. Doak and A. H. Corwin, *J. Amer. Chem. Soc.* **71**, 159 (1949).
[95] N. V. Sidwick, "The Organic Chemistry of Nitrogen," pp. 482–490. Oxford Univ. Press, London and New York, 1937.
[96] J. H. Ridd, *J. Chem. Soc.* 1238 (1955).
[97] J. H. Ridd, *Proc. Chem. Soc.* 256 (1958).
[98] A. Grimison and J. H. Ridd, *J. Chem. Soc.* 3019 (1959).
[99] L. Shutte, P. Provó Kluit, and E. Havinga, *Tetrahedron Suppl.* **7**, 295 (1966).
[100] D. M. Brouwer, M. J. Van der Vlugt, and E. Havinga, *Kon. Ned. Akad. Wetenschap, Proc., Ser. B* **62**, 93 (1959). *Chem. Abstr.* **54**, 1335g (1960).

of 2,4,5-trideuterioimidazole (4.5), 4,5-dideuterioimidazole (3.7), and 2-deuterioimidazole (about 1) indicates that the C–H bond at the 4(5) position is weakened in the rate-determining step and implies, in contrast with previous reports,[101] that the initial substitution is mainly at the 4-position.

Only in strongly alkaline solution is the orientation of the first attack possibly different.[101]

The main kinetic features of the iodination of imidazole—which show a suggestive resemblance to those of the iodination of phenols[102, 103]—have been substantially confirmed by a study of Havinga and co-workers[99, 100] on the kinetics of iodination of histidine and some other imidazole derivatives.

Although several different rate laws have been drawn for different substrates, the kinetic equations are generally consistent with a mechanism involving an electrophilic attack by $I_2$ on an anionic imidazole ring system, followed by proton abstraction from the sigma complex as the rate-determining step.

The data permit the conclusion that the reactivity of the dissociated imidazole anion exceeds that of the neutral molecule by at least a factor of $10^8$.

Some peculiar aspects of the kinetic behavior of histidine are ascribed to the formation of a tautomeric form of the anion (11), favored by the ammonium group in the side chain.

(11)

Vaughan and co-workers[104] have studied the iodination of pyrazole. The kinetic features are similar to those of the iodination of imidazole. Also in this case, the rate could be separated into parallel "uncatalyzed" and "base-catalyzed" reactions and the mechanism proposed

---

[101] C. Pauly and E. Arauner, *J. Prakt. Chem.* **118**, 33 (1928).
[102] E. Berliner, *J. Amer. Chem. Soc.* **73**, 4307 (1951).
[103] E. Grovenstein and N. S. Apprahanian, *J. Amer. Chem. Soc.* **84**, 212 (1962).
[104] J. D. Vaughan, D. G. Lambert, and V. L. Vaughan, *J. Amer. Chem. Soc.* **86**, 2857 (1964).

(*vide infra*) involves the attack of the conjugate base of pyrazole by the electrophile (the iodine molecule).

$$I_3^- \rightleftarrows I_2 + I^-$$

$$\underset{\text{substrate}}{HS} + H_2O \rightleftarrows S^- + H_3O^+$$

$$S^- + I_2 \rightleftarrows S\diagup_H^I + I^-$$

$$S\diagup_H^I + H_2O \xrightarrow{\text{uncatalyzed}} SI^- + H_3O^+$$

$$S\diagup_H^I + B \xrightarrow{\text{base-catalyzed}} SI^- + BH^+$$

(the base may be pyrazole itself)

The reactivities of pyrazole and imidazole in their uncatalyzed reactions were found to be comparable in magnitude ($k_I/k_P = 1.3$).

Afterwards, the kinetics of iodination of some 1-alkyl-substituted pyrazoles in buffered solutions have been studied by the same research group.[105] In these cases dissociation is, of course, not possible and the rate law observed is the same as that of iodination of aniline.[106]

The reactivities of 1-methyl-, 1-ethyl-, and 1-isopropylpyrazoles follow the inductive order of the alkyl group. The rate of iodination of the pyrazole neutral molecule was estimated from those of the 1-alkyl derivatives and the relative rate $k_{P^-}/k_P$ for the uncatalyzed reaction was estimated to be in the range $3 \times 10^9$ to $7 \times 10^{12}$, or even larger if it is assumed that the anionic substrate undergoes attack by molecular iodine (the probability of encounter between the anion and $IOH_2^+$ is very small) and the neutral substrate by either $I_2$ or $IOH_2^+$.

The kinetics of iodination of indole and some of its derivatives in aqueous ethanol have been studied over a wide range of iodide ion concentration ($4 \times 10^{-4}$ to $10^{-1}$ $M$) and at different pH values (6.50–7.86).[107–109] The observed second-order rate constants are inversely

---

[105] J. D. Vaughan, G. L. Lewett, and V. L. Vaughan, *J. Amer. Chem. Soc.* **89**, 6218 (1967).
[106] E. Berliner, *J. Amer. Chem. Soc.* **72**, 4003 (1950).
[107] E. C. R. De Fabrizio, *Ann. Chim. (Rome)* **58**, 651 (1968).
[108] E. C. R. De Fabrizio, *Ann. Chim. (Rome)* **58**, 1435 (1968).
[109] P. Amat di San Filippo and E. C. R. De Fabrizio, *Ann. Chim. (Rome)* **59**, 799 (1969).

proportional to the square of the initial iodide concentration and insensitive to change in pH. The reaction exhibits a positive deuterium isotope effect[108] ($k_H/k_D \simeq 2$ for indole and $\simeq 3$ for 2-methylindole) which remains practically constant on varying the iodide ion concentration. These experimental results suggest a mechanism involving a rapid preequilibrium in which the iodinating agent (likely the molecular iodine) attacks the indole molecule, followed by a rate-determining proton transfer.

## D. Nitration

Since the classic papers by Ingold and his co-workers,[110, 111] nitration has for a long time been considered as the "standard" electrophilic substitution. Many orientation and relative rate data on the nitration of both carbocyclic and heterocyclic substrates have been accumulated and the results have been generalized as valid for all electrophilic substitutions. As a matter of fact, this popularity is partially undeserved; nitration is a complicated reaction, which can occur by a multiplicity of parallel mechanisms.[112] In particular, in the case of the very reactive substrates that five-membered heterocycles are, two complications may make meaningless both kinetic measurements and competitive experiments.[113] (*i*) Due to the great reactivity of both partners the encounter limiting rate may be achieved; in this case, of course, all the substrates react at the same rate and the effect of structure on the reactivity cannot be studied. (*ii*) Nitrous acid, always present in traces, may exert an anticatalytic effect in some cases and a markedly catalytic effect in others; with a very reactive substrate, nitration may proceed essentially via nitrosation, followed by oxidation. For these reasons, the nitration data must be handled with much caution.

The nitration of furan by nitric acid in organic solvents takes a peculiar course. It has been known for a long time[114–116] that the nitration of furan in acetic acid proceeds through the formation of a

---

[110] G. A. Benford and C. K. Ingold, *J. Chem. Soc.* 929 (1938).
[111] E. D. Hughes, C. K. Ingold, and R. I. Reed, *J. Chem. Soc.* 2400 (1950).
[112] J. G. Hoggett, R. B. Moodie, and K. Schofield, *Chem. Commun.* 605 (1969).
[113] J. G. Hoggett, R. B. Moodie, and K. Schofield, *J. Chem. Soc. B* 1 (1969).
[114] R. Marquis, *C. R. Acad. Sci.* **132**, 140 (1091); **134** 776 (1902).
[115] I. J. Rinkes, *Rec. Trav. Chim. Pay-Bas* **49**, 1169 (1930).
[116] H. Gilman and G. Wright, *J. Amer. Chem Soc.* **52**, 2550 (1930).

quite stable intermediate, which is subsequently converted into nitrofuran by the action of a mild base. Various structures were proposed for the nitration intermediate, before the structure of 2-acetoxy-5-nitro-2,5-dihydrofuran, (**12**) first suggested by Freure and Johnson[117] was firmly established by Clauson-Kaas and Fakstorp.[118]

$$O_2N \diagdown \diagup OCOCH_3$$
$$H \diagdown O \diagup H$$

(**12**)

Afterward, analogous intermediates have been isolated by several authors in the nitration of other furan derivatives bearing electron-withdrawing substituents.[119, 120] Michels and Hayes[119] studied the kinetics of the conversion of the addition intermediate into the final nitro-substituted derivative by the action of several bases.

Many quantitative and semiquantitative studies on the nitration of thiophene have been carried out[113, 121-123] and the rate of nitration of thiophene relative to benzene has been determined under different experimental conditions. However, these relative rate values probably do not reflect the true relative reactivities, for the reasons stated above.

Östman[122] has determined the kinetic isotope effect in the nitration of thiophene by benzoyl nitrate in acetonitrile; the observed effect ($k_T/k_H = 1.14 + 0.10$) is opposite to that expected for a primary isotope effect. It seems likely that it is a secondary α-isotope effect, ascribable to a rate-determining nature of the bond-formation step and similar to that observed by Olah et al. in the nitration of benzene derivatives with nitronium fluoroborate in tetramethylenesulfone.[124]

[117] B. T. Freure and J. R. Johnson, *J. Amer. Chem. Soc.* **53**, 1142 (1931).
[118] N. Clauson-Kaas and J. Fakstorp, *Acta Chem. Scand.* **1**, 210 (1947).
[119] J. G. Michels and K. J. Hayes, *J. Amer. Chem. Soc.* **80**, 1114 (1958).
[120] K. Venters, J. Stradins, and S. Hillers, *Tr. Soveshch. Fiz. Metod. Issled. Org. Soedin. Khim.*; *Akad, Nauk Kirg. SSR, Inst. Org. Khim. Frunze* **20** (1962); *Chem. Abstr.* **62**, 2677e (1965).
[121] R. Motoyama, E. Imoto, and J. Ogawa, *Nippon Kagaku Zasshi* **78**, 962 (1957); *Chem. Abstr.* **54**, 14224g (1960).
[122] B. Östman, *Ark. Kemi* **19**, 499 (1962).
[123] G. A. Olah, S. J. Kuhn, and A. Mlinkò, *J. Chem. Soc.* 4257 (1956).
[124] G. A. Olah, S. J. Kuhn, and S. H. Flood, *J. Amer. Chem. Soc.* **83**, 4571, 4581 (1961).

A fundamental investigation of the kinetics and the mechanism of nitration of five-membered heterocycles containing nitrogen is due to Ridd and his co-workers.[125, 126] They studied the reaction of pyrazole and imidazole with nitric acid in 90–99% sulfuric acid.

Since imidazole and pyrazole are almost completely protonated in the reaction medium, the reaction could, in principle, occur either through the conjugated acids or through the much smaller concentration of the much more reactive neutral molecules.

The observed kinetic law, the type of rate profile (plot of log $k$ vs. sulfuric acid concentration), the values of the Arrhenius parameters, the comparison of the observed reaction rates with the calculated encounter rates, and the agreement with the features of the nitration of quinoline[127] are in favor of a reaction of nitronium ions with the azolium cations. Only at lower acidities ($< 90\%$ $H_2SO_4$) can the reaction of the neutral azole molecules become important.

## E. Acylation and Related Reactions

### 1. Acylation

There are some differences between the experimental conditions used in acylation of heteroaromatic five-membered rings and of benzene derivatives.

(1) The use of acid anhydrides is generally preferred to that of acyl chlorides, since a weaker acid is liberated in the course of the reaction; this reduces the decomposition of acid-sensitive starting materials.

(2) Since aluminum chloride often induces polymerization, it is generally replaced by other milder catalysts, such as tin tetrachloride,[128–131] zinc chloride,[131, 132] magnesium perchlorate,[133] ferric

---

[125] M. W. Austin and J. H. Ridd, *Chem. Ind. (London)* 1057 (1962).
[126] M. W. Austin, V. R. Blackborow, J. H. Ridd, and B. V. Smith, *J. Chem. Soc.* 1051 (1965).
[127] M. W. Austin and J. H. Ridd, *J. Chem. Soc.* 4204 (1963).
[128] M. W. Farrar and R. Levine, *J. Amer. Chem. Soc.* **72**, 4433 (1950).
[129] Ya. L. Gol'dfarb and L. M. Smorgonskii, *J. Gen. Chem. USSR* **8**, 1523 (1938); *Chem. Abstr.* **33**, 4593 (1939).
[130] P. Linda and G. Marino, *Tetrahedron* **23**, 1739 (1967).
[131] W. R. Edwards and R. J. Eckert, *J. Org. Chem.* **31**, 1283 (1966).
[132] S. Hillers and I. Berklava, *Latv. PSR Zinat. Akad. Vestis* 53 (1956); *Chem. Abstr.* **51**, 5747d (1957).
[133] G. N. Dorofeenko and V. I. Dulenko, *Zh. Vses. Khim. Obschest. im. D.I. Mendeleeva* **7**, 120 (1962); *Chem. Abstr.* **56**, 13596 (1962).

chloride,[134] phosphoric acid,[135] and iodine.[130, 136]

(3) In contrast with the acylation of benzene derivatives, which generally requires stoichiometric amounts of the Lewis acid,[137] the acylation of five-membered heterocycles requires only catalytic amounts of the catalyst (generally 0.01 mole of catalyst for each mole of acid anhydride is sufficient). This is probably due either to the greater reactivity of the substrates or to the lower ability of the ketones formed to coordinate the catalyst.

(4) When the substrate has a basic center (imidazole, pyrazole) or is able to form complexes with Lewis acids (pyrrole),[138] the use of Friedel–Crafts type catalyst is completely prevented.[139] It is interesting to note, in this context, that while pyrrole is acetylated in the absence of any added catalyst,[140] it is *not* acetylated in the presence of tin tetrachloride.[139] Furthermore, the presence of pyrrole inhibits also the tin tetrachloride-catalyzed acetylation of furan and thiophene.[139]

In the reaction with acetic anhydride in an inert solvent, pyrrole gives a mixture of 1- and 2-acetyl derivatives.[139] The substitution at the 2-carbon seems to involve the neutral molecule of pyrrole, whereas that at nitrogen probably involves the dissociated anion. In fact, the C/N isomer ratio is decreased by adding sodium acetate (which favors ionization) and increased by adding acetic acid (which opposes it).

The relative reactivities of many five-membered heterocycles in acylation reactions[130, 141–143] have been determined by the competitive method, which has the advantage over the kinetic one of not requiring a complete knowledge of the reaction kinetics.

---

[134] G. G. Galust'yan and I. P. Tsukervanik, *Z. Obschch. Khim.* **34**, 1478 (1964); *Chem. Abstr.* **61**, 55906 (1964).
[135] H. D. Hartough and A. I. Kosak, *J. Amer. Chem. Soc.* **69**, 3093 (1947).
[136] H. D. Hartough and A. I. Kosak, *J. Amer. Chem. Soc.* **68**, 2639 (1946).
[137] P. H. Gore, in "Friedel–Crafts and Related Reactions" (G. A. Olah, ed), Vol. III, Part 1, p. 82 ff. Wiley (Interscience), New York, 1964.
[138] O. Schmitz-Dumont, *Chem. Ber.* **62**, 226 (1929).
[139] P. Linda and G. Marino, *Ric. Sci.* **37**, 424 (1967).
[140] G. L. Ciamician and M. Dennstedt, *Gazz. Chim. Ital.* **13**, 455 (1883); **15**, 9 (1885).
[141] Ya. L. Gol'dfarb, V. P. Litvinov, and V. I. Shvedov, *Zh. Obshch. Khim.* **30**, 534 (1960); *Chem. Abstr.* **54**, 24638g (1960).
[142] S. Clementi, F. Genel, and G. Marino, *Chem. Commun.* 498 (1967).
[143] S. Clementi and G. Marino, *Tetrahedron* **25**, 4599 (1969).

Relative rates and isomer distributions of the catalyzed acetylation of five-membered heterocycles are not affected by change in catalyst (tin tetrachloride or iodine).[130] This is an indication that the electrophilic species involved in the reaction is the acetylium ion $CH_3CO^+$ rather than the oxonium complex.

Trifluoroacetylation[142, 143, 143a] by trifluoroacetic anhydride in an inert solvent has proved to be particularly suitable for such studies of comparative reactivity. Trifluoroacetic anhydride is, in fact, able to acylate reactive aromatic substrates in the absence of any added Friedel–Crafts catalyst,[144] a circumstance which avoids the above-mentioned complications arising from the interactions of many heterocycles with such catalysts.[139]

Another interesting acylating reagent which does not need a catalyst is the mixed anhydride of acetic and trifluoroacetic acid, often called acetyl trifluoroacetate (ATFA).[145] Anisole and phenetole react with ATFA to give the acetylated products exclusively.[146] However, Marino and co-workers[147, 148] have observed that the five-membered heterocycles, when treated with ATFA, give mixtures of acetylated and trifluoroacetylated products, the composition of which depends on the reactivity of the substrate and the experimental conditions. The product distributions observed in 1,2-dichloroethane at 75° are reported in Table III. The data indicate that the amount of the trifluoroacetylated product increases as the reactivity of the substrate increases. Thiophene and 2-methylpyrrole represent the limit cases; the former yields the 2-acetyl derivative exclusively, whereas the latter forms over 90% 5-trifluoroacetylated compound.

Temperature, solvents of high polarity and Lewis acids favor acetylation over trifluoroacetylation, while adding lithium trifluoroacetate increases the amount of trifluoroacetylated product.[148] On the basis of these facts and of a comparison of the reactivity of this with other electrophilic reagents, a mechanism has been proposed,[148]

[143a] S. Clementi and G. Marino, *Chem. Commun.* 1642 (1970).
[144] W. D. Cooper, *J. Org. Chem.* **23**, 1382 (1958).
[145] P. H. Gore, in "Friedel–Crafts and Related Reactions" (G. A. Olah, ed.), Vol. 3, Part 1, p. 33. Wiley (Interscience), New York, 1964.
[146] E. J. Bourne, M. Stacey, J. C. Tatlow, and R. Worrall, *J. Chem. Soc.* 2006 (1954).
[147] S. Clementi, F. Genel, and G. Marino, *Ric. Sci.* **37**, 418 (1967); *Chem. Abstr.* **68**, 29509 (1968).
[148] S. Clementi and G. Marino, *Gazz. Chim. Ital.* **100**, 556 (1970).

Sec. II. E.] SUBSTITUTIONS OF FIVE-MEMBERED RINGS 259

according to which the trifluoroacetylated products are formed by action of the undissociated anhydride and the acetylated ones by action of the acetylium ions $CH_3CO^+$ (as free cations or ion pairs).

TABLE III

REACTION OF 5-MEMBERED HETEROCYCLES WITH ACETYL TRIFLUOROACETATE[a]

| Compound | Trifluoro-acetylated product (%) | Acetylated product (%) |
|---|---|---|
| Thiophene | 0.0 | 100 |
| Furan | 0.9 | 99.1 |
| 2-Methylthiophene | 1.4 | 98.6 |
| 2-Methylfuran | 15.9 | 84.1 |
| Pyrrole | 46.9 | 53.1 |
| 2-Methylpyrrole | 91.9 | 8.1 |

[a] Reaction in dichloroethane at 75°. Data from Clementi et al.[147] The compositions of the reaction products are given in mole%.

## 2. Formylation

Five-membered heteroaromatic rings can be formylated under a variety of conditions,[149-155] a behavior analogous to that of phenols and anilines.

A kinetic study of the Vilsmeier–Haak formylation of thiophene derivatives in dichloroethane solution has been recently reported.[156] Reactions of thiophene and 2-methylthiophene follow third-order kinetics, first-order in substrate, dimethylformamide (DMF), and phosphorus oxychloride. These results are in agreement with a mechanism involving a rapid preequilibrium step leading to an

[149] E. J.-H. Chu and T. C. Chu, J. Org. Chem. **19**, 266 (1954).
[150] V. J. Traynelis, J. J. Miskel, and J. R. Sowa, J. Org. Chem. **22**, 1269 (1957).
[151] I. L. Finar and G. H. Lord, J. Chem. Soc. 3314 (1957).
[152] M. Dezelic and K. Grom-Dursum, Glas. Drus. Hem. Technol. NR Bosne Hercegovine **9**, 49 (1960); Chem. Abstr. **58**, 2423h (1963).
[153] A. Rieche, H. Gross, and E. Hoft, Ber. **93**, 88 (1960).
[154] A. Ermili, A. J. Castro, and P. A. Westfall, J. Org. Chem. **30**, 339 (1965).
[155] B. A. Tertov and A. V. Koblik, Khim. Geterotsikl. Soedin. 1123 (1967); Chem. Abstr. **69**, 59158 (1968).
[156] P. Linda, G. Marino, and S. Santini, Tetrahedron Lett. 4223 (1970).

adduct[157, 158] $DMF \cdot POCl_3$, followed by a rate-determining step in which the latter attacks the heteroaromatic substrate.

(1) $DMF + POCl_3 \underset{k_{-1}}{\overset{k_1}{\rightleftarrows}}$ Electrophilic adduct

(2) Electrophilic adduct + Heterocycle $\overset{k_2}{\longrightarrow}$ Intermediate

(3) (Intermediate $\overset{H_2O}{\longrightarrow}$ Formyl derivative)

With very reactive substrates, such as 2-methoxythiophene, step (2) is very fast and step (1) becomes rate-determining; the reaction then follows second-order kinetics and the observed rate constant is equal to the rate constant $k_1$ for the formation of the electrophilic adduct.

Rate measurements on 2-methyl-5-deuteriothiophene have shown the existence of a small kinetic isotope effect ($k_H/k_D = 1.18 \pm 0.02$). There appears to be no way of deciding if such a small effect is a primary (i.e., indicating a partial weakening of the carbon–hydrogen bond in the transition state) or a secondary one.

## F. Other Reactions

### 1. Diazonium Coupling

This is another reaction typical of aromatic amines and phenols, which is successful with many five-membered heterocycles.

Imidazole is reported[159] to be coupled at position 2, in contrast with the other electrophilic substitutions which occur preferentially at position 4. This peculiar orientation, together with the observation that 1-methylimidazole does not couple with diazonium compounds,[160] supports the view that ionization of the N–H bond is a prerequisite for coupling.

A kinetic study of the coupling of imidazole with diazotized sulfanilic acid in buffered aqueous solution has been performed.[161]

[157] H. H. Boshard and H. Zollinger, *Helv. Chim. Acta* **42**, 1653 (1959).
[158] G. Martin and M. Martin, *Bull. Soc. Chim. Fr.* 1637 (1963).
[159] R. G. Fargher and F. L. Pyman, *J. Chem. Soc.* **115**, 217 (1919).
[160] R. Burian, *Ber.* **37**, 696 (1904).
[161] R. D. Brown, H. C. Duffin, J. C. Maynard, and J. H. Ridd, *J. Chem. Soc.* 3937 (1953).

There is a significant correspondence between the reactions of imidazole and phenol[162]; the kinetic equations have the same form and in both reactions there is no isotope effect.[98, 163] The reaction is second-order and plots of log $k$ against pH confirm that the anion is involved in the reaction.

Binks and Ridd[164] have made a complete kinetic study of the reaction of indole with several diazotized amines ($p$-nitroaniline, $p$-chloroaniline, sulfanilic acid, and aniline). Only the reaction with $p$-nitrodiazonium salt exhibits a simple kinetic form (pseudo first-order reaction); in the other cases the kinetics appear to be due to the superposition of two reactions, a normal azo-coupling reaction and an autocatalytic side reaction that removes diazonium ions, but does not form azo compounds.

In the pH range 3.7–6.2, the kinetic equation does not contain a pH-dependent term; this is consistent with a direct reaction of diazonium ions with neutral molecules of indole. At higher pH (6.2–8.6) an initial increase in rate is observed, in spite of the partial conversion of the diazonium chloride into the unreactive diazohydroxide. Probably an alternative path involving the conjugate base of indole also becomes operative.

As in the reaction of imidazole, no major isotope effect has been observed; 3-deuterioindole reacts at about the same rate as indole. The diazo coupling of five-membered heteroaromatic substrates therefore appears similar to that of homocyclic compounds where, in the absence of steric hindrance, slow attack of the diazonium ion is followed by a fast proton loss.[165]

## 2. Mercuration

The mercuration of thiophene shows some peculiarities. The rate of acetoxymercuration of thiophene is extraordinarily high,[166] the relative rate $k_T/k_B$ being $6.2 \times 10^5$. The $\alpha:\beta$ reactivity ratio is likewise very high and no amount of $\beta$ isomer was detected in the reaction mixture.[167] These facts are in contrast with the normally relatively

---

[162] H. Zollinger, *Chem. Rev.* **51**, 347 (1952).
[163] A. Grimison and J. H. Ridd, *Proc. Chem. Soc.* 256 (1958).
[164] J. H. Binks and J. H. Ridd, *J. Chem. Soc.* 2398 (1957).
[165] H. Zollinger, *Experientia* **12**, 165 (1956).
[166] Motoyama *et al.*[121] report for the relative rate $k_T/k_B$ a value of $7 \times 10^4$, apparently at variance with the reported rate constants.
[167] P. Linda and G. Marino, unpublished results.

low selectivity of the mercuration reaction[168]; reactions far more selective than mercuration exhibit lower values for the $k_T/k_B$ and $k_\alpha/k_\beta$ ratios[75] (see Section III,A,3). It is very probable that the mercuration of thiophene proceeds through a mechanism different from that of benzene, presumably *via* a preliminary coordination of mercury with sulfur atom.[169]

Kinetic studies on the mercuration of selenophene and its derivatives have also been reported.[170]

## 3. *Replacement Reactions*

By replacement reactions is generally meant those reactions in which groups besides hydrogen are substituted by an electrophile, often the proton itself. These reactions are very frequent in the chemistry of five-membered heteroaromatic compounds, and kinetic studies on some of them have been reported.

Brown, Buchanan, and Humffray have studied kinetically the protodemercuration[171] of 2- and 3-furyl, 2-thienyl, and 2-selenophenyl mercuric chlorides, and the protodeboronation[172] and the iododeboronation[173] of 2- and 3-thiophenboronic acids.

Eaborn and his co-workers have measured the rates of protodesilylation of 2- and 3-trimethylsilylthiophene under different experimental conditions[174,175] and of 2-trimethylsilylfuran[175] in methanol–perchloric acid solution. The kinetic behavior and the mechanism of the reactions do not present any peculiarity with respect to the analogous reactions of homocyclic derivatives.

---

[168] H. C. Brown and G. Goldman, *J. Amer. Chem. Soc.* **84**, 1650 (1962).
[169] M. Jones, *Advan. Chem. Ser.* **37**, 121 (1963).
[170] Yu. K. Yur'ev, M. A. Gal'bershtam, and I. I. Kandror, *Khim. Geterotsikl. Soedin.* 897 (1966); *Chem. Abstr.* **66**, 94453 (1967).
[171] R. D. Brown, A. S. Buchanan, and A. A. Humffray, *Aust. J. Chem.* **18**, 1513 (1965).
[172] R. D. Brown, A. S. Buchanan, and A. A. Humffray, *Aust. J. Chem.* **18**, 1521 (1965).
[173] R. D. Brown, A. S. Buchanan, and A. A. Humffray, *Aust. J. Chem.* **18**, 1527 (1965).
[174] F. B. Deans and C. Eaborn, *J. Chem. Soc.* 2303 (1959).
[175] C. Eaborn and J. A. Sperry, *J. Chem. Soc.* 4921 (1961).

## III. Relative Reactivity of Rings and Ring Positions

### A. MONOCYCLIC SYSTEMS WITH ONE HETEROATOM

#### 1. Overall Relative Reactivities of the Fundamental Rings

Until a few years ago, no quantitative comparison of overall reactivities in electrophilic substitution of the fundamental five-membered rings was available. Only the reactivity of thiophene relative to benzene had been measured quantitatively in several electrophilic substitutions: bromination,[72] chlorination,[72] nitration,[121] hydrogen exchange,[57] protodesilylation,[174] and mercuration.[121]

Concerning the comparison among thiophene, furan, and pyrrole, the reactivity order was established mainly on the basis of qualitative criteria.

(1) The comparison of the diverse experimental conditions (temperature, amount and nature of the catalyst, time of reaction, etc.) necessary to obtain the products in similar reactions of the various systems.

(2) The orientation of the substitution in mixed systems containing two different rings. For instance, the higher reactivity of furan with respect to thiophene was deduced from the orientation in the nitration of 2-furyl-2-thienylketone (13)[176] and in the acetylation of 2-furyl-2-thienylmethane[177] (14).

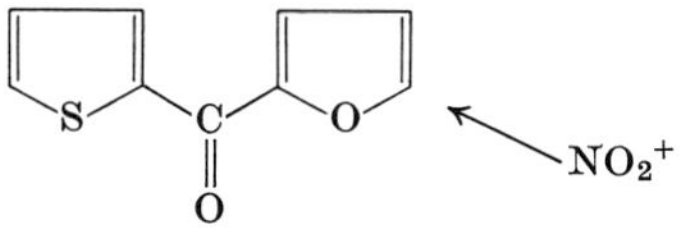

(13) 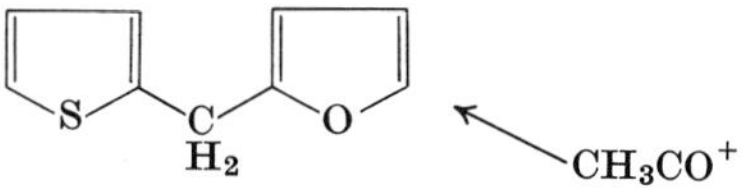    (14)

This method is of uncertain value since the reactivity of each ring is, at least partially, altered by the electronic effects of the other ring acting as substituent.

On the basis of these qualitative criteria the following order of reactivity was universally established: benzene < thiophene < furan. Concerning the comparison of furan and pyrrole, the interpretation of

[176] H. Gilman and R. V. Young, *J. Amer. Chem. Soc.* **56**, 464 (1934).
[177] Ya. L. Gol'dfarb and Ya. L. Danyushevskii, *Zh. Obshch. Khim.* **31**, 3654 (1961); *Chem. Abstr.* **57**, 9776 (1962).

the preparative data was not unequivocal; some textbooks[35, 38] report that pyrrole is more reactive than furan in electrophilic substitution, others[178, 179] the opposite.

Only in recent years has the reactivity sequence of the five-membered rings (including selenophene) been placed on a quantitative basis, by the use of more rigorous kinetic and competitive procedures.

In the tin tetrachloride-catalyzed acetylation by acetic anhydride in dichloroethane, furan is more reactive than thiophene by a factor of 11.9 as determined by a competitive method. The use of iodine as catalyst does not substantially alter the reactivity ratio.[130] The comparison could not be extended to pyrrole, because of the interaction of this substrate with Friedel–Crafts catalysts[139] (see discussion in Section II, E).

Later, Linda and Marino[84, 90, 180] were able to compare the relative reactivities of all four fundamental systems (furan, thiophene, selenophene, and pyrrole) toward bromination by molecular bromine in acetic acid. Unfortunately, the comparison could not be made on the unsubstituted rings for the following reasons: first, the rates of substitution for furan and pyrrole were too high to be followed by standard kinetic techniques; second, furan and pyrrole undergo ring fission and/or polymerization under the influence of the hydrobromic acid formed in the reaction; finally, furan tends to give addition as well as substitution products in the reaction with bromine.[181a]

For these reasons, the rates of bromination of the 2-methoxycarbonyl derivatives were determined: it is well known that the introduction of an electron-withdrawing substituent slows down considerably the rate of substitution, stabilizes the rings toward the action of mineral acids,[181b] and reduces the tendency of the furan ring to undergo addition reactions[181c] However, owing to the large difference in reactivity among the substrates examined, it was necessary to use two different solvents (50% aqueous acetic acid for the less reactive substrates and anhydrous acetic acid for the more reactive ones) and to establish the relative rates for the whole series with the aid of a substrate exhibiting intermediate reactivity, which was examined in both solvents (Table IV).

[178] J. D. Roberts and M. C. Caserio, "Basic Principles of Organic Chemistry," p. 987. Benjamin, New York, 1964.
[179] See Paquette,[37] p. 115.
[180] P. Linda and G. Marino, *Chem. Commun.* 499 (1967).
[181] See Dunlop and Peters,[80] (a) p. 80, (b) p. 62, (c) p. 494.

## TABLE IV

### Rate Data from Bromination[a]

| Substrate (solvent)[b] | | | | | | | | | | | $k^c$ |
|---|---|---|---|---|---|---|---|---|---|---|---|
| LiBr ($M$): | 0.200 | 0.175 | 0.150 | 0.125 | 0.100 | | | | | | |
| LiClO$_4$ ($M$): | 0.000 | 0.025 | 0.050 | 0.075 | 0.100 | | | | | | |
| 2-Methoxycarbonylthiophene (A) | $1.62 \times 10^{-4}$ | — | $2.10 \times 10^{-4}$ | $2.54 \times 10^{-4}$ | $2.88 \times 10^{-4}$ | $1.69 \times 10^{-3}$ |
| 2-Methoxycarbonylfuran (A) | $2.22 \times 10^{-2}$ | $2.75 \times 10^{-2}$ | $3.21 \times 10^{-2}$ | $3.70 \times 10^{-2}$ | $1.97 \times 10^{-1}$ |
| Thiophene (A) | $5.46 \times 10^{2}$ | $6.07 \times 10^{2}$ | $6.60 \times 10^{2}$ | — | $8.79 \times 10^{2}$ | $4.36 \times 10^{3}$ |
| Thiophene (B) | $1.98 \times 10^{-2}$ | — | $2.59 \times 10^{-2}$ | $3.10 \times 10^{-2}$ | $3.85 \times 10^{-2}$ | $6.27 \times 10^{-1}$ |
| 2-Methoxycarbonylpyrrole (B) | 20.8 | 24.0 | 27.1 | — | — | $6.20 \times 10^{2}$ |

[a] At 25°C. Data from Linda and Marino.[84]
[b] Solvent: A, 50% AcOH; B, 100% AcOH.
[c] The true rate constants obtained from the plots of $k_{obs} (1 + K[\text{Br}^-])$ against [Br$^-$] by extrapolation to [Br$^-$] = 0. ($K$ is the equilibrium constant for Br$_2$ + Br$^-$ ⇌ Br$_3^-$.)

Rate and isomer distribution data were used to calculate the relative rates for the α positions (Table V).

TABLE V

RELATIVE RATES OF SUBSTITUTION OF THE
FUNDAMENTAL 5-MEMBERED RINGS

| Compound | Trifluoro-acetylation[a] | Bromination[b] |
|---|---|---|
| Thiophene | 1 | 1 |
| Furan | $1.4 \times 10^2$ | $1.2 \times 10^2$ |
| Pyrrole | $5.3 \times 10^7$ | $5.9 \times 10^8$ |

[a] Relative rates of substitution of the unsubstituted rings at 75°C. See Clementi and Marino.[143]
[b] Relative rates of substitution at 25°C at position 5 of the 2-methoxycarbonyl derivatives, obtained from the rate constants of Table IV, corrected for isomer distributions. Linda and Marino.[84]

Since the rate of bromination of thiophene relative to benzene is known,[72] it was possible to calculate the partial rate factors (with respect to a single position of benzene) for the α position of the four rings: thiophene, $5 \times 10^9$; selenophene, $2.4 \times 10^{11}$; furan, $6 \times 10^{11}$; pyrrole, $3 \times 10^{18}$. So far, these values, although approximate, represent the sole comparison existing between the reactivities of all five-membered rings with respect to benzene.

The relative reactivities of all four *unsubstituted* rings have been subsequently determined in another electrophilic substitution: trifluoroacetylation by trifluoroacetic anhydride in dichloroethane.[142, 143] The relative rates, obtained by a competitive procedure, are in good agreement with the bromination data (Table V) and confirm, in particular, the big jump in reactivity from furan to pyrrole. The great reactivity of pyrrole cannot be ascribed to a reaction involving the anion $C_4H_4N^-$, since $N$-methylpyrrole is still more reactive than pyrrole by a factor of about 2.

The very high reactivity of pyrrole compared with furan in electrophilic substitutions is also confirmed by rate measurements of hydrogen–deuterium exchange in methanol–water–sulfuric acid mixtures[53] (Table VI). The rate of exchange of pyrrole-2-*d* in 0.5%

$H_2SO_4$ is about 4000 times greater than the rate of exchange of furan-2-d in 21.6% sulfuric acid. Although the exchange rates have been

TABLE VI

KINETIC DATA FOR PROTODEDEUTERIATION IN METHANOL–WATER[a]

| Compound | $k_1 \times 10^6$ (sec$^{-1}$)[c] | | | | |
|---|---|---|---|---|---|
| | I | II | III | IV | V |
| Benzene-d | 0.916 | — | — | — | — |
| Thiophene 3-d | 8610 | 0.500 | — | — | — |
| Thiophene 2-d | — | 972 | 2.86 | 0.469 | — |
| Furan-3-d[b] | — | < 50 | < 0.3 | — | — |
| Furan-2-d[b] | — | 472 | 2.83 | 0.500 | — |
| Pyrrole-2,3,4,5-d$_4$ | | | | | |
| 3-Substitution | — | — | — | — | 800 |
| 2-Substitution | — | — | — | — | 1900 |

[a] Data from Schwetlick et al.[53] Solvent: methanol–water (1.67:1.00 by weight).
[b] Partial cleavage of the furan ring occurs under the experimental conditions used.
[c] In $H_2SO_4$ (%): I, 78.5; II, 56.0; III, 31.0; IV, 21.6; and V, 0.5.

measured at different acidities and a direct rate ratio cannot then be calculated, a factor of about $10^5$ for the relative rate of pyrrole to furan may be estimated for this reaction.

The reactivity sequence: furan > selenophene > thiophene ≫ benzene has also been observed in the *nucleophilic* substitutions of the halogenonitro derivatives of these rings.[21, 22] This shows that the observed trend does not depend on the effectiveness of lone-pair conjugation of the heteroatoms NH, O, Se, and S and the $\pi$-electron density at the carbon atoms. It is interesting to note that a good correlation is observed between molecular ionization potentials (determined from electron impact measurements) and reactivity data in electrophilic substitution, in that higher reactivities correspond to lower ionization potentials[182]: pyrrole ≪ furan < selenophene < thiophene ≪ benzene (see Table VII). This is expected in view of a

[182] P. Linda, G. Marino, and S. Pignataro, *Ric. Sci.* **39**, 666 (1969).

## TABLE VII
### Ionization Potentials and Reactivity Data

| Compound | I.P. (eV)[a] | Log $\alpha_f$ (bromination)[b] |
|---|---|---|
| Benzene | 9.53 | 0 |
| Thiophene | 9.12 | 9.7 |
| Selenophene | 9.01 | 11.4 |
| Furan | 8.99 | 11.8 |
| Pyrrole | 8.40 | 18.5 |

[a] Electron impact technique (Linda et al.[182]).
[b] Data from Linda and Marino.[84]

certain similarity between the molecular ion obtained in the first ionization process (**15**) with the transition state in the substitution process (**16**).

(**15**)      (**16**)

### 2. The α:β Ratios

Electrophilic attack occurs more readily at the α than at the β position in all five-membered rings. The preference for α substitution may be rationalized by comparing the energies of the transition states leading to α- and β-substituted products. Taking the Wheland intermediates as models for the transition states, it is possible to see that, while for α attack, three limiting resonance structures (**17–19**)

(**17**)      (**18**)      (**19**)

may be written, only two such structures are possible for the β attack (**20, 21**).

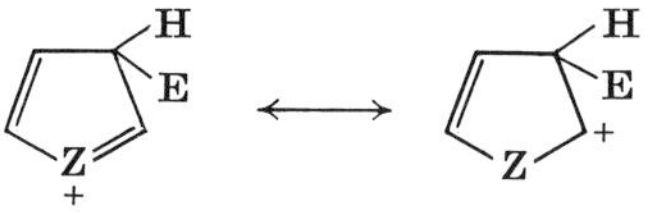

Isomer distributions and α:β ratios depend strongly on the electrophilic reagent. Data referring to fourteen substitutions of thiophene are summarized in Table VIII. In acetylation, the isomer distribution

TABLE VIII

Isomer Distributions and α:β Reactivity Ratios in Electrophilic Substitutions of Thiophene[a]

| Reaction | Conditions | β (%) | α:β | Ref. |
|---|---|---|---|---|
| Protodedeuteriation | In $CF_3COOH$ | — | 3400 | 55 |
| Protodedeuteriation | In aq. $H_2SO_4$ | — | 1045 | 58 |
| Protodetritiation | In aq. $H_2SO_4$ | — | 955 | 57 |
| Vilsmeier formylation | — | <0.1 | >1000 | 75 |
| Trifluoroacetylation | — | <0.1 | >1000 | 75 |
| Bromination | By $Br_2$ | 0.205 | 490 | 75 |
| Acetylation | By acetyl trifluoroacetate | 0.30 | 330 | 75 |
| Acetylation | By $Ac_2O$ and $I_2$ | 0.50 | 200 | 75 |
| Acetylation | By $Ac_2O$ and $SnCl_4$ | 0.50 | 200 | 75 |
| Benzoylation | By $Bz_2O$ and $SnCl_4$ | 0.70 | 140 | 75 |
| Bromination | By $Br^+$ | 0.82 | 120 | 75 |
| Chlorination | By $Cl_2$ | 1.00 | 100 | 75 |
| Protodesilylation | — | — | 43.5 | 174 |
| Nitration | By $BzNO_3$ (at 0°) | 13.8 | 6.2 | 122 |

[a] At 25°C.

is not influenced by the nature of the catalyst (tin tetrachloride or iodine); this indicates that the same electrophile, probably $CH_3CO^+$ is involved. On the other hand, the isomer distribution in acylation is strongly affected by temperature, as is illustrated by the data of Table IX).

## TABLE IX

EFFECT OF TEMPERATURE ON $\alpha{:}\beta$ RATIO IN ACYLATION REACTIONS OF THIOPHENE[a]

| Reaction | Conditions | Temp. (°C) | $\beta$ (%) | $\alpha{:}\beta$ |
|---|---|---|---|---|
| Acetylation | By Ac$_2$O and SnCl$_4$ | 25 | 0.5 | 200 |
|  |  | 75 | 1.2 | 82 |
| Acetylation | With acetyl trifluoroacetate | 25 | 0.3 | 330 |
|  |  | 75 | 1.4 | 71 |
| Benzoylation | By Bz$_2$O and SnCl$_4$ | 25 | 0.7 | 140 |
|  |  | 75 | 1.5 | 66 |

[a] Data from Clementi et al.[75]

A similar effect is also observed in the hydrogen exchange; for this reaction, the $\alpha{:}\beta$ ratio is strongly dependent on the nature of the solvent and the acidity of the medium also (Table X).

The $\alpha{:}\beta$ ratio is also strongly dependent on the heteroatom. Data for a homogeneous comparison under strictly equivalent experimental conditions are not available. The existing data (Table XI) and other qualitative reports seem to confirm the general validity of the following order for the $\alpha$-directing power: furan $\gg$ thiophene $\simeq$ selenophene $\gg$ pyrrole.

Many reasons may be responsible for the different $\alpha{:}\beta$ ratios observed in the four rings. In particular, the small $\alpha{:}\beta$ ratios for pyrrole [as well as the low sensitivity of this ring to substituent effects (Section IV, B)] may be due to the fact that in this case the Wheland intermediate is not a good model for the transition state. This hypothesis is in keeping with the Hammond postulate,[183] according to which the transition state approaches closer to the unperturbed starting molecule as its reactivity increases.

### 3. *Partial Rate Factors and Linear Free Energy Relationships*

*Partial rate factor* is defined as the rate of substitution at a single position of the heteroaromatic ring relative to a single position in benzene.

[183] G. Hammond, *J. Amer. Chem. Soc.* **77**, 334 (1955).

## TABLE X

INFLUENCE OF SOLVENTS, ACIDITY, AND TEMPERATURE ON THE $\alpha:\beta$ RATIO IN HYDROGEN EXCHANGE

| Reaction | Conditions | $H_0$ | Temp. (°C) | $\alpha:\beta$ | Ref. |
|---|---|---|---|---|---|
| Protodetritiation | In aq. $H_2SO_4$ | −6.0 | 25 | 440 | 56 |
| Protodetritiation | In aq. $H_2SO_4$ | −4.0 | 25 | 900 | 56 |
| Protodetritiation | In aq. $H_2SO_4$ | −4.0 | 2 | 1200 | 56 |
| Protodedeuteriation | In aq. $H_2SO_4$ (heterogeneous) | — | 25 | 1045 | 58 |
| Protodedeuteriation | In methanol/56% $H_2SO_4$ | — | 20 | 1945 | 53 |
| Protodedeuteriation | In $CF_3COOH$ (20%) and $CH_3COOH$ (80%) | — | 25 | 2500 | 55 |
| Protodedeuteriation | In $CF_3COOH$ (50%) and $CH_3COOH$ (50%) | — | 25 | 3000 | 55 |
| Protodedeuteriation | In $CF_3COOH$ | — | 25 | 3500 | 55 |

## TABLE XI

COMPARISON OF THE $\alpha:\beta$ RATIOS IN ELECTROPHILIC SUBSTITUTIONS OF FURAN, THIOPHENE, AND PYRROLE

| Reaction | Conditions | Furan | Thiophene | Pyrrole | Ref. |
|---|---|---|---|---|---|
| Benzoylation | $Bz_2O$, $SnCl_4$, DCE, 75° | 250 | 65.6 | — | a |
| Acetylation | ATFA, DCE, 75° | >1000 | 71.4 | 6 | b |
| Protodedeuteriation | — | — | 1940 | 2.4 | c |

[a] S. Clementi, P. Linda, and G. Marino.[75, 77]
[b] Reactions with acetyl trifluoroacetate (ATFA) give mixtures of acetylated and trifluoroacetylated products. The reported figures refer to the $\alpha:\beta$ acetylation ratios (S. Clementi and G. Marino, unpublished data).
[c] Reactions in methanol/water/sulfuric acid. The ratio for thiophene was determined in 78% $H_2SO_4$ and that for pyrrole at 0.5% $H_2SO_4$.[53]

For displacement reactions the partial rate factors $\alpha_f$ and $\beta_f$ are calculated directly from the ratios of the rate constants of the heterocyclic and the benzene derivatives. For the electrophilic substitutions in which the proton is replaced, they may be easily calculated from the overall relative rates and the isomer distributions:

$$\alpha_f = \frac{k_{\text{Het}}}{k_\text{B}} \cdot \frac{\alpha\%}{100} \cdot \frac{6}{2}$$

$$\beta_f = \frac{k_{\text{Het}}}{k_\text{B}} \cdot \frac{\beta\%}{100} \cdot \frac{6}{2}$$

a. *Thiophene.* The available partial rate factors for the electrophilic substitutions of thiophene, together with some derived quantities, are summarized in Table XII. This table also includes data referring to some side-chain reactions, for which a resonance-stabilized carbonium ion is formed in the transition state. These reactions resemble, in certain aspects, the electrophilic substitutions and they have been utilized by many authors for determining aromatic reactivities.[184-187]

The partial rate factors vary within wide limits. Electrophilic species of lower activity, such as molecular bromine, are more selective, i.e., more capable of discriminating either between thiophene and benzene or between positions $\alpha$ and $\beta$ of thiophene. Quantitatively, a linear trend is observed (Fig. 1) between $\log \alpha/\beta$ and $\log \alpha_f$. This is a correlation formally analogous to the "selectivity relationship" proposed by Brown and Nelson[188] for the reactions of monosubstituted benzenes.

The constancy of the quantity $\log \alpha_f / \log \beta_f$ is remarkably good: $1.54 \pm 0.13$, if only the electrophilic substitutions are taken into account and $1.61 \pm 0.17$, if also the side-chain reactions are included. The constancy is even better than that observed for substitutions of toluene[189] ($\log p_f / \log m_f = 3.98 \pm 0.53$).

The data of Table XII can be utilized for a test of the applicability of a linear free energy relationship to electrophilic reactions of thiophene.

[184] M. J. S. Dewar and R. J. Sampson, *J. Chem. Soc.* 2946 (1957).
[185] H. C. Brown and Y. Okamoto, *J. Amer. Chem. Soc.* **79**, 1913 (1957).
[186] M. Planchen, P. J. C. Fierens, and R. H. Martin, *Helv. Chim. Acta* **42**, 517 (1959).
[187] A Streitwieser, "Molecular Orbital Theory for Organic Chemistry," p. 370. Wiley, New York, 1961.
[188] H. C. Brown and K. L. Nelson, *J. Amer. Chem. Soc.* **75**, 6292 (1953).
[189] L. M. Stock and H. C. Brown, *Advan. Phys. Org. Chem.* **1**, 92 (1963).

TABLE XII PARTIAL RATE FACTORS AND RELATED QUANTITIES FOR ELECTROPHILIC REACTIONS AT THE THIOPHENE RING

| Reactions | $\rho^a$ | $\alpha_t$ | $\beta_t$ | $\dfrac{\log \alpha_t}{\log \beta_t}$ | Ref. |
|---|---|---|---|---|---|
| Electrophilic substitutions | | | | | |
| 1. Bromination by $Br_2$ | −12.1 | $5.1 \times 10^9$ | $1.05 \times 10^7$ | 1.38 | 72, 75 |
| 2. Chlorination | −10.0 | $3.9 \times 10^7$ | $3.9 \times 10^5$ | 1.36 | 72, 75 |
| 3. Acetylation | −9.1 | $2.7 \times 10^6$ | $1.35 \times 10^4$ | 1.55 | 75 |
| 4. Protodedeuteriation | — | $8.7 \times 10^6$ | $8.3 \times 10^3$ | 1.77 | 58[b] |
| 5. Protodetritiation | −8.2 | $5.3 \times 10^6$ | $5.8 \times 10^3$ | 1.78 | 58[b] |
| 6. Bromination by $Br^+$ | −6.2 | $1.9 \times 10^5$ | $1.6 \times 10^3$ | 1.65 | 75 |
| 7. Nitration | — | $(7.1 \times 10^2)$ | $(1.15 \times 10^2)$ | 1.40 | 122 |
| 8. Protodeboronation | — | $8.5 \times 10^5$ | $7.1 \times 10^3$ | 1.54 | 172 |
| 9. Iododeboronation | −4.8 | $9.7 \times 10^3$ | $7.0 \times 10^2$ | 1.40 | 173 |
| 10. Protodesilylation | −4.6 | $5.0 \times 10^3$ | $1.15 \times 10^2$ | 1.78 | 174 |
| 11. Mercuration | −4.0 | $1.85 \times 10^6$ | — | — | 121, 75 |
| 12. Protodemercuration | −2.4 | $1.7 \times 10^3$ | — | — | 171 |
| Side-chain carbonium ion reactions | | | | | |
| 13. Solvolysis of 1-arylethyl-$p$-nitrobenzoates | −6.0 | $6.3 \times 10^4$ | $1 \times 10^3$ | 1.60 | c |
| 14. Solvolysis of 1-arylethylacetates | −5.7 | $5.4 \times 10^4$ | $4.8 \times 10^2$ | 1.76 | 239 |
| 15. Isomerization of $cis$-1-aryl-2-phenylethenes | −3.3 | $3.5 \times 10^2$ | — | — | c |
| 16. Rearrangement of arylpropenyl carbinols | −2.9 | 50.2 | — | — | d |
| 17. Pyrolysis of 1-arylethylacetates | −0.66 | 3.3 | 1.8 | 2.08 | e |

[a] Reaction constants $\rho$ for benzene derivatives [L. M. Stock and H. C. Brown, *Advan. Phys. Org. Chem.* **1**, 35 (1963)].
[b] See also B. Östman, *FOA* (*Foersvarets Forskningsanst.*) *Rep.* **3**, 1 (1969).
[c] See D. S. Noyce, C. A. Lipinski, and G. M. Loudon, *J. Org. Chem.* **35**, 1718 (1970).
[d] See E. A. Braude and E. S. Stern, *J. Chem. Soc.* 1097 (1947); E. A. Braude and J. S. Fawcett, *J. Chem. Soc.* 4158 (1952).
[e] See R. Taylor, *J. Chem. Soc. B* 1397 (1968).

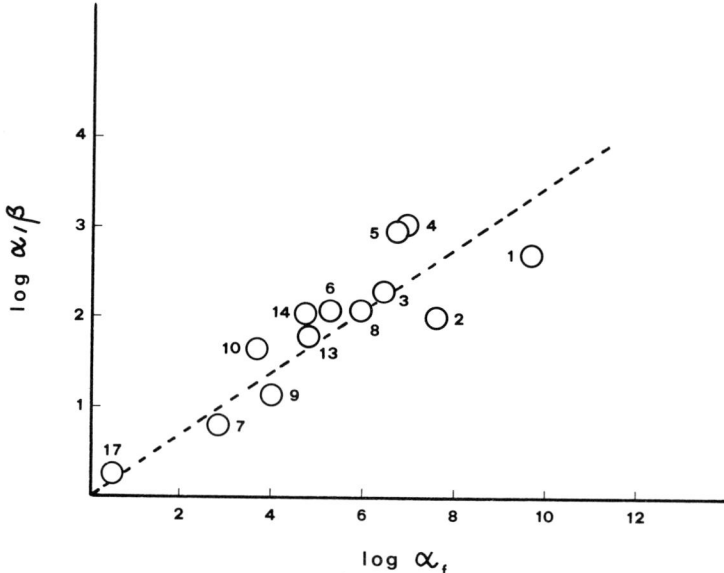

FIG. 1. The relationship between $\log \alpha/\beta$ and $\log \alpha_f$ for electrophilic reactions at the thiophene ring. The numbers identify the reactions; see Table XII.

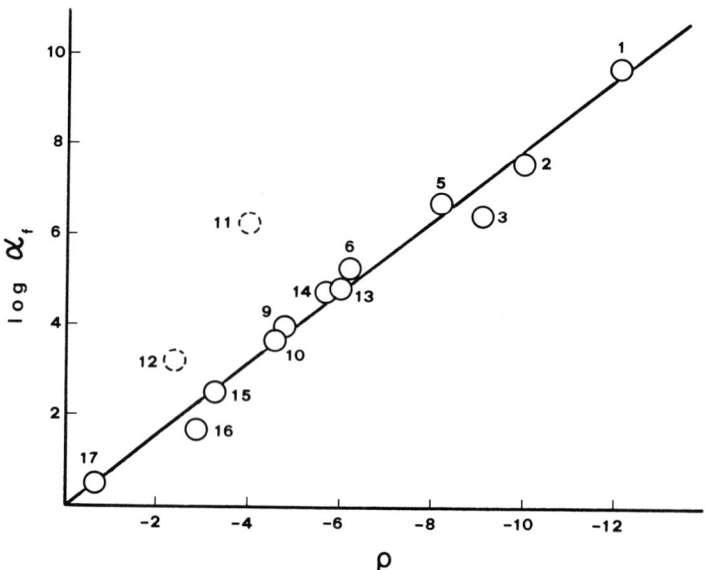

FIG. 2. The "extended selectivity relationship" for electrophilic reactions at the $\alpha$ position of thiophene. The numbers identify the reactions; see Table XII.

Following the procedure suggested by Brown and Stock[190] (termed "the extended selectivity treatment") $\log \alpha_f$ and $\log \beta_f$ have been plotted against the reaction constants $\rho$ for benzene derivatives (Figs. 2 and 3).

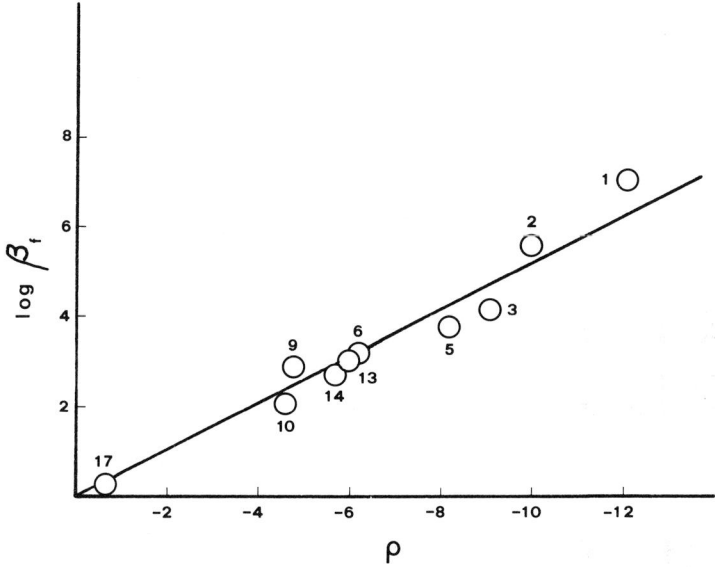

Fig. 3. The "extended selectivity relationship" for electrophilic reactions at the $\beta$ position of thiophene. The numbers identify the reactions; see Table XII.

Values relative to nitration and protodeboronation have not been recorded in the graph since in the literature $\rho$ values for these reactions have been determined under completely different experimental conditions. Further, the values of $k_T/k_B$ for nitration is questionable for the reasons discussed in Section II, D.

Figure 2 reveals two major discrepancies, mercuration and protodemercuration, which seriously disagree with the predictions of the relationship. It is very probable that these reactions proceed through a mechanism different from that in benzene, presumably via a preliminary coordination of mercury with the sulfur atom (for a more extensive discussion, see Section II, F, 2).

The data for the remaining reactions (12 at the $\alpha$ and 10 at the $\beta$ position) closely follow the linear rleationship. A least-squares

[190] H. C. Brown and L. M. Stock, *J. Amer. Chem. Soc.* **84**, 3298 (1962).

analysis shows that the correlation is equal to or even better than that observed in the reactions of monosubstituted benzenes.[189]

The satisfactory correlation of the data argues for a linear response of the resonance contributions of the thiophene ring. Substituent constants[191] $\sigma_\alpha^+$ and $\sigma_\beta^+$ relative to the perturbation produced by the substitution of a S atom for a CH=CH moiety in the benzene ring, can be easily calculated from the slopes of the plots, $-0.79$ and $-0.52$, respectively. (Slopes refer to least-square regression lines with the theoretical origin.)

The linearity observed in the $\alpha$-reactivity plot implies the absence of significant steric effects exerted by the lone pairs of the sulfur atom on the reactivity at the $\alpha$ position.

b. *Furan.* The $\alpha$-orienting power is much more marked in furan than in thiophene (Section III, A, 2). In most electrophilic substitutions it is reported that only the $\alpha$ isomer has been isolated.[80] As a consequence of this, very few partial rate factors are available for the $\beta$ position, most of them referring to replacement and side-chain reactions. All the existing data are summarized in Table XIII.

Those for reactions at the $\alpha$ position are numerous enough for testing the applicability of the "extended selectivity treatment[192] (Fig. 4). The points nicely fit a straight line with the theoretical origin ($r = 0.995$, $s = 0.79$). From the slope a substituent constant $\sigma_\alpha^+$ relative to the structural modification caused by the substitution of an oxygen atom for a CH=CH in the benzene ring may be calculated. The value obtained $-0.93$, is noticeably more negative than $\sigma_\alpha^+$ for thiophene, $-0.79$.

c. *Selenophene and Pyrrole.* The relative rates of substitution for selenophene to thiophene, have been determined for a number of reactions using kinetic and competitive methods. The amount of $\beta$ isomer formed is always less than a few percent[90] and, therefore, the relative rates can be considered as referring to $\alpha$ reactivities. Since for all the reactions examined (except trifluoroacetylation) the

---

[191] The terms $\sigma_{a\text{-thienyl}}$, $\sigma_{a\text{-pyrrolyl}}$, etc., have often been used to measure the effects caused by the "replacement" of a CH=CH–group in the benzene ring with a S or a NH group. This use is incorrect, and terms such as $\sigma_{a\text{-S}}$ or $\sigma_{a\text{-NH}}$ (and $\sigma_{a\text{-S}}^+$, etc.) should be used. The use of the terms $\sigma_{a\text{-thienyl}}$, etc., should be limited only to the effects caused by a substitution of an H atom with $C_4H_3S$, $C_4H_3NH$, etc., groups (see F. Fringuelli, G. Marino, and A. Taticchi, *J. Chem. Soc. B* 1595 (1970).

[192] S. Clementi, P. Linda, and G. Marino, *Tetrahedron Lett.* 1389 (1970).

## TABLE XIII
### Partial Rate Factors and Related Quantities for Electrophilic Reactions at the Furan Ring

| Reactions | $\rho$ | $\alpha_f$ | $\beta_f$ | Ref. |
|---|---|---|---|---|
| Electrophilic substitutions | | | | |
| 1. Bromination by $Br_2$ | −12.1 | $6.1 \times 10^{11}$ | — | 84 |
| 2. Chlorination | −10.0 | $1.9 \times 10^9$ | — | 192 |
| 3. Acetylation | −9.1 | $3.2 \times 10^7$ | $4.7 \times 10^3$ | 130 |
| 4. Protodesilylation | −4.3 | $1.7 \times 10^4$ | $1.2 \times 10^2$ | 175[a] |
| 5. Protodemercuration | −2.4 | $4.0 \times 10^3$ | $1.5 \times 10^2$ | 171 |
| Side-chain carbonium ion reactions | | | | |
| Solvolysis | | | | |
| 6. Of 1-arylethyl-$p$-nitrobenzoates | −6.0 | $1.0 \times 10^5$ | $2.9 \times 10^2$ | [b] |
| 7. Of 1-arylethyl acetates | −5.7 | $2.1 \times 10^5$ | $6.7 \times 10^2$ | 239 |
| 9. Rearrangement of arylpropenylcarbinols | −2.9 | 90 | — | [c] |
| 8. Pyrrolysis of 1-arylethylacetates | −0.66 | 3.8 | 1.9 | [d] |

[a] R. Taylor, *J. Chem. Soc. B*, 1364 (1970).
[b] In 80% aqueous ethanol at 75° (H. Jorge and D. S. Noyce, private communication.)
[c] E. A. Braude and J. S. Fawcett, *J. Chem. Soc.* 4158 (1952).
[d] R. Taylor, *J. Chem. Soc. B*, 1397 (1968).

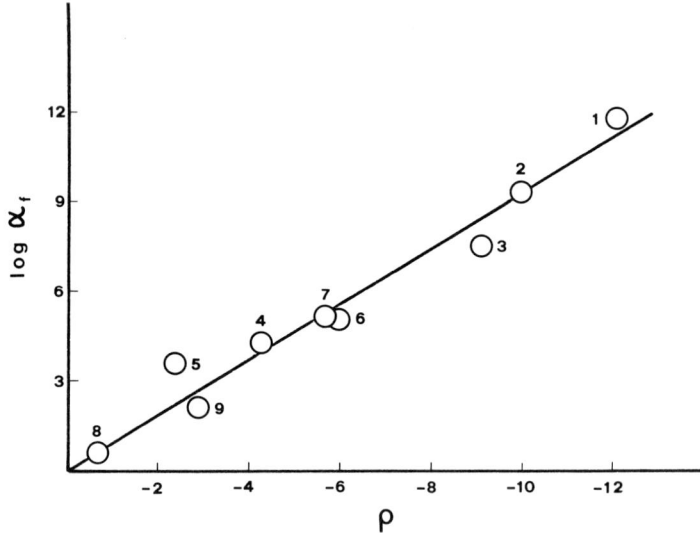

Fig. 4. The "extended selectivity relationship" for electrophilic reactions at the α position of furan. The numbers identify the reactions; see Table XIII.

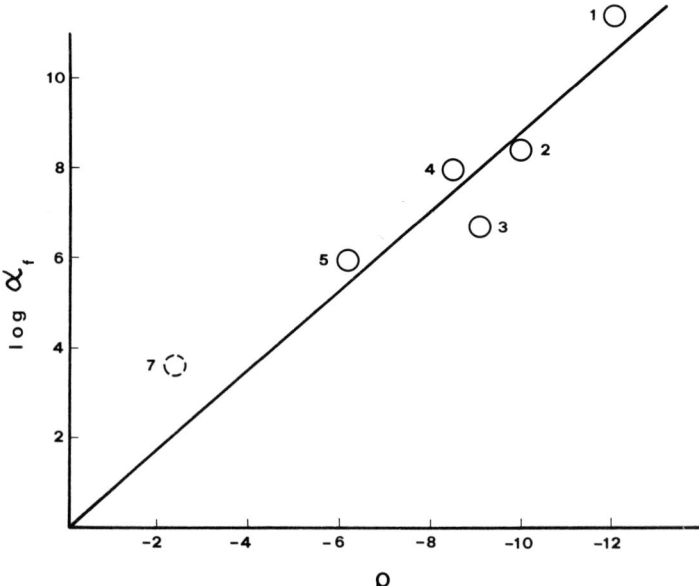

Fig. 5. The "extended selectivity relationship" for electrophilic reactions at the α position of selenophene. The numbers identify the reactions; see Table XIV.

rates of thiophene relative to benzene are also known, it is possible to calculate the partial rate factors for substitution at the α position of selenophene. The relevant values are summarized in Table XIV.

TABLE XIV

RELATIVE REACTIVITIES OF SELENOPHENE TO THIOPHENE AND PARTIAL RATE FACTORS FOR SUBSTITUTION AT THE SELENOPHENE RING

| Reaction | $\rho^a$ | Relative rate selenophene to thiophene | $\alpha_f$ | Ref. |
|---|---|---|---|---|
| 1. Bromination by $Br_2$ | −12.1 | 47.5 | $2.4 \times 10^{11}$ | 90 |
| 2. Chlorination | −10.0 | 6.5 | $2.5 \times 10^8$ | 90 |
| 3. Acetylation | −9.1 | 1.9 | $5.1 \times 10^6$ | 90 |
| 4. Protodedeuteriation | $(-8.2)^b$ | 11 | $9.6 \times 10^7$ | 59[c] |
| 5. Bromination by $Br^+$ | −6.2 | 4.5 | $8.5 \times 10^5$ | 90 |
| 6. Trifluoroacetylation | — | 6.6 | — | 90 |
| 7. Protodemercuration | −2.4 | 2.2 | $3.7 \times 10^3$ | 171 |

[a] Reaction constants for benzene derivatives [L. M. Stock and H. C. Brown, *Advan. Phys. Chem.* **1**, 35 (1963)].

[b] $\rho$ value for the protodetritiation reaction.

[c] In Linda and Marino[90] an erroneous value for the relative rate of protodedeuteriation has been reported.

The plot of $\log \alpha_f$ vs. $\rho$ is not as good ($r = 0.994$) as the corresponding plot for thiophene (Fig. 5). From the slope a value of $\sigma^+_{\alpha-Se}$ equal to −0.88 is obtained.

For pyrrole, the only partial rate factor available refers to the bromination by molecular bromine in acetic acid, and this was calculated from reactivity data on the 2-methoxycarbonyl derivatives.[84] For the trifluoroacetylation reaction there exists a value for the rate relative to thiophene, but it is not possible to calculate the partial rate factor since data on the reactivity of benzene under the same conditions are not obtainable.

Partial rate factors are also available for two α-carbonium ion side-chain reactions of 1-methylpyrrole derivatives. The relevant values are summarized in Table XV.

TABLE XV

PARTIAL RATE FACTORS FOR ELECTROPHILIC REACTIONS
OF PYRROLE AND $N$-METHYLPYRROLE

| Reaction | Compound[a] | $\alpha_f$ | $\sigma_a^{+}$ [b] | Ref. |
|---|---|---|---|---|
| Bromination | P | $3 \times 10^{18}$ | $-1.53$ | 84 |
| Solvolysis | | | | |
| of 1-arylethyl $p$-nitro-benzoates | MP | $6 \times 10^{10}$ | $-1.89$ | c |
| of 1-arylethylacetates | MP | $5.7 \times 10^{10}$ | $-1.88$ | 239 |

[a] P, pyrrole; MP, $N$-methylpyrrole.
[b] $\sigma_{\alpha}^{+}$-$_{NH}$ = log $\alpha_f/\rho$.
[c] In 70% dioxane at 75° (G. M. Fraser and D. S. Noyce, private communication).

## B. MONOCYCLIC SYSTEMS WITH TWO OR MORE HETEROATOMS

Oxazole (**22**), thiazole (**23**), and imidazole (**24**) are formally derived from furan, thiophene, and pyrrole by replacement of a $\beta$-CH group

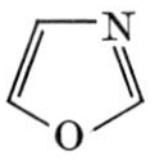

(**22**) 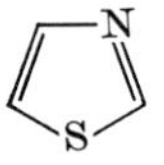    (**23**) 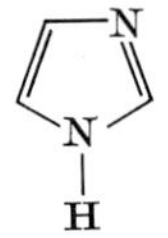    (**24**)

of the ring by an aza function. Replacement of an $\alpha$-CH by an aza group leads to isoxazole (**25**), isothiazole (**26**), and pyrazole (**27**), respectively.

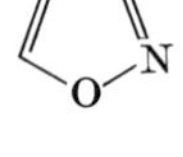

(**25**) 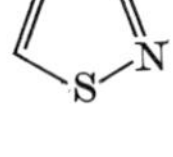    (**26**) 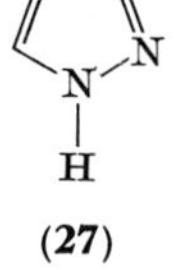    (**27**)

All these rings retain aromatic character and react with a variety of electrophilic reagents, but require, of course, somewhat more vigorous conditions.

As concerns the orientation of the substitution, in compounds **22–24**, both the meta-directing effect of the aza group and the alpha-directing effect of O, S, and NH converge the substitution to the 5-position. This is what is actually observed in substitutions of thiazole,[193] oxazole,[194] and imidazole[195, 196] (note that in imidazole positions 4 and 5 are equivalent because of the prototropic equilibrium).

In compounds **25–27** the directing effects of the two heteroatoms are conflicting. The effect of the "aza" group appears to be prevalent and the molecules of isoxazole,[197, 198] isothiazole,[199] and pyrazole[200–202] are substituted preferentially at position 4.

These orientations of substitution are observed when either the neutral or the protonated molecules are the reacting species. On the other hand, different orientations can be observed in those reactions of imidazole (and perhaps pyrazole) which involve the dissociated anions $C_3H_3N_2^-$. For example, diazo coupling[96] and deuterium exchange[203] of imidazole in alkaline media give rise to 2-substitution.

Concerning the relative reactivities, the presence of an electron-withdrawing aza group obviously reduces the reactivity toward electrophiles with respect to the parent heterocycles having a single heteroatom. Many authors[204] regard these azoles even less reactive than benzene itself, since they generally do not undergo Friedel–Crafts reactions. This argument is, however, fallacious, since the unreactivity in these reactions is due to the formation of complexes between these substrates and the Lewis acid. According to an approximate calculation based on the values of the $\sigma$ constants for the heteroatoms, considered as substituents, the reactivity of all six azoles

[193] J. M. Sprague and A. H. Land, *in* "Heterocyclic Compounds" (R. C. Elderfield, ed.), Vol. 5, p. 484. Wiley, New York, 1957.
[194] J. W. Cornforth, *in* "Heterocyclic Compounds" (R. C. Elderfield, ed.), Vol. 5, p. 298. Wiley, New York, 1957.
[195] K. Hoffmann, *in* "The Chemistry of Heterocyclic Compounds" (A. Weissberger, ed.), Vol. 6. Wiley (Interscience), New York, 1953.
[196] F. Moller, *in* "Methoden der Organischen Chemie" (J. Houben and T. Weyl, eds.), Vol. 11, Part 1, p. 9. Thieme, Stuttgart, 1957.
[197] A. Quilico and G. Speroni, *in* "The Chemistry of Heterocyclic Compounds" (A. Weissberger, ed.), Vol. 17. Wiley (Interscience), New York, 1962.
[198] N. K. Kochetkov and S. D. Sokolov, *Advan. Heterocycl. Chem.* **2**, 365 (1963).
[199] R. Slack and K. R. H. Wooldridge, *Advan. Heterocycl. Chem.* **4**, 107 (1965).
[200] T. L. Jacobs, *in* "Heterocyclic Compounds" (R. C. Elderfield, ed.), Vol. 5, p. 45. Wiley, New York, 1957.
[201] R. Fusco, *in* "The Chemistry of Heterocyclic Compounds" (A. Weissberger, ed.), Vol. 22. Wiley (Interscience), New York, 1967.
[202] A. N. Kost and I. I. Grandberg, *Advan. Heterocycl. Chem.* **6**, 247 (1966).
[203] R. J. Gillespie, A. Grimison, J. H. Ridd, and R. F. M. White, *J. Chem. Soc.* 3228 (1958).
[204] See, for instance, Albert[38] pp. 260, 275, 276.

(**22–27**) *when reacting as neutral molecules* should still be greater than that of benzene. This expectation is confirmed by the orientation of the substitution in mixed systems (although this criterion of relative reactivity, as underlined in Section II, A, is open to criticism). For instance, bromination of 5-phenylisoxazole gives 4-bromo-5-phenylisoxazole[205] and acylation of 1-phenylpyrazole gives 4-acetyl-1-phenylpyrazole.[206]

Some side-chain reactivity data substantiate the above statement. Thus, the rates of hydrolysis of compounds **28** and **29** are, respectively,

$2.6 \times 10^4$ and $3.6 \times 10^5$ times larger than that of the phenyl analog.[207] Further, in the hydrolysis of *t*-cumyl chlorides, the 5-thiazolyl derivative is about 8.5 times more reactive than the phenyl under the same conditions.[207]

The situation changes if the reactions are carried out in strongly acidic solutions; in this case, the substrates are completely protonated and the reactivities greatly decrease.

A quantitative study has been carried out by Ridd and co-workers[126] on the nitration of imidazole and pyrazole in 98% sulfuric acid; the partial rate factors for the 4-positions of imidazolium and pyrazolium cations are $3.0 \times 10^{-9}$ and $2.1 \times 10^{-10}$, respectively. Thiazole and isoxazole cations are also far less reactive than benzene. As a consequence, phenyl derivatives give products substituted in the benzene ring, on sulfonation or nitration.[208–210]

[205] N. K. Kochetkov and E. D. Khomutova, *Zh. Obshch. Khim.* **28**, 359 (1958); *Chem. Abstr.* **52**, 13710b (1958).
[206] C. A. Royan, *Ber.* **55**, 291 (1922); A. Michaels and C. A. Royan, *ibid.* **50**, 737 (1917).
[207] D. S. Noyce, S. Fike, E. Gordon, and G. Stowe, unpublished results.
[208] C. Musante, *Farmaco* (*Ed. Sci. Tec.*) **6**, 32 (1951); *Chem. Abstr.* **45**, 5879 (1951).
[209] B. S. Friedman, M. Sparks, and R. Adams, *J. Amer. Chem. Soc.* **59**, 2262 (1937).
[210] E. Ochiai, T. Kakuda, I. Nakayama, and G. Masuda, *J. Pharm. Soc. Jap.* **59**, 228, 462 (1939); *Chem. Abstr.* **34**, 101 (1940).

On the other hand, the reactivity of the anions (obtained in the acid dissociations of imidazole and pyrazole) is, of course, very much higher. On iodination in water, the reactivity of the pyrazole anion $C_3H_3N_2^-$ has been estimated to be greater than that of the neutral molecule by a factor in the range of $3 \times 10^9$ to $7 \times 10^{12}$, or perhaps even larger.[105]

Very few data are available for a quantitative comparison of the reactivities of the azoles, among themselves. It is probable that, in view of the high values for the $\alpha : \beta$ ratios observed in furan, thiophene, and pyrrole, the overall reactivities of the $\beta$-azalogs always exceed those of the $\alpha$-azalogs: oxazole > isoxazole; thiazole > isothiazole; imidazole > pyrazole. The only existing data refer to the imidazole–pyrazole pair and confirm this hypothesis. Imidazole is more reactive than pyrazole in iodination[104] by a factor of 1.3 and in nitration[126] (the conjugate acids being the reacting species) by a factor of 14.3.

It is also possible that the order of activating power NH > O > S observed in the rings with only one heteroatom, is also retained in the corresponding azalogs, i.e., imidazole > oxazole > thiazole; pyrazole > isoxazole > isothiazole. So far, however, this is a pure speculation since there are no data for a serious homogeneous comparison.

The only available data concern some $\alpha$-carbonium ion side-chain reactions (hydrolysis of $t$-cumyl chlorides and $p$-nitrobenzoates); they reveal that the substitution of nitrogen for a CH group, in furan, thiophene, or pyrrole, leads to a fairly uniform reduction in reactivity at the 2-position.[211]

The introduction of a second aza group into the 5-membered rings reduces further the susceptibility to electrophilic attack. Triazoles,[212, 213] oxadiazoles[214, 215] and thiadiazoles[216–218] are practically completely resistant to electrophilic substitution unless powerful electron-releasing substituents are present. No quantitative studies on the reactivities of these rings have been made.

[211] D. S. Noyce, private communication.
[212] J. H. Boyer, in "Heterocyclic Compounds" (R. C. Elderfield, ed.), Vol. 7, p. 384. Wiley, New York, 1961.
[213] K. T. Potts, Chem. Rev. **61**, 87 (1961).
[214] L. C. Behr, in "The Chemistry of Heterocyclic Compounds" (A. Weissberger, ed.), Vol. 17, p. 235. Wiley (Interscience), New York, 1962.
[215] A. Hetzheim and K. Möckel, Advan. Heterocycl. Chem. **7**, 183 (1966).
[216] L. M. Weinstock and P. I. Pollack, Advan. Heterocycl. Chem. **9**, 107 (1968).
[217] F. Kurzer, Advan. Heterocycl. Chem. **5**, 119 (1965).
[218] J. Sandström, Advan. Heterocycl. Chem. **9**, 165 (1968).

## C. Polycyclic Systems

### 1. Benzofuran and Benzothiophene

a. *Orientation of Substitution.* A striking difference between the two systems is the different orientation in substitution by electrophiles. It is well known that benzofuran, like the parent furan, is substituted almost exclusively at the α position. Nitration,[219] acylation[77, 220-222] chlorination,[223] Vilsmeier formylation,[224] and chloromethylation[225] are reported to give the 2-substituted derivatives, alone or together with minor amounts of the 3-substituted isomers.

In contrast with benzofuran (and with thiophene itself), benzothiophene is substituted preferentially at the β position. Bromination,[77, 226, 227] formylation,[228, 229] acylation,[77, 221, 230-232] chlorination,[233] chloromethylation,[234] sulfonation,[235] mercuration,[236] and iodination[237] give exclusively or prevalently the 3-substituted isomers. Anomalous orientation (in position 2) has been observed in isopropylation,[238] but it is possible that isomerization occurs under the conditions reported.

In both systems all the positions of the benzene ring are much less reactive than either of those of the heterocyclic ring and only very

[219] R. Stormer and B. Kahlert, *Ber.* **35**, 1640 (1902).
[220] N. P. Buu-Hoi, N. D. Xuong, and N. V. Bac, *J. Chem. Soc.* 173 (1964).
[221] M. W. Farrar and R. Levine, *J. Amer. Chem. Soc.* **72**, 4433 (1950).
[222] R. Magnusson, *Acta Chem. Scand.* **17**, 2358 (1963).
[223] See Palmer,[36] p. 322.
[224] M. Bisagni, N. P. Buu-Hoi, and R. Roger, *J. Chem. Soc.* 3688 (1955).
[225] A. L. Mndzhoyan and A. A. Aroyan, *Izv. Akad. Nauk. Arm. SSSR, Khim. Nauki* **14**, 591 (1961); *Chem. Abstr.* **58**, 5606h (1963).
[226] J. Szmuszkovicz and E. J. Modest, *J. Amer. Chem. Soc.* **72**, 571 (1950).
[227] N. P. Buu-Hoi and J. Lecocq, *C. R. Acad. Sci.* **222**, 1441 (1946).
[228] V. V. Ghaisas, *J. Org. Chem.* **22**, 703 (1957).
[229] A. W. Weston and R. J. Michaels, *J. Amer. Chem. Soc.* **72**, 1422 (1950).
[230] G. Komppa, *J. Prakt. Chem.* **122**, 322 (1929).
[231] C. Hansch and H. G. Lindwall, *J. Org. Chem.* **10**, 381 (1945).
[232] H. D. Hartough and A. I. Kosak, *J. Amer. Chem. Soc.* **68**, 2639 (1946).
[233] A. H. Schlesinger and D. T. Mowry, *J. Amer. Chem. Soc.* **73**, 2614 (1951).
[234] F. F. Blicke and D. G. Sheets, *J. Amer. Chem. Soc.* **70**, 3768 (1948).
[235] R. Weissgerber, *German Patent* 353932 (1921).
[236] F. Challenger and S. A. Miller, *J. Chem. Soc.* 1005 (1939).
[237] R. Gaertner, *J. Amer. Chem. Soc.* **74**, 4950 (1952).
[238] S. F. Bedell, E. C. Spaeth, and J. M. Bobbitt, *J. Org. Chem.* **27**, 2026 (1962).

small amounts, if any, of the isomers monosubstituted in the benzene ring have been isolated in the reaction mixtures.

Recently, more accurate determinations of isomer distributions and $\alpha:\beta$ reactivity ratios in electrophilic substitutions of benzofuran and benzothiophene have been determined using the gas chromatographic technique.[77] The data are summarized in Table XVI which also includes $\alpha:\beta$ reactivity ratios referring to electrophilic displacement reactions such as protodesilylation and to side-chain reactions related to electrophilic substitution, such as the solvolysis of 1-arylethyl-acetates[239] and 1-arylethyl-$p$-nitrobenzoates.[240]

The reactivity ratios vary over a wide range, depending on the nature of the electrophile and the conditions used. Temperature appears to be an important factor in determining the isomer distribution, at least in acetylation. The amounts of $\alpha$-substituted isomer in benzothiophene and $\beta$-substituted isomer in benzofuran increases, as the temperature increases (Table XVII). Plots of $\log \alpha/\beta$ against $1/T$ are linear for the acetylation of both systems over the range of temperature examined. It is possible that at higher temperatures ($> 250°$) an inversion of the orientation pattern could be observed;

TABLE XVII

Effect of Temperature on Isomer Distribution of Tin Tetrachloride-Catalyzed Acetylation of Benzofuran and Benzothiophene in Dichloroethane[a]

| Temp. (°C) | Benzofuran | | Benzothiophene | |
|---|---|---|---|---|
| | $\beta$ (%) | $\alpha:\beta$ | $\alpha$ (%) | $\beta:\alpha$ |
| 25 | 12.2 | 7.2 | 13.0 | 6.7 |
| 75 | 22.2 | 3.5 | 21.9 | 3.6 |
| 130 | 37.0 | 1.7 | 35.1 | 1.8 |
| 200 | — | — | 41.2 | 1.4 |

[a] Data from S. Clementi, P. Linda, and G. Marino.[77]

[239] E. A. Hill, M. L. Gross, M. Stasiewicz, and M. Manion, *J. Amer. Chem. Soc.* **91**, 7381 (1969).
[240] D. S. Noyce, C. A. Lipinski, and R. W. Nichols, unpublished results.

TABLE XVI

α:β REACTIVITY RATIOS IN ELECTROPHILIC SUBSTITUTIONS AND RELATED REACTIONS IN BENZOFURAN AND BENZOTHIOPHENE

| Reaction | Conditions[a] | Temp (°C) | α:β (Benzofuran) | β:α (Benzothiophene) | Ref. |
|---|---|---|---|---|---|
| Bromination | $Br_2$, AcOH | 25 | — | 90.9 | 77 |
| Bromination | HOBr, $HClO_4$, 50% dioxane | 25 | — | 76.9 | 77 |
| Chlorination | $Cl_2$, AcOH | 25 | — | 43.5 | 77 |
| Acetylation | $Ac_2O$, $SnCl_4$, DCE | 75 | 3.5 | 3.6 | 77 |
| Acetylation | ATFA, DCE | 75 | 8.0 | 3.7 | 77 |
| Benzoylation | $Bz_2O$, $SnCl_4$, DCE | 75 | 8.4 | 1.9 | 77 |
| Protodedeuteriation | $CF_3COOH$, AcOH | 25 | — | 1.6 | [b] |
| Protodesilylation | $HClO_4$, $CH_3OH$ | 50 | — | 1.03 | 175 |
| Nitration | $HNO_3$, AcOH | — | — | 4.7–6 | [c] |
| Solvolysis | | | | | |
| Of 1-arylethyl acetates | — | 25 | 1.19 | 3.09 | 239 |
| Of 1-arylethyl-OPNB | — | 75 | 0.97 | 2.7 | 211 |

[a] DCE, Dichloroethane; ATFA, acetyl trifluoroacetate; OPNB, p-nitrobenzoate.
[b] See A. I. Shatenshtein et al., Zh. Obshch. Khim. **40**, 1622 (1970).
[c] See G. Van Zyl, C. J. Bredewey, R. H. Rynbrandt, and D. C. Neckers, Can. J. Chem. **44**, 2283 (1966); K. J. Armstrong, M. Martin-Smith, N. M. D. Brown, G. C. Brophy, and S. Sternhell, J. Chem. Soc. C 1766 (1969).

however, at high temperatures these molecules undergo extensive decomposition under the reaction conditions and the hypothesis could not be proved.

These differences in orientation between benzofuran and benzothiophene have long puzzled chemists; however, as will be discussed below, the effect of the "annelation" on the reactivities of the $\alpha$ and $\beta$ positions of furan and thiophene rings is substantially similar, and the apparent difference in behavior originates in the different $\alpha:\beta$ reactivity ratios in the two monocyclic systems.

b. *Relative Reactivities.* The benzo derivatives are less reactive than the parent compounds, and generally require more severe conditions to achieve substitution.

Recently the overall reactivities relative to the monocyclic rings have been determined for a number of reactions[77] by kinetic or competitive procedures. The data, reported in Table XVIII, show that fusion with a benzene ring produces an overall decrease in reactivity in both systems. The decrease is much more pronounced for furan than for thiophene ring. As a consequence of this, the overall reactivities of benzofuran and benzothiophene are nearly equal in all the substitutions for which quantitative data are available (column 3 of Table XVIII; for a useful comparison the relative reactivities of the monocyclic rings in the same reactions are also reported in column 4).

c. *Partial Relative Rates and "Annelation" Effects.* The relative rates and isomer distributions in monocyclic and bicyclic systems permit the calculation, by simple mathematical operations, of the *partial relative rates* of the various single nonequivalent positions of these molecules. Table XIX reports the values in reference to the $\beta$ position of thiophene. An inspection of the data reveals some discrepancies and inversions in reactivity order for the various reactions. Many reasons may account for these deviations: among others, the steric effects exerted by the *peri*-H atoms of the bicyclic systems on the reactivities of the $\beta$ positions, the possibility of interaction between some electrophiles and the S atom in benzothiophene,[76] specific solvation effects, etc. Nevertheless, these facts do not obscure the general picture of the relative reactivities, which is as follows:
$\beta\text{-T} \simeq \beta\text{-F} < \beta\text{-BF} \simeq \alpha\text{-BT} < \beta\text{-BT} \simeq \alpha\text{-BF} < \alpha\text{-T} < \alpha\text{-F}$.

The effects caused by the annelation on the reactivity of the $\alpha$ and $\beta$ positions of furan and thiophene are of particular interest. The

TABLE XVIII

OVERALL RELATIVE RATES OF SUBSTITUTIONS IN BENZOFURAN AND BENZOTHIOPHENE[a,b]

| Reaction | Conditions | Temp. (°C) | BF/F | BT/T | BF/BT | F/T |
|---|---|---|---|---|---|---|
| Bromination | Br$_2$, AcOH | 25 | — | 0.19 | — | (120)[c] |
| Bromination | HOBr, HClO$_4$ | 25 | — | 0.05 | — | — |
| Chlorination | Cl$_2$, AcOH | 25 | — | 0.64 | — | (50)[c] |
| Acetylation | Ac$_2$O, SnCl$_4$, DCE | 25 | 0.003 | 0.04 | 0.90 | 11.9 |
| Acetylation | Ac$_2$O, SnCl$_4$, DCE | 75 | 0.008 | 0.07 | 1.00 | 8.55 |
| Acetylation | ATFA, DCE | 75 | 0.012 | 0.12 | 0.97 | 9.84 |
| Benzoylation | Bz$_2$O, SnCl$_4$, DCE | 75 | 0.021 | 0.15 | 1.44 | 9.22 |

[a] Data from S. Clementi, P. Linda, and G. Marino.[77]
[b] F, Furan; T, thiophene; BF, benzofuran; BT, benzothiophene; DCE, dichloroethane; ATFA, acetyl trifluoroacetate.
[c] Relative rate of halogenation of the 2-methoxycarbonyl derivatives.

## TABLE XIX

Electrophilic Substitutions and Related Reactions of Furan, Thiophene, Benzofuran, and Benzothiophene. Partial Relative Rates[a,b]

| Reaction | Conditions | Temp. (°C) | β-T | β-F | β-BF | α-BT | β-BT | α-BF | α-T | α-F |
|---|---|---|---|---|---|---|---|---|---|---|
| Bromination | Br$_2$, AcOH | 25 | 1 | — | — | 2.07 | 189 | — | 490 | (58,000)[c] |
| Chlorination | Cl$_2$, AcOH | 25 | 1 | — | — | 2.90 | 126 | — | 100 | (5,000)[c] |
| Bromination | HOBr, HClO$_4$, 50% dioxane | 25 | 1 | — | — | 0.16 | 12.1 | — | 120 | — |
| Benzoylation | Bz$_2$O, SnCl$_4$, DCE | 75 | 1 | 2.43 | 2.97 | 6.66 | 12.8 | 25.0 | 65.6 | 609 |
| Acetylation | Ac$_2$O, SnCl$_4$, DCE | 25 | 1 | <2.40 | 1.78 | 2.00 | 13.3 | 12.9 | 200 | 2,380 |
| Acetylation | ATFA, DCE | 75 | 1 | <0.71 | 1.86 | 3.57 | 13.3 | 14.8 | 71.4 | 704 |
| Protodedeuteriation[d] | CF$_3$COOH, AcOH | 25 | 1 | — | — | 6.25 | 10.0 | — | 3,000 | — |
| Protodesilylation | HClO$_4$, CH$_3$OH | 50 | 1 | 1.02[e] | — | 0.35 | 0.36 | — | 42.8 | 153 |
| Solvolysis Of 1-arylethyl acetates[f] | — | 25 | 1 | 1.41 | 0.87 | 0.69 | 2.14 | 1.04 | 113 | 448 |
| Of 1-arylethyl-OPNB[g] | — | 75 | 1 | 0.84 | 2.06 | 1.27 | 3.43 | 2.06 | 62.1 | 293 |

[a] Data from S. Clementi, P. Linda, and G. Marino.[77]
[b] T, Thiophene; F, furan; BT, benzothiophene; BF, benzofuran; DCE, dichloroethane; ATFA, acetyl trifluoroacetate; OPNB, p-nitrobenzoate.
[c] Data from the 2-methoxycarbonyl derivatives.
[d] A. I. Shatenshtein et al., Zh. Obshch. Khim. **40**, 1622 (1970).
[e] R. Taylor, J. Chem. Soc. B 1364 (1970).
[f] See Hill et al.[239]
[g] D. S. Noyce, private communication.

elaboration of the data (Table XX) leads to a rather surprising result. Although the orientation of substitution in the two bicyclic systems is different, nevertheless, the effect of fusion of a benzene ring is substantially similar in the two rings. The reactivity of the α position is always *decreased* by a similar factor, and the reactivity of the β position is—with two exceptions—*increased* in both systems.

2. *Indole*

Indole, like benzothiophene, reacts with electrophiles to give preferentially β-substituted products. Bromination[241-243] in dioxane or pyridine, nitration by ethyl nitrate,[244] chlorination by sulfuryl chloride,[245] iodination in aqueous solution,[246, 247] Vilsmeier and Reimer–Tiemann formylations,[248-251] diazo coupling,[252] thiocyanation,[253] and nitrosation[254] all give the 3-substituted indoles, practically free from other isomers.

The position 3 is so much preferred in electrophilic substitution, that it is the site for primary attack even when it is already substituted. Jackson and co-workers[255] have provided a considerable body of evidence that electrophilic substitution at the 2-position of 3-substituted indoles is an indirect process involving prior attack at the 3-position to give an indolenine derivative (**30**), which subsequently rearranges to a 2,3-disubstituted indole.

[241] K. Piers, C. Meimaroglou, R. V. Jardine, and R. K. Brown, *Can. J. Chem.* **41**, 2399 (1963).
[242] L. A. Yanovskaya, *Dokl. Akad. Nauk SSSR* **71** 693 (1950); *Chem. Abstr.* **44**, 8354 (1950).
[243] A. P. Terent'ev, L. I. Belen'kii, and L. A. Yanovskaya, *Zh. Obshch. Khim.* **24**, 1265 (1954); *Chem. Abstr.* **49**, 12327d (1955).
[244] F. Angelico and G. Velardi, *Atti. Accad. Naz. Lincei* **13** Classe Scienze Fis. Mat. Nat. (1) 241 (1904).
[245] R. Weissgerber, *Ber.* **46**, 651 (1913).
[246] H. Pauly and K. Gundermann, *Ber.* **41**, 3999 (1908).
[247] A. Oswald, *Z. Phys. Chem.* **73**, 128 (1911).
[248] A. C. Shabica, E. E. Howe, J. B. Ziegler, and M. Tishler, *J. Amer. Chem. Soc.* **68**, 1156 (1946).
[249] F. T. Tyson and J. T. Shaw, *J. Amer. Chem. Soc.* **74**, 2273 (1952).
[250] W. J. Boyd and W. Robson, *Biochem. J.* **29**, 555 (1935).
[251] G. F. Smith, *J. Chem. Soc.* 3842 (1954).
[252] W. Madelung and O. Wihelm, *Ber.* **57B**, 234 (1924).
[253] M. S. Grant and H. R. Snyder, *J. Amer. Chem. Soc.* **82**, 2742 (1966).
[254] P. Seidel, *Ber.* **77B**, 797 (1944).
[255] A. H. Jackson and A. Smith, *Tetrahedron* **21**, 898 (1965) and subsequent papers of the series.

TABLE XX

Effect of "Annelation" on the Reactivities of $\alpha$ and $\beta$ Positions of Furan and Thiophene[a]

| Reaction | Conditions | Temp. (°C) | $\dfrac{\alpha\text{-BF}}{\alpha\text{-F}}$ | $\dfrac{\beta\text{-BF}}{\beta\text{-F}}$ | $\dfrac{\alpha\text{-BT}}{\alpha\text{-T}}$ | $\dfrac{\beta\text{-BT}}{\beta\text{-T}}$ | Ref. |
|---|---|---|---|---|---|---|---|
| Bromination | $Br_2$, AcOH | 25 | — | — | $4.2 \times 10^{-3}$ | 189 | 77 |
| Chlorination | $Cl_2$, AcOH | 25 | — | — | $2.9 \times 10^{-2}$ | 126 | 77 |
| Bromination | HOBr, $HClO_4$, aq. dioxane | 25 | — | — | $1.3 \times 10^{-3}$ | 12.1 | 77 |
| Acetylation | $Ac_2O$, $SnCl_4$, DCE | 75 | $1.3 \times 10^{-2}$ | > 3.6 | $3.0 \times 10^{-2}$ | 9.1 | 77 |
| Acetylation | ATFA, DCE | 75 | $2.1 \times 10^{-2}$ | > 2.6 | $5.0 \times 10^{-2}$ | 13.3 | 77 |
| Benzoylation | $Bz_2O$, $SnCl_4$, DCE | 75 | $4.1 \times 10^{-2}$ | 1.22 | $1.0 \times 10^{-1}$ | 12.8 | 77 |
| Protodedeuteriation | $CF_3CO_2H$, AcOH | 25 | — | — | $2.0 \times 10^{-3}$ | 10.0 | [b] |
| Protodesilylation | $HClO_4$, $CH_3OH$ | 50 | — | — | $8.2 \times 10^{-3}$ | 0.36 | 175 |
| Solvolysis |  |  |  |  |  |  |  |
| Of 1-arylethyl acetates | — | 25 | $2.3 \times 10^{-3}$ | 0.62 | $6.1 \times 10^{-3}$ | 2.1 | 239 |
| Of 1-arylethyl OPNB | — | 75 | $7.0 \times 10^{-3}$ | 2.53 | $2.0 \times 10^{-2}$ | 3.4 | [c] |

[a] For explanation of symbols and abbreviations, see footnotes to Table XIX.
[b] See A. I. Shatenshtein et al., Zh. Obshch. Khim. **40**, 1622 (1970).
[c] D. C. Noyce, private communication.

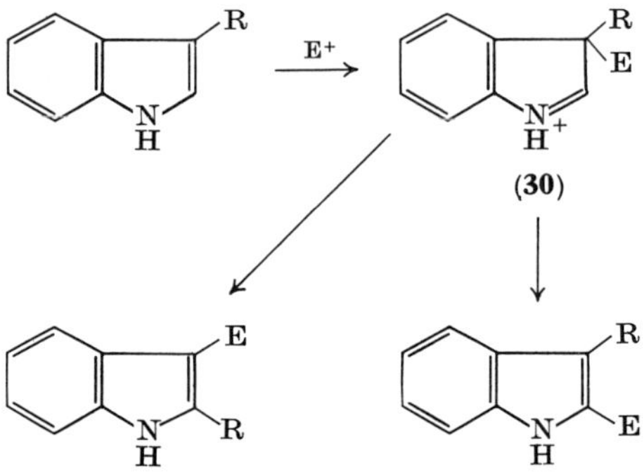

[Diagram: indole + E⁺ → intermediate (30) → 2-substituted or 3-substituted indole products]

(30)

(31)

When sulfonated, indole behaves anomalously and gives the 2-sulfonic acid as the main product.[256] Under the conditions used, the reaction could be thermodynamically rather than kinetically controlled.

Another anomalous orientation is observed in the nitration of 2-methylindole in sulfuric acid, which affords the 5-nitro derivative almost exclusively.[257, 258] However, nitration in acetic acid yields the expected 3-nitro isomer. It has been shown[258a] that whereas nitration in acetic acid involves free indole, nitration in sulfuric acid proceeds through the protonated form (31).

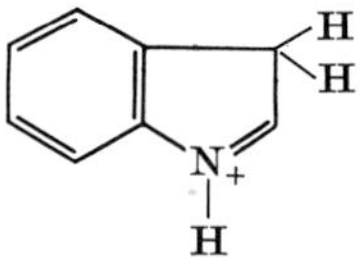

There are no quantitative studies to permit direct comparison of the reactivity of indole in electrophilic substitution with that of pyrrole. However, the available data indicate that the reactivity at position 3 of indole does not differ very much from that of the 2-position of pyrrole. For instance, indole is reported to be 50 times more reactive than aniline in the iodination reaction.[107]

A direct comparison between 1-methylindole and 1-methylpyrrole is possible only for an α-carbonium ion reaction, the solvolysis of

[256] A. P. Terent'ev, S. K. Golubeva, and L. V. Tsymbal, *Zh. Obshch. Khim.* **19**, 781 (1949); *Chem. Abstr.* **44**, 1095 (1950).
[257] W. E. Noland and R. D. Rieke, *J. Org. Chem.* **27**, 2250 (1962).
[258] W. E. Noland, L. R. Smith, and D. C. Johnson, *J. Org. Chem.* **28**, 2262 (1963).
[258a] K. Brown, and A. R. Katritzky, *Tetrahedron Lett.* 803 (1964).

1-arylethylacetates,[239] in which the 1-methyl-3-indolyl derivative is about 1.3 times as reactive as the 1-methyl-2-pyrrolyl derivative.

## IV. Effects of Substituents

### A. Directing Effects

#### 1. *Monocyclic Rings*

Gronowitz first applied the modern electronic theories of organic chemistry to elucidating the directing effects of substituents in the thiophene ring[259] and has summarized concepts and experimental data in a recent review.[2] The same rules can be extended without appreciable modifications to the other monocyclic systems: furan, selenophene, and pyrrole.

a. *Alkyl Substituents.* 2-Alkylthiophenes orient the substitution (chlorination,[260] bromination,[85, 261] acetylation,[130, 141] trifluoroacetylation,[143] hydrogen exchange,[55] *t*-butylation,[262] formylation[263]) almost exclusively to the other free α position (5). Small amounts of 2-alkyl 3-substituted isomers are also formed in some cases, but only in the nitration of 2-methylthiophene are appreciable amounts (30%) of 3-substitution obtained.[264]

Selenophene is very similar in chemical behavior to thiophene[4a, 90]; therefore, it is not surprising that also here a 2-alkyl group directs the substitution preferentially to C-5. This has been observed in acetylation,[265] chloromethylation,[266] aminomethylation,[267] and formylation[268] of 2-methylselenophene.

[259] S. Gronowitz, *Ark. Kemi* **13**, 295 (1958).
[260] H. L. Coonradt, H. D. Hartough, and G. C. Johnson, *J. Amer. Chem. Soc.* **70**, 2564 (1948).
[261] P. Cagniant and D. Cagniant, *Bull. Soc. Chim. Fr.* 713 (1952).
[262] Ya. L. Gol'dfarb and I. S. Korsakova, *Bull. Acad. Sci. USSR Div. Chem.* 481 (1954); *Chem. Abstr.* **49**, 9615d (1955).
[263] P. Linda, G. Marino, and S. Santini, unpublished results.
[264] R. A. Hoffman and S. Gronowitz, *Ark. Kemi* **16**, 563 (1960).
[265] E. G. Kataev and M. V. Palkina, *Uch. Zap. Kazan. Gos. Univ.* **113**, 115 (1953); *Chem. Abstr.* **50**, 937i (1956).
[266] Yu. K. Yur'ev, N. K. Sadovaya, and M. A. Gal'bershtam, *Zh. Obshch. Khim.* **32**, 259 (1962); *Chem. Abstr.* **57**, 16536 (1962).
[267] Yu. K. Yur'ev, N. K. Sadovaya, and A. B. Ibraginova, *Zh. Obshch. Khim.* **29**, 3647 (1959); *Chem. Abstr.* **54**, 19644a (1960).
[268] Yu. K. Yur'ev, N. N. Mezentsova, and V. E. Vas'Kovskii, *Zh. Obshch. Khim.* **27**, 3155 (1957); *Chem. Abstr.* **52**, 9065g (1958).

Since the α-directing power is more pronounced in furan than in thiophene (see Section III, A, 2), the preference for substitution at C-5 is here reinforced. Substitution in 2-alkylfurans actually occurs at this position in all cases and no detectable formation of other isomers is observed.[1]

In pyrrole the α:β reactivity ratio is much smaller than in the other 5-membered rings (see Section III, A, 2); here the formation of a relatively larger amount of 3-substituted isomer could be expected. Actually, the 5-substituted 2-alkyl derivative appears to be the main product in all the electrophilic substitutions of the 2-alkylpyrroles[3]; in some cases, as in the reactions with trifluoroacetic anhydride and acetyl trifluoroacetate,[147] the 5-substituted isomer is apparently the only product formed.

If the alkyl substituent is in position 3, the 2-position is the most favored from the electronic point of view, being ortho to the alkyl group and alpha to the heteroatom. The other α position (5) is, however, free from steric effects. Consequently, reactions of 3-alkyl thiophenes, -furans, and -pyrroles yield a mixture of 2- and 5-substituted isomers, the composition of which depends on the nature of the heteroatom, the electrophile, and the size of the alkyl group. Table XXI reports the isomer distribution in the acylation of 3-methyl-, 3-isopropyl-, and 3-t-butylthiophenes and shows the importance of the steric factor in determining the orientation.

TABLE XXI

Isomer Distribution in Acetylation of 3-Alkylthiophenes

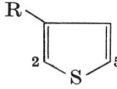

| R | Yield of acetylated products | | | Ref. |
|---|---|---|---|---|
| | Total | 2 | 5 | |
| Methyl | 92 | 74 | 18 | 135 |
| Isopropyl | 79 | 31 | 48 | 276 |
| t-Butyl | 82 | 0 | 82 | a |

[a] M. Sy, N. P. Buu-Hoi, and N. D. Xuong, *J. Chem. Soc.* 21 (1955).

b. *Substituents with* $-I$, $+M$ *Effects.* Substituents belonging to this class (i.e., halogens; oxy, alkoxy, and thioalkoxy; amino and alkylamino, etc.) show an orientation pattern similar to that of the alkyl groups. If the substituent is in position 2, substitution occurs at C-5; if it is in position 3, substitution occurs preferentially at C-2 or, if the steric requirements of the group or the electrophile predominate, at C-5. Most of the available information concerns the substituted thiophenes,[2] since the corresponding derivatives of furan and pyrrole are generally unstable, but the few reported reactions of furans and pyrroles bearing $-I$, $+M$ substituents (for instance, acylation of the halogenofurans) also follow the above orientation rules.

c. *Substituents with* $-I$, $-M$ *Effects.* If a $-I$, $-M$ substituent (such as $NO_2$, $CN$, $COR$, etc.) is in position 3, the effects of the substituent and the heteroatom combine to direct the substitution to position 5. This is what is actually observed, without exception.

A different situation follows when the electronegative substituent is in the 2-position. While the 5-position is the most activated by the heteroatom, the 4-position is the least deactivated by the substituent; the orientation of the substitution will be the result of competition between these two effects.

In furan, where the preference for α substitution is very pronounced (see Section III,A,2), the orienting effect of the heteroatom predominates and the main (often the sole) product formed is the 2,5-disubstituted isomer, in all cases.[1]

In thiophene the α-directing power of the sulfur atom still dominates in many cases. Halogenation of 2-thiophenecarboxaldehyde,[269, 270] 2-thiophenecarboxylic acid,[271] and 2-methoxycarbonylthiophene[85] gives mainly or exclusively the 5-substituted isomer. Chloromethylation gives similar results.[272] However, nitration of electronegatively 2-substituted thiophenes[273-275] gives a mixture of 4- and 5-nitro 2-substituted isomers, in different ratios (Table XXII).

[269] S. Gronowitz, *Ark. Kemi.* **8**, 87 (1955).
[270] J. Lamy, D. Lavit, and N. P. Buu-Hoi, *J. Chem. Soc.* 4202 (1958).
[271] J. F. Bunnett, D. M. Bachmann, L. P. Snipper, and J. H. Maloney, *J. Amer. Chem. Soc.* **71**, 1493 (1949).
[272] R. Lukeš, M. Janda, and K. Kefurt, *Collect. Czech. Chem. Commun.* **25**, 1058 (1960).
[273] J. Tirouflet and P. Fournari, *C. R. Acad. Sci.* **246**, 2003 (1958).
[274] B. Östman, *Acta Chem. Scand.* **22**, 2754 (1968).
[275] A. J. De Dominicis, Ph.D. Thesis, New York Univ., 1963.

## TABLE XXII

Isomer Distributions in Nitration of Negatively 2-Substituted Thiophenes

| 2-R | Conditions | Amounts of nitro isomers[a] | | | Ref. |
|---|---|---|---|---|---|
| | | 5-NO$_2$ | 4-NO$_2$ | 3-NO$_2$ | |
| CHO | HNO$_3$ in AcOH | 25 | 75 | — | 273 |
| COCH$_3$ | HNO$_3$ in AcOH | 48 | 52 | — | 273 |
| CN | HNO$_3$ in AcOH | 57 | 43 | — | 273 |
| CO$_2$H | HNO$_3$ in AcOH | 69 | 31 | — | 273 |
| CHO | HNO$_3$ in CF$_3$COOH | 42.3 | 57.2 | 0.5 | 274 |
| CN | HNO$_3$ in CF$_3$COOH | 59.9 | 39.9 | 0.2 | 274 |
| NO$_2$ | HNO$_3$ in CF$_3$COOH | 42.5 | 57.4 | 0.1 | 274 |
| COCH$_3$ | HNO$_3$ in H$_2$SO$_4$ | 25.9 | 73.2 | 0.8 | 275 |
| CO$_2$H | HNO$_3$ in H$_2$SO$_4$ | 40.4 | 55.9 | 2.0 | 275 |
| CO$_2$CH$_3$ | HNO$_3$ in H$_2$SO$_4$ | 42.0 | 50.9 | 7.1 | 275 |
| NO$_2$ | HNO$_3$ in H$_2$SO$_4$ | 36.7 | 63.3 | — | 275 |

[a] The sum of the substituted products equals 100.

Finally, in some other reactions (e.g., isopropylation of 2-acetylthiophene[276]) the 4-substituted isomer is practically the only product formed. Overall, the situation is not easily rationalizable; often anomalous orientation may be the result of a peculiar mechanism.

In pyrrole, the α-directing effect of the heteroatom is much weaker than in the other two fundamental rings; consequently, it is overcome by the effect of the substituent and the substitution occurs mainly at C-4. For example, nitration of 2-acetylpyrrole[277] gives 4-nitro and 5-nitro derivatives in ratios of roughly 2:1; bromination of 2-methoxycarbonylpyrrole gives a mixture of 4-bromo and 5-bromo derivatives, of composition varying according to the conditions used, but in which the 4-bromo isomer is always predominant.[278, 84]

d. *Disubstituted Rings.* When two substituents are present in the ring, the position of further substitution can be deduced from the directing effects of the heteroatom and each substituent. The orientation rules summarized in Scheme 2 can tentatively be predicted.

[276] E. C. Spaeth and C. B. Germain, *J. Amer. Chem. Soc.* **77**, 4066 (1955).
[277] I. J. Rinkes, *Rec. Trav. Chim.* **53**, 1167 (1934).
[278] P. Hodge and R. W. Rickards, *J. Chem. Soc.* 459 (1965).

Sec. IV.A.] SUBSTITUTIONS OF FIVE-MEMBERED RINGS 297

(In formulas **32–41**, methyl and nitro groups have been chosen as typical ortho-para and meta-directing substituents, respectively.)

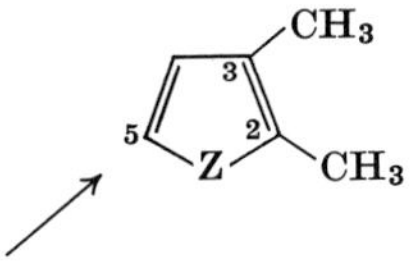

**(32)** 2,3-di-CH₃

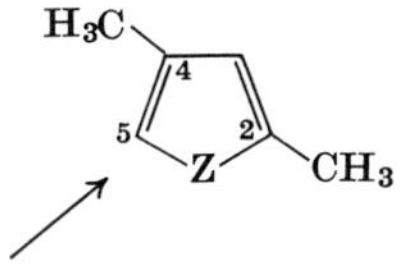

**(33)** 3-CH₃, 2-CH₃ (with 4-position marked)

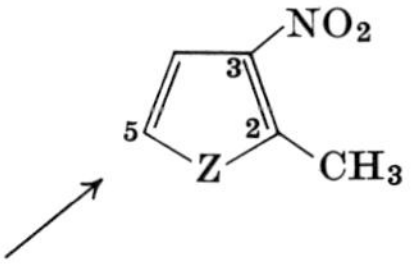

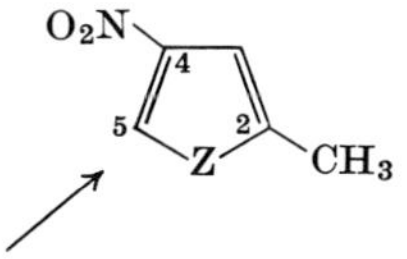

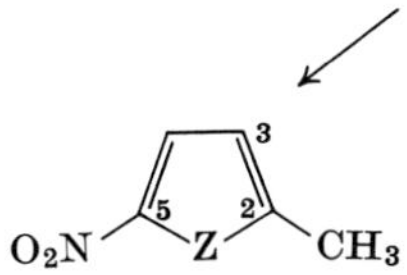

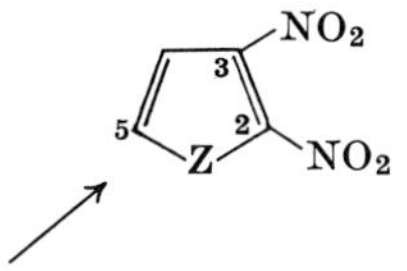

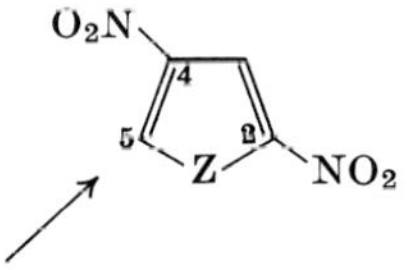

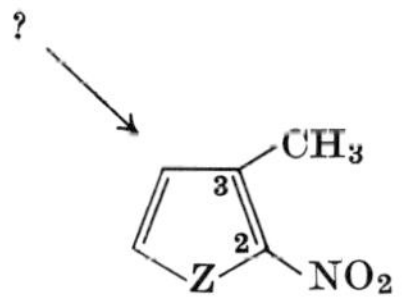

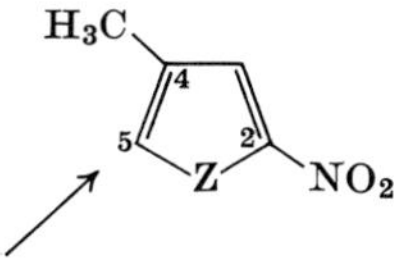

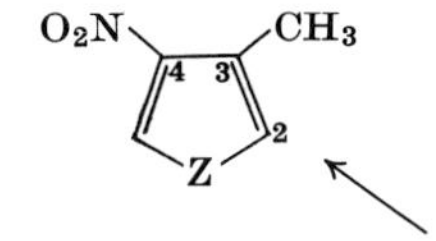

**(34)** 3-NO₂, 2-CH₃

**(35)** 4-NO₂, 2-CH₃

**(36)** 5-O₂N, 2-CH₃

**(37)** 3,2-di-NO₂

**(38)** 4-O₂N, 2-NO₂

**(39)** 3-CH₃, 2-NO₂

**(40)** 4-H₃C, 2-NO₂

**(41)** 4-O₂N, 3-CH₃

SCHEME 2

The scattered information available in the literature seems to confirm the general validity of these predictions. However, it must be remembered that, in many cases, especially in 2,5-disubstituted compounds, an α substituent already present can be replaced in preference to a β proton.

e. *Substituted Benzo Derivatives.* No systematic study on the effects of substituents in benzo derivatives is available. On the basis of fragmentary experimental data, the following general rules can be drawn: (*i*) an activating substituent in positions 2 or 3 directs the entering group in the other free position of the heterocyclic ring; (*ii*) a deactivating substituent in the five-membered ring generally directs the entering group into the benzene ring. (*iii*) a strongly activating substituent (such as hydroxy or amino) in the benzene ring directs the substitution into the same ring (ortho or para); and (*iv*) with a deactivating or a weakly activating substituent (such as an alkyl group) in the benzene ring, the substitution occurs in the heterocyclic ring.

## B. Hammett-Type Correlations

### 1. *General Aspects*

The problem of the application of the Hammett equation to the electronic transmission of substituent effects through five-membered rings has recently been discussed by Jaffé and Jones.[279]

In principle, two distinct procedures may be adopted. One might use a single $\rho$ value for any type of reaction (e.g., $\rho = 1$ for the ionization of carboxylic acids in water at 25°) and define new $\sigma$ values for each substituent, depending on the nature of the ring and the relative position with respect to both the reaction center and the heteroatom. This procedure has rarely been used since it would require extremely extensive experimental material before any reasonable test of the correlation could be made.[279] Recently, it has been used by Butler[280] to calculate a large number of $\sigma$ constants for the thiophene ring from the ionization constants of substituted thiophene carboxylic acids. The values obtained were found to be in excellent agreement with those calculated by the method of Dewar and Grisdale,[281] indicating that the sulfur atom has very little effect upon the transmission of substituent effects.

[279] H. H. Jaffé and H. L. Jones, *Advan. Heterocycl. Chem.* **3**, 238 (1963).
[280] A. R. Butler, *J. Chem. Soc. B* 867 (1970).
[281] M. J. S. Dewar and P. J. Grisdale, *J. Amer. Chem. Soc.* **84**, 3539 (1962).

In the alternative procedure, originally proposed by Hammett in treating the furoic acids,[282] the pairs of positions 2–5 (of "conjugative" type) and 2–4 ("nonconjugative") are assimilated to the pairs of positions 1–4 ("para") and 1–3 ("meta") respectively, of the benzene ring. Accordingly, the conventional $\sigma_p$ and $\sigma_m$ constants, derived from benzene chemistry, are employed and separate $\rho$ constants are derived from the reactions at each different ring.

This procedure has been extensively applied in correlating side-chain reactivities[283-289] and spectroscopic quantities.[290] In correlating electrophilic substitution reactivity data, the Hammett $\sigma$ constants must, of course, be replaced by the Brown and Okamoto $\sigma^+$ constants.[185]

### 2. Thiophene

Relative rates of substitution for a number of substituted thiophenes sufficient to test the applicability of $\rho\sigma^+$ relationships are available for seven reactions: bromination by molecular bromine, chlorination by molecular chlorine, protodetritiation, protodedeuteriation, acetoxymercuration, tin tetrachloride-catalyzed acetylation, and trifluoroacetylation. The relevant data are assembled in Table XXIII.

The relative rates of bromination of some 2-substituted thiophenes (R = H, $CH_3$, Cl, Br, I, $CO_2H$, $CO_2Et$) by molecular bromine in anhydrous acetic acid have been determined by the author,[85] comparing the times necessary to achieve 10% reaction. All the thiophenes examined gave the 5-bromo derivative on substitution, without appreciable formation of any other isomer. Only 2-iodothiophene

---

[282] W. E. Catlin, *Iowa State Coll. J. Sci.* **10**, 65 (1935).
[283] E. Imoto and R. Motoyama, *Bull. Naniwa Univ.* **2A**, 127 (1954); *Chem. Abstr.* **49**, 9614e (1955).
[284] E. Imoto, Y. Otsuji, and T. Hirai, *Nippon Kagaku Zasshi* **77**, 804 (1956); *Chem. Abstr.* **52**, 9066e (1958).
[285] E. Imoto and Y. Otsuji, *Bull. Univ. Osaka Prefect., Ser. A* **6**, 115 (1958); *Chem. Abstr.* **53**, 3027h (1959).
[286] Y. Otsuji, Y. Koda, M. Kubo, M. Furukawa, and E. Imoto, *Nippon Kagaku Zasshi* **80**, 1293 (1959); *Chem. Abstr.* **55**, 6476g (1961).
[287] D. Spinelli, G. Guanti, and C. Dell'Erba, *Ric. Sci.* **38**, 1048 (1968).
[288] F. Fringuelli, G. Marino, and G. Savelli, *Tetrahedron* **25**, 5815 (1969).
[289] J. Tirouflet and J. P. Chané, *C. R. Acad. Sci.* **245**, 80, 500 (1957).
[290] Y. Otsuji and E. Imoto, *Nippon Kagaku Zasshi* **80**, 1199 (1959); *Chem. Abstr.* **55**, 3194 (1961).

TABLE XXIII

Relative Rates of α Substitution of Substituted Thiophenes[a]

| Substituent | Bromination[b] | Chlorination[c] | Detritiation[d] | Dedeuteriation[e] | Acetylation[f] | Trifluoroacetylation[g] | Mercuration[h] |
|---|---|---|---|---|---|---|---|
| 2-OCH$_3$ | — | — | — | $1.5 \times 10^6$ | — | $1.8 \times 10^6$ | — |
| 2-SCH$_3$ | — | — | — | $1.5 \times 10^3$ | — | $5.2 \times 10^3$ | — |
| 2-CH$_3$ | $6.3 \times 10^2$ | — | $2.1 \times 10^2$ | $2.3 \times 10^2$ | 33.7 | $3.8 \times 10^2$ | — |
| 2-C$_2$H$_5$ | — | — | — | — | — | $5.2 \times 10^2$ | — |
| 2-$t$-C$_4$H$_9$ | — | — | $2.3 \times 10^2$ | — | 48.5 | $5.4 \times 10^2$ | — |
| 2-C$_6$H$_5$ | — | — | 15.5 | — | — | $1.1 \times 10^2$ | — |
| 3-CH$_3$ | — | — | — | 12 | 3.40 | — | — |
| H | 1 | 1 | 1 | 1 | 1 | 1 | 1 |
| 2-Cl | $5.2 \times 10^{-1}$ | 0.42 | $1.7 \times 10^{-1}$ | — | 0.14 | $5.8 \times 10^{-1}$ | — |
| 2-Br | $3.8 \times 10^{-1}$ | 0.33 | $1.3 \times 10^{-1}$ | — | — | $4.6 \times 10^{-1}$ | $2.6 \times 10^{-1}$ |
| 2-I | $(9.4 \times 10^{-1})$ | 0.53 | $1.9 \times 10^{-1}$ | — | — | — | — |
| 3-SCH$_3$ | — | — | — | 0.1 | — | — | — |
| 2-COCH$_3$ | — | — | — | — | — | — | $2.8 \times 10^{-3}$ |
| 2-CO$_2$H | $3.3 \times 10^{-5}$ | $1.1 \times 10^{-4}$ | — | — | — | — | — |
| 2-CO$_2$C$_2$H$_5$ | $1.1 \times 10^{-5}$ | $1.3 \times 10^{-4}$ | — | — | — | — | $3.4 \times 10^{-3}$ |
| 2-NO$_2$ | — | $1.1 \times 10^{-6}$ | — | — | — | — | — |

[a] Relative rates at position 5 of thiophene ring. The values are corrected for the statistical factor.
[b] Reaction with Br$_2$ in anhydrous acetic acid at 25°.[85]
[c] Reaction with Cl$_2$ in acetic acid at 25°. Values from different sources; see text.
[d] Reaction in CF$_3$CO$_2$H–CH$_3$CO$_2$H mixture at 24.8°.[297]
[e] Reaction in CF$_3$CO$_2$H–CH$_3$CO$_2$H mixture at 25°.[55, 299, 300]
[f] SnCl$_4$-catalyzed acetylation by Ac$_2$O in dichloroethane at 25°. See Linda and Marino,[130] and unpublished results.
[g] Reaction with (CF$_3$CO)$_2$O in dichloroethane at 75°. See Clementi and Marino,[143, 302]
[h] Acetoxymercuration in acetic acid at 50°.[121]

gave a complex reaction mixture, containing 2-iodo- (15%), 2-bromo- (14%), 2,5-dibromo- (16%), and 2-bromo-5-iodothiophene (55%). It is clear that 2-bromothiophene was formed in an I–Br exchange reaction,[291] whereas 2,5-dibromothiophene could be formed either by further substitution on 2-bromothiophene or in a halogen-exchange reaction of 2-bromo-5-iodothiophene.

A plot of $\log k/k_0$ against $\sigma_p^+$ is linear (Fig. 6). The main deviations concern the 2-halogenothiophenes and are in the same direction as is

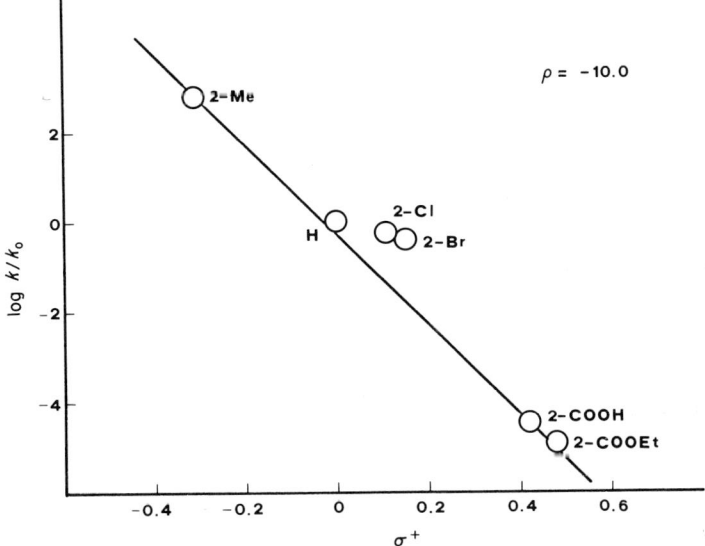

FIG. 6. Plot of $\log k/k_0$ vs $\sigma_p^+$ constants for the bromination of 2-substituted thiophenes in acetic acid; see Marino.[85]

observed in the noncatalyzed halogenation of the halogenobenzenes.[292] These deviations can easily be explained by the known interactions between halogen substituents and halogen molecules.[293] They are so strong in 2-iodothiophene that they lead to the breakage of the C–I bond (see previous paragraph). The reaction constant $\rho$, derived from the graph, has a value of $-10.0$.

[291] Similar reactions have been described in other iodo derivatives of thiophenes. See, for instance, W. Steinkopf, H. F. Schmitt, and H. Friedler, *Ann.* **527**, 237 (1936).
[292] L. M. Stock and F. W. Baker, *J. Amer. Chem. Soc.* **84**, 1661 (1962).
[293] L. J. Andrews, *Chem. Rev.* **54**, 713 (1954).

Later, Butler and Hendry[294] repeated this work, measuring the rates of bromination of some 2-substituted thiophenes (the same derivatives used by Marino[85] plus 2-ethyl, 2-t-butyl, and 2-phenyl) in 85% aqueous acetic acid and in the presence of lithium bromide. The $\rho$ value obtained was identical ($-10.0$), showing that it is not affected at all by passing from anhydrous to aqueous acetic acid.

In Fig. 7 an analogous plot for the uncatalyzed chlorination of 2-substituted thiophenes in anhydrous acetic acid is reported. The

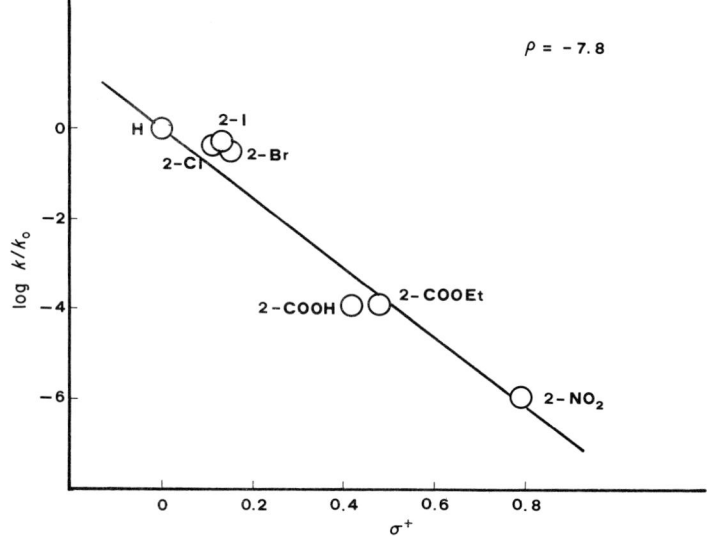

FIG. 7. Plot of log $k/k_0$ vs. $\sigma_p^+$ constants for the chlorination of 2-substituted thiophenes in acetic acid.

graph assembles values from different sources: unsubstituted thiophene,[72] 2-chloro-,[295] 2-bromo-,[295] 2-carboxy,-[294] 2-carbomethoxy-,[294] 2-iodo-,[295] and 2-nitrothiophene.[295] The reaction of 2-nitrothiophene has been carried out in 85% acetic acid and the relative rate is obtained by a comparison with the rate of chlorination of 2-thiophenecarboxylic acid in the same solvent. A $\rho$ value of $-7.8$ is so obtained.

Imoto and his co-workers[121] have measured the rate of acetoxymercuration of some 2-R-substituted thiophenes (R=$CH_3$, H, Br, $COCH_3$, $CO_2Et$) in acetic acid. Substitution occurred exclusively

[294] A. R. Butler and J. B. Hendry, *J. Chem. Soc. B* 848 (1970).
[295] G. Marino, unpublished results.

at position 5. These authors have tested the applicability of a linear free energy relationship using the Hammett σ constants. A much better fit is obtained if $\sigma_p^+$ constants are used (Fig. 8). The ρ value[296] is then −5.3.

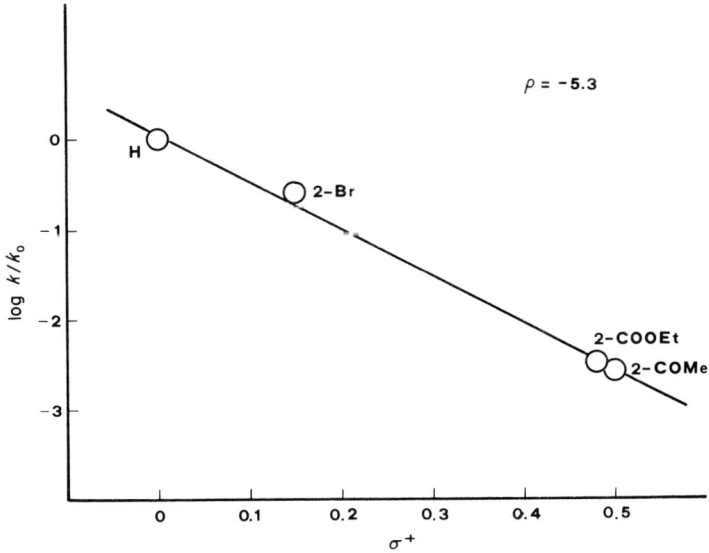

FIG. 8. Plot of log $k/k_0$ vs. $\sigma_p^+$ constants for the mercuration of 2-substituted thiophenes in acetic acid. Data from Motoyama et al.[121]

Another study of the effects of substituents in an electrophilic substitution at the thiophene ring has been made recently by Butler and Eaborn,[297] who measured the rates of protodetritiation of a number of 5-substituted 2-tritiothiophenes. The substituents were $OCH_3$, $CH_3$, $t$-$C_4H_9$, $C_6H_5$, H, Cl, Br, I, $CO_2C_2H_5$, and $NO_2$. The compounds having the last two substituents were too unreactive and it was impossible to follow the rate of exchange. 2-Methoxythiophene gave a bright red color when added to acid, possibly as a result of a side reaction. The other thiophene derivatives behaved normally. Because of the large range in reactivity, it was impracticable to study all the compounds in the same medium. Accordingly, three mixtures of trifluoroacetic and acetic acids having different compositions were

[296] Motoyama et al.[121] report erroneously a ρ value of −3.7. See Yur'ev et al.[304]
[297] A. R. Butler and C. Eaborn, J. Chem. Soc. B 370 (1968).

used. The rates of H/T exchange of the unsubstituted thiophene 2-*t* were determined in all three solvents. Since the relative rates of 5-methyl- and 5-*t*-butyl-2-tritiothiophenes are similar in two different solvents, it was concluded to be likely, although not certain, that changes in the acidity of the medium do not affect the relative reactivities. In Fig. 9 a plot of log $k/k_0$ against $\sigma_p^+$ is reported. The correlation is again satisfactory; the $\rho$ value is $-7.2$. Only the *t*-butyl group deviates substantially; however, it is interesting to note that the order of reactivity observed—*t*-butyl > methyl—is the same as is observed in the protodetritiation of *p*-alkyltritiobenzenes in trifluoroacetic acid.[298]

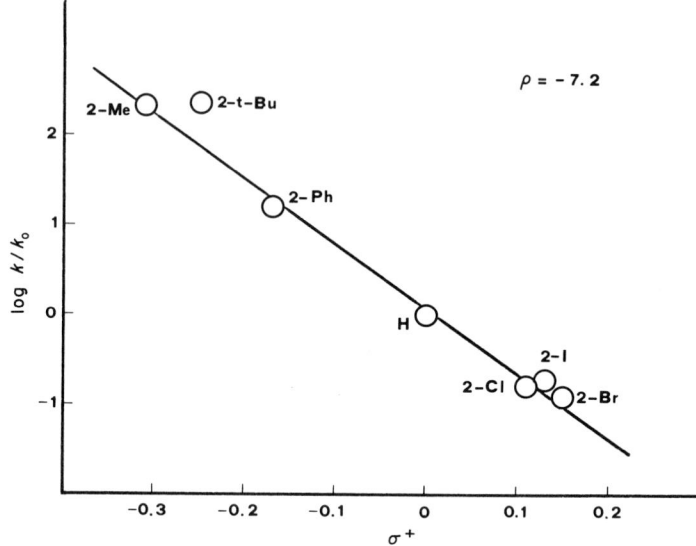

FIG. 9. Plot of log $k/k_0$ vs. $\sigma_p^+$ constants for the protodetritiation of 5-X-2-tritiothiophenes; see Butler and Eaborn.[297] (Reprinted by permission of the Chemical Society.)

The kinetics of protodedeuteriation of a number of 4- and 5-substituted 2-deuteriothiophenes in acetic–trifluoroacetic acid mixtures have been studied by Shatenshtein and his co-

---

[298] R. Baker, C. Eaborn, and R. Taylor, *J. Chem. Soc.* 4927 (1961).

workers.[55, 299, 300] The data adhere to the linear correlation with good precision (Fig. 10), yielding a $\rho$ value of $-7.5$, not very dissimilar from that obtained by Butler and Eaborn for the T/H exchange. The only important deviation from linearity concerns the 5-thiomethoxy group. However, it has been noted that the $p$-SCH$_3$ group almost never fits on the $\rho\sigma^+$ plot; the reasons for this have recently been discussed.[301]

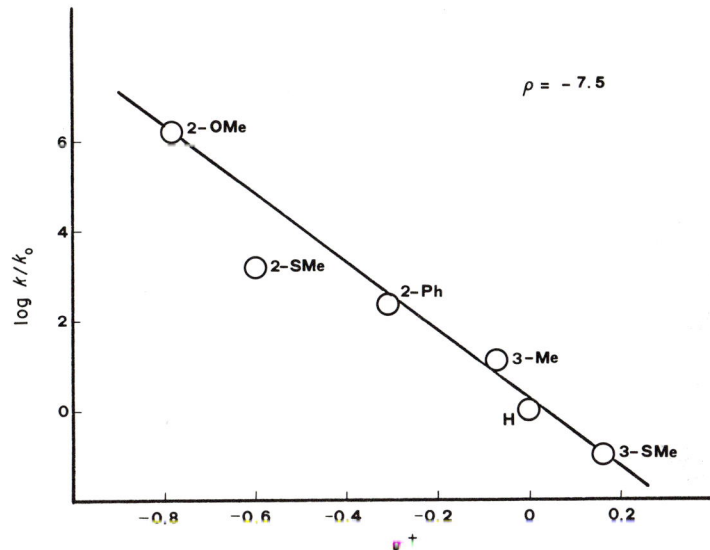

FIG. 10. Plot of log $k/k_0$ vs. $\sigma_p^+$ and $\sigma_m^+$ constants for the protodedeuteriation of substituted thiophenes. Data from Shatenshtein et al.[55, 299, 300]

The plot of Fig. 10 is interesting since it also includes $\beta$-substituted thiophenes and permits a test of the applicability of the benzene $\sigma_m^+$ constants.

Marino and Linda[130] have measured the relative rates of 2-chlorothiophene, thiophene, and 2-methylthiophene in the reaction with acetic anhydride and tin tetrachloride in dichloroethane. Additional measurements on two other thiophene derivatives have been made

[299] A. I. Shatenshtein, E. A. Gvodzeva, and A. G. Kamrad, *Isotopenpraxis* **2**, 462 (1966).
[300] A. I. Shatenshtein, Ya. L. Gol'dfarb, I. O. Shapiro, E. N. Zvyagintseva, and L. I. Belen'kii, *Dokl. Akad. Nauk SSSR* **180**, 117 (1968); *Chem. Abstr.* **69**, 66577 (1968).
[301] S. Clementi and P. Linda, *Tetrahedron* **26**, 2869 (1970).

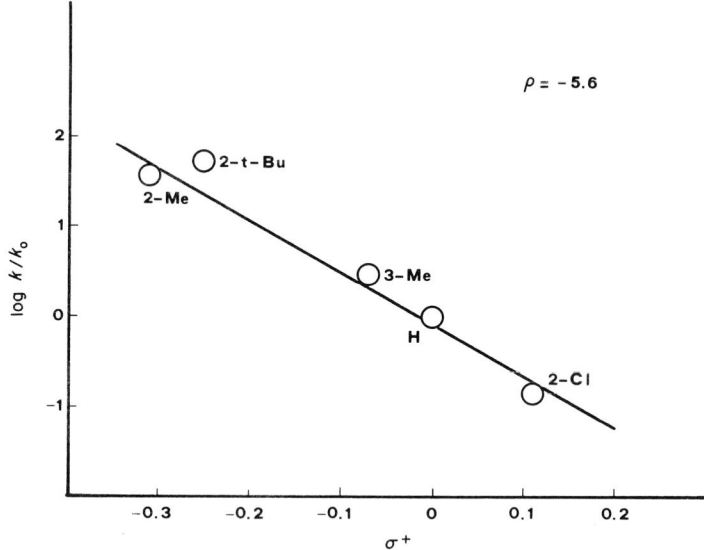

FIG. 11. Plot of log $k/k_0$ vs. $\sigma^+$ constants for the tin tetrachloride-catalyzed acetylation of substituted thiophenes. Data from Marino et al.[130, 302]

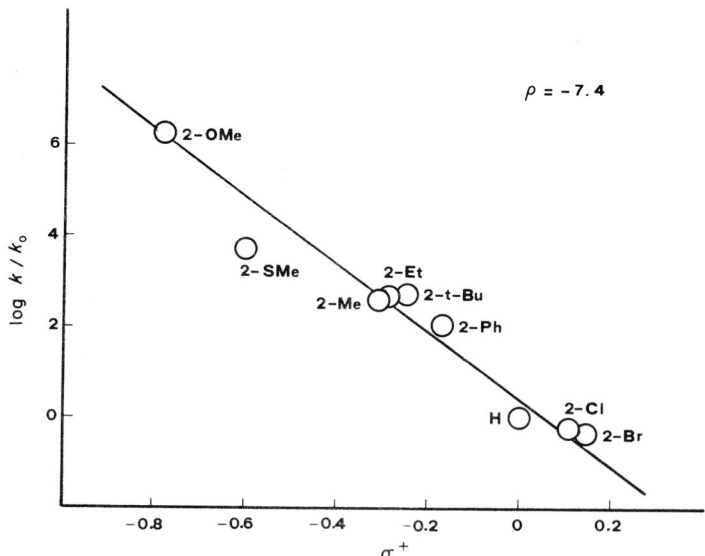

FIG. 12. Plot of log $k/k_0$ vs. $\sigma^+$ constants for the trifluoroacetylation of substituted thiophenes. Data from Clementi and Marino.[143, 302]

[302] S. Clementi and G. Marino, *Chem. Commun.* 1642 (1970).

subsequently[314]; the results yield a reasonable correlation with the $\sigma_p^+$ and $\sigma_m^+$ constants (Fig. 11).

The numerous data gathered in recent studies of the trifluoroacetylation reaction in dichloroethane[143, 302] are presented graphically in Fig. 12; again, the 2-thiomethoxy group falls below the line.[301]

Table XXIV provides a summary of the $\rho$ values for the electrophilic substitutions of thiophene derivatives; when available, the $\rho$ values for the corresponding reactions of benzene derivatives are also recorded for comparison.

TABLE XXIV

REACTION CONSTANTS $\rho$ FOR SUBSTITUTIONS AT THIOPHENE AND BENZENE RINGS

| Reaction | $\rho$ | |
|---|---|---|
| | Thiophenes[a] | Benzenes[b] |
| Bromination | −10.0 | −12.1 |
| Chlorination | −7.8 | −10.0 |
| Protodedeuteriation | −7.5 | — |
| Protodetritiation | −7.2 | −8.2 |
| Acetylation | −5.6 | −9.1 |
| Trifluoroacetylation | −7.4 | — |
| Mercuration | −5.3 | −4.0 |

[a] See footnotes to Table XXII for original references.
[b] L. M. Stock and H. C. Brown, *Advan. Phys. Org. Chem.* **1**, 35 (1963).

In all the examined reactions (it must be noted, however, that the conditions for acetylation and protodetritiation were not the same for the thiophene and the benzene derivatives) the $\rho$ values for reaction at the thiophene nucleus are always somewhat lower than the corresponding $\rho$ values for reaction at the benzene ring.

The only exception to this rule concerns the mercuration reaction, but there are good reasons to believe that mercuration of thiophene derivatives occurs by a mechanism different from that in benzene derivatives (see Section II, F, 2).

Also for the reactions in which a direct comparison of the $\rho$ constants is not possible, the data indicate that the activating or deactivating effects of the substituents are smaller in thiophene than in benzene. For instance, the partial rate factor for the para position of toluene in protodetritiation in trifluoroacetic acid[303] is over 300, compared with an increase in reactivity of about 200 times caused by a α-methyl group on the rate of protodetritiation of the other α position of the thiophene ring.[297] A possible explanation for this fact is that thiophene being much more reactive than benzene, the transition state lies further from the Wheland intermediate, and thus a smaller degree of positive charge is present on the ring in the transition state.[297]

### 3. Selenophene

Rate constants for the mercuration of selenophene and its 2-bromo, 2-carboethoxy, 2-acetyl, and 2-nitro derivatives under the conditions employed by Motoyama et al.,[121] have been determined at various temperatures by Yur'ev and his co-workers.[304] The selenophene derivatives were mercurated from 1.5 to 3 times more rapidly than the corresponding thiophene derivatives. A straight line is obtained in the Hammett plot with $\sigma_p^+$ constants, with a $\rho$ value of $-5.7$ at $25°$.

### 4. Furan

The only study regarding substituent effects on the rate of an electrophilic substitution in the furan ring is that on the trifluoroacetylation reaction[302] (Table XXV and Fig. 13).

TABLE XXV

Relative Rates of Trifluoroacetylation of 2-Substituted Furans[a]

| Substituent | $k/k_H$ |
|---|---|
| Br | $1.8 \times 10^{-2}$ |
| H | 1 |
| $CH_3$ | $8.6 \times 10^2$ |
| $C_2H_5$ | $6.9 \times 10^2$ |

[a] Reaction with trifluoroacetic anhydride in dichloroethane at 75° (S. Clementi and G. Marino).[302]

[303] C. Eaborn and R. Taylor, *J. Chem. Soc.* 247 (1961).
[304] Yu. K. Yur'ev, M. A. Gal'bershtam, and I. I. Kandror, *Khim. Gerotsikl. Soedin.* 897 (1966); *Chem. Abstr.* **66**, 94453 (1967).

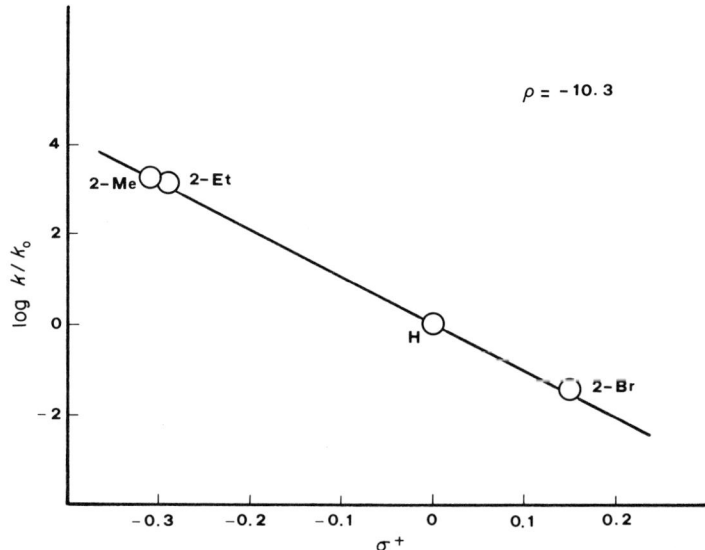

FIG. 13. Plot of log $k/k_0$ vs. $\sigma_p^+$ constants for the trifluoroacetylation of substituted furans (S. Clementi and G. Marino).[302]

The $\rho$ constant is remarkably more negative ($-10.3$) than that observed for the same reaction of thiophene ($-7.4$). This is a rather surprising result, in view of the higher reactivity of the furan ring (higher reactivity should lead to lower selectivity). Although data on other electrophilic substitutions would be desirable before this observation is generalized, it may be interesting to note that the same trend (i.e., greater sensitivity of the furan ring to the effects of substituents) is seen in two other reactions, which can be considered in certain respects similar to electrophilic substitution: the molecular ionization in the gas phase[305] and the solvolysis of the arylmethylcarbinol $p$-nitrobenzoates.[306]

Figure 14 shows a plot of the ionization potentials of substituted furans against those of the corrresponding thiophenes; the slope is, remarkably, greater than unity (1.28).

The solvolysis of furylmethylcarbinol $p$-nitrobenzoates[306] (a sidechain reaction in which a positive charge is developed in the transition state and can be delocalized in the ring) has a reaction constant

[305] P. Linda, G. Marino, and S. Pignataro, *J. Chem. Soc. B* (in press).
[306] D. S. Noyce and G. V. Kaiser, *J. Org. Chem.* **34**, 1008 (1969).

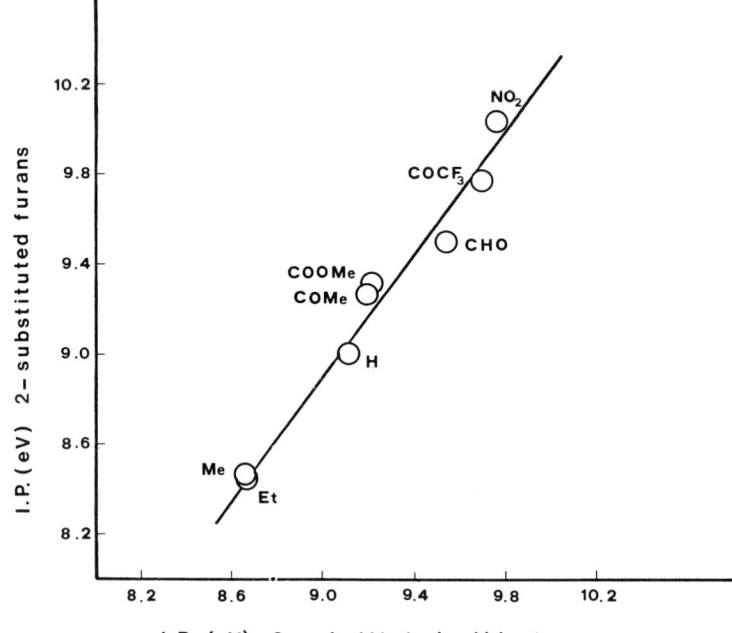

Fig. 14. Plot of ionization potentials (I.P.) of substituted furans vs. ionization potentials of substituted thiophenes (P. Linda, G. Marino, and S. Pignataro, J. Chem. Soc. B, in press).

of $-7.44$, substantially more negative than the $\rho$ value for similar reactions of benzyl derivatives[307, 308] (in the range between $-6$ and $-6.5$).

It must be recalled that furan behaves abnormally also in its "selectivity of position" (very large $\alpha:\beta$ ratios, see Section III, B).

It is not easy to find convincing explanations for these facts, but it is possible that a connection exists with the greater "bond fixation" in the furan ring.

5. *Pyrrole*

No systematic studies similar to those reported, in the previous paragraphs, for thiophene, selenophene, and furan have been made for pyrrole.

[307] V. J. Shiner, A. E. Buddenbaum, B. L. Murr, and G. Lamatz, J. Amer. Chem. Soc. **90**, 418 (1968).

[308] H. C. Brown, R. Bernheimer, G. J. Kim, and S. E. Shepple, J. Amer. Chem. Soc. **89**, 370 (1967).

The only available data concern the effects of alkyl groups on the trifluoroacetylation reaction (Table XXVI). A methyl group in position 2 increases (taking into account the statistical factor) the reactivity of the other α position by a factor of 23.8 (compared with factors of 380 and 1720 observed in thiophene and furan, respectively).[143]

TABLE XXVI

Relative Rates of Trifluoroacetylation of Substituted Pyrroles[a]

| Compound | $k/k_{II}$ |
|---|---|
| Pyrrole | 1 |
| 1-Methylpyrrole | 1.88 |
| 2-Methylpyrrole | 11.9 |
| 2-Ethylpyrrole | 12.4 |
| 2-t-Butylpyrrole | 12.4 |

[a] Reaction with trifluoroacetic anhydride in dichloroethane at 75° (S. Clementi and G. Marino).[143, 302]

This low "substrate selectivity" and the low "positional selectivity" (see Section III, B) are in keeping with a transition state little perturbed with respect to the initial state. This was to be expected, on the basis of the Hammond postulate,[183] in view of the very great reactivity of the pyrrole ring.

A 1-methyl group increases the reactivity only by a factor of less than 2. It must be considered that from position 1 the hyperconjugative effect is not operating; further, there is a steric effect which slows the rate of substitution at the adjacent position. Similar effects were observed in iodination.[94]

### 6. Benzo Derivatives

Amat di San Filippo and De Fabrizio[109] studied the effects of several substituents in the benzene ring on the reactivity at C-3 of indole on iodination. The effects of a methyl group at positions 1 and 2 were also studied. The results are summarized in Table XXVII. There is no possibility of resonance interaction between a substituent in position 5 (or 6) and the reaction center in 3. Heteronuclear "nonconjugative"

## TABLE XXVII

RELATIVE RATES OF IODINATION OF SUBSTITUTED INDOLES[a]

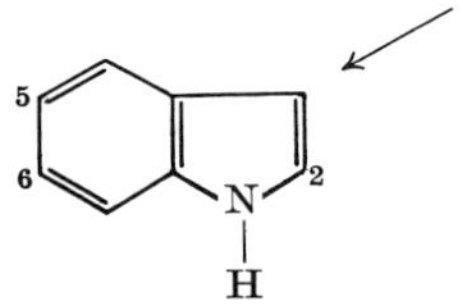

| Substituent | Relative rate |
|---|---|
| 2-$CH_3$ | 358 |
| 5-$CH_3$ | 3.07 |
| 5-$OCH_3$ | 2.25 |
| 6-$OCH_3$ | 2.18 |
| 1-$CH_3$ | 1.47 |
| H | 1 |
| 5-Br | 0.071 |
| 5-Cl | 0.069 |
| 5-$NO_2$ | [b] |

[a] In 40% ethanol. Temp. 25°C; [KI] = $7 \times 10^{-3}$ $M$; [$I_2$] ≃ $10^{-5}$ $M$; $\mu = 0.10$ $M$. Data from Amat di San Filippo and De Fabrizio.[109]

[b] 5-Nitroindole is unreactive under the reaction conditions.

$\sigma$ constants have been determined for the quinoline system (the $\sigma_{epi}$ constants).[309] In Fig. 15 the values of log $k_{rel}$ for the iodination of substituted indoles are plotted against the $\sigma_{epi}$ constants. Also the value for 2-methylindole is reported against the $\sigma_{ortho}^+$ for methyl, determined by Charton[310] (−0.276). The correlation among the seven points is very good ($r = 0.997$) and cannot be fortuitous. [The fit is even better ($r = 0.999$) if the $\sigma_p^+$ for 2-methyl is used, a circumstance not surprising in view of the smaller steric effects expected in a five-membered ring.] This good correlation illustrates the value of extending the use of $\sigma_{epi}$ (and possibly $\sigma_{cata}$) derived from quinoline to other bicyclic systems.

[309] E. Baciocchi, G. Illuminati, and G. Marino, *J. Amer. Chem. Soc.* **80**, 2270 (1958).

[310] M. Charton, *J. Amer. Chem. Soc.* **91**, 6649 (1969).

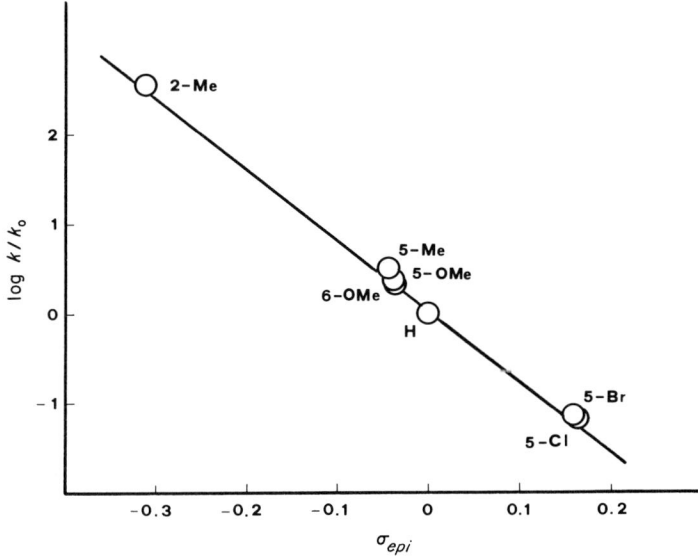

FIG. 15. Plot of log $k/k_0$ vs. $\sigma_{epi}$ constants for the iodination of substituted indoles. Data from Amat di San Filippo and De Fabrizio.[109]

The use of other substituent constants (e.g., $\sigma_m$) does not lead to linear correlations.

Similar studies on benzothiophene and benzofuran are lacking and it is not possible, therefore, to verify the general validity of the treatment.

NOTE ADDED IN PROOF

Since the completion of this review, a monography on the chemistry of indoles[311] and a review on the chemistry of imidazole[312] have been published.

The nitration of pyrrole has been investigated under a variety of experimental conditions.[313] The reaction with nitric acid in acetic anhydride proceeds smoothly and both spectroscopic and kinetic results indicate that the acetyl nitrate is the effective electrophilic agent. The $\alpha : \beta$ ratio is about 4 for nitration in acetic anhydride over a wide temperature range. By using competitive procedures, the partial rate factors for $\alpha$- and $\beta$-positions of the pyrrole ring in the nitration were estimated by these authors to be $1.3 \times 10^5$ and $3 \times 10^4$, respectively.

The relative importance of primary steric effects in benzene and thiophene rings has been determined by a critical comparison of the partial rate factors for the tin

[311] R. J. Sundberg, "The Chemistry of Indoles." Academic Press, New York, 1970.
[312] M. R. Grimmett, *Advan. Heterocycl. Chem.* **12**, 104 (1970).
[313] M. R. Cooksey, M. J. Morgan, and D. P. Morrey, *Tetrahedron* **26**, 5101 (1970).

tetrachloride-catalyzed acetylation of toluene, *t*-butylbenzene, 3-methyl- and 3-*t*-butylthiophene.[314]

The nitration and hydrogen exchange of 1,3,5-trimethylpyrazole, 3,5-dimethyloxazole, and 3,5-dimethylisothiazole have been studied kinetically.[315] The main features of this work are the following: for 3,5-dimethylisoxazole, both hydrogen exchange and nitration occur exclusively on the free base. For 1,3,5-trimethylpyrazole and 3,5-dimethylisothiazole, nitration occurs on the conjugate acids, while hydrogen exchange undergoes a changeover from free base to conjugate acid reaction as the acidity is increased.

The comparison among the reactivities towards electrophiles of the five-membered rings has been extended to tellurophen.[316] In both the reactions examined (tin tetrachloride catalyzed acetylation and trifluoroacetylation), tellurophen exhibits a reactivity intermediate between those of selenophen and furan.

---

[314] S. Clementi, P. Linda, and M. Vergoni, *Tetrahedron Lett.* 611 (1971).
[315] A. G. Burton, P. P. Forsythe, C. D. Johnson, and A. R. Katritzky, *J. Chem. Soc. B* (in press).
[316] F. Fringuelli, G. Marino, and A. Taticchi, unpublished results.

# Recent Developments in Phenanthridine Chemistry

B. R. T. KEENE

*Medway and Maidstone College of Technology, Kent, England*

AND

P. TISSINGTON

*Ciba-Geigy Chemicals Ltd., Grimsby, England*

|       |                                                                 |     |
|-------|-----------------------------------------------------------------|-----|
| I.    | Introduction and Nomenclature                                   | 316 |
| II.   | General Methods of Synthesis                                    | 317 |
|       | A. From Biphenyl and Fluorene Derivatives                       | 317 |
|       | B. From Systems Not Containing the Biphenyl Link                | 340 |
|       | C. Miscellaneous Methods                                        | 359 |
|       | D. The Conversion of Phenanthridones into Phenanthridines       | 368 |
| III.  | Physical Properties                                             | 369 |
|       | A. Dipole Moments, Molecular Geometry, and Electronic Structure | 369 |
|       | B. Spectra                                                      | 371 |
|       | C. Ionization Constants                                         | 376 |
|       | D. Polarographic Behavior                                       | 377 |
| IV.   | Reactions of the Phenanthridine Nucleus                         | 378 |
|       | A. Reactions at Nitrogen                                        | 378 |
|       | B. Electrophilic Substitution Reactions at Ring Carbon Atoms    | 387 |
|       | C. Nucleophilic Substitution Reactions                          | 390 |
|       | D. Other Reactions                                              | 397 |
| V.    | The Properties of Functional Groups                             | 402 |
|       | A. Alkyl and Substituted Alkyl Groups                           | 402 |
|       | B. Amino and Substituted Amino Groups                           | 404 |
|       | C. Carbonyl, Carboxyl, and Related Groups                       | 406 |
|       | D. Halogen Atoms                                                | 408 |
|       | E. Hydroxyl and Substituted Hydroxyl Groups                     | 409 |
|       | F. Cyano Groups                                                 | 410 |
|       | G. *N*-Oxides                                                   | 410 |
|       | H. Quaternary Salts                                             | 412 |

## I. Introduction and Nomenclature

Although phenanthridine was discovered in the late nineteenth century[1,2] neither the parent base nor its derivatives attracted attention until useful therapeutic activity was established in certain quaternary phenanthridinium compounds.[3] A substantial number of substituted phenanthridines (and many phenanthridinium salts) have now been described and lists of compounds appearing in the literature from 1884 until 1955 are available.[4,5] Phenanthridines have attracted surprisingly little systematic attention, although the system is of considerable theoretical interest and, with its nine nonequivalent carbon atoms, may be expected to provide a rigorous test of molecular orbital reactivity correlations. Naturally occurring derivatives include several *Amaryllidaceae* and *Papaveraceae* alkaloids, notably, lycorine, haemanthamine, and chelidonine; the chemistry of the phenanthridine alkaloids has been reviewed.[6]

Of the two ring-numbering systems in use for phenanthridine, (**1**) is favored by *Chemical Abstracts*, the "Ring Index", and the Chemical Society and will be used here. The older alternative (**2**) is encountered less frequently.

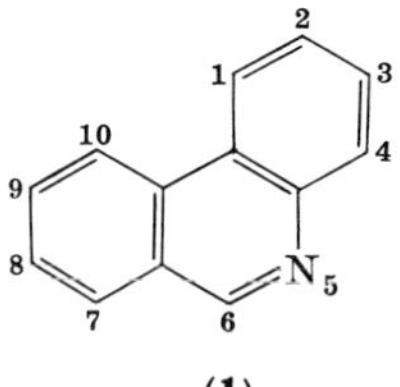

(**1**) 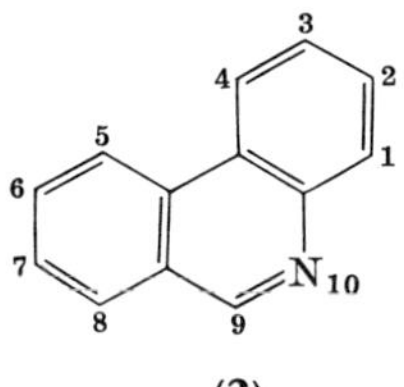    (**2**)

---

[1] A. Pictet and H. J. Ankersmit, *Chem. Ber.* **22**, 3339 (1889).
[2] C. Gräbe, *Chem. Ber.* **17**, 1370 (1884).
[3] See, e.g., G. T. Morgan and L. P. Walls, *J. Chem. Soc.* 2447 (1931); C. H. Browning, G. T. Morgan, J. V. M. Robb, and L. P. Walls, *J. Pathol. Bacteriol.* **46**, 203 (1938); J. Carmichael and F. R. Bell, *Vet. Record* **56**, 496 (1944); A. Albert, S. D. Rubbo, and M. I. Burvill, *Brit. J. Exp. Pathol.* **30**, 159 (1949).
[4] R. S. Theobald and K. Schofield, *Chem. Rev.* **46**, 171 (1950).
[5] J. Eisch and H. Gilman, *Chem. Rev.* **57**, 525 (1957).
[6] W. C. Wildman, *Alkaloids* **6**, 289 (1960) and references therein.

## II. General Methods of Synthesis

Methods of synthesis of the phenanthridine system reported up to 1950 have been surveyed in an article by Walls.[7] Since the later[5] of the two reviews mentioned above did not include synthetic work and the earlier article[4] dealt largely with the ring closure of 2-acylaminobiphenyls, an attempt has been made here to cover the synthetic literature from 1950 while largely restricting the discussion of properties to material not dealt with by Eisch and Gilman.[5]

The main methods of synthesis can conveniently be classified into two groups, depending on whether the tricyclic system is formed from a suitably substituted biphenyl or fluorene derivative, or from a compound not containing the biphenyl link. In this review the major syntheses are treated under these headings, subdivided appropriately.

### A. From Biphenyl and Fluorene Derivatives

#### 1. *The Cyclization of 2-Substituted Biphenyls*

The cyclization of 2-acylaminobiphenyls is a well-established route to phenanthridine derivatives, and a number of interesting examples of this procedure have been described since the last review[4] on synthesis. Originally, Morgan and Walls[3] used phosphoryl chloride to effect ring closure, either alone or in the presence of a high-boiling, polar solvent such as nitrobenzene.[8] The carbonium ion nature of the reaction was first recognized by Ritchie[9] and has been confirmed by Barber,[10] who established that the iminochloride (3) is an intermediate in the cyclization of 2-benzamido-4,4'-dinitrobiphenyl. Compound 3 cyclized readily only in the presence of Friedel–Crafts catalysts, of which several (e.g., stannic chloride) were found to be more efficient than phosphoryl chloride. Good yields of the phenanthridine (4) were obtained by converting 2-benzamido-4,4'-dinitrobiphenyl to the iminochloride (3) and cyclizing *in situ* in the presence of stannic chloride.

[7] L. P. Walls, *in* "Heterocyclic Compounds" (R. C. Elderfield, ed.), Vol. 4, Chapter IV. Wiley, 1952.
[8] L. P. Walls, *J. Chem. Soc.* 294 (1945).
[9] E. Ritchie, *J. Proc. Roy. Soc. N.S. Wales* **78**, 147 (1945).
[10] H. J. Barber, L. Bretherick, E. M. Eldridge, S. J. Holt, and W. R. Wragg, *J. Soc. Chem. Ind., London* **69**, 82 (1950).

More recent examples of the use of the Morgan–Walls method have provided a wide range of 6-alkyl-[11–15] and 6 arylphenanthridines,[14–17] and other isolated examples have been reported. Although the original procedure fails with 2-formamidobiphenyl, presumably because the formation of the unsubstituted nitrilium salt is energetically less favorable, cyclization of the amide has been achieved by fusion with zinc chloride,[18] and more recently 4-phenylphenanthridine has been obtained from 2′-formamido-m-terphenyl by the same technique.[19] However, the cyclization of 2-formamidobiphenyl is achieved more conveniently by the action of a mixture of phosphoryl chloride and stannic chloride,[20] or on stirring in polyphosphoric acid at 140–160°.[21]

Surprisingly, in the absence of stirring, the latter reaction leads only to material which has been tentatively identified as $N,N'$-bis(o-biphenylyl)formamide.[21] Nevertheless, polyphosphoric acid usually provides an attractive alternative to the "classical" procedure; thus, 1-methylphenanthridine has been obtained from 2-formamido-6-

[11] K. Mitsuhashi, *J. Pharm. Soc. Japan* **71**, 1232 (1951); *Chem. Abstr.* **46**, 5593 (1952).
[12] V. M. Rodionov, N. N. Suvorov, and L. V. Shagalov, *Dokl. Akad. Nauk SSSR* **82**, 731 (1952); *Chem. Abstr.* **47**, 4339 (1953).
[13] N. P. Buu-Hoi, P. Jacquignon, and C. T. Long, *J. Chem. Soc.* 505 (1957).
[14] B. L. Hollingsworth and V. Petrow, *J. Chem. Soc.* 3771 (1961).
[15] G. S. Matvelashvili, S. F. Belevskii, O. Y. Fedotova, and G. S. Kolesnikov, *Khim. Geterotsikl. Soedin* 1044 (1969); *Chem. Abstr.* **72**, 111265 (1970).
[16] A. E. Fairfull, V. Petrow, and W. F. Short, *J. Chem. Soc.* 3549 (1955).
[17] C. K. Bradsher and K. B. Moser, *J. Org. Chem.* **24**, 592 (1959).
[18] A. Pictet and A. Hubert, *Chem. Ber.* **29**, 1182 (1896).
[19] A. Risaliti and S. Bozzini, *Ann. Chim. (Rome)* **54**, 685 (1964).
[20] D. W. Ockenden and K. Schofield, *J. Chem. Soc.* 717 (1953).
[21] E. C. Taylor and N. W. Kalenda, *J. Amer. Chem. Soc.* **76**, 1699 (1954).

methylbiphenyl[22] and 6-phenylphenanthridine from 2-benzamidobiphenyl.[23] In a simpler alternative to the latter method, the 2-benzamidobiphenyl is formed *in situ* by heating a mixture of 2-aminobiphenyl and benzoic acid in polyphosphoric acid,[23] and an acceptable yield of 6-methylphenanthridine is obtained when 2-aminobiphenyl and acetic acid are heated together in "polyphosphate ester" (PPE) at 100°.[24]

Badger and Sasse have described the preparation of 2-, 3-, and 8-bromophenanthridine by the cyclization of the appropriate bromoformamidobiphenyl with polyphosphoric acid.[25] In the case of 2-bromo-2'-formamidobiphenyl a higher concentration of phosphorus pentoxide in the acid was necessary to effect ring closure, and a simple steric effect was invoked.[26] Nevertheless, the Morgan–Walls reaction has been used to obtain several "overcrowded" compounds in which unfavorable steric factors operate. Following the original report of the preparation of the 1,10-dimethylphenanthridine (**5a**) by this procedure,[27] several other examples have been described, notably the synthesis of the related phenanthridine (**5b**),[28] the 1,2-(**6**)[29] and 9,10-benzophenanthridines (**7**),[30] and the 1,2 : 9,10-dibenzophenanthridine (**8**).[29]

(**5a**) R = H
(**5b**) R = Me

(**6**)

[22] B. R. T. Keene and P. Tissington, *J. Chem. Soc.* 3032 (1965).
[23] B. Staskun, *J. Org. Chem.* **29**, 2856 (1964).
[24] A. Banger and C. Halls, private communication.
[25] G. M. Badger and W. H. F. Sasse, *J. Chem. Soc.* 4 (1957).
[26] G. S. Chandler, J. L. Huppatz, R. A. Jones, and W. H. F. Sasse, *Aust. J. Chem.* **20**, 2037 (1967).
[27] E. Ritchie, *J. Proc. Roy. Soc. N.S. Wales* **78**, 159 (1945).
[28] P. M. Everitt, S. M. Loh, and E. E. Turner, *J. Chem. Soc.* 4587 (1960).
[29] T. R. Govindachari, K. Nagarajan, B. R. Pai, and V. N. Sundararajan, *Chem. Ber.* **91**, 2053 (1958).
[30] B. Mills and K. Schofield, *J. Chem. Soc.* 4213 (1956).

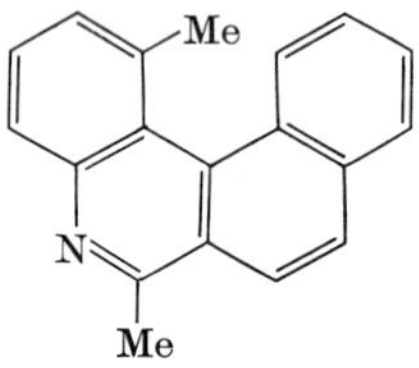

Other examples of annelated derivatives which have been obtained by the Morgan–Walls synthesis are 3,4-benzophenanthridine,[31] 2,3-benzophenanthridine,[32] a number of its methoxy and methylenedioxy derivatives,[33, 34] and dihydronitidine (9c).[35, 36] It is interesting that in one of the last syntheses the oxime acetate (9a) undergoes spontaneous ring closure to 9b during Semmler–Wolff aromatization; this represents the first instance of a Morgan–Walls cyclization proceeding without the use of conventional reagents.[36]

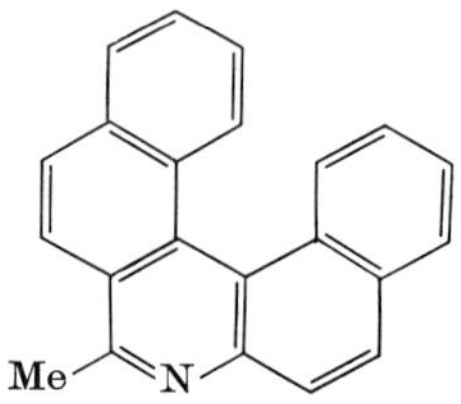

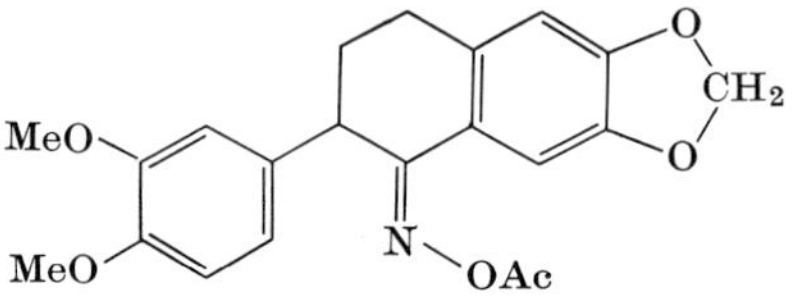

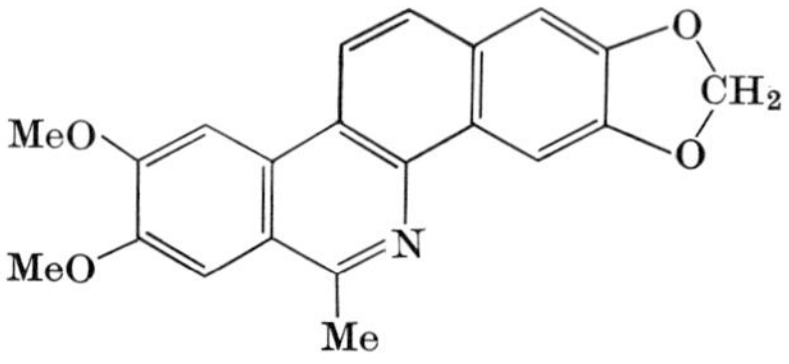

[31] W. M. Whaley and M. Meadow, *J. Org. Chem.* **19**, 661 (1954).
[32] D. N. Brown, D. H. Hey, and C. W. Rees, *J. Chem. Soc.* 3873 (1961).
[33] K. W. Gopinath, T. R. Govindachari, K. Nagarajan, and N. Viswanathan, *J. Chem. Soc.* 4761 (1957).
[34] K. W. Gopinath, T. R. Govindachari, K. Nagarajan, and K. K. Purushothaman, *J. Chem. Soc.* 504 (1958).
[35] H. R. Arthur and Y. L. Ng, *J. Chem. Soc.* 4010 (1959).
[36] K. W. Gopinath, T. R. Govindachari, P. C. Parthasarathy, and N. Viswanathan, *J. Chem. Soc.* 4012 (1959).

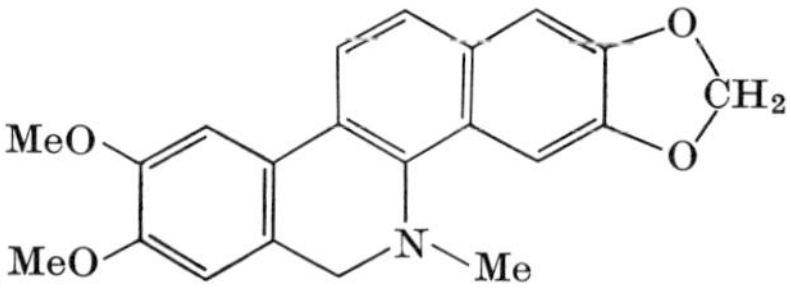

Double cyclization of the bisamide (**10**) (R = Me) occurs almost quantitatively in polyphosphoric acid, although the bisformamido compound (**10**) (R = H) undergoes deformylation in this reagent. However, 6,13-diazabenz[a,h]anthracene is formed in good yield on heating **10** (R = H) at 285° in a melt of sodium chloride and aluminum chloride,[37a] a reagent which had been used previously to prepare 4,9-diazachrysene (and its 5,10-dimethyl derivative) by a similar double cyclization procedure.[37b]

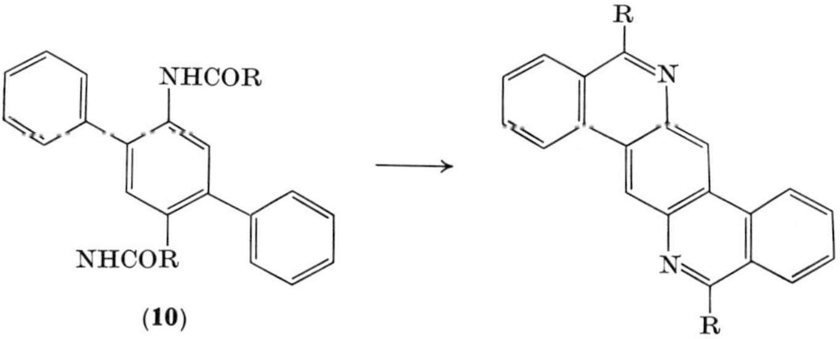

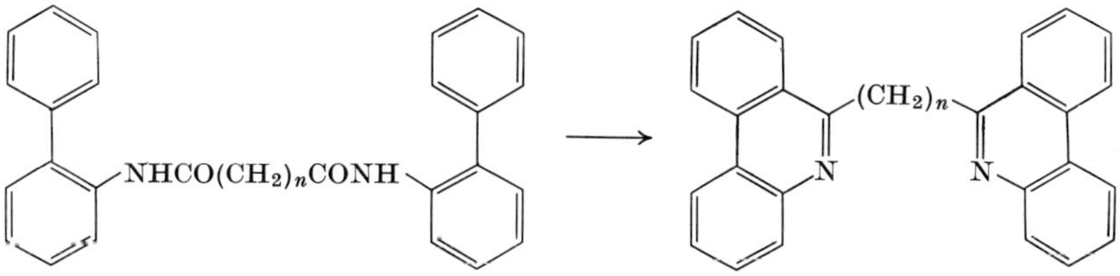

Following the original report of the preparation of 1,4-bis(phenanthridin-6-yl)butane (**12**) ($n = 4$) from the adipamide (**11**) ($n = 4$),[38] Hollingsworth and Petrow have obtained a series of α,ω-bisphenanthridinylalkanes (**12**) ($n = 4$–8, 10) by the same general method, although the authors were unable to cyclize the glutaramide (**11**) ($n = 3$).[39]

6-Styrylphenanthridine is formed as a minor product (23%) in the isomerization of 2-cinnamoylaminobiphenyl to 4,8-diphenyl-3,4-dihydrocarbostyril in polyphosphoric acid at 138°.[40]

[37] (a) L. H. Klemm and A. Weisert, *J. Heterocycl. Chem.* **2**, 140 (1965). (b) R. F. Robbins, *J. Chem. Soc.* 2553 (1960).
[38] E. Ritchie, *J. Proc. Roy. Soc. N.S. Wales* **78**, 155 (1945).
[39] B. L. Hollingsworth and V. Petrow, *J. Chem. Soc.* 3664 (1961).
[40] K. M. Johnston, *J. Heterocycl. Chem.* **6**, 847 (1969).

The Morgan–Walls cyclization has also been employed in cases where the acylamino group is attached to a reduced ring. For example, the amides (**13a–c**,[41] and **d**[42]) are readily cyclized by phosphoryl chloride and the tetrahydro-1,2-benzophenanthridine (**14**) has been prepared

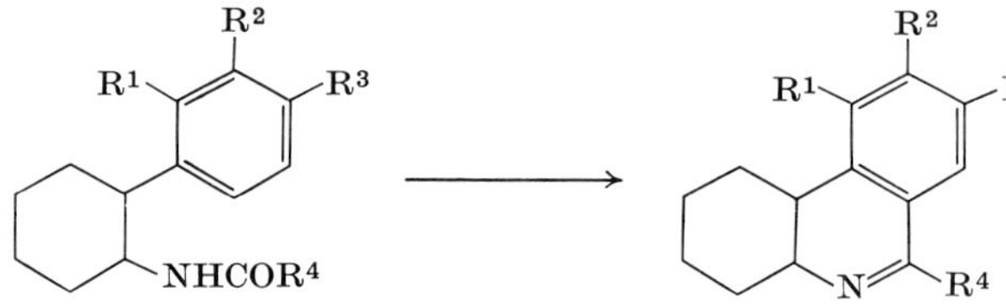

(**13a**) $R^1 = R^2 = R^3 = H$; $R^4 = Me$
(**13b**) $R^1 = R^3 = H$; $R^2 = OMe$; $R^4 = Me$
(**13c**) $R^1 = R^2 = H$; $R^3 = OMe$; $R^4 = CH_2Cl$
(**13d**) $R^1 = H$; $R^2 + R^3 = OCH_2O$; $R^4 = Me$
(**13e**) $R^1 = R^2 = R^3 = H$; $R^4 = Ph$
(**13f**) $R^1 = R^2 = R^3 = H$; $R^4 = CH_2NC_5H_{10}$
(**13g**) $R^1 = Me$; $R^2 = R^3 = R^4 = H$

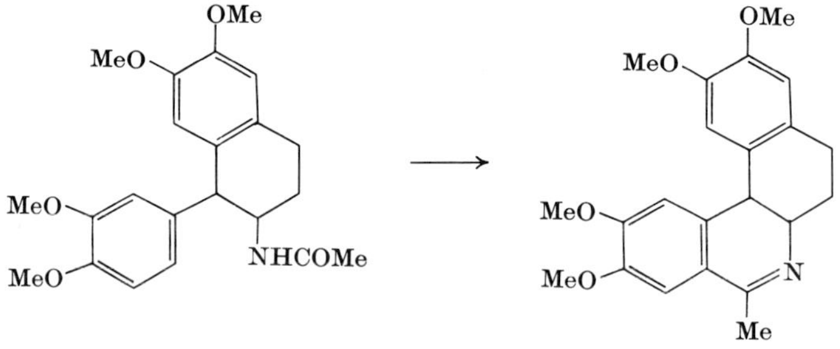

(**14**)

in a similar way.[43] Govindachari[44] has employed rather severe conditions (a mixture of phosphoryl chloride and phosphorus pentoxide in boiling xylene) to cyclize the benzamido derivative (**13e**), but later

---

[41] T. Nomura, *Yakugaku Zasshi* **77**, 270 (1957); *Chem. Abstr.* **51**, 11351 (1957).
[42] Hoffmann-La Roche and Co., *Belgian Patent* 628,614; *Chem. Abstr.* **61**, 646 (1964).
[43] G. N. Walker, *J. Amer. Chem. Soc.* **76**, 3999 (1954).
[44] T. R. Govindachari, K. Nagarajan, B. R. Pai, and N. Arumugan, *J. Chem. Soc.* 4280 (1956).

work showed that phosphoryl chloride alone is sufficient.[45] On the other hand, the piperidinomethyl derivative (**13f**) can only be cyclized with polyphosphoric acid, both phosphorus pentoxide and phosphoryl chloride being ineffective.[46] The formamidobiphenyl (**13g**) gives the corresponding hexahydrophenanthridine in poor yield on treatment with phosphorus pentoxide in xylene.[47]

Successful cyclizations involving attack on reduced rings have also been reported; in the presence of polyphosphoric acid the originally formed hexahydrophenanthridines (**15**) all undergo spontaneous partial dehydrogenation to the corresponding 7,8,9,10-tetrahydrophenanthridines (**16**).[48]

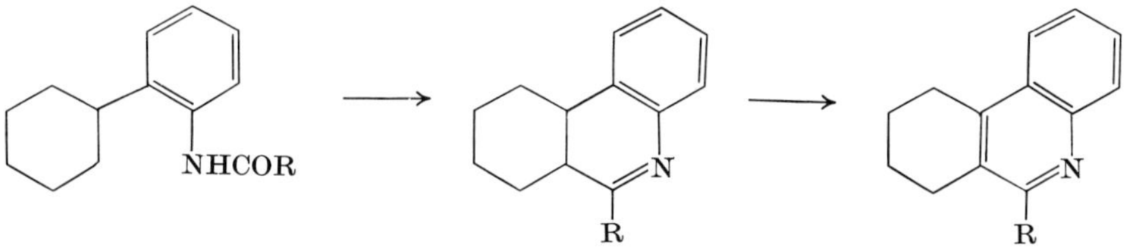

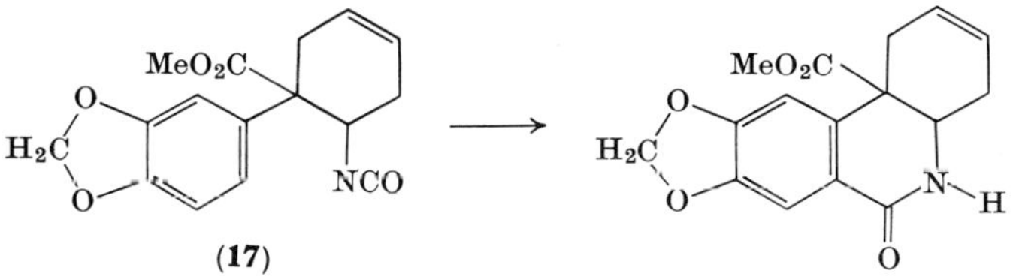

[45] T. R. Govindachari, B. R. Pai, and V. N. Sundararajan, *J. Chem. Soc.* 1715 (1958).
[46] A. C. Das Gupta, S. S. Chakravorti, B. P. Das, and U. P. Basu, *J. Indian Chem. Soc.* **46**, 1085 (1969).
[47] M. S. Gibson, *J. Chem. Soc.* 2249 (1961).
[48] D. A. Denton, R. K. Smalley, and H. Suschitzky, *J. Chem. Soc.* 2421 (1964).

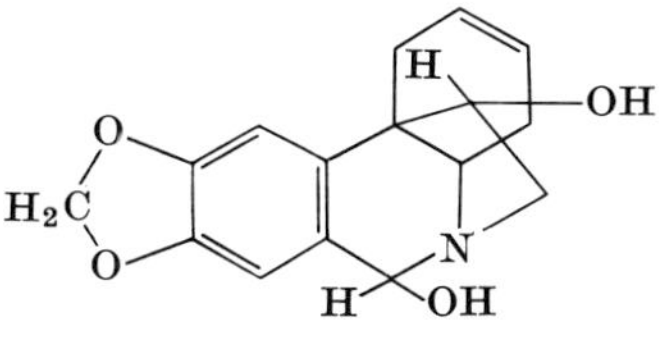

Phenanthridone has been obtained by the action of carbon dioxide on 2-lithioaminobiphenyl[49] and by cyclizing 2-biphenylyl isocyanate with aluminum chloride,[50] the latter reaction appearing to be a general one. Thus, cyclization of 5-bromo-2-biphenylyl isocyanate gives 2-bromophenanthridone,[51] while reaction of the isocyanate (17) with trifluoroacetic acid is a key stage in the synthesis of dihydroxycrinene (18).[52]

The Friedel–Crafts cyclization of the acid chloride (19a) occurs normally at $-70°$, with the formation of the azacycloheptenone (20), but the sole product at $5°$ is 5,6-dihydro-5-tolylsulfonylphenanthridine (21). Similarly, the related acid (19b) gives phenanthridine when heated with polyphosphoric acid at $60°$.[53] Presumably a 5,6-dihydrophenanthridine is formed which undergoes dehydrogenation in the reaction medium (cf. Das Gupta et al.[46]).

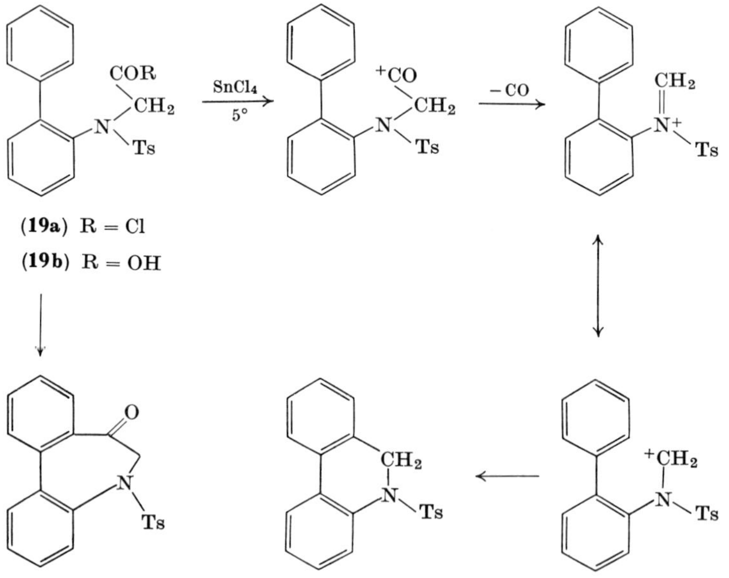

(19a) R = Cl
(19b) R = OH

(20)        (21)

[49] N. S. Narasimhan and R. H. Alukar, *Indian J. Chem.* **7**, 1280 (1969).
[50] J. M. Butler, *J. Amer. Chem. Soc.* **71**, 2578 (1949).
[51] W. L. Mosby, *J. Amer. Chem. Soc.* **76**, 936 (1954).
[52] J. B. Hendrikson, C. Foote, and N. Yoshimura, *Chem. Commun.* 165 (1965).
[53] W. Paterson and G. R. Procter, *J. Chem. Soc.* 3468 (1962).

3,4,4a,10b-Tetrahydrophenanthridone has been prepared by heating 2-phenyl-3-cyclohexenyl isocyanate in polyphosphoric acid,[54] and 2-biphenylyl isocyanate is believed to be an intermediate in the preparation of phenanthridone when a mixture of biphenyl-2-carboxylic acid and hydroxylamine is heated in polyphosphoric acid.[55] More recently, the same isocyanate has been shown to undergo photochemical ring closure, though in poor yield,[56] while the photolysis of ethyl N-2-biphenylyl carbamate gives phenanthridone in better yield.[57] (The carbamate also gives phenanthridone when heated with zinc chloride in diethylene glycol, and 3-methylphenanthridone has been prepared by the same method.[14]) 8-Phenylphenanthridone is believed to be a by-product when p-terphenyl-2-carboxylic acid undergoes the Curtius reaction.[58]

6-Alkylphenanthridines (and 6-aryl) have been obtained by passing a mixture of 2-nitrobiphenyl and the appropriate alcohol over a copper–alumina catalyst at 320–360°.[59]

o-Phenylbenzophenone oxime rearranges to a mixture of 6-phenylphenanthridine and 9-fluorenone anil (1 : 4) in polyphosphoric acid.[60] The formation of the two products has been accounted for on the basis of aryl migration in the mixture of isomeric oximes (22) and (24) (R = H), the resulting nitrilium salts (23) and (25) (R = H) undergoing cyclization in preference to hydrolysis.[60, 61] The rearrangement of

[54] R. Dran and M. Hill, *C. R. Acad. Sci.* **261**, 770 (1965).
[55] H. R. Snyder, C. T. Elston, and D. B. Kellom, *J. Amer. Chem. Soc.* **75**, 2014 (1953).
[56] J. S. Swenton, *Tetrahedron Lett.* 2855 (1967).
[57] N. C. Yang, A. Shani, and G. R. Lenz, *J. Amer. Chem. Soc.* **88**, 5369 (1967).
[58] G. R. Ames and W. Davey, *J. Chem. Soc.* 3480 (1957).
[59] N. S. Kozlov, B. I. Kiselov, and T. A. Skorokhodova, *Izobret. Prom. Obraztsy, Tovarnye Znaki* **46**, 32 (1969); *Chem. Abstr.* **71**, 49801 (1969).
[60] P. A. S. Smith, *J. Amer. Chem. Soc.* **76**, 431 (1954).
[61] P. T. Lansbury and S. P. Spitz, *J. Org. Chem.* **32**, 2623 (1967).

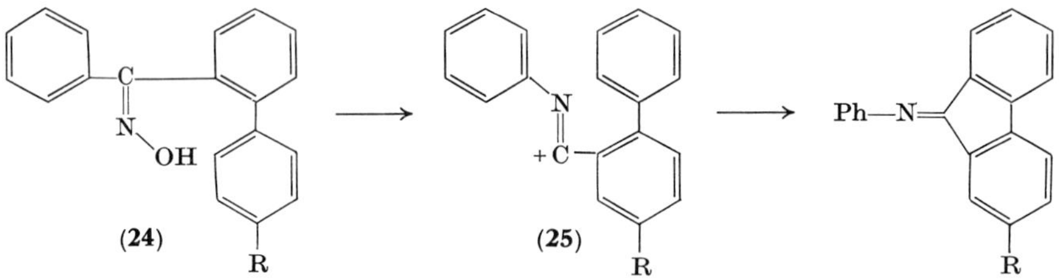

[Structures (24), (25), and Ph—N=fluorenyl shown with R substituents]

the related *o*-(*p*-tolyl)benzophenone oxime [presumably a mixture of 22 and 24 (R = Me)] is similar.[61] The Schmidt reaction with *o*-phenylbenzophenone in sulfuric acid gives 6-phenylphenanthridine as the sole product, suggesting preferential migration of the *o*-biphenylyl group.[60]

6,6-Dimethyl-5,6-dihydrophenanthridine (29) is one of the minor products obtained from the thermolysis of 1-(biphenyl-2-yl)-1-methylethyl azide (26). This appears to be the first authentic example of an intramolecular substitution by an *alkyl* nitrene (27). The formation of 29 was also observed in the photolysis of 26, but evidence was brought forward suggesting that 29 arises here from the photolysis of the imine (28); 28 also formed 29 on treatment with polyphosphoric acid (PPA).[62]

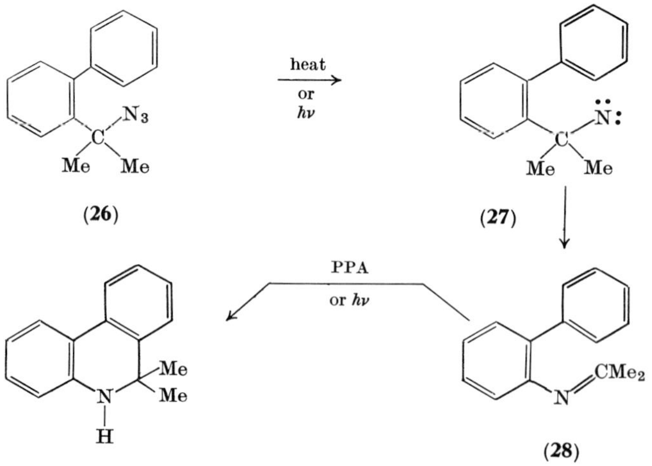

[Scheme showing (26) → (27) via heat or hν; (27) → (28) with CMe$_2$; (28) → (29) via PPA or hν; (29) is 6,6-dimethyl-5,6-dihydrophenanthridine]

[62] R. A. Abramovitch and E. P. Kyba, *Chem Commun.* 265 (1969).

The thermolysis of biphenyl-2-yl-diphenylmethyl azide provides the 6,6-diphenyl analog of **29** in small yield.[62] On the other hand, intramolecular attack is not observed in either pyrolytic or photolytic decomposition of the simpler 2-azidomethylbiphenyl.[63]

The formation of a phenanthridine lactone by the persulfate oxidation of 2′-cyanobiphenyl-2-carboxylic acid is interesting in that little is known of the addition of radicals to the C≡N system. The mechanism proposed below (Scheme 1) also accounts for the occurrence of fluorenone as a by-product.[64]

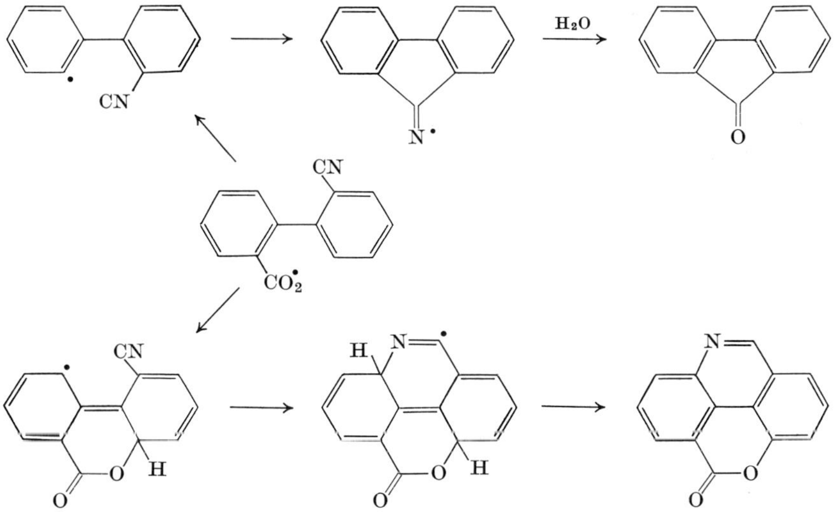

SCHEME 1

## 2. The Cyclization of 2,2′-Disubstituted Biphenyls

Phenanthridines are obtained from 2,2′-disubstituted biphenyls in which the substituents react intramolecularly to form the heterocyclic ring. The oldest examples of syntheses of this type involve Hofmann[65] and Curtius[66,67] reactions on biphenic acid, and have been explained

[63] B. Coffin and R. F. Robbins, *J. Chem. Soc.* 1252 (1965).
[64] P. M. Brown, J. Russell, R. H. Thomson, and A. G. Wylie, *J. Chem. Soc. C* 842 (1968).
[65] L. Oyster and H. Adkins, *J. Amer. Chem. Soc.* **43**, 208 (1921).
[66] R. A. Labriola, *J. Org. Chem.* **5**, 329 (1940).
[67] R. A. Labriola and A. Felitte, *J. Org. Chem.* **8**, 536 (1943).

in terms of the intermediate formation of 2-aminobiphenyl-2'-carboxylic acid, which undergoes spontaneous cyclization to phenanthridone. The reduction of 2'-nitrobiphenyl-2-carboxylic acid with ammoniacal ferrous sulfate[68] or with zinc and hydrochloric acid[69] also gives phenanthridone, but reduction of the nitro acid with zinc and ammonium chloride, or with boiling tetralin, gives $N$-hydroxyphenanthridone (**31**) probably via the hydroxylamino derivative (**30**).[69] Rather unexpectedly, boiling tetralin converts 2,2'-dinitro-6,6'-biphenic acid into the lactone (**32**), the reaction involving nucleophilic displacement of the nitro groups by carboxylate anion. The reaction is catalyzed by quinoline.[70] Mixtures of phenanthridone

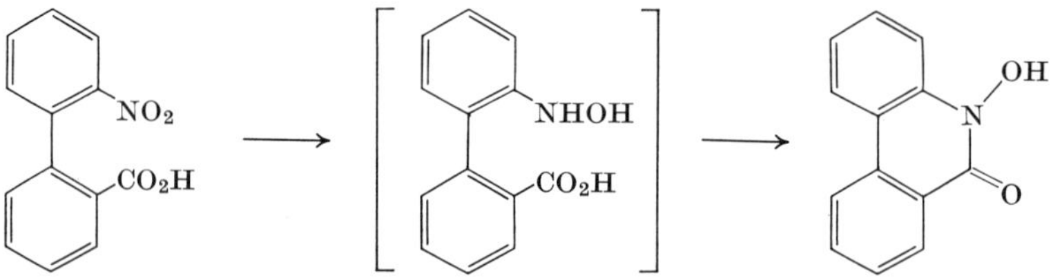

and $N$-hydroxyphenanthridone are obtained by the catalytic hydrogenation of derivatives of 2'-nitrobiphenyl-2-carboxylic acid; the relative amounts of hydroxylamino trapping (leading to **31**) are dependent both on pH and on the nature of the acid function.[71]

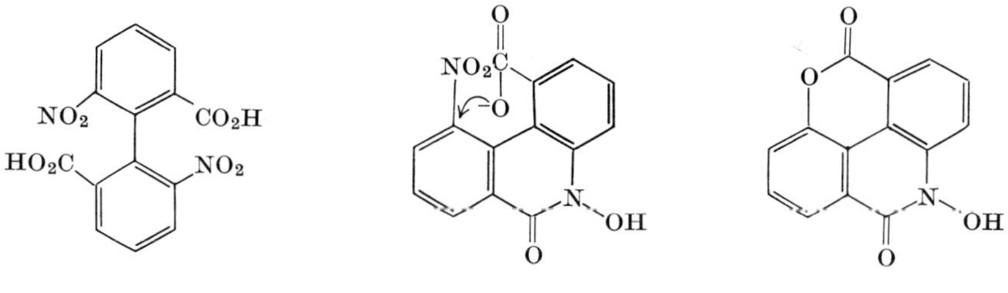

The course of the reduction of 6-nitrodiphenic acid with zinc and hydrochloric acid is not clear. Hey[69] has proposed that the product

---

[68] C. Angelini, *Ann. Chim.* (*Rome*) **49**, 879 (1957).
[69] D. H. Hey, J. A. Leonard, and C. W. Rees, *J. Chem. Soc.* 4579 (1962).
[70] M. Gawlak and R. F. Robbins, *J. Chem. Soc.* 5135 (1964).
[71] C. W. Muth, J. R. Elkins, M. L. DeMatte, and S. T. Chiang, *J. Org. Chem.* **32**, 1106 (1967).

originally isolated by Bell[72] was N-hydroxyphenanthridone-1-carboxylic acid, but other authors[73] have claimed that phenanthridone-1-carboxylic acid is formed. Catalytic hydrogenation of the 2'-nitro-1,2,3,4,-tetrahydrobiphenyl-2,3-dicarboxylic acid (33) gives the corresponding phenanthridone (34)[74, 75]; a similar procedure converts 2-formyl-1,2,3,4,-tetrahydro-2'-nitrobiphenyl (35) into the hexahydrophenanthridine (36).[75, 76]

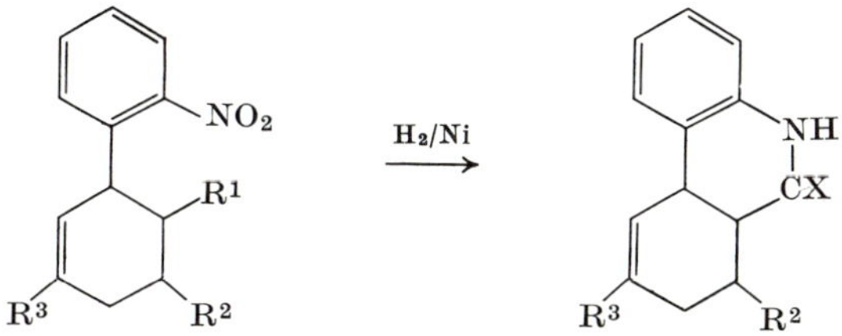

(33) $R^1 = R^2 = CO_2H$; $R^3 = Me$    (34) $R^2 = CO_2H$; $R^3 = Me$; $X = O$
(35) $R^1 = CHO$; $R^2 = R^3 = H$    (36) $R^2 = R^3 = H$; $X = H_2$

The base-catalyzed cyclization of certain 2'-substituted 2-nitrobiphenyls (37)→(38) provides an interesting example of the participation of an aromatic nitro group in an aldol-type condensation[77, 78];

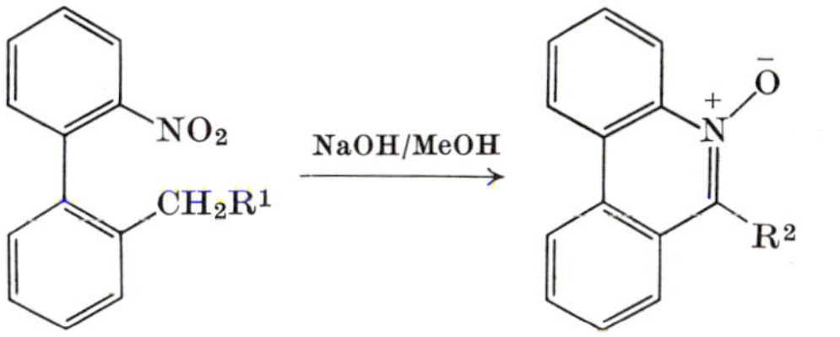

(37)  (38)

(37a) $R^1 = CO_2Me$    (38a) $R^2 = CO_2H$
(37b) $R^1 = CONH_2$    (38b) $R^2 = CONH_2$
(37c) $R^1 = CN$    (38c) $R^2 = CN$
(37d) $R^1 = COPh$    (38d) $R^2 = H$
(37e) $R^1 = SO_2Ph$    (38e) $R^2 = H$

[72] F. Bell, *J. Chem. Soc.* 835 (1934).
[73] B. M. Krasovitskii, D. G. Pereyaslova, and N. K. Kobyak, *Ukr. Khim. Zh.* **18**, 97 (1952); *Chem. Abstr.* **48**, 11422 (1954).
[74] E. C. Taylor and E. J. Strojny, *J. Amer. Chem. Soc.* **78**, 5104 (1956).
[75] T. Masamune, M. Takasugi, H. Suginome, and M. Yokoyama, *J. Org. Chem.* **29**, 681 (1964).
[76] E. A. Braude and J. S. Fawcett, *J. Chem. Soc.* 3113 (1951).
[77] C. W. Muth, J. C. Ellers, and O. F. Folmer, *J. Amer. Chem. Soc.* **79**, 6500 (1957).
[78] C. W. Muth, N. Abraham, M. L. Linfield, R. B. Wotring, and E. A. Pacofsky, *J. Org. Chem.* **25**, 736 (1960).

the 2'-methylene group is activated to facilitate carbanion formation.[79]

Photochemical reactions involving ortho-ortho interactions have been reported.[80] The anil (**39**) gives phenanthridone (25%) when irradiated in ethanol, and 2-(cyanoanilinomethyl)-2'-nitrobiphenyl (**40**) forms 6-cyanophenanthridine under similar conditions. The same product is also formed when the biphenyl (**40**) is treated with sodium ethoxide.[80]

(**39**)  (**40**)

The formation of 8,10-dimethylphenanthridine (**44**), 2'-amino-2,4,6-trimethylbiphenyl, and 2,4,9-trimethylcarbazole during the thermal decomposition of 2'-azido-2,4,6-trimethylbiphenyl (**41**) ($R = N_3$) in hexadecane at 230° has been attributed to the participation of the corresponding nitrene (**42**),[81] probably in its triplet state.[82] Because of the absence of a substituent at C-6, the thermal decomposition of 2-azido-2'-methylbiphenyl gives 4-methylcarbazole exclusively.[63] Deoxygenation of 2'-nitro-2,4,6-trimethylbiphenyl (**41**) ($R = NO_2$) with triethyl phosphite is believed also to give the nitrene (**42**), but formation of the adduct (**43**) with the excess of phosphite occurs to the exclusion of the energetically less favorable insertion reaction leading to the phenanthridine.[83, 84] However, when the reaction is carried out under high dilution in either isopropylbenzene or *t*-butylbenzene, the undesired coupling reaction is suppressed sufficiently to allow the formation of 8,10-dimethylphenanthridine, though in only 12–14% yield.[84] Abramovitch has proposed a nitrene

---

[79] J. D. Loudon and G. Tennant, *Quart. Rev. Chem. Soc.* **18**, 391 (1964).
[80] E. C. Taylor, B. Furth, and M. Pfau, *J. Amer. Chem. Soc.* **87**, 1400 (1965).
[81] G. Smolinsky, *J. Amer. Chem. Soc.* **82**, 4717 (1960).
[82] G. Smolinsky, *J. Amer. Chem. Soc.* **83**, 2489 (1961).
[83] G. Smolinsky and B. I. Feuer, *J. Org. Chem.* **31**, 3882 (1966).
[84] J. I. G. Cadogan and M. J. Todd, *Chem. Commun.* 178 (1967).

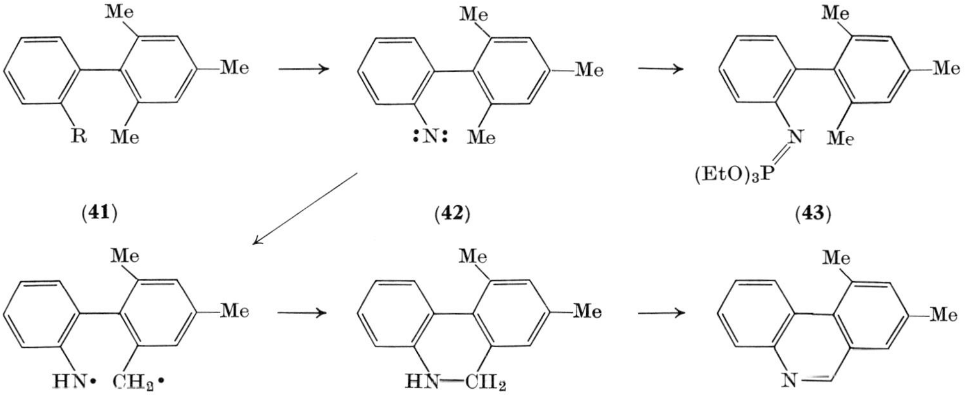

mechanism for the conversion of the nitrobiphenyl (41) (R=NO₂) to 8,10-dimethylphenanthridine by oxidation with ferrous oxalate at 300°[85, 86]; however, since the same product is formed at 350° in diphenyl ether in the absence of oxalate, Smolinsky has suggested an alternative mechanism involving the *aci* form (45) of the nitro compound.[83]

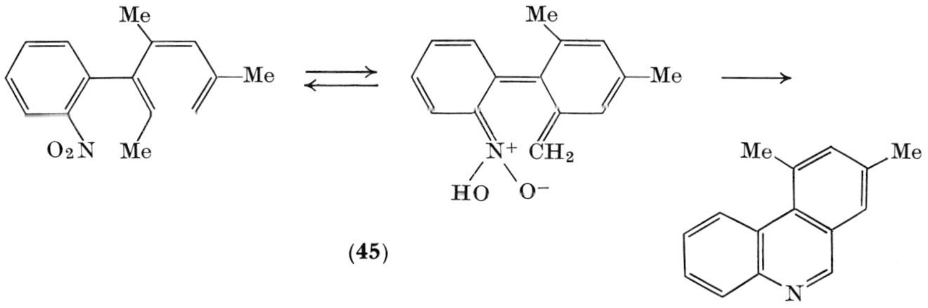

The ferrous oxalate procedure has been employed successfully to prepare 8,10-dimethyl-3,4-benzophenanthridine from 2-mesityl-1-nitronaphthalene.[85] However, an attempted preparation of 4,9-diazachrysene from 2,2'-dimethyl-6,6'-dinitrobiphenyl failed, while treatment with an excess of triethyl phosphite gave 1,10-dimethylbenzo[c]cinnoline (11%) as the only recognizable product. Failure

[85] R. A. Abramovitch, Y. Ahmad, and D. Newman, *Tetrahedron Lett.* 752 (1961).
[86] R. A. Abramovitch, D. Newman, and G. Tertzakian, *Can. J. Chem.* 41, 2390 (1963).

has also been reported in the case of the thermolysis and photolysis of 2,2′-dimethyl-6,6′-diazidobiphenyl.[63]

The decomposition of 2,2′-bis(azidomethyl)biphenyl in boiling biphenyl ether gives a mixture of phenanthridine and 5H-dibenz[c,e]-azepine (46). Hydrazoic acid is evolved and the following mechanism has been proposed (Scheme 2) (although attempts to confirm carbene formation were not successful).[63]

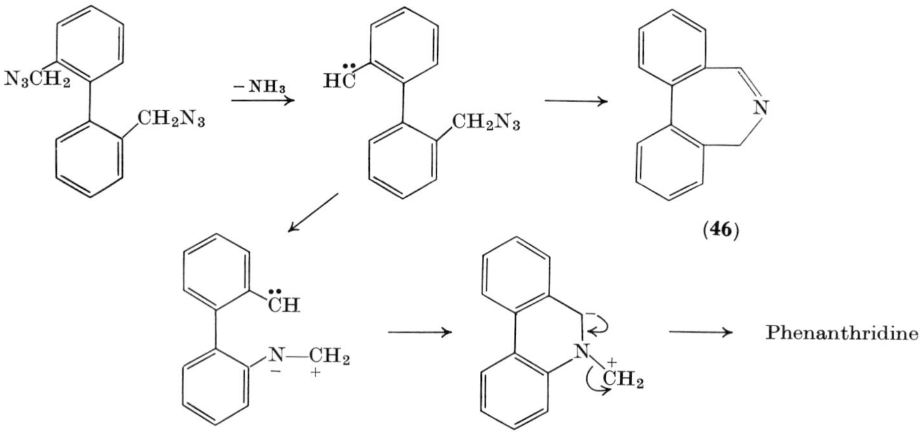

SCHEME 2

Following the original report that hydrazoic acid converts phenanthraquinone into phenanthridone,[87] Stephenson has shown that a 10 M excess of hydrazoic acid is required to avoid increasing the amount of diphenamic acid at the expense of phenanthridone.[88] Even this excess of acid converts 4-nitrophenanthraquinone into a mixture of nitrodiphenamic acids, with only a trace of 5-nitrophenanthridone.[89] Both 3,4-benzophenanthridone (48) and 7,8-benzophenanthridone (49) have been prepared from chrysenequinone (47) by the scheme outlined below.[90]

5-Keto-5,7-seco-6-norcholestan-7-oic- acid (50) reacts with ethanolic ammonia under pressure to give 6-aza-4-cholesten-7-one (51) (R = H),[91] whereas treatment of the same acid (or its methyl ester)

[87] G. Caronna, Gazz. Chim. Ital. 71, 483 (1941).
[88] E. F. M. Stephenson, J. Chem. Soc. 2620 (1949).
[89] A. G. Caldwell and L. P. Walls, J. Chem. Soc. 2156 (1952).
[90] G. M. Badger and J. H. Seidler, J. Chem. Soc. 2329 (1954).
[91] T. L. Jacobs and R. B. Brownfield, J. Amer. Chem. Soc. 82, 4033 (1960).

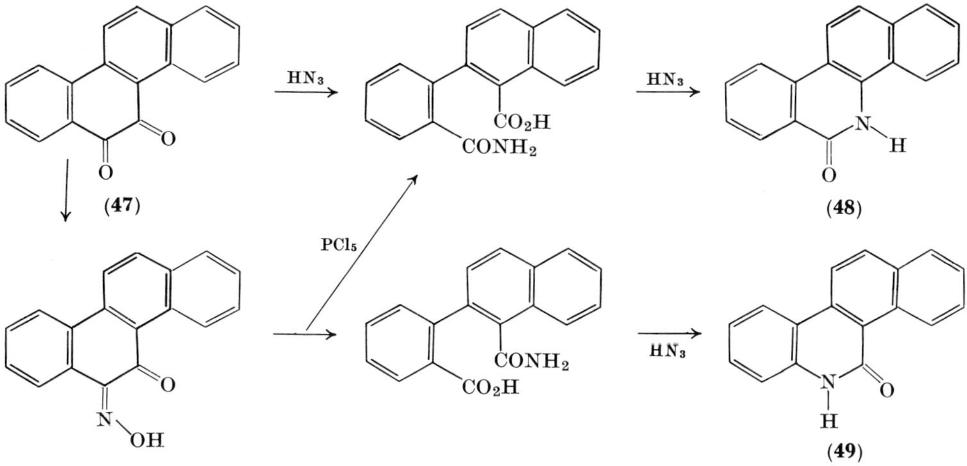

with benzylamine gives the corresponding $N$-benzylenamine (**51**) (R = CH$_2$Ph).[91, 92] Other azasteroids have been obtained in the same way.[92-94]

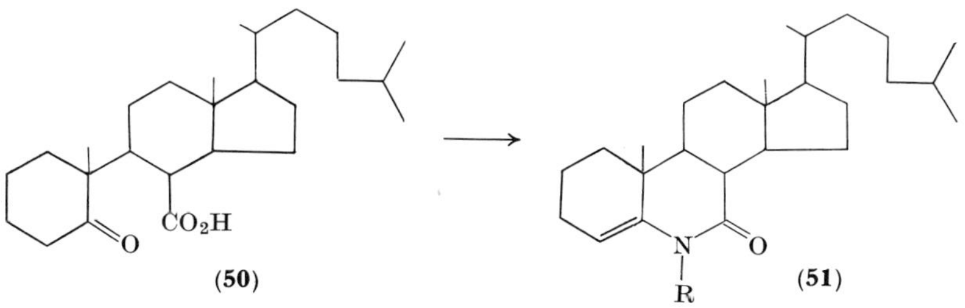

Catalytic hydrogenation of the ketoaldoxime (**52**) gives a hydrophenanthridine to which structure **53** has been assigned.[95]

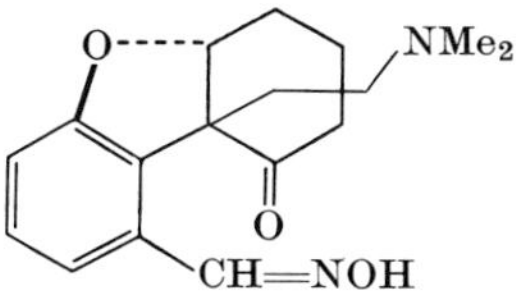

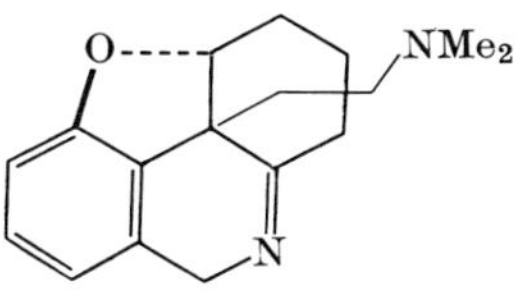

[92] J. P. Kutney and R. A. Johnson, *Chem. Ind.* (*London*) 1713 (1961).
[93] J. P. Kutney, R. A. Johnson, and I. Vlattas, *Can. J. Chem.* **41**, 613 (1963).
[94] J. P. Kutney, G. Eigendorf, and J. E. Hall, *Tetrahedron* **24**, 845 (1968).
[95] H. Mishima, M. Kurabayashi, and I. Iwai, *J. Org. Chem.* **28**, 2621 (1963).

## 3. Ring Expansion Procedures

Phenanthridone has been obtained from both fluorenone and its oxime by Schmidt and Beckmann procedures, respectively. Starting from unsymmetrically substituted fluorenones, mixtures of isomeric phenanthridones are frequently formed.

The conversion of fluorenone oxime into phenanthridone was first achieved using a mixture of phosphorus pentachloride and phosphoryl chloride,[96] although reactions with this reagent are complicated by chlorination in the 6-position.[97] Phenanthridone has been obtained almost quantitatively from fluorenone oxime by the action of polyphosphoric acid at 180°,[98] and the same reagent converts 4,5-dimethylfluorenone oxime into 1,10-dimethylphenanthridone,[22] although in this case the yield is lower, presumably for steric reasons. The rearrangement of 1,3-dimethylfluorenone oxime (54) has been claimed to give a single amide, but the authors were unable to distinguish between the two possible alternatives (55) and (56).[99] Since the oxime probably exists preferentially in the *anti*-dimethylphenyl configura-

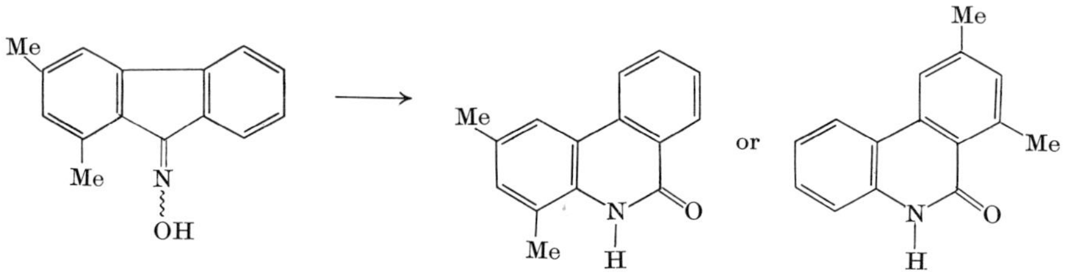

(54)            (55)            (56)

tion, the product is probably 2,4-dimethylphenanthridone (55), assuming a normal trans migration.

3-Nitrofluorenone oxime rearranges to a single amide, 2-nitrophenanthridone, in a mixture of phosphoryl chloride and phosphorus pentachloride. The same reagents convert 2-nitrofluorenone oxime (of apparent stereochemical purity) into a mixture of 3- and 8-nitrophenanthridone,[97] though only the first product is formed in

[96] E. Beckman and P. Wegerhoff, *Ann. Chem.* **252**, 35 (1889).
[97] A. J. Nunn, K. Schofield, and R. S. Theobald, *J. Chem. Soc.* 2797 (1952).
[98] E. C. Horning, V. L. Stromberg, and H. A. Lloyd, *J. Amer. Chem. Soc.* **74**, 5153 (1952).
[99] L. Chardonnens and A. Würmli, *Helv. Chim. Acta* **33**, 1338 (1950).

polyphosphoric acid.[100] The steric factors present in 1-nitro- and 1-iodofluorenone oxime suggest that these compounds should be stereochemically pure; however, on heating in polyphosphoric acid they merely undergo deoximation in preference to rearrangement.[100] Various 2,7-disubstituted fluorenone oximes have been shown to give 3,8-disubstituted phenanthridones in good yield on heating in polyphosphoric acid at above 180°.[100–102]

Examples of the preparation of hydrogenated phenanthridones via the Beckmann rearrangement have also been reported. Thus, the action of thionyl chloride on the oxime (**57**) gives a hexahydrophenanthridone which must be **58**, since on dehydrogenation 4-methylphenanthridone (**59**) is formed.[103] Perhydrofluorenone is a minor product of the carbonylation of cyclohexene; evidence for the identity of the product was obtained by conversion to a mixture of stereoisomers of perhydrophenanthridone on heating in polyphosphoric acid.[104] The same reagent also converts 2,3-benzofluorenone oxime (**60**) into

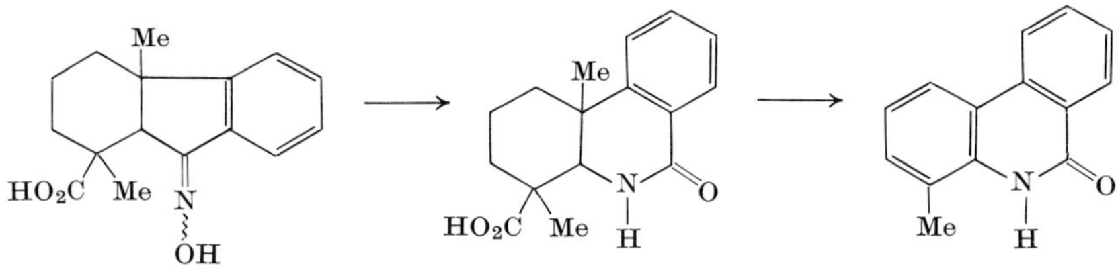

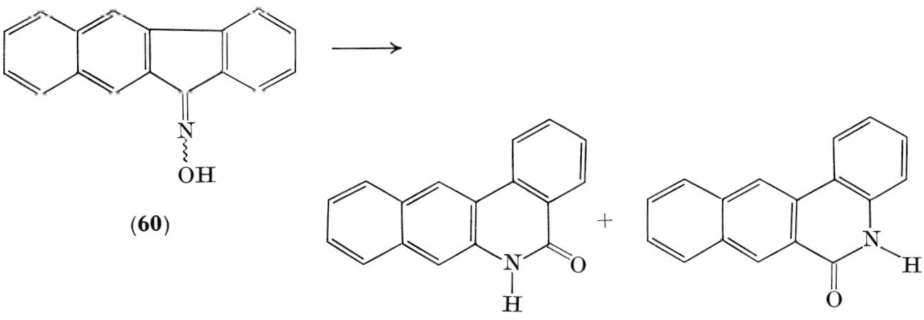

[100] H.-L. Pan and T. L. Fletcher, *J. Med. Chem.* **12**, 822 (1969).
[101] H.-L. Pan and T. L. Fletcher, *J. Heterocycl. Chem.* **7**, 313,597 (1970).
[102] B. R. T. Keene and G. L. Turner, *Chem. Commun.* 221 (1967) and unpublished work.
[103] M. Ohta, *Chem. Pharm. Bull.* **5**, 256 (1957); *Chem. Abstr.* **52**, 6285 (1958).
[104] P. T. Lansbury and R. W. Meschke, *J. Org. Chem.* **24**, 104 (1959).

an approximately equimolar mixture of 2,3- (**61**) and 8,9-benzophenanthridone (**62**) in an overall yield of 21%,[105] although only 9,10-benzophenanthridone was isolated from the action of polyphosphoric acid on 3,4-benzofluorenone oxime.[22] The Beckmann ring expansion of 1,2:7,8-dibenzofluorenone oxime has also been employed in an unambiguous synthesis of 3,4:7,8-dibenzophenanthridine (see page 350).

Unsuccessful attempts to obtain phenanthridone carboxylic acids by the Beckmann route have also been reported. The oxime of fluorenone-1-carboxylic acid (**63**) gave only the oxazinone (**64**) on

(63) → (64)

treatment with polyphosphoric acid at 175°, and attempts using other reagents failed.[106] (A successful application of the Schmidt reaction is described below.)

A mixture of polyphosphoric acid and nitromethane converts fluorenone into phenanthridone at 250°. The reaction probably proceeds via the oxime, which can be isolated when the reaction is carried out at 190°.[107] The same conversion can be achieved, though in lower yield, by the action of sodamide on fluorenone in liquid ammonia or *t*-butylamine. In this reaction biphenyl-2-carboxamide (isolated in 5% yield) is probably a precursor of phenanthridone, since the conversion of the amide to the latter in 40% yield has been demonstrated in a second experiment. The major product of the reaction with fluorenone in liquid ammonia is, not surprisingly, fluorenimine.[108]

Fluorenone forms phenanthridone almost quantitatively on treatment with hydrazoic acid (the Schmidt reaction) in either

[105] L. H. Klemm and A. Weisert, *J. Heterocycl. Chem.* **2**, 15 (1965).
[106] A. Resplandy and P. Le Roux, *Bull. Soc. Chim. Fr.* 4975 (1968).
[107] F. A. L. Anet, P. M. G. Bavin, and M. J. S. Dewar, *Can. J. Chem.* **35**, 180 (1957).
[108] G. W. Kenner, M. J. T. Robinson, C. M. B. Taylor, and B. R. Webster, *J. Chem. Soc.* 1756 (1962).

sulfuric[109] or polyphosphoric acid.[110] The influence of electronic effects in determining migratory aptitudes is in dispute. The ratios of isomeric phenanthridones obtained by the action of hydrazoic acid on 2-nitro-, 2-methoxy-, and 3-nitrofluorenone in sulfuric acid were originally interpreted as suggesting that there is no simple correlation between the electronic character or bulk of a biphenylene system and its migratory aptitude.[111] More recently several instances have been quoted in which strongly electron-withdrawing substituents appear to promote fission between the carbonyl group and the substituent-containing ring, resulting in preferential formation of one of the isomeric phenanthridones.[101] Thus, the nitro-substituted phenylene rings of 2-nitro-, 3-nitro-,[111] and 4-nitrofluorenone and a series of bromo- and chloronitrofluorenones invariably migrate preferentially; the amino group (which under these conditions is presumably protonated and hence electron-withdrawing) behaves similarly.[101] On the other hand, 2-methoxyfluorenone gives approximately equal amounts of 3- and 8-methoxyphenanthridone,[111] possibly because the adverse electronic effect is compensated by the greater bulk of the substituted ring. That such a bulk effect exists (even when the substituent is remote from the carbonyl carbon atom of the original ketone) is suggested by the rearrangement of 2-chlorofluorenone to 3-chlorophenanthridone.[101]

Hydrazoic acid attacks the carbonyl and carboxy groups of fluorenone-1-carboxylic acid simultaneously giving mainly 4-amino phenanthridone together with some acidic material, presumably phenanthridone-4-carboxylic acid.[89] The latter is best prepared by prior protection of the carboxy group with pyrrolidine.[106, 112] The increase in strain associated with the conversion of 4,5-dimethyl-fluorenone into 1,10-dimethylphenanthridone is reflected in the relatively low yield (50%) obtained under Schmidt conditions.[22]

2,3-Benzofluorenone undergoes the Schmidt reaction in trichloroacetic acid to give an approximately equimolecular mixture of 2,3- and 8,9-benzophenanthridone.[105] Under similar conditions 3,4-benzofluorenone (65) forms a mixture of 1,2- (66) and 9,10-benzophenanthridone (67) in which the former predominates.[113] The failure of

[109] P. A. S. Smith, *J. Amer. Chem. Soc.* **70**, 320 (1948).
[110] R. T. Conley, *J. Org. Chem.* **23**, 1330 (1958).
[111] C. L. Arcus, M. M. Coombs, and J. V. Evans, *J. Chem. Soc.* 1498 (1956).
[112] A. Resplandy and P. Le Roux, *C. R. Acad. Sci.* **265** (21), 1181 (1967).
[113] B. R. T. Keene and K. Schofield, *J. Chem. Soc.* 2609 (1958).

1,2-benzofluorenone to undergo the Schmidt reaction in sulfuric, trichloroacetic, or polyphosphoric acids has been attributed to steric factors.[114]

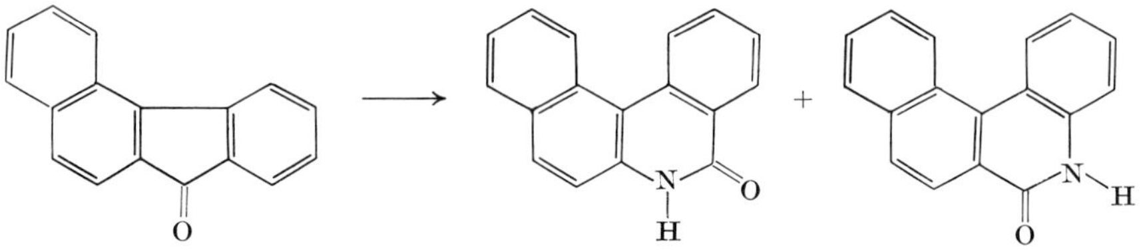

(65)   (66)   +   (67)

The conversion of the phenanthridones obtained by both Beckmann and Schmidt procedures into the parent bases is described in Section II, D. Alternatively, phenanthridines (and their 6-alkyl derivatives) can be obtained directly by a variation of the Schmidt reaction

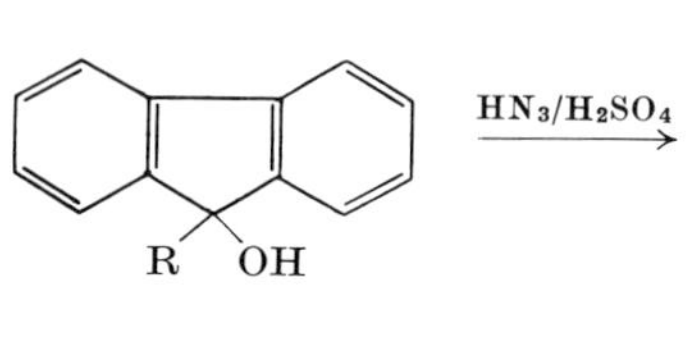

employing fluoren-9-ols[115] and 9-alkylfluoren-9-ols, respectively [Eq. (1)].[116] The reaction has been applied successfully to the ring expansion of methyl-,[22, 116] nitro-, amino-, and methoxyfluorenols.[117] The ratios of phenanthridines obtained suggest that the ease of migration of a biphenylene system is related directly to its capacity for electron release at the point of attachment to C-9.[117] Kinetic studies have shown that the protonated azide is involved, and aryl migration and loss of nitrogen are probably simultaneous.[118]

The conversion of 9-methylfluorenol into 6-methylphenanthridine proceeds in abnormally low yield owing to the formation of a "neutral" by-product (25%),[117] which has been identified as the phenanthridine

---

[114] C. L. Arcus, R. E. Marks, and M. M. Coombs, *J. Chem. Soc.* 4064 (1957).
[115] C. L. Arcus and R. J. Mesley, *J. Chem. Soc.* 178 (1953).
[116] C. L. Arcus and E. A. Lucken, *J. Chem. Soc.* 1634 (1955).
[117] C. L. Arcus and M. M. Coombs, *J. Chem. Soc.* 4319 (1954).
[118] C. L. Arcus and J. V. Evans, *J. Chem. Soc.* 789 (1958).

(68).[119] 9-(9-Fluorenyl)-9-fluorenol (69) has been converted into 6-(9-fluorenyl)phenanthridine (70) in low yield.[120]

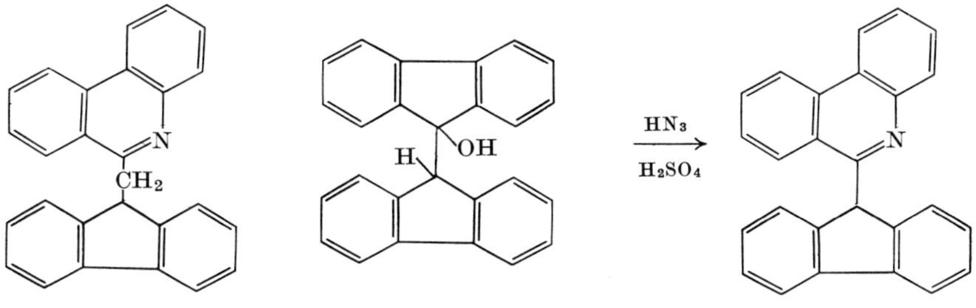

Moore and Snyder[121] have employed the Schmidt reaction to determine the structure of the hydrocarbon (71), which is one of the products of the self-condensation of acetophenone in polyphosphoric acid. Treatment with hydrogen bromide and dimethyl sulfoxide (DMSO) gives the fluorenol (72), which yields a mixture of 1-methyl-2,4,6-triphenylphenanthridine (73) and 10-methyl-6,7,9-triphenylphenanthridine (74) when treated with hydrazoic acid.

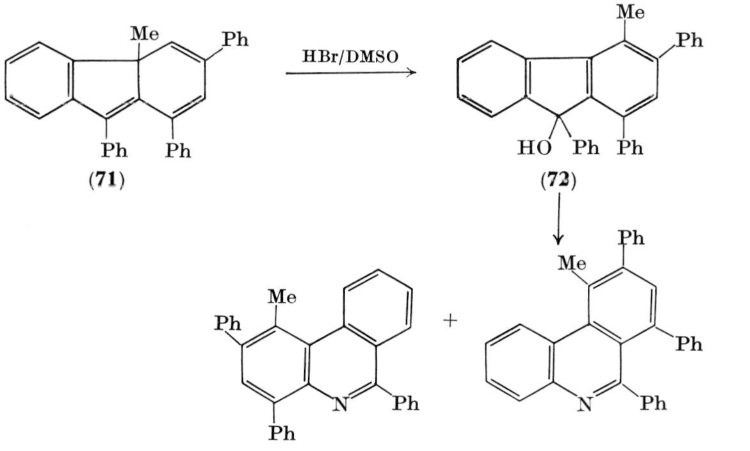

[119] M. M. Coombs, *J. Chem. Soc.* 3454 (1958).
[120] P. M. G. Bavin, *Can. J. Chem.* **43**, 2919 (1965).
[121] H. W. Moore and H. R. Snyder, *J. Org. Chem.* **28**, 535 (1963).

Although 1,2-benzofluoren-9-ol does not react directly under conditions normally employed in the Schmidt reaction, it may be converted to 9-azido-1,2-benzofluorene, which in trifluoroacetic acid rearranges after protonation (75) to a mixture of 3,4-benzophenan-

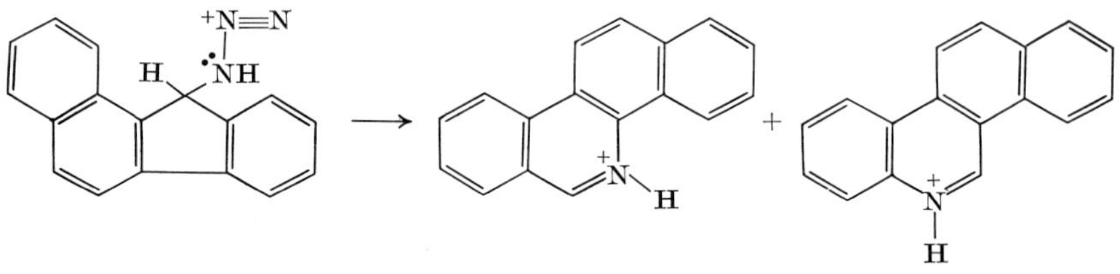

(75) (76) (77)

thridine (76) and 7,8-benzophenanthridine (77), in which the latter predominates.[114]

### B. From Systems Not Containing the Biphenyl Link

#### 1. The Pschorr and Related Reactions

*N*-Methylphenanthridone (79) can be obtained from the diazonium sulfates of the aminobenzanilides (78) and (80) by heating in aqueous solution or by allowing the sulfate (or better, the corresponding

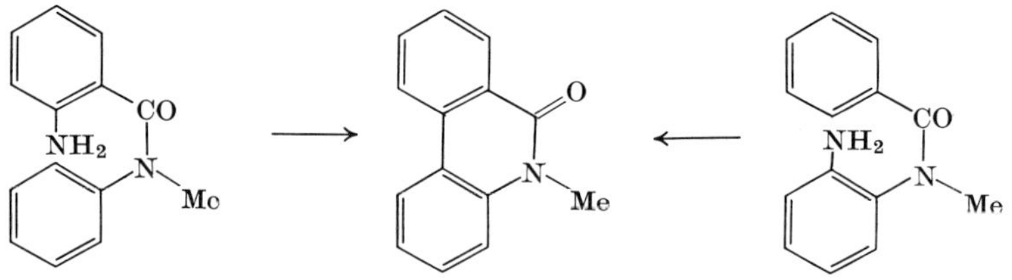

(78) (79) (80)

fluoroborate) to decompose in acetone in the presence of copper powder.[122] Reactions of this type, however, frequently give rise to a number of other compounds, sometimes to the virtual exclusion of the desired product. Thus Hey and Turpin[123] have shown that the ortho-substituted derivatives (81a) decompose mainly by deamination and demethylation. The same authors also demonstrated that decomposition by these routes occurs with the ortho-substituted benzanilides

---

[122] R. A. Heacock and D. H. Hey, *J. Chem. Soc.* 1508 (1952).
[123] D. H. Hey and D. G. Turpin, *J. Chem. Soc.* 2471 (1954).

(82) and (83) which had previously been claimed to undergo cyclization under Pschorr conditions.[99, 124] Deamination and demethylation have also been observed in the case of the amides (81b) and (81c),

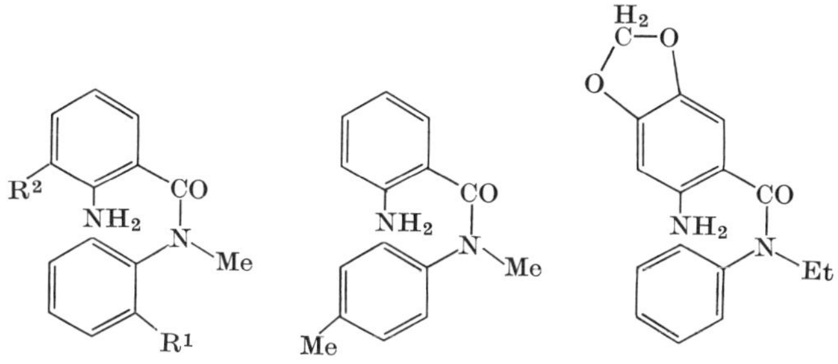

(81a) $R^1$ = Me, Et, Cl, $NO_2$, $CO_2H$, $CO_2Me$, or α-naphthyl;
     $R^2$ = H in each case
(81b) $R^1$ = H; $R^2$ = OMe
(81c) $R^1$ = H; $R^2$ = Me

exclusively in the latter, in which a methyl substituent is ortho to the amino group.[125] On the other hand, thermal decomposition of the diazonium sulfate (84) obtained from 2-amino-N-ethyl-4'-methoxybenzanilide gives the spirodienone (85), probably by the mechanism shown[126, 127]; the 2'-methoxy isomer undergoes a similar reaction. In phosphoric acid at 170° the spirodienone (85) undergoes the dienone–phenol rearrangement giving 2-hydroxy-N-ethylphenanthridone (86). Similarly, a mixture of acetic anhydride and sulphuric acid converts 85 to 2-acetoxy-N-ethylphenanthridone,[126] and equal amounts of 2,6-dichlorophenanthridine and 2-chloro-5-ethylphenanthridone are formed when 85 is heated with phosphorus pentachloride at 175°.[128] Alternatively, reduction to the dienol (87) followed by elimination and rearrangement gives N-ethylphenanthridone (88).[126]

---

[124] R. B. Kelly, W. I. Taylor, and K. Wiesner, *J. Chem. Soc.* 2094 (1953).
[125] D. H. Hey and R. A. J. Long, *J. Chem. Soc.* 4110 (1959).
[126] D. H. Hey, J. A. Leonard, T. M. Moynehan, and C. W. Rees, *J. Chem. Soc.* 232 (1961).
[127] D. H. Hey, J. A. Leonard, C. W. Rees, and A. R. Todd, *J. Chem. Soc. C* 1513 (1967).
[128] D. H. Hey, J. A. Leonard, and C. W. Rees, *J. Chem. Soc.* 5251 (1963).

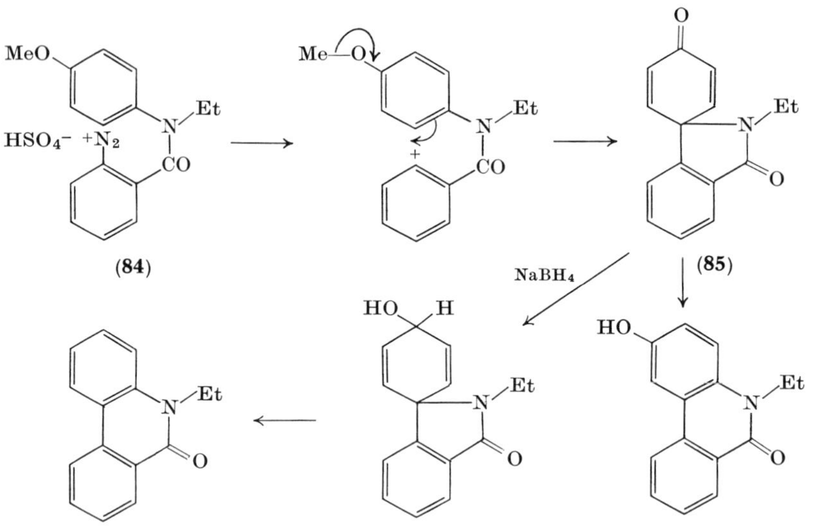


(84) (85) (86) (87) (88)

Thermal decomposition in aqueous solution of the diazonium chloride (**89a**) gives *N*-phenylphenanthridone (**90a**) and *N*-*o*-hydroxybenzoyldiphenylamine. On the other hand, catalytic decomposition gives *N*-phenylphenanthridone and biphenyl-2-carboxyanilide (**91a**), and it has been shown that the substituted derivatives (**89b–d**) decompose in a similar way. The absence of substituent effects in the

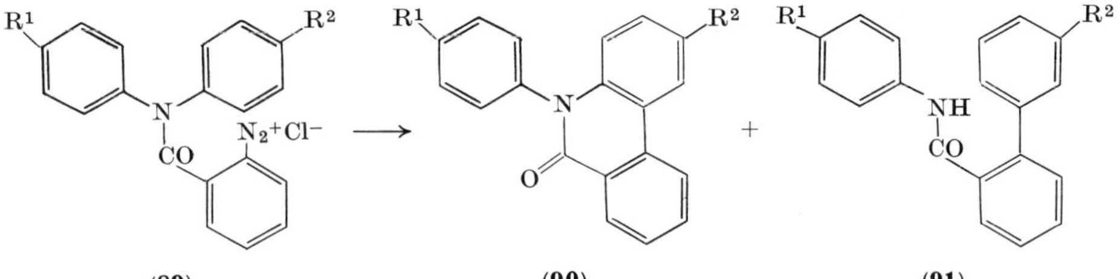


(**89**) (**90**) (**91**)

(**a**) $R^1 = R^2 = H$
(**b**) $R^1 = R^2 = Me$
(**c**) $R^1 = Me$
(**d**) $R^1 = Cl$; $R^2 = H$

anilide-forming reactions of the unsymmetrically substituted derivatives (**89c, d**) suggest that a radical mechanism is responsible.[129]

[129] D. H. Hey and T. M. Moynehan, *J. Chem. Soc.* 1563 (1959).

A careful reexamination of the Pschorr synthesis has shown that, in the copper-catalyzed reaction, products are formed which are not found when the diazonium salt is decomposed thermally. Thus, in a typical example, the fluoroborate (92) of o-amino-N-methylbenzanilide forms both the spirocyclohexadienone (93) and the spirohexadiene dimer (94) as well as N-methylphenanthridone, and the proposed radical mechanism[130] is now established. In an interesting extension of the reaction, N-methylbenzanilide-2-diazonium fluoroborate in oxygen-free methylene chloride has been shown to undergo rapid conversion into N-methylphenanthridone (35%) and the spirodiene (95) when hydrogen iodide is bubbled through at room temperature. The initial, molecule-induced homolysis of the diazonium salt (below) probably involves an electron-transfer process.[131]

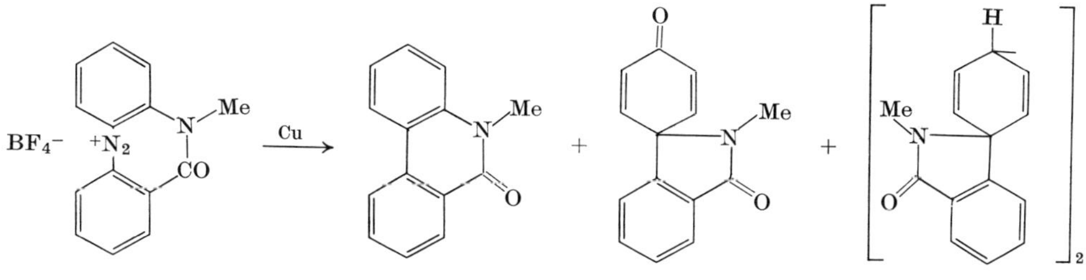

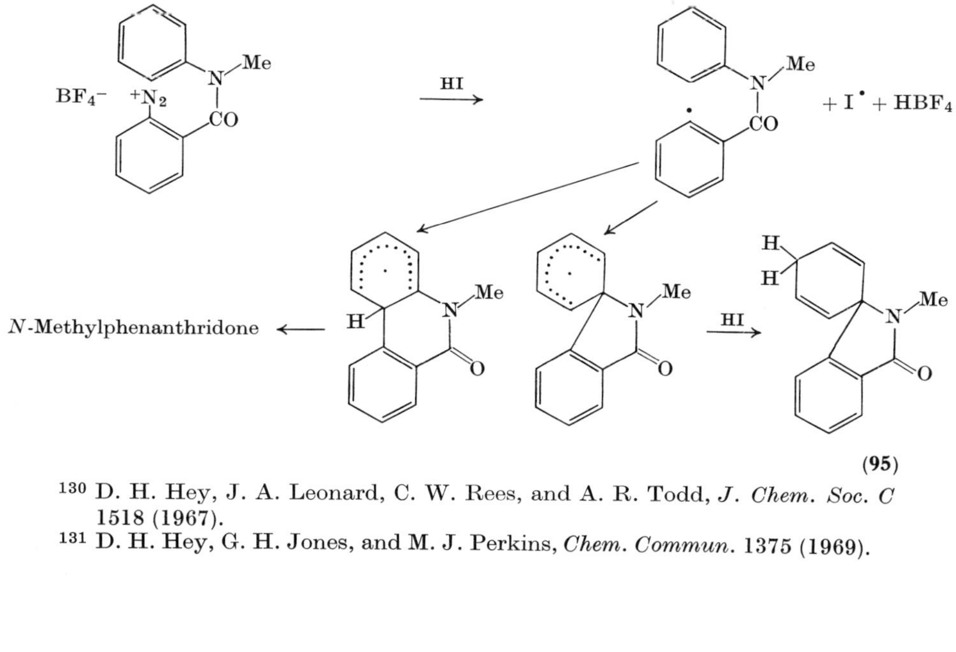

[130] D. H. Hey, J. A. Leonard, C. W. Rees, and A. R. Todd, *J. Chem. Soc. C* 1518 (1967).
[131] D. H. Hey, G. H. Jones, and M. J. Perkins, *Chem. Commun.* 1375 (1969).

The formation of by-products appears to predominate with annelated derivatives, and the desired phenanthridone is, at best, only a minor product of the copper-catalyzed decomposition of the appropriate diazonium salt. Typically (and contrary to an earlier report[123]), the decomposition of N-methyl-2-naphthanilide-1-diazonium fluoroborate (**96**) gives poor yields of four products: 2-naphthanilide (2%), N-methyl-9,10-benzophenanthridone (**97**) (3%), 2-methyl-6,7-benzisoindoline-1-spirocyclohexa-2′,5′-diene-3,4′-dione (**98**) (5%), and bi-(2-methyl-3-oxo-6,7-benzisoindoline-1-spirocyclohexa-2′,5′-dien-4′-yl) (**99**) (15%).[132, 133]

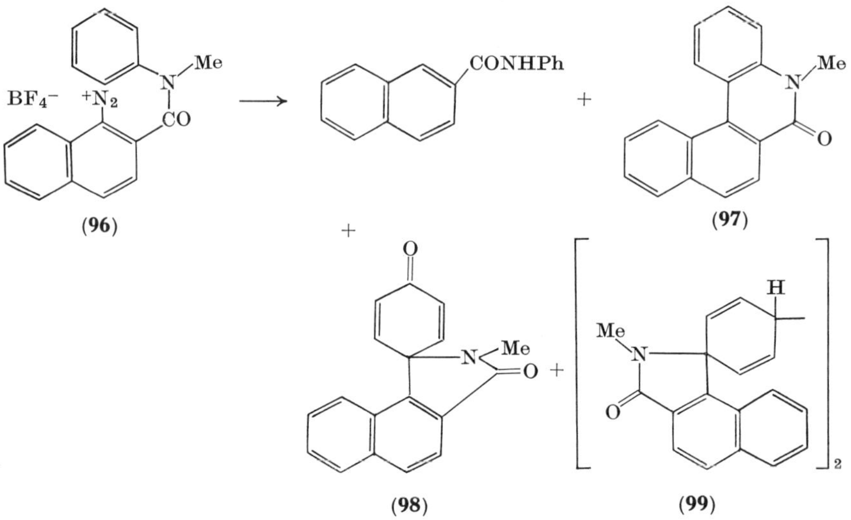

The spirohexadiene dimers dissociate and rearrange when heated to give the corresponding N-methylphenanthridones; in the case of the unsubstituted compound **100** the yield is quantitative, and acceptable yields of N-methyl-1,2-, N-methyl-3,4-, and N-methyl-8,9-benzophenanthridone have been obtained from the appropriate dimers.[133]

---

[132] D. M. Collington, D. H. Hey, and C. W. Rees, *J. Chem. Soc. C* 1017 (1968).
[133] D. M. Collington, D. H. Hey, C. W. Rees, and E. le R. Bradley, *J. Chem. Soc. C* 1021 (1968).

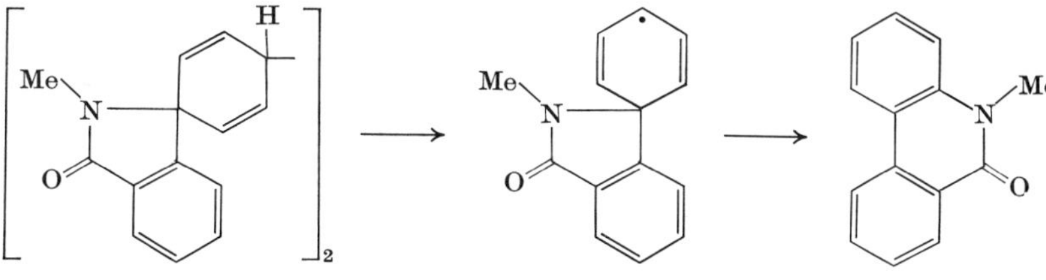

Successful cyclizations (albeit in low yields) have been reported in the case of the amides (**101**)[134] and (**102**),[135] although no benzophenanthridone was obtained from (**103**).[136] 7-Methoxycarbonyl-phenanthridone and its 2-chloro derivative have recently been obtained by a standard Pschorr procedure.[137]

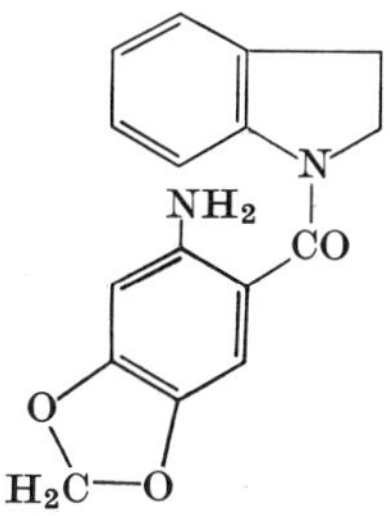

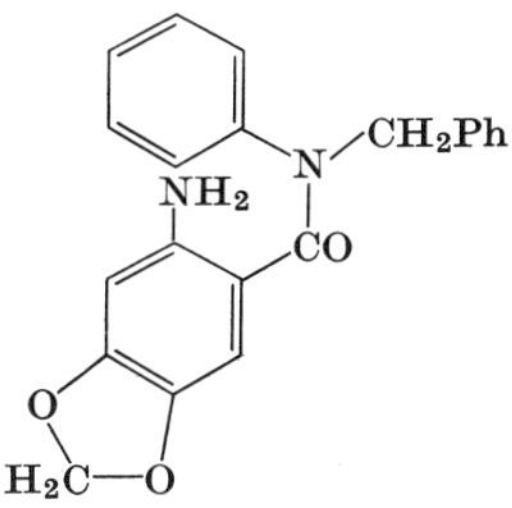

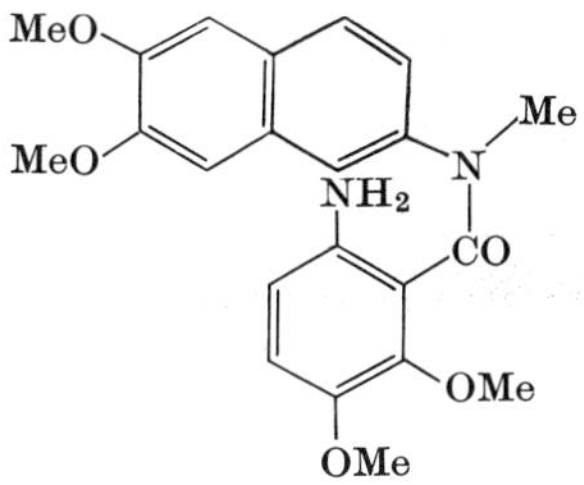

In Pschorr reactions involving anilides, amido groups must be protected to avoid the formation of triazinones[122]; *N*-alkylation has commonly been employed, but the resulting *N*-alkylphenanthridones are difficult to dealkylate. Similarly, protection by *N*-alkylation has been employed in the case of the benzylamine (**104** (R = Me).[138] In a procedure designed to circumvent this difficulty, the *N*-sulfonylated diazonium chlorides, e.g., **104a–c**, are decomposed catalytically; acid hydrolysis, followed by oxidation, of the 5,6-dihydro derivative

---

[134] J. W. Cook, J. D. Loudon, and P. McCloskey, *J. Chem. Soc.* 4176 (1954).
[135] K. Mitsuhashi, *J. Pharm. Soc. Japan* **72**, 344 (1952); *Chem. Abstr.* **47**, 6419 (1953).
[136] T. R. Govindachari and N. Arumugam, *J. Chem. Soc.* 2534 (1955).
[137] A. Resplandy and P. Le Roux, *Bull. Soc. Chim. Fr.* 4947 (1968).
[138] R. P. Patel and H. R. Patel, *Indian J. Pharm.* **18**, 334 (1956); *Chem. Abstr.* **51**, 11351 (1957).

(105) yields the parent base.[139] 4-Bromo-,[140] 7-bromo, and 9-bromophenanthridine[141] have all been obtained in this way.

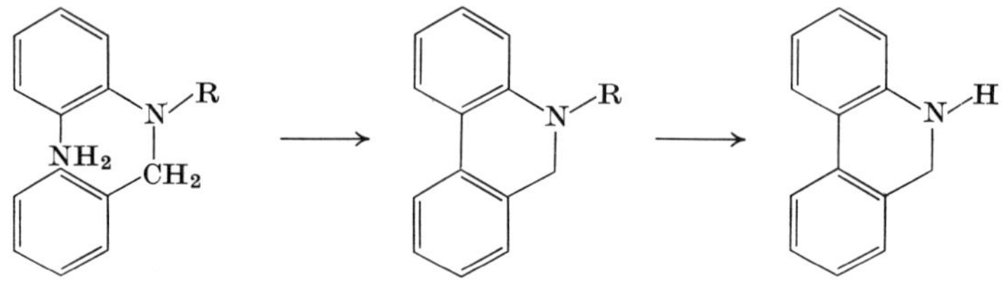

(104a) R = Ph·SO$_2$
(104b) R = p-Me·C$_6$H$_4$·SO$_2$
(104c) R = Me·SO$_2$

(105)

An attempted "biosynthetic-type" approach to lycorine was carried out without prior protection of the amido nitrogen of the fluoroborate (106) and gave, not unexpectedly, the triazinone (107).

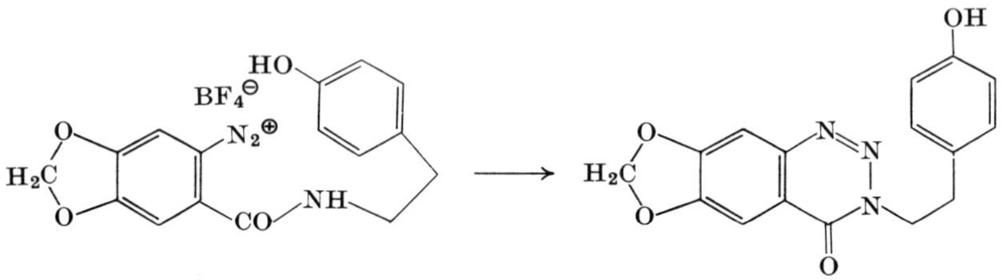

(106)   (107)

Although, in principle, this triazinone is in equilibrium with the diazonium salt in strong acid solution, it was unaffected by anhydrous fluoroboric acid, boron trifluoride, or polyphosphoric acid (cf. below).[142]

Catalytic decomposition of the diazonium chloride of α-(4-isoquinolyl)-o-aminocinnamic acid (108a) gives 3,4-benzophenanthridine-1-carboxylic acid (109a)[143] and the substituted derivatives (109b–d) have been prepared by the same general route.[144]

[139] J. L. Huppatz and W. H. F. Sasse, *Aust. J. Chem.* **16**, 417 (1963).
[140] J. L. Huppatz and W. H. F. Sasse, *Aust. J. Chem.* **18**, 206 (1965).
[141] J. L. Huppatz and W. H. F. Sasse, *Aust. J. Chem.* **17**, 406 (1964).
[142] J. B. Hendrickson, R. W. Alder, D. R. Dalton, and D. G. Hey, *J. Org. Chem.* **34**, 2667 (1969).
[143] R. A. Abramovitch and G. Tertzakian, *Can. J. Chem.* **41**, 2265 (1963).
[144] S. F. Dyke, M. Sainsbury, and B. J. Moon, *Tetrahedron* **24**, 1467 (1968).

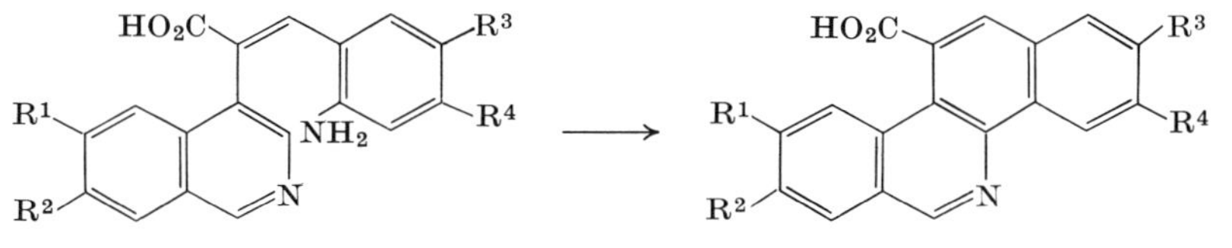

(a) $R^1 = R^2 = R^3 = R^4 = H$
(b) $R^1 + R^2 = R^3 + R^4 = OCH_2O$
(c) $R^1 = R^2 = OMe; R^3 + R^4 = OCH_2O$
(d) $R^1 = R^2 = R^3 = R^4 = OMe$

The decomposition of 3,4-dihydro-4-oxo-3-phenyl-1,2,3-benzotriazines (**110**) in phosphoric acid is mechanistically similar to the Pschorr reaction in the sense that it involves intramolecular arylation

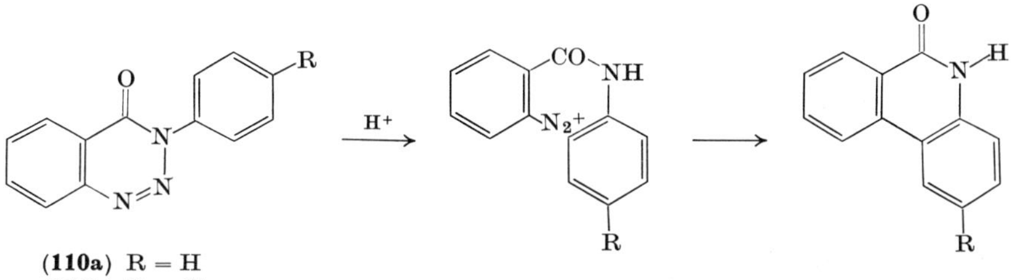

(**110a**) R = H
(**110b**) R = Cl
(**110c**) R = Br

with loss of nitrogen.[145, 146] 2-Chloro- and 2-bromophenanthridone have both been prepared by this method.[147] Thermal decomposition of the triazinone (**110a**) gives phenanthridone (14%) together with 31% acridone.[143, 149]

Iminotriazines, e.g., **111**, also decompose in phosphoric acid to give the corresponding 6-aminophenanthridines; similar decompositions can be carried out catalytically using copper powder in 30% sulfuric acid.[150]

---

[145] T. Mitsuhashi, *J. Pharm. Soc. Japan* **71**, 1235 (1951); *Chem. Abstr.* **46**, 5593 (1952).
[146] M. S. Gibson, *Chem. Ind. (London)* 698 (1962).
[147] M. S. Gibson, *J. Chem. Soc.* 3539 (1963).
[148] D. H. Hey, C. W. Rees, and A. R. Todd, *Chem. Ind. (London)* 1332 (1962).
[149] D. H. Hey, C. W. Rees, and A. R. Todd, *J. Chem. Soc. C* 1028 (1968).
[150] M. W. Partridge and M. F. G. Stevens, *J. Chem. Soc.* 3663 (1964).

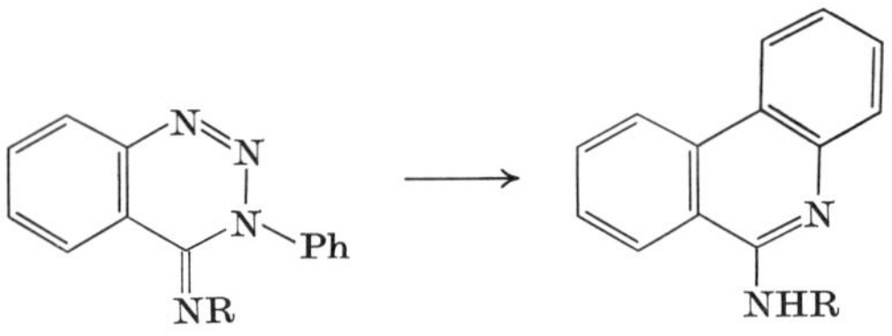

(111)

R = H, Me, Et, or Ph

## 2. Photochemical Coupling Reactions

Recently, several syntheses of phenanthridines have been described in which the biphenyl bond is formed by photochemical coupling. The chief advantage of this method is that relatively simple intermediates are required; on the other hand, since dilute solutions must be employed to minimize dimerization of the reactant, the technique is usually convenient only for the preparation of relatively small amounts of material.

In the simplest case, the conversion of benzylideneaniline to phenanthridine, ring closure occurs satisfactorily in sulfuric acid,[151] and both 6-methyl- and 3,6-dimethylphenanthridine have been prepared in the same way, albeit in low yield, from the appropriate acetophenone anil.[152] However, Mallory and Woods[153] have shown that no phenanthridine is formed when benzylideneaniline is irradiated in either cyclohexane, benzene, or ethanol at 30°–40°, and have ascribed this to the short half-life (ca. 1 second at room temperature[154, 155]) of the cis isomer [although a successful cyclization has been reported in the case of the dimethylamino derivative (112b)[156]].

The observation that phenanthridine is formed in 2% yield when the anil (112a) is irradiated at 10°, whereas benzophenone anil (113),

---

[151] G. M. Badger, C. P. Joshua, and G. E. Lewis, *Tetrahedron Lett.* 3711 (1964).
[152] A. V. El'tsov, O. P. Studzinskii, and N. V. Ogol'tsova, *Zh. Obshch. Khim.* **6**, 405 (1970).
[153] F. B. Mallory and C. S. Wood, *Tetrahedron Lett.* 2643 (1965).
[154] E. Fischer and Y. Frei, *J. Chem. Phys.* **27**, 808 (1957).
[155] D. G. Anderson and G. Wettermark, *J. Amer. Chem. Soc.* **87**, 1433 (1965); G. Wettermark, J. Weinstein, J. Sousa, and L. Dogliotti, *J. Phys. Chem.* **69**, 1584 (1965).
[156] S. Searles and R. A. Classen, *Tetrahedron Lett.* 1627 (1965).

(**112a**) R = H
(**112b**) R = $NMe_2$

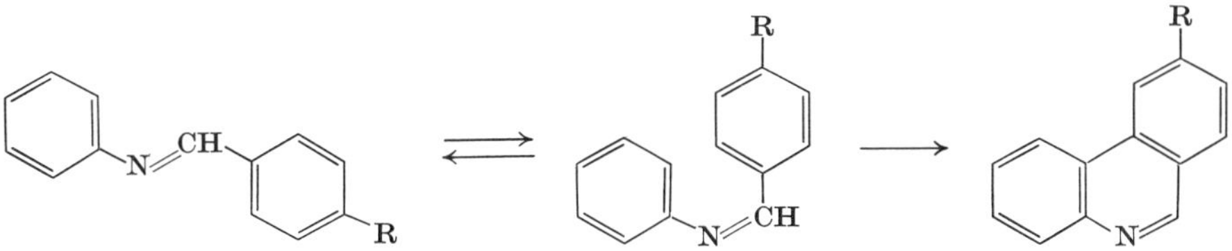

which is not subject to the same stereochemical prohibition, gives 6-phenylphenanthridine (**114**) in good yield is consistent with this explanation. The presence of oxygen or iodine is essential, and the dihydro intermediate shown is probably involved.[153] Irradiation of a

(**113**) (**114**)

solution of anhydro-4,5-diphenyl-2-mercapto-1,2,4-thiadiazolium hydroxide (in acetonitrile) gives the cyclized product **115** (R = H) and the analogous chloro compound (R = Cl) can be prepared similarly.[157]

(**115**)

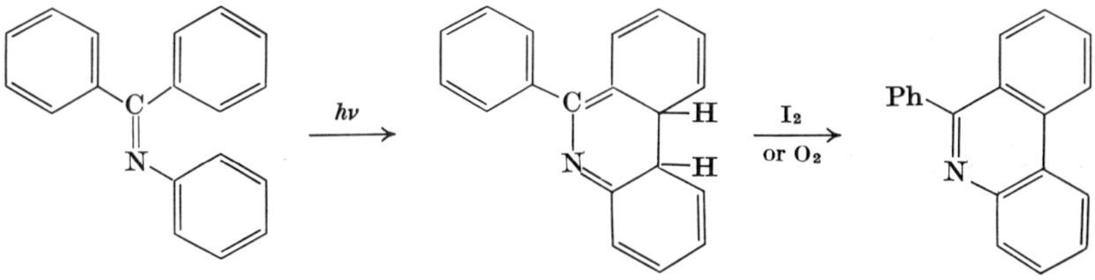

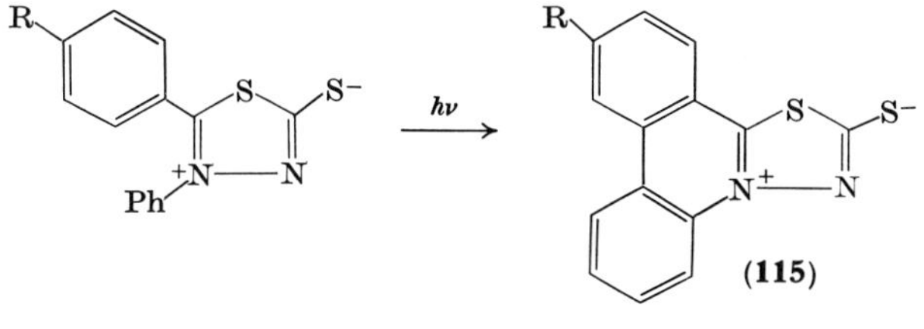

Oxidative photochemical coupling of the anil (**116**) gives 3,4:7,8-dibenzophenanthridine (**117**),[158] the structure of which was confirmed (as shown) by an independent synthesis from 1,2:7,8-dibenzofluorenone oxime (**118**).[158]

[157] R. Moriarty, J. Kliegman, and R. B. Desai, *Chem. Commun.* 1255 (1967).
[158] M. P. Cava and R. H. Schlessinger, *Tetrahedron Lett.* 2109 (1964).

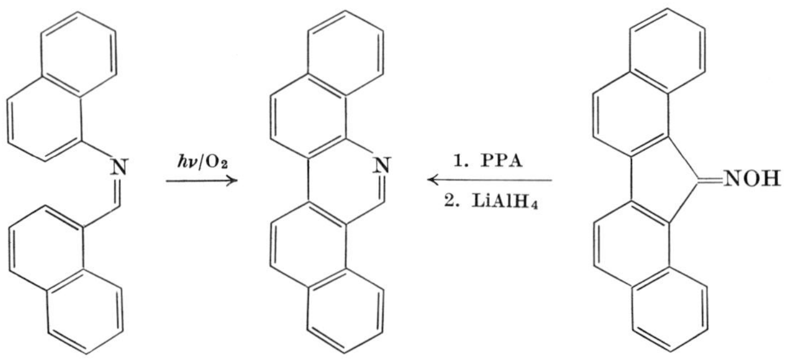

(116)   (117)   (118)

Irradiation of benzanilide in the presence of iodine gives phenanthridone, which has also been obtained from 2- and 2'-iodobenzanilide (**119a, b**) in the absence of added oxidant.[159]

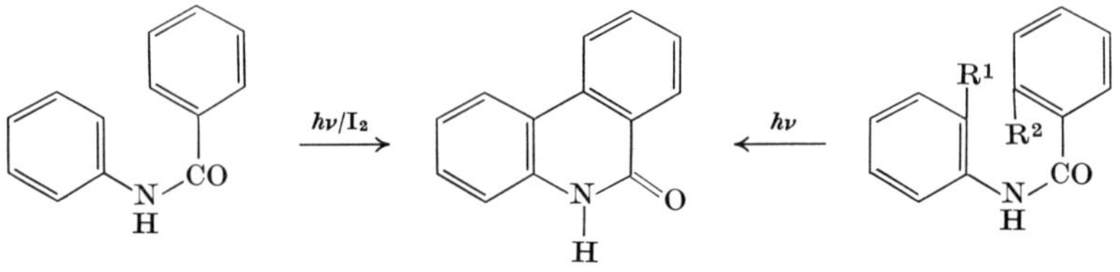

(119a) $R^1 = H; R^2 = I$
(119b) $R^1 = I; R^2 = H$

The *N*-substituted 1a,3,4,4a-tetrahydro-1,2-benzophenanthridones (**121a–d**) have been obtained by irradiation of the corresponding *N*-benzoylamines (**120a–d**) in ether; in the case of the *N*-methyl

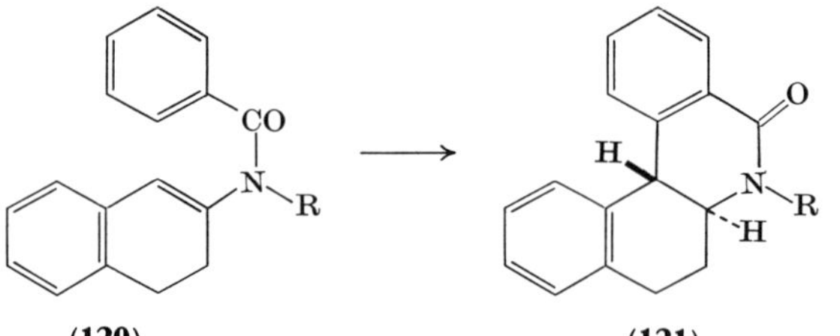

(120)   (121)

(a) $R = CH_2-CH=CH_2$
(b) $R = Me$
(c) $R = CH_2Ph$
(d) $R = n\text{-Bu}$

[159] B. S. Thyagarajan, N. Kharasch, H. B. Lewis, and V. Wolf, *Chem. Commun.* 614 (1967).

derivative (**121b**), the structure was verified by dehydrogenation to the known $N$-methyl-1,2-benzophenanthridone.[160]

Loader and Timmons have described a number of photocyclizations which involve coupling with a heterocyclic ring in the presence of atmospheric oxygen. Irradiation of *trans*-4-styrylquinoline (**122**) in hexane or cyclohexane gives 7,8-benzophenanthridine (**124**).[161–163] In the latter solvent a small amount of 6-cyclohexyl-7,8-benzophenanthridine, resulting from the photochemical reaction between 7,8-benzophenanthridine and the solvent, has been isolated; alkylation also occurs to a limited extent when the reaction is performed in hexane.[162] It has been suggested that a dihydro intermediate (**123**), similar to that proposed for the cyclization of stilbene to phenanthrene,[164, 165] is involved, although no evidence for its existence has been obtained.[162]

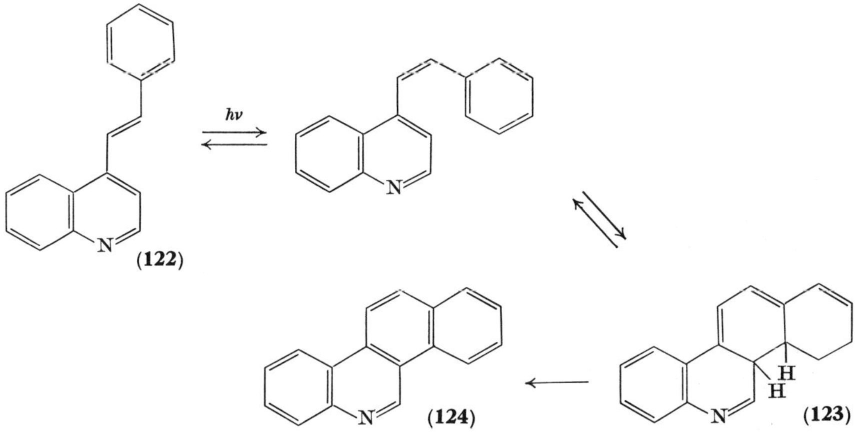

The same photochemical procedure merely converts *trans*-2-styrylquinoline into polymeric material, although *trans*-3-styrylisoquinoline (**125**) and *trans*-4-styrylisoquinoline (**127a**) give moderate yields of 1,2- (**126**) and 3,4-benzophenanthridine (**128a**), respectively.[162]

[160] I. Ninomiya, T. Naito, and T. Moxi, *Tetrahedron Lett.* 2259 (1969).
[161] C. E. Loader, M. V. Sargent, and C. J. Timmons, *Chem. Commun.* 127 (1965).
[162] C. E. Loader and C. J. Timmons, *J. Chem. Soc. C* 1457 (1967).
[163] C. E. Loader and C. J. Timmons, *J. Chem. Soc. C* 330 (1968).
[164] M. V. Sargent and C. J. Timmons, *J. Amer. Chem. Soc.* **85**, 2186 (1963); *J. Chem. Soc.* 5544 (1964).
[165] F. B. Mallory, C. S. Woods, and J. T. Gordon, *J. Org. Chem.* **29**, 3373 (1964).

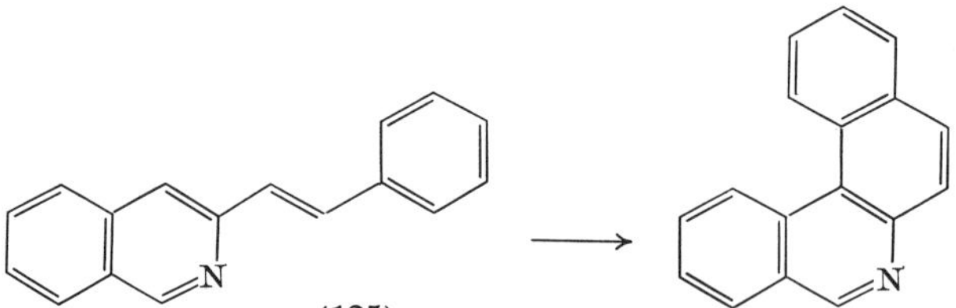

(a) R = H
(b) R = OMe

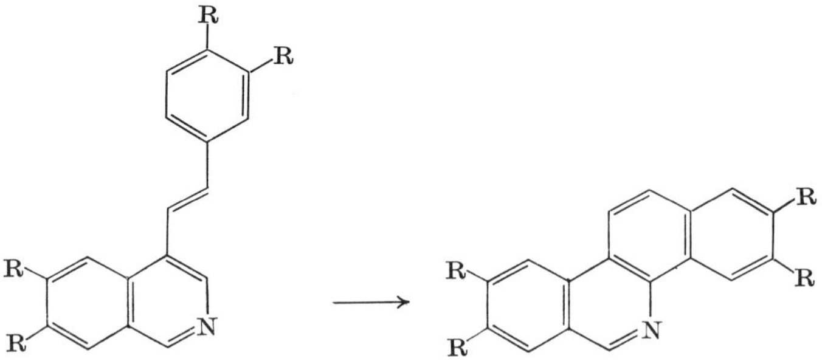

In the alkaloid field, Dyke and Sainsbury[166a] have prepared 2′,3′,8,9-tetramethoxy-3,4-benzophenanthridine (**128b**) from 4-(3,4-dimethoxystyryl)-6,7-dimethoxyisoquinoline (**127b**); the relatively high yield (ca. 50%) was attributed to the insolubility of the product which protected it from further photochemical attack, and the more soluble 2′,3′,7,8-tetramethoxy-3,4-benzophenanthridine was obtained only in minute yield.

The recent synthesis [Eq. (2)] of sanguinarine chloride, involving the irradiation of anhydroprotopine, provides an interesting variant of the more usual coupling procedure.[166b]

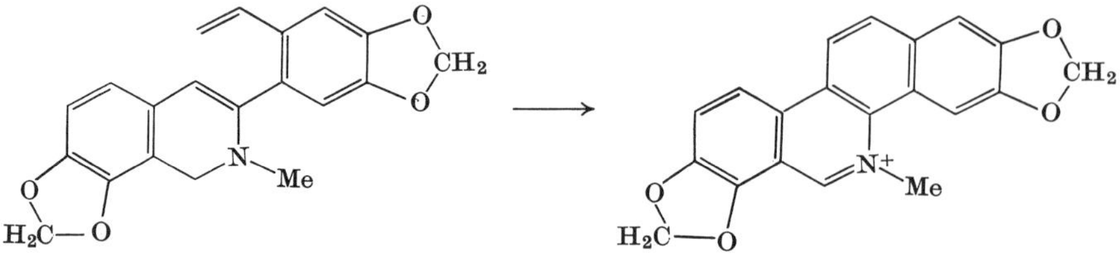

[166] (a) S. F. Dyke and M. Sainsbury, *Tetrahedron* **23**, 3161 (1967); (b) M. Onda, K. Yonezwa, and K. Abe, *Chem. Pharm. Bull.* **17**, 404 (1969).

### 3. *Cyclodehydration Reactions Leading to Hydrophenanthridines*

Reduced phenanthridines have been prepared from anilines and derivatives of cyclohexanone. In the first reported example, Borsche[167] obtained 6,9-dimethyl-7,8,9,10-tetrahydrophenanthridine (**130a**) in 15% overall yield by condensing 6-acetyl-3-methylcyclohexanone with aniline and cyclizing the resultant anil (**129a**) with sulfuric acid. Although the anil (**129b**) of 2-acetylcyclohexane-1,3-dione cyclizes in

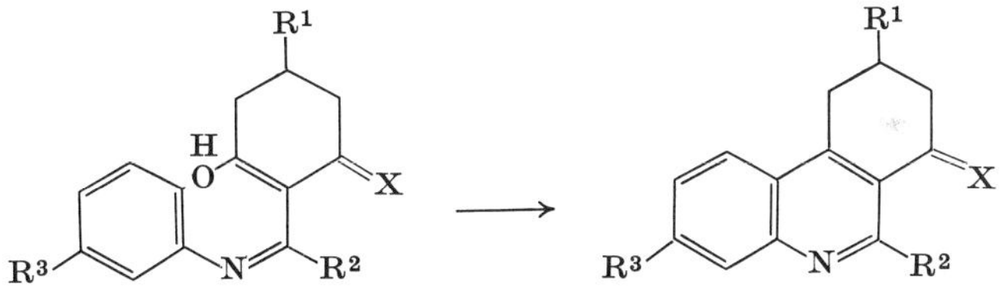

(a) $R^1 = R^2 = Me$; $R^3 = H$; $X = H_2$
(b) $R^1 = R^3 = H$; $R^2 = Me$; $X = O$
(c) $R^1 = R^2 = H$; $R^3 = OMe$; $X = O$

polyphosphoric acid,[168] the reaction fails with the simpler anil (**129**) ($R^1 = R^2 = R^3 = H$; $X = O$). Infrared spectroscopy suggests that a trans configuration is responsible, although at the time failure of the anil to cyclize in formic acid[169] (a reagent which was considered to be effective in these circumstances[170, 171]) was believed to be evidence against this. However, it was shown later (see p. 355) that this work had not been carried out with authentic anils. On the other hand, the anil (**129c**) cyclizes readily in polyphosphoric acid,[172] and the same procedure has been applied successfully in the synthesis of the phenanthridine (**131**)[173] and the 3,4-benzophenanthridines (**132a–c**).[174]

[167] W. Borsche, *Ann. Chem.* **377**, 70 (1910).
[168] H. Smith, *J. Chem. Soc.* 803 (1953).
[169] N. A. J. Rogers and H. Smith, *J. Chem. Soc.* 341 (1955).
[170] H. B. Hollingsworth and V. Petrow, *J. Chem. Soc.* 1537 (1948).
[171] H. B. Hollingsworth and V. Petrow, *J. Chem. Soc.* 263 (1960).
[172] W. Kirkor and T. Baranowicz, *Lodz. Tow. Nauk. Wydz. 3 Acta Chim.* **8**, 69 (1962); *Chem. Abstr.* **59**, 7415 (1963).
[173] P. E. Cross and E. R. H. Jones, *J. Chem. Soc.* 5916 (1964).
[174] S. V. Kessar, I. Singh, and A. Kumar, *Tetrahedron Lett.* 2207 (1965).

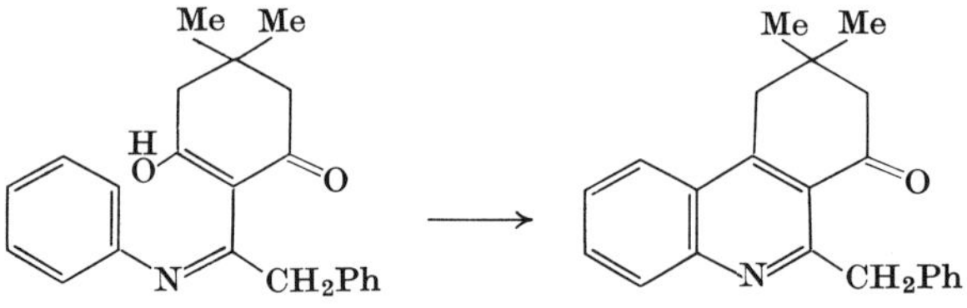

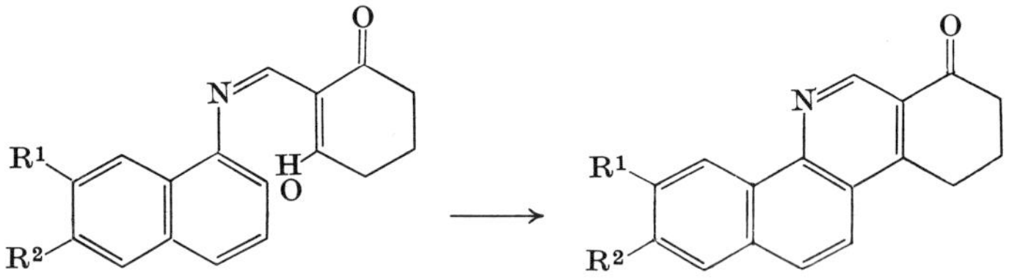

(132a) R¹ = R² = H
(132b) R¹ = H; R² = OMe
(132c) R¹ = R² = OMe

A series of partly reduced 6-aryl-1,2-benzophenanthridin-7-ones (133) have been prepared by heating anils (which may be prepared *in situ*) of 2-naphthylamine with dimedone in boiling ethanol, and similar reactions were also carried out with other 5,5-disubstituted cyclohexa-1,3-diones.[175] The most notable feature of these reactions is the ease of cyclodehydration in the absence of an acid catalyst.

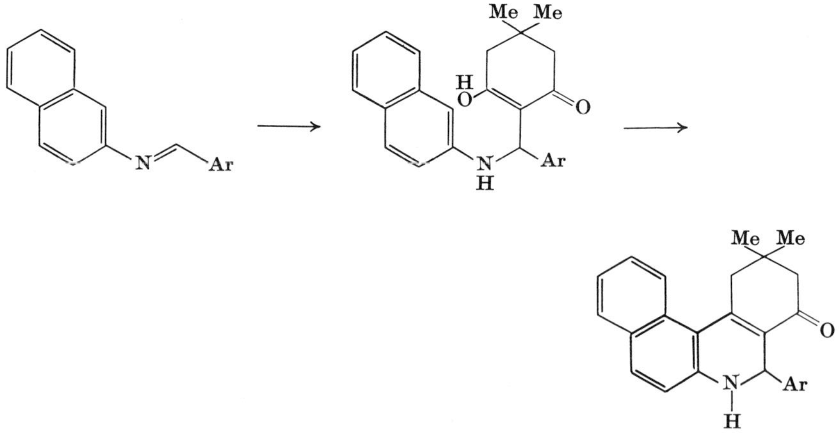

[175] I. Lielbriedis, V. V. Chirkova, and E. Gudriniece, *Latv. PSR Zinat. Akad. Vestis. Kim. Ser.* 197 (1969); 193 (1969); *Chem. Abstr.* **71**, 61179–80 (1969).

Hollingsworth and Petrow, who assumed the *trans*-anil (**134**) structure for the condensation product of aniline and 2-hydroxymethylenecyclohexanone, found that cyclization to 7,8,9,10-tetrahydrophenanthridine could be achieved by heating with formic acid. Since this reagent reduces azomethine bonds, it was argued that favorable stereochemistry is achieved by reduction[170] and, in fact, the secondary amine (**135**) can be isolated.[171] However, later work has shown that the "anil" is, in fact, *cis* (**136**) and that cyclization to

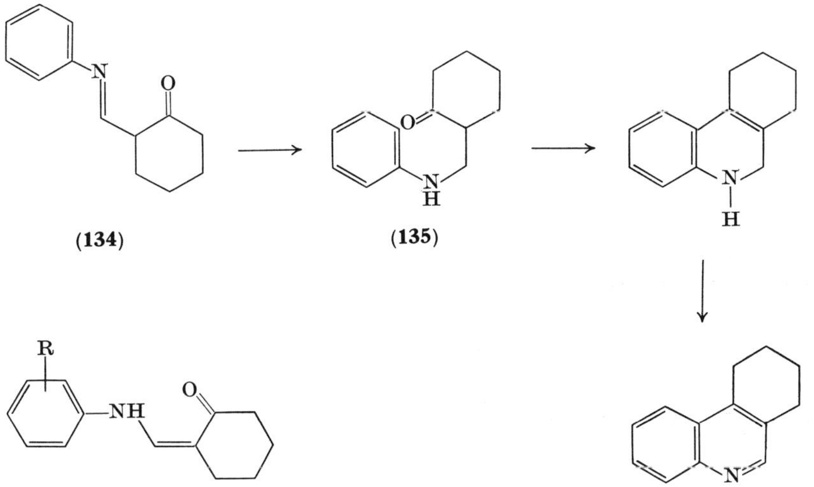

(**134**)   (**135**)

(**136**)

7,8,9,10-tetrahydrophenanthridine also occurs in polyphosphoric acid. A number of methyl-, methoxy-, and chlorosubstituted derivatives can be made to react in the same way, whereas 1,2,3,4-tetrahydroacridines are formed on heating in a mixture of zinc chloride and the appropriate arylamine hydrochloride (corresponding to the substituted aryl-NH moiety) in ethanol.[176]

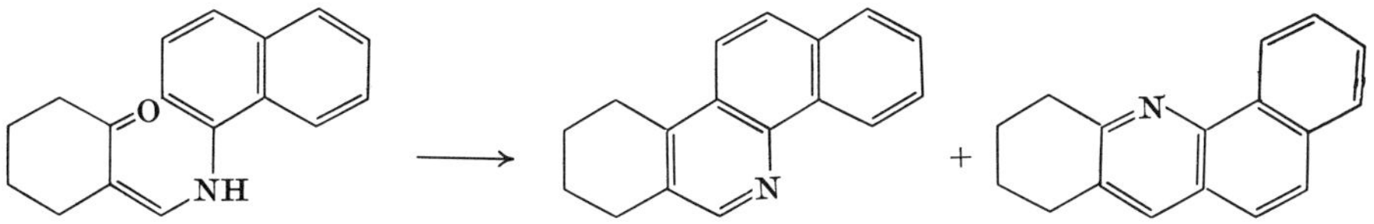

(**137**)   (**138**)   (**139**)

[176] B. D. Tilak, H. Berde, V. N. Gogte, and T. Ravindranathan, *Indian J. Chem.* **8**, 1 (1970).

The cyclization of 2-(1-naphthylaminomethylene)cyclohexanone (**137**) in formic acid has been shown to provide 6,7,8,9-tetrahydro-1,2-benzacridine (**139**) (17%), in addition to the 7,8,9,10-tetrahydro-3,4-benzophenanthridine (**138**) (28%)[177] which was claimed earlier to be the only cyclization product.[170]

Hall and Walker[177] consider that since the cyclization is formally an acid-catalyzed cyclodehydration reaction it should owe nothing to the known reducing properties of formic acid, as suggested by Hollingsworth and Petrow.[170, 171] However, none of the other acids examined (except polyphosphoric) gave any appreciable yield of the phenanthridine (**138**) although most gave good yields of the acridine (**139**), perhaps via a Hofmann–Martius rearrangement prior to cyclization. Similarly, 2-(2-naphthylaminomethylene)cyclohexanone (**140**) has been shown to give 7,8,9,10-tetrahydro-1,2-benzophenanthridine (**141**) (39%) and 6,7,8,9-tetrahydro-3,4-benzacridine (**142**) (12%) in hot formic acid, whereas treatment with lactic acid at 130° provides the benzacridine (**142**) (60%) and no benzophenanthridine (**141**).[177]

Treatment of 2-(1-naphthylaminomethylene)-1-tetralone (**143**) with formic acid provides 7,8-dihydro 3,4:9,10-dibenzophenanthridine (**144**),[177] although the simpler 2-anilinomethylene-1-tetralone (**145**) does not give a phenanthridine derivative under similar conditions.[30, 177] The observation[30] that ring closure of **145** with hot formic acid, followed by treatment with ammonia, gives 5,6,8,9-tetrahydrodibenz[c,h]acridine has been confirmed, and is rationalized in terms of an intermediate xanthylium formate. The nitrogen atom in the final product comes from the ammonia used in the reaction.[177]

2-Hydroxymethylenecyclohexanone and o-nitroaniline give 4-nitro-7,8,9,10-tetrahydrophenanthridine when heated together below 120° in a mixture of arsenic and phosphoric acids, without isolation of the uncyclized intermediate.[178]

[177] G. E. Hall and J. Walker, *J. Chem. Soc. C* 2237 (1968).
[178] F. H. Chase, *J. Org. Chem.* **21**, 1069 (1956).

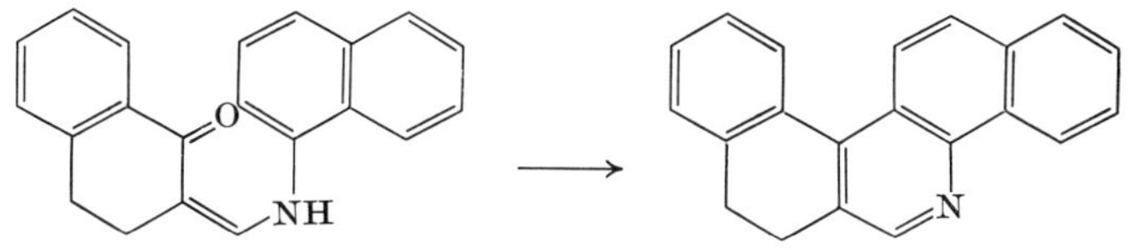

(143) → (144)

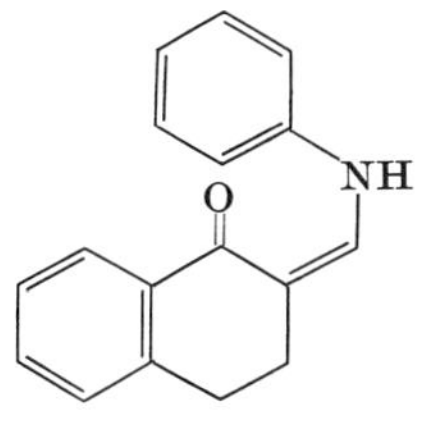

(145)

Amino ketones of the type **146** cyclize to the corresponding 7,8-dihydrophenanthridines (**147**) in the presence of aniline hydrochloride[179, 180] and the amino ketone **148** likewise gives 3-methoxy-

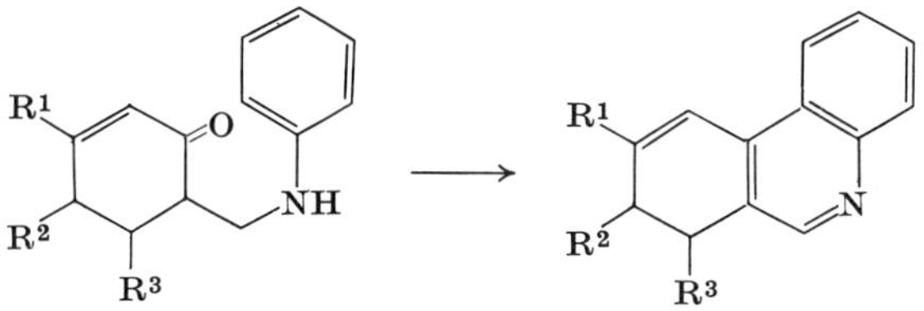

(146) → (147)

$R^1$ = Me; $R^2$ = $R^3$ = H
$R^1$ = Et; $R^2$ = Me; $R^3$ = H
$R^1$ = Me; $R^2$ = H; $R^3$ = Ph

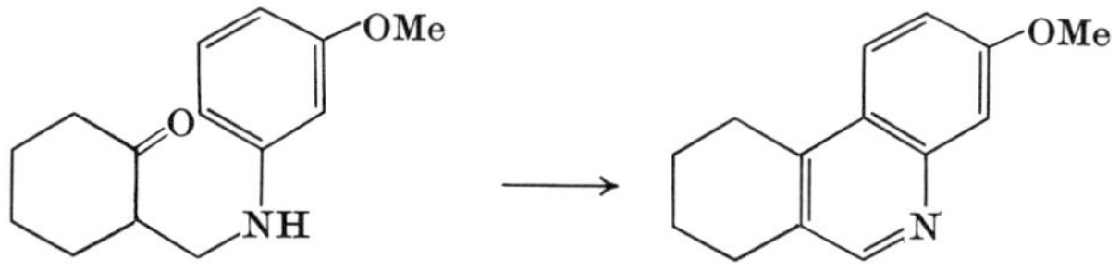

(148) → (149)

---

[179] F. Boyer and J. Decombe, *C. R. Acad. Sci.* **262**, 145 (1966); *Bull. Soc. Chim. Fr.* 2373 (1967).
[180] B. D. Tilak, T. Ravindranathan, and K. N. Subbaswami, *Tetrahedron Lett.* 1959 (1966).

7,8,9,10-tetrahydrophenanthridine (**149**) when heated with polyphosphoric acid in the presence of triphenylmethyl chloride as a hydride ion abstractor.[180] On the other hand, attempts to cyclize the amino alcohol (**150**) to octahydrophenanthridine in sulfuric acid at room temperature were unsuccessful,[75] although at 100°, 7,8,9,10-tetrahydrophenanthridine was formed.[75]

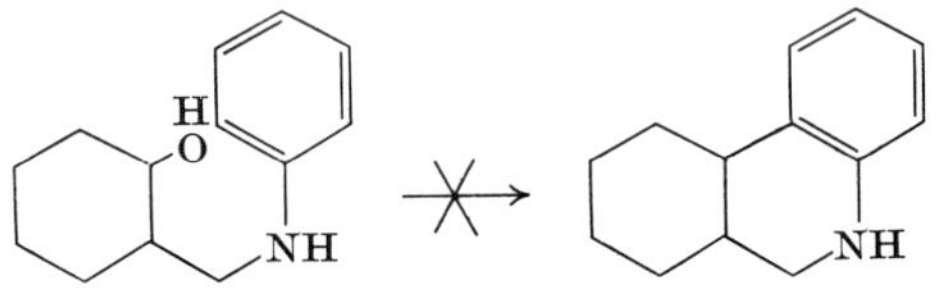

(**150**)

The azasteroid (**152**) has been prepared from (**151**) by cyclodehydration with *p*-toluenesulfonic acid.[181]

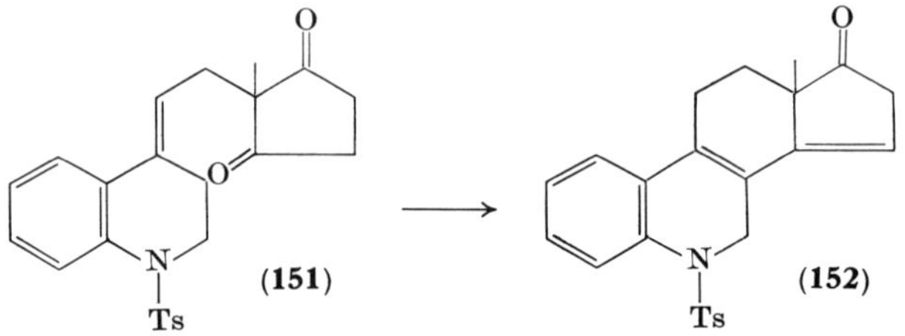

(**151**) → (**152**)

Both anilides and thioanilides of cyclohexanone-2-carboxylic acid cyclize satisfactorily under acid conditions; thus, 7,8,9,10-tetrahydrophenanthridone (**154a**) and the corresponding tetrahydrophenanthridine thione (**154b**) and its 2-methyl derivative (**154c**) have been obtained by the action of sulfuric acid on (**153a–c**), respectively.[182]

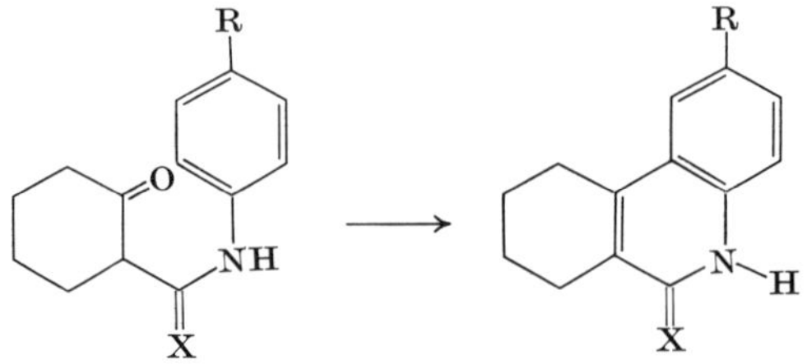

(**153**) → (**154**)

(a) R = H; X = O
(b) R = H; X = S
(c) R = Me; X = S

[181] H. O. Huisman, W. N. Speckamp, and U. K. Pandit, *Rec. Trav. Chim.* **82**, 898 (1963).
[182] W. Reid and W. Kaeppeler, *Ann. Chem.* **673**, 132 (1964); **688**, 177 (1965).

Finally the amido ketones (**155a, b**) cyclize in polyphosphoric acid giving good yields of, respectively, 7,10-dihydro-8,9-benzophenanthridone (**156a**) and its 2-methyl derivative (**156b**).[183]

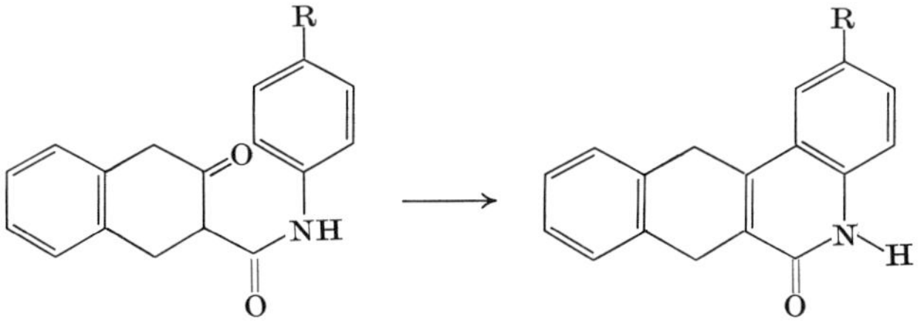

(155) (156)

(a) R = H
(b) R = Me

## C. Miscellaneous Methods

Several reactions involving benzyne intermediates have been employed in the synthesis of phenanthridines. Benzyne itself reacts with phenyl isocyanate to give phenanthridone, probably via the Diels–Alder adduct (**157**). A second product is 6-phenoxyphenanthri-

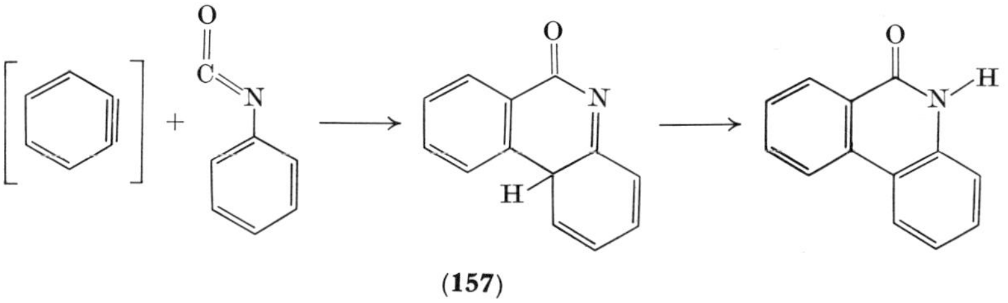

(157)

dine, formed by the attack of a second molecule of benzyne on the adduct, or perhaps on phenanthridone.[184] 3,4-Dehydroquinoline (**159**) has been postulated as an intermediate in the formation of phenanthridine (9%) from the reaction of 3-bromo-4-chloroquinoline (**158**) with lithium amalgam in furan.[185] 5,6-Dihydrophenanthridine

---

[183] N. M. Przhiyalgovskaya, V. F. Schner, and V. N. Belov, *Zh. Obshch. Khim.* **33**, 3690 (1963).
[184] J. C. Sheehan and G. D. Daves, *J. Org. Chem.* **30**, 3247 (1965).
[185] T. Kauffmann, F. P. Boettcher, and J. Hansen, *Angew. Chem.* **73**, 341 (1961); *Ann. Chem.* **659**, 102 (1962).

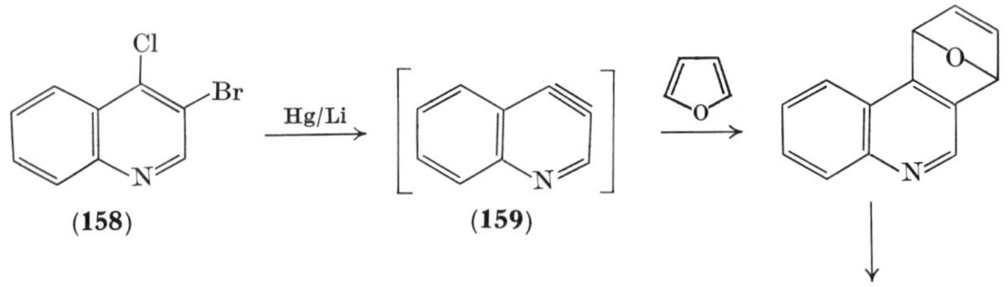

has been prepared in almost quantitative yield by the action of potassamide on the chloroamine (160) (R = H) in liquid ammonia, the reaction probably involving the aryne intermediate (161).

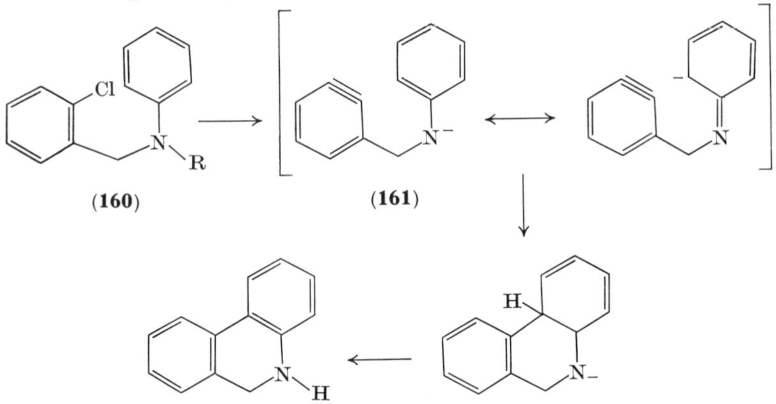

The N-methylamine (160) (R = Me) gives a complex mixture of products when treated under the same conditions.[186] It was shown subsequently that prior reduction of the anils used as starting materials is, in fact, unnecessary since in the presence of potassamide in liquid ammonia (o-chlorobenzylidene)aniline forms phenanthridine in excellent yield, and the analogous anils from p-toluidine and α-naphthylamine undergo ring closure under similar conditions. In view of the known trans geometry of the anils and the unlikelihood of trans–cis equilibration under these conditions this reaction deserves further study.[187] Potassamide in liquid ammonia also converts 2-bromo-N-ethyl-3'-hydroxybenzanilide (162) into a mixture of N-ethyl-3-(164) and N-ethyl-1-hydroxyphenanthridone (165), probably via the aryne (163).[188]

[186] S. V. Kessar, R. Gopal, and M. Singh, *Tetrahedron Lett.* 71 (1969).
[187] S. V. Kessar and M. Singh, *Tetrahedron Lett.* 1155 (1969).
[188] D. H. Hey, C. W. Rees, and J. A. Leonard, *J. Chem. Soc.* 5266 (1963).

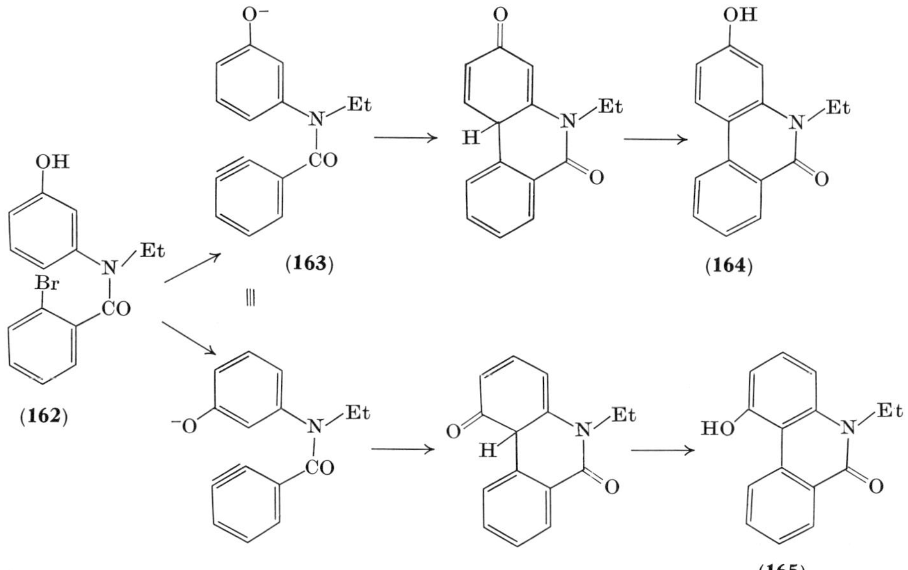

Several pyrolytic methods of preparation have been reported. The thermal decomposition of 1-benzylbenzotriazole in the presence of copper at 350–400° gives phenanthridine in low yield,[189] together with some N-benzylaniline.[190] The pyrolysis of 11-phenyldibenzothiazepine (166) ($R^1 = R^2 = H$) in diethyl phthalate in the presence of copper bronze provides 6-phenylphenanthridine in 94% yield,[191]

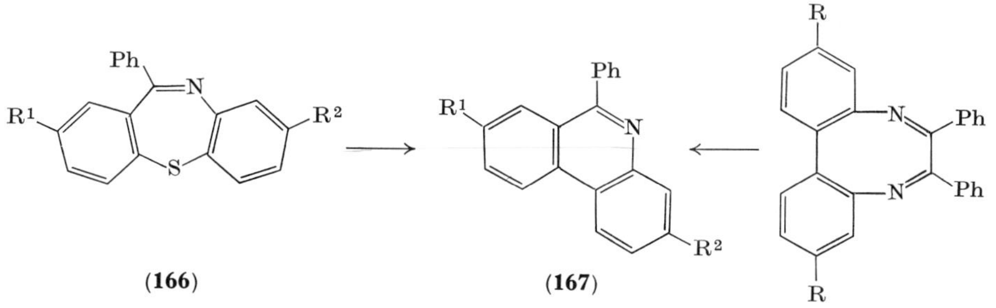

(168a) R = H
(168b) R = $CO_2Me$

[189] M. S. Gibson, *J. Chem. Soc.* 1076 (1956).
[190] B. W. Ashton and H. Suschitzky, *J. Chem. Soc.* 4559 (1957).
[191] R. H. B. Galt, J. D. Loudon, and A. D. B. Sloan, *J. Chem. Soc.* 1588 (1958).

and a number of methyl, methoxy, chloro, and nitro derivatives (167)* have been obtained by similar sulfur extrusion processes.[192, 193] 6-Phenylphenanthridine is also formed when 6,7-diphenyldibenzo-[e,g][1,4]diazocine (168a) is pyrolyzed; the bismethoxycarbonyl derivative (168b) undergoes a similar reaction.[194]

Ring contraction reactions leading to phenanthridines have also been observed with azadibenzocycloheptenones. Thus, treatment of 5,6-dihydro-5-tolylsulfonyldibenz[b,d]azepin-7-one (169a) suspended in sodium methoxide in toluene with oxygen gives phenanthridone (35%) and the hydroxylamine (169b) (42%), which, in turn, can be converted to phenanthridone by heating with palladized charcoal.[53]

(169a) $R^1 = H$; $R^2 = Ts$
(169b) $R^1 = H$; $R^2 = OH$
(169c) $R^1 = Me$; $R^2 = Ts$

(170a) $R = H$
(170b) $R = OEt$

Ring contraction (to 6-methylphenanthridine) also occurs when the ketone (169a) is detosylated with sodium in liquid ammonia, whereas 6-formylphenanthridine is the product of acid hydrolysis catalyzed by zinc chloride. Bromination of the same ketone with a molar proportion of bromine or with N-bromosuccinimide gives phenanthridine-6-carboxylic acid (as its hydrobromide), whereas the 2-bromo derivative of the latter is the probable product when an excess of bromine is employed.[195] Thermal decomposition of the azatropone (170a), or heating with lithium aluminum hydride in tetrahydrofuran, gives phenanthridone; since the yield in the former reaction is always about 50%, disproportion may be involved. 6-Ethoxycarbonyl-phenanthridine is formed when the ethoxy compound (170b) is

* The products are described,[192] apparently incorrectly, as 4-substituted rather than 3-substituted phenanthridines.

[192] R. H. B. Galt and J. D. Loudon, *J. Chem. Soc.* 885 (1959).
[193] A. D. Jarrett and J. D. Loudon, *J. Chem. Soc.* 3818 (1957).
[194] N. L. Allinger, W. Szkrybalo, and M. A. DaRooge, *J. Org. Chem.* **28**, 3007 (1963).
[195] G. R. Procter and W. C. Peaston, *J. Chem. Soc. C* 2151 (1969).

treated with sodium methoxide; the same reagent effects the ring contraction of the 6-methyl derivative (**169c**) to 6-acetylphenanthridine.[196]

Passage of a mixture of cyclohexanone and ammonia over a variety of catalysts, e.g., thoria–alumina[197] or silica–magnesia,[198] at 200°–420° provides 6-pentyl-1,2,3,4,7,8,9,10-octahydrophenanthridine, which has also been obtained in a similar way from 2-(1-cyclohexenyl)-cyclohexanone, $n$-hexaldehyde, and ammonia at 350°.[199]

Povarov has reported that the addition of ethyl-1-cyclohexenyl ether to benzylideneaniline and a number of its derivatives in the presence of boron trifluoride etherate provides the adducts (**171a–d**). **171a** can be converted into 6-phenyl-7,8,9,10-tetrahydrophenanthridine either by treatment with $p$-toluenesulfonic acid or oxidation with potassium permanganate.[200]

(**171a**) $R^1 = R^2 = R^3 = H$
(**171b**) $R^1 = OH; R^2 = R^3 = H$
(**171c**) $R^1 = R^2 = H; R^3 = Br$
(**171d**) $R^1 = R^3 = H; R^2 = CO_2H$

The condensation of the sodio derivative of 1,3-diformyl-4-keto-1,2,3,4-tetrahydroquinoline (**172**) with ethyl acetonedicarboxylate has been reported to give the 5-formyl-9-hydroxy-5,6-dihydrophenanthridine-8,10-dicarboxylate (**173**). (However, the material obtained from **173** by hydrolysis, decarboxylation, and dehydrogenation[201]

---

[196] W. C. Peaston and G. R. Procter, *J. Chem. Soc. C* 2481 (1968).
[197] W. E. Kuhr and C. C. Nathan (to Texaco Inc.), U.S. Patent 3,250,706 (1966); *Chem. Abstr.* **65**, 412 (1966).
[198] H. Chafetz and R. C. Anderson (to Texaco Inc.), U.S. Patent 3,349,092 (1967); *Chem. Abstr.* **68**, 105026 (1968).
[199] H. Chafetz and R. C. Anderson (to Texaco Inc.), U.S. Patent 3,311,631 (1967); *Chem. Abstr.* **67**, 90697 (1967).
[200] L. S. Povarov, *Iz. Akad. Nauk. SSSR., Ser. Khim.* 337 (1966); *Chem. Abstr.* **64**, 17539 (1966).
[201] W. E. Edmiston and K. Wiesner, *Can. J. Chem.* **29**, 105 (1951).

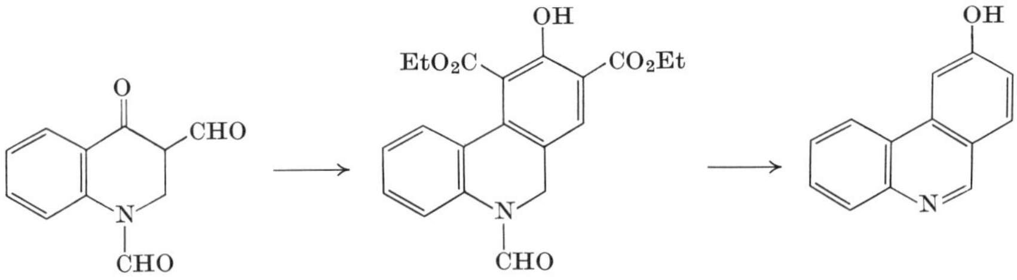

(172) → (173) →

melted considerably lower than 9-hydroxyphenanthridine prepared later by the hydrolysis of the diazonium salt of 9-aminophenanthridine.[117]

Treatment of the dibenzpyrone (174) with methylamine under pressure gives 7,8,9,10-tetrahydro-1-hydroxy-5,9-dimethyl-3-$n$-pentylphenanthridone (175).[202] Similarly, reaction of the tetrahydro-

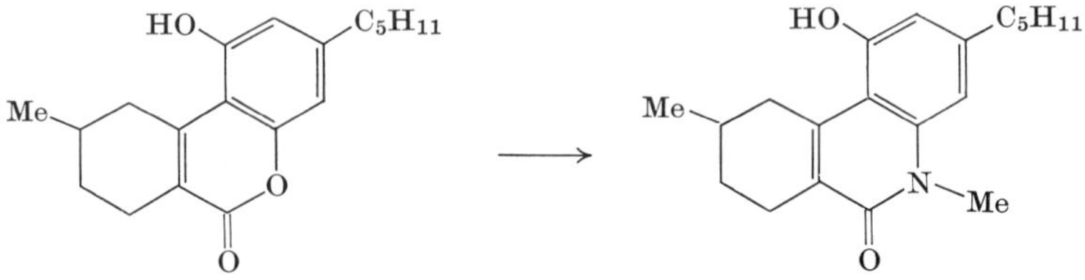

(174) (175)

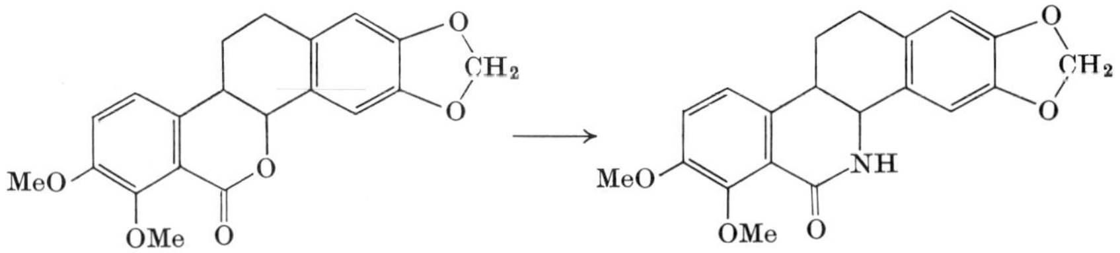

(176) (177)

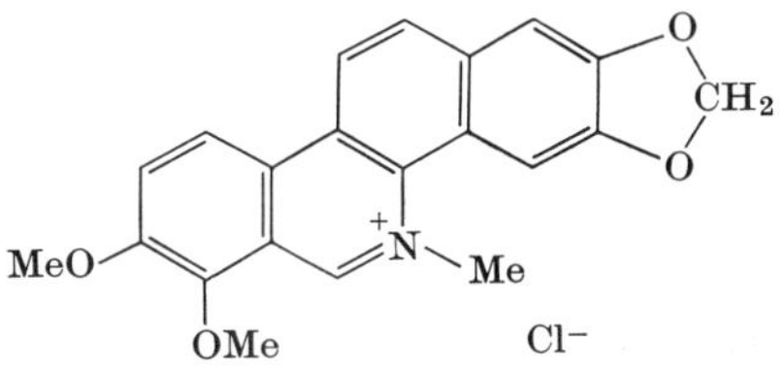

(178)

[202] J. F. Hoops, H. Bader, and J. H. Biel, *J. Org. Chem.* **33**, 2995 (1968).

naphthoisocoumarin (**176**) with ammonia at 210° gives the phenanthridone (**177**) which has been converted to the alkaloid chelerythrine [as the methochloride (**178**)].[203]

The indolophenanthridine (**179**) was the unexpected product when 1,4-diphenylcarbazole was treated with a mixture of *N*-methylformanilide and phosphoryl chloride.[204] Presumably the Vilsmeier–Haack complex attacks at nitrogen, but no explanation was offered to account for the oxidation (of either the Vilsmeier intermediate or 9-formyl-1,4-diphenylcarbazole) which must be involved in the formation of the phenanthridone.

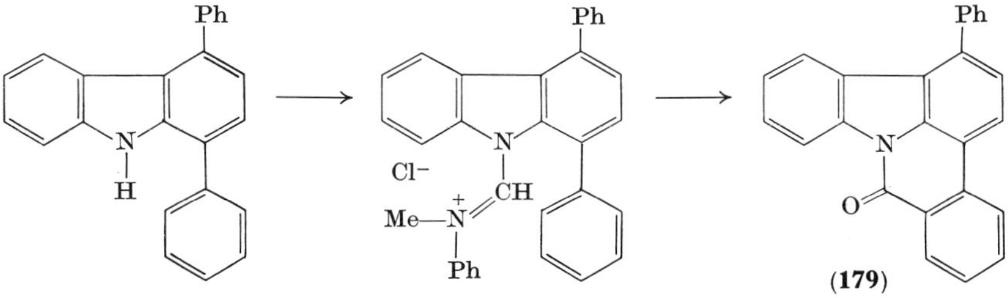

Phosphoryl chloride at 130° converts spiro(homophthalimide-4,2′-indan) (**180**) into 6-chloro-2,3-benzophenanthridine (**181**) (41%), possibly by the mechanism indicated. A second product is 2,3-

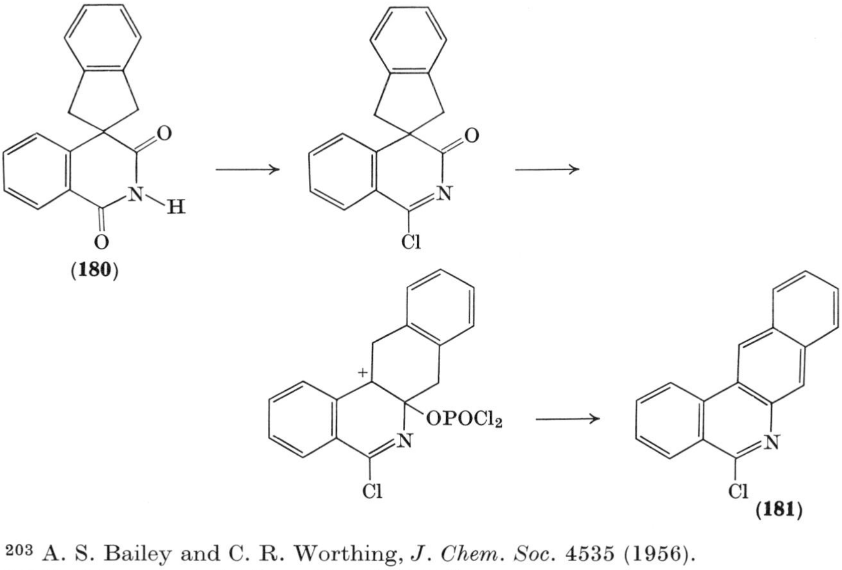

[203] A. S. Bailey and C. R. Worthing, *J. Chem. Soc.* 4535 (1956).
[204] T. Teitei, *Aust. J. Chem.* **23**, 185 (1970).

benzophenanthridone (11%) which may be formed by hydrolysis of **181** during reaction.[205]

The action of ethylamine on the dienone lactone (**182**) gives *N*-ethyl-3-ethylaminophenanthridone (**183**) instead of the expected *N*-ethyl lactam. The yield of the phenanthridone rises with increasing temperature and at 210° it is the sole product. The following mechanism has been suggested:

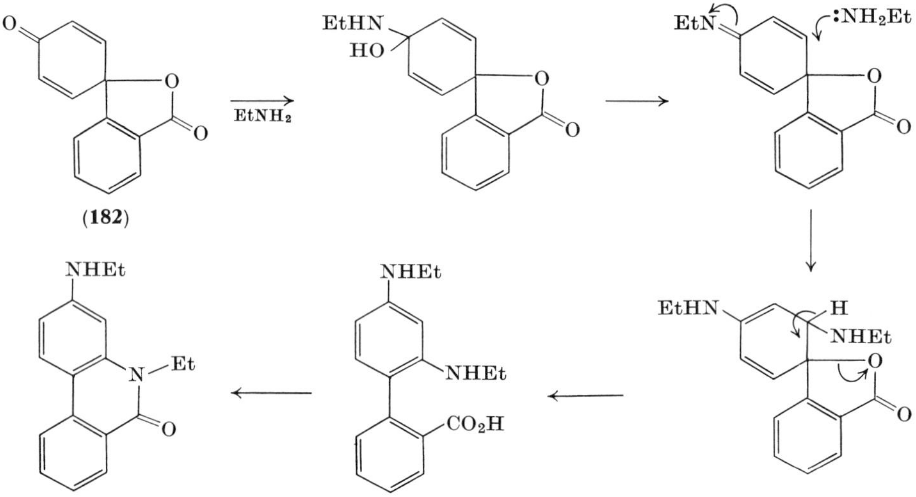

Similarly, the dienone lactone (**182**) gives 3-aminophenanthridone on heating with ammonia (formed by the thermal decomposition of urea *in situ*) and the same conversion probably occurs on treatment with sodamide in liquid ammonia.[128]

Oxidation of 2,3-cycloheptenoindole (**184**) over platinum provides the lactam (**185**) which undergoes a transannular condensation when subjected to chromatography on alumina, giving 7,8,9,10-tetrahydrophenanthridone.[206]

The norbelladin derivative (**186**) undergoes oxidative coupling in ferric chloride solution, the product (**187**) suffering further coupling on hydrolysis to give the crinin derivative (**188**).[207] That these

[205] W. J. Gensler, M. Vinovskis, and N. Wang, *J. Org. Chem.* **34**, 3664 (1969).
[206] B. Witkop, J. B. Patrick, and M. Rosenblum, *J. Amer. Chem. Soc.* **73**, 2641 (1951).
[207] B. Franck and H. J. Lubs, *Angew. Chem. Int. Ed.* **7**, 223 (1968).

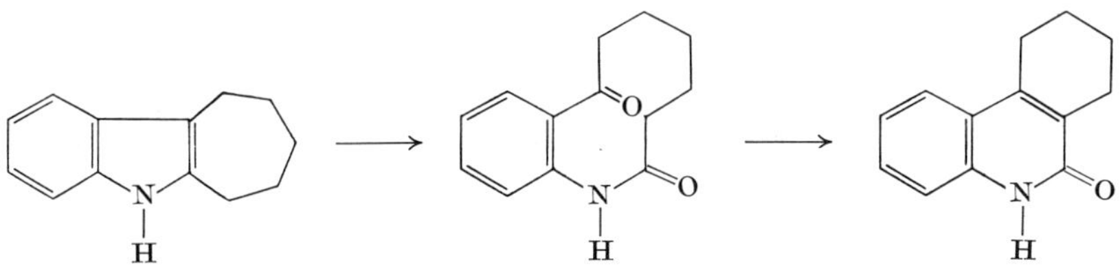

Sec. II. C.]  PHENANTHRIDINES  367

(184)  (185)

reactions proceed at room temperature is of particular interest, since it has been demonstrated that norbelladin is a biogenetic precursor of crinin.[208]

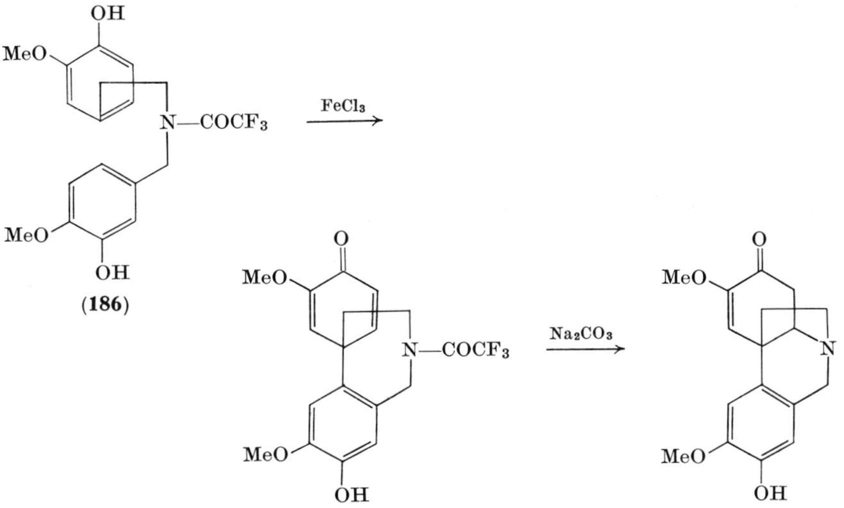

(186)

(187)  (188)

Examples of the application of the Mannich reaction also deserve mention. The amine (189) condenses with aqueous formaldehyde in the presence of sodium bicarbonate to give the phenanthridine (190).[209] Similarly, phenanthridines have been obtained from formaldehyde and the amines (191)[210] and (192),[142] although in these cases cyclization was carried out in hydrochloric acid.

[208] D. H. R. Barton, G. W. Kirby, J. D. Taylor, and G. M. Thomas, *Proc. Chem. Soc.* 254 (1961); A. R. Battersby, H. M. Fales, and W. C. Wildman, *J. Amer. Chem. Soc.* **83**, 4098 (1961).
[209] K. Okada, *Chem. Pharm. Bull.* **10**, 852 (1962); *Chem. Abstr.* **59**, 2788 (1963).
[210] J. S. Bindra and N. Anand, *Indian J. Chem.* **5**, 344 (1967); *Chem. Abstr.* **68**, 114416 (1968).

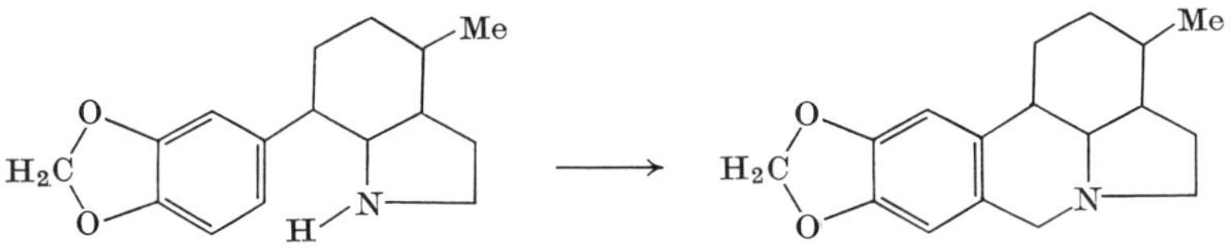

(189) (190)

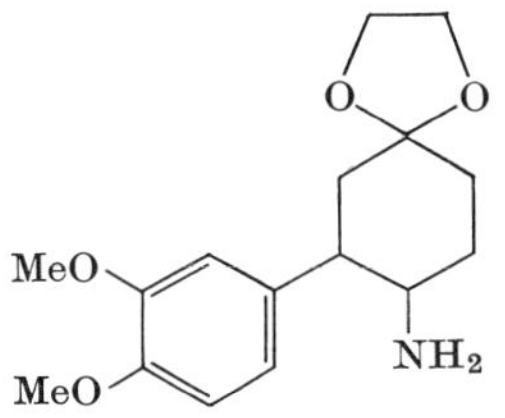

(191) 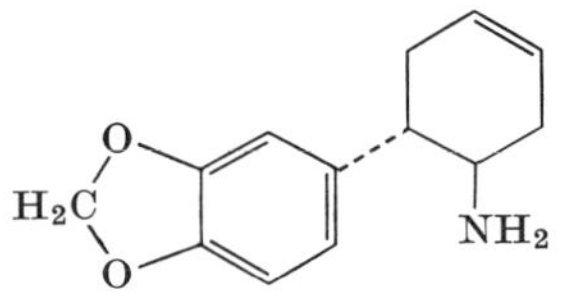 (192)

## D. The Conversion of Phenanthridones into Phenanthridines

Since several of the general synthetic methods described above give rise to phenanthridones, the conversion of the latter compounds into the corresponding phenanthridines is of some importance.

In the case of phenanthridine itself, this has been achieved by reduction with lithium aluminum hydride[211]; presumably stoichiometric quantities were employed since the hydride can reduce phenanthridine to its 5,6-dihydro derivative.[212] In a procedure which has been adopted more widely, an excess of lithium aluminum hydride is employed and the resultant 5,6-dihydrophenanthridine is dehydrogenated catalytically or,[213] better, with an excess of chloranil.[102, 200] Oxidation with aqueous potassium permanganate has been employed for the second stage,[139] but the method does not appear to be general.[22]

A convenient procedure which has been recommended for large-scale work entails conversion of the lactam to the phenanthridine-thione by heating with phosphorus pentasulfide, followed by desulfurization with Raney nickel[214] or preferably Raney cobalt.[215]

[211] P. De Mayo and W. Rigby, *Nature* **166**, 1075 (1950).
[212] W. C. Wooton and R. L. McKee, *J. Amer. Chem. Soc.* **71**, 2946 (1949).
[213] G. M. Badger, J. H. Seidler, and B. Thomson, *J. Chem. Soc.* 3207 (1951).
[214] E. C. Taylor and A. E. Martin, *J. Amer. Chem. Soc.* **74**, 6295 (1952).
[215] G. M. Badger, N. Kowanko, and W. H. F. Sasse, *J. Chem. Soc.* 440 (1957).

## III. Physical Properties

### A. Dipole Moments, Molecular Geometry, and Electronic Structure

Phenanthridine, m.p. 106°–107°, crystallizes readily from petroleum and aqueous ethanol. The long-known mercuric chloride complex[216] (B, HCl, $HgCl_2$) provides a convenient means of purification.[217] An early value for the dipole moment (1.5 D)[218] differs markedly from that obtained more recently (2.93 D).[217] The higher figure is more consistent with the moments of related systems and the original value is probably erroneous, perhaps because of insufficient experimental data.[217] The rather lower moment of 6-phenylphenanthridine (2.27 D) is appreciably less than those of 2-phenylpyridine and 2-phenylquinoline, indicating that in the phenanthridine derivative delocalization is sterically restricted in a nonplanar structure.[217]

The geometry of the phenanthridine molecule has not been determined accurately, but there seems no reason to doubt that the parent system is planar. The simplest derivative for which data are available is bis(6-phenanthridinyl)methane (**193a**⇌**193b**). The red tautomer is planar[219] and isomorphous with tetrabenz[*a,c,h,j,*]anthracene, consistent with the hydrogen-bonded structure(**193b**); the crystals are monoclinic, space group $P2_1/c$[220, 221] with $a = 9.52$, $b = 1.54$, $c = 6.66$

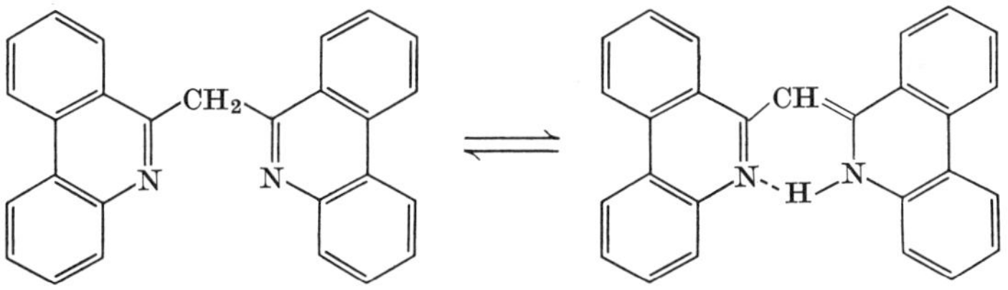

(**193a**)   (**193b**)

---

[216] A. Pictet and A. Hubert, *Chem. Ber.* **29**, 1813 (1896).
[217] C. W. Cumper, R. F. Ginman and A. I. Vogel, *J. Chem. Soc.* 4518 (1962).
[218] V. de Gauock and R. J. W. Le Fèvre, *J. Chem. Soc.* 1392 (1939).
[219] H. J. Friedrich and G. Scheibe, *Z. Electrochem.* **65**, 767 (1961).
[220] J. Van Thuijl and C. Romers, *Rec. Trav. Chim.* **87**, 5 (1968).
[221] H. Poppe, and W. Hoppe. *Z. Kristallogr., Kristallgeometrie, Kristallphys.* **122**, 298 (1965).

A°, $\beta = 100°$, and $z = 2$.[220] The information available is not sufficiently accurate for bond length and bond angle determination and an accurate crystallographic analysis of phenanthridine itself is needed.

Early calculations by Longuet-Higgins and Coulson[222] have been followed by a number of attempts to employ more refined theoretical treatments to predict bond parameters[223-225] and $\pi$-electron densities[223, 225, 226] and to calculate reactivity indices[223, 227] which accord more satisfactorily with even the limited experimental information available. The $\pi$-electron densities and bond lengths given in (**194**) are those obtained recently by Tinland using a SCF-ASMO-CI (variable $\beta$) method.[225] The alternative bond lengths (in parentheses) were calculated using a modified Pople–Pariser–Parr SCF-LCAO approach for which results generally superior to those obtained previously have been claimed.[224] In related systems for which experimental data are available, observed and calculated bond lengths are generally in satisfactory agreement and the geometry of the phenanthridine molecule is probably fairly represented by the values given in **194**. The same authors have also calculated values for the "resonance energy" (2.940 eV, 67.8 kcal/mole) and the heat of formation (119.2 ± 0.07 eV, 2747.6 ± 0.2 kcal/mole),[224] but experimental data are lacking.

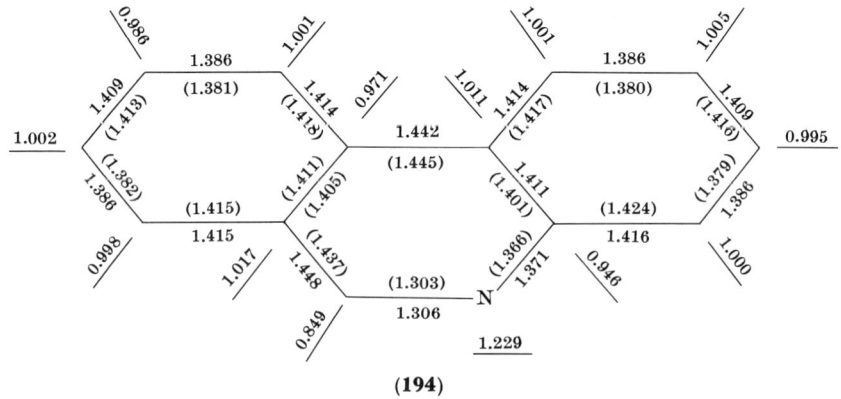

(**194**)

[222] H. C. Longuet-Higgins and C. A. Coulson, *J. Chem. Soc.* 971 (1949).
[223] R. Zahradnik and C. Parkanyi, *Collect. Czech. Chem. Commun.* **30**, 355 (1965).
[224] M. J. S. Dewar and G. J. Gleicher, *J. Chem. Phys.* **44**, 759 (1966).
[225] B. Tinland, *Tetrahedron* **25**, 583 (1969).
[226] G. Coppens and J. Nasielski, *Tetrahedron* **18**, 507 (1962).
[227] M. J. S. Dewar and P. M. Maitlis, *J. Chem. Soc.* 2521 (1957).

Calculated reactivity indices are at one with qualitative electronic theory in accounting for ready nucleophilic attack at C-6 in the phenanthridine molecule. However, the limited data available on positional reactivities in electrophilic substitutions is not accounted for satisfactorily by any of the available treatments (see later) and it has been pointed out that the simple Hückel treatment used by some authors,[223, 226] is generally inapplicable in heteroaromatic systems.[228]

## B. Spectra

Strong bands arising from $\pi, \pi^*$ transitions (the $\alpha$, $\beta$, and $p$ bands of Clar's classification)[229] appear in the absorption spectrum of phenanthridine at 346, 250, and 295 nm (in cyclohexane) and the corresponding extinction coefficients (1950, 45000, and 5800) fall in the expected order.[230] A weak $\pi,\pi^*$ triplet absorption reported at 450.5 nm has been studied (making use of the perturbing effect of oxygen under pressure to increase the band intensity).[231] No $n,\pi^*$ absorption is apparent in the phenanthridine spectrum (even in cyclohexane solution)[232, 233] and the overall spectral shifts induced by changing from a nonpolar to a polar solvent are therefore small.[232] The suggestion that a weaker singlet $n,\pi^*$ band probably lies under the strong $\alpha$ band has received experimental support. Thus, the addition of small amounts of proton donors to hexane solutions of azaaromatics generally has a hyperchromic effect on bands arising from $\pi,\pi^*$ transitions as well as causing the disappearance of $n,\pi^*$ absorption, and from a quantitative examination of the intensity changes produced by adding various donors Coppens has concluded that the $n,\pi^*$ band of phenanthridine appears at 327 nm. This value is in good agreement with an LCAO prediction.[234] The results of similar studies with a series of carboxylic acids in hexane solution have been interpreted in favor of a tautomeric equilibrium between an associated donor base pair and the fully protonated species (**195b**).

[228] M. J. S. Dewar, *Rev. Mod. Phys.* **35**, 586 (1963).
[229] E. Clar, "Aromatische Kohlenwasserstoffe," 2nd ed., Springer, Berlin, 1952.
[230] S. F. Mason, *in* "Physical Methods in Heterocyclic Chemistry" (A. R. Katritzky, ed.), Vol. 1, p. 1. Academic Press, 1963.
[231] D. F. Evans, *J. Chem. Soc.* 2753 (1959).
[232] G. M. Badger and I. S. Walker, *J. Chem. Soc.* 122 (1956).
[233] W. Slough and A. R. Ubbelohde, *J. Chem. Soc.* 911 (1957).
[234] G. Coppens, C. Gillet, J. Nasielski, and E. Vander Donckt, *Spectrochim. Acta* **18**, 1441 (1962).

(195a) ⇌ (195b)

The equilibrium composition was calculated for phenanthridine and a number of other azaaromatics, but no simple relationship existed between these figures and the corresponding $pK_a$ values and no quantitative significance could be attached to the presence or absence of *peri*-hydrogen atoms,[235] which have a controlling influence in other reactions of phenanthridine (*vide infra*). (Infrared studies on hydrogen bonding are considered below.)

The position of the $^1L_a$ transition in the phenanthridine spectrum is in good agreement with the LCAO transition energy (as in other monoazaaromatics). The agreement between calculated and observed intensities is less satisfactory.[236]

Phenanthridine, like other monoazaaromatics which fluoresce only weakly in nonpolar solvents, is subject to marked fluorescence activation by hydroxylic solvents. Recent studies have shown this to result from the effects of solvent on vibronic interactions between $n,\pi^*$ and $\pi,\pi^*$ electronic states, and the effect of solvent changes on the phosphorescence half-life has been similarly explained.[237] Measurements of fluorescence and, more especially, phosphorescence characteristics have been proposed as analytical methods for mixtures containing phenanthridines[238, 239] and detailed studies of the emission spectra of phenanthridine,[240, 241] its cation,[241] 9-methylphenanthridine,[242] and phenanthridine-$N$-oxide[243] have been reported. A

[235] J. Nasielski and E. Vander Donckt, *Spectrochim. Acta* **19**, 1989 (1963).
[236] J. Nasielski and E. Vander Donckt, *Bull. Soc. Chim. Belg.* **72**, 725 (1963).
[237] E. C. Lim and J. M. H. Yu, *J. Chem. Phys.* **47**, 3270 (1967); **45**, 4742 (1966).
[238] B. L. Van Duuren, *Anal. Chem.* **32**, 1436 (1960).
[239] H. V. Drushel and A. L. Sommers, *Anal. Chem.* **38**, 10 (1966).
[240] H. H. Perkampus and K. Kortüm, *Z. Phys. Chem.* **56**, 73 (1967).
[241] H. Gropper and F. Doerr, *Ber. Bunsenges. Phys. Chem.* **67**, 46 (1963).
[242] A. P. Kilimov and L. N. Zvegintseva, *Izv. Sibirsk. Otd. Akad. Nauk. SSSR, Ser. Khim. Nauk* 14 (1964).
[243] T. Kubota and H. Miyazaki, *Chem. Pharm. Bull.* **9**, 948 (1961).

systematic examination of the effects of substituents on ultraviolet absorption has yet to be made, but the spectra of a limited number of alkyl-,[22, 244, 245] substituted alkyl-,[245] alkoxy-,[246, 248] amino-,[245-247] carboxy-,[245] halogeno-,[26, 140, 245] and hydroxyphenanthridines[245, 248, 249] (together with some $N$-oxides[243] and quaternary salts[246]), have been reported. The predominance of the amino (**196**) and lactam (**197**) forms in 6-amino- and 6-hydroxyphenanthridine, respectively, was established by the usual methods (see below) including spectrophotometric comparison with model compounds.[246]

(**196**)  (**197**)

Charge-transfer (C-T) bands have been located in the spectra of solutions of phenanthridine in 1,2-dimethoxyethane containing bromine; solutions containing up to a 2:1 mole ratio of halogen to base were examined, but the structure of the species involved is not clear.[233] Phenanthridine satisfies the conditions necessary for both $n$ and $\pi$ donation[250] and $n$ donation is apparently involved in the charge-transfer interaction with iodine. The equilibrium constant for this reaction has been determined spectrophotometrically, but the claimed correlation (for a series of $N$-heteroaromatic bases) between $pK_a$ values (in 50% ethanol) and these C–T equilibrium constants[251] appears to be an unsatisfactory one and in any case lacks theoretical justification, since it is doubtful whether dissociation constants provide, in general, an accurate measure of $n$-ionization potentials. In particular, the excellent correlation in the case of phenanthridine is probably fortuitous, since the authors report that $n$-halogen interactions are markedly sensitive to steric factors[251] which are almost

---

[244] B. R. T. Keene and P. Tissington, *J. Chem. Soc.* 4426 (1965).
[245] R. M. Acheson and A. O. Plunkett, *J. Chem. Soc.* 3758 (1962).
[246] C. B. Reese, *J. Chem. Soc.* 895 (1958).
[247] A. R. Osborn, K. Schofield, and L. N. Short, *J. Chem. Soc.* 4191 (1956).
[248] A. Albert and J. N. Phillips, *J. Chem. Soc.* 1294 (1956).
[249] S. F. Mason, *J. Chem. Soc.* 5010 (1957).
[250] J. J. Eisch, *Advan. Heterocycl. Chem.* **7**, 1 (1966).
[251] J. N. Chaudhuri and S. Basu, *Trans. Faraday Soc.* **55**, 898 (1959).

certainly operative where *peri*-hydrogen atoms are present. The "spectroscopic difference method" has been used to determine the position of the C-T band in solutions of $N$-methylphenanthridinium iodide in chloroform (425 nm) and acetonitrile (393 nm). The addition of polycyclic aromatic hydrocarbons (e.g., anthracene, phenanthrene, and pyrene) to chloroform solutions of $N$-methylphenanthridinium bromide results in each case in the appearance of an additional, sharp C–T band.[252] The charge-transfer behavior of a series of polynuclear azaaromatics (including phenanthridine) with benzoquinone[253] and tetrachlorophthalic anhydride[254] closely parallels that of the analogous hydrocarbons, suggesting that the heteroaromatic compounds function here as $\pi$ donors rather than $n$ donors. The energies associated with the C–T bands observed in the complexes of, *inter alia*, phenanthridine with chloranil and tetracyanoethylene have been correlated with the LCAO-MO energies of the highest occupied orbitals of the donors.[255]

The infrared spectrum of phenanthridine in carbon disulfide shows characteristic bands (C–H out-of-plane bending) at 746 (4H, strong) and 888 cm$^{-1}$ (1H, medium), with other medium intensity bands at 716 and 765 cm$^{-1}$.[256] The thin film (solid phase) spectrum shows essentially similar features, and the spectrum of the hydrochloride under these conditions closely resembles that of phenanthrene.[257] A C–H (stretching) band appears at 3086 cm$^{-1}$ (CCl$_4$).[256] A number of bands in the 1650–1220 cm$^{-1}$ region have been assigned tentatively to C–H (in-plane) skeletal vibrations in both phenanthridine and the bromophenanthridines.[26] The equilibrium constant for the phenanthridine–methanol interaction in carbon tetrachloride has been obtained by infrared spectroscopy; the value (4.7 liter mole$^{-1}$) is higher than that for the corresponding complexes of quinoline (3.8 liter mole$^{-1}$) and pyridine (3.1 liter mole$^{-1}$).[258]

The infrared spectra of a limited number of derivatives have been reported. The spectrum of 6-aminophenanthridine (**196**) shows

[252] J. Nasielski and E. Vander Donckt, *Theoret. Chim. Acta* **2**, 22 (1964).
[253] M. Chowdhury, *Trans. Faraday Soc.* **57**, 1482 (1961).
[254] M. Chowdhury and S. Basu, *Trans. Faraday Soc.* **56**, 335 (1930).
[255] M. Nepras and R. Zahradnik, *Collect. Czech. Chem. Commun.* **29**, 1555 (1964).
[256] G. Coppens and J. Nasielski, *Bull. Soc. Chim. Belg.* **70**, 136 (1961).
[257] H. H. Perkampus and E. Baumgarten, *Z. Electrochem.* **64**, 951 (1960).
[258] G. Coppens, J. Nasielski, and N. Sprecher, *Bull. Soc. Chim. Belg.* **72**, 626 (1963).

(198)        (199)

characteristic N–H (sym) and N–H (antisym) stretching bands at 3409 and 3513 cm$^{-1}$ (carbon tetrachloride) and hence the compound exists predominantly in the aminoform.[259] Similar bands (at 3405 and 3511 cm$^{-1}$) appear in the spectrum of 1,10-dimethyl-6-aminophenanthridine (198) (chloroform) so that the relief of strain which would be involved in forming the imino tautomer (199) clearly fails to outweigh the overriding tendency for α-amino-$N$-heteroaromatic compounds to exist in the amino form.[244] In sharp (but predictable) contrast the spectrum of phenanthridone in carbon tetrachloride shows bands at 3402 (N–H stretching) and 1669 cm$^{-1}$ (C=O stretching) and the lactam form therefore predominates. Under the same conditions the spectra of 3-, 8-, and 9-hydroxyphenanthridine show normal O–H stretching absorptions near 3600 cm$^{-1}$.[260] Brief details have been given of the spectra of a number of phenanthridones.[127]

The infrared spectra of some perhydrophenanthridines have been reported in connection with stereochemical studies.[261]

The pmr spectrum of phenanthridine has been recorded in carbon tetrachloride,[22, 262] deuteriochloroform,[263] and hexadeuteriobenzene.[263] In common with other $N$-heteroaromatic systems the proton at C-6 (adjacent to the ring nitrogen) appears at lowest field in all three solvents (0.88 τ in carbon tetrachloride). In carbon tetrachloride the C-1 and C-10 protons (which are deshielded by the adjacent rings) give rise to a doublet (ortho coupling to C(2)–H and C(9)–H, respectively) at slightly higher field (1.59 τ)[22]; the spectrum in deuteriochloroform is similar.[263] An interesting difference appears in the spectrum in hexadeuteriobenzene, however; the C-4 proton (peri to

[259] S. F. Mason, *J. Chem. Soc.* 1281 (1959).
[260] S. F. Mason, *J. Chem. Soc.* 4874 (1957).
[261] T. Masamune, *Chem. Commun.* 244 (1968).
[262] E. Vander Donckt, R. H. Martin, and F. Geerts-Evrard, *Tetrahedron* **20**, 1495 (1964).
[263] R. H. Martin, N. Defay, F. Geerts-Evrard, and D. Bogaert-Verhoogen, *Tetrahedron, Suppl.* **8**, Part I, 181 (1966).

the ring nitrogen) resonates downfield of the C(1)–H and C(10)–H signals, although the remaining protons of the carbocyclic rings give rise to the usual unresolved multiplet at higher field.[263] Methyl groups at C-1 and C-10 are deshielded by the adjacent rings and give rise to signals near 7.0 $\tau$,[22, 121] but in 1,10-dimethyl derivatives enforced deviation from planarity shifts the methyl proton signals to higher field, e.g., to 7.75 and 7.77 $\tau$ in 1,10-dimethyl-5,6-dihydrophenanthridine.[102] An attempt has been made to establish a correlation between the chemical shifts of methyl protons in $\alpha$-methyl-$N$-heteroaromatics (including 6-methylphenanthridine) and rates of exchange with EtOD; a general parallelism is observed.[264] Pmr studies of suitable 6-aryl- and 6-aralkylphenanthridinium vinyl oxides have revealed the first instance of stable rotamers arising from hindrance to rotation of cis substituents on a double bond. The two forms of **200** are separable chromatographically, and have identical ultraviolet and infrared absorptions, but different pmr and mass spectra.[265] The probable

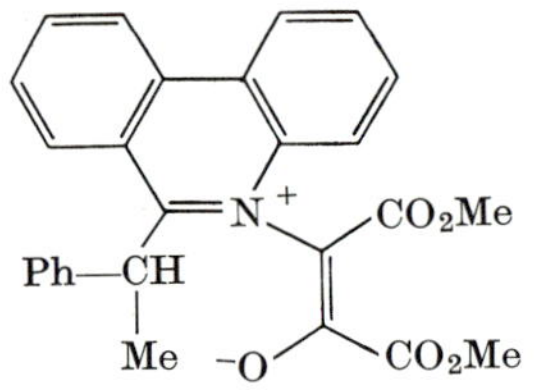

(200)

conformations of some octahydrophenanthridines have been deduced from pmr data.[266]

## C. Ionization Constants

Phenanthridine is a fairly weak base (p$K_a$, in water, 4.52 ± 0.01 at 20°).[247] The effect of substituents has not been examined systematically (and predictions based on a Hammett treatment appear somewhat unreliable in the phenanthridine series).[267] The available data are in line with qualitative predictions based on the electronic

---

[264] N. N. Zatsepina, I. F. Tupitsyn, and L. S. Efros, *Zh. Obshch. Khim.* **34**, 4072 (1964).
[265] R. M. Acheson and I. A. Selby, *Chem. Commun.* 62 (1970).
[266] T. Masamune, S. Ohuchi, S. Shimokawa, and H. Booth, *Tetrahedron* **22**, 773 (1966).
[267] D. D. Perrin, *J. Chem. Soc.* 5590 (1965).

character of the groups concerned. Thus, alkyl,[245] hydroxyl,[248] and, more particularly, amino[247, 268] substituents in the carbocyclic rings are generally base-strengthening and bromine atoms exert a base-weakening effect.[26] Annelation appears to be virtually without effect.[244] Amino substitution at C-6 produces the expected marked increase in $pK_a$ (to 7.31 in the case of 6-aminophenanthridine itself)[269] and comparable increases have been observed in alkyl homologs; the corresponding increments are smaller in the benzophenanthridines.[244]

Spectrophotometric measurements have confirmed that 6-aminophenanthridine is protonated at the ring nitrogen atom, and the $pK_a$ values (in 70% ethanol) of 6-amino-(6.30), 6-methylamino-(6.70), 6-dimethylamino-(4.91), and 5,6-dihydro-6-imino-5-methylphenanthridine (9.10) were used to establish the predominance of the amino tautomer, the relatively low value for 6-dimethylaminophenanthridine being ascribed to steric inhibition of delocalization of the exocyclic lone pair.[246] The $pK_a$ (in water) of phenanthridone is below $-1.5$, whereas that of 6-methoxyphenanthridine is 2.38, confirming the lactam structure of the former.[248] Enhanced basic strengths in excited states have been reported for phenanthridine and 6-methylphenanthridine.[270] The $pK_a$ values of the parent base in aqueous solution in the first singlet and first triplet states have been estimated to be 9.6 and 5.7, respectively.[271]

### D. Polarographic Behavior

Phenanthridine was reported initially to exhibit a single (probably two-electron)[272] reduction wave in dimethylformamide containing tetraalkylammonium salts as supporting electrolytes.[272–274] Polarographic half-wave potentials for the reduction of $N$-heteroaromatics in dimethylformamide (DMF) are controlled by electron transfer

[268] K. A. Allen, J. Cymerman-Craig, and A. A. Diamantis, *J. Chem. Soc.* 234 (1954).
[269] A. Albert, R. Goldacre, and J. Phillips, *J. Chem. Soc.* 2240 (1948).
[270] M. Das and S. Basu, *Proc. Nat. Inst. Sci. India, Sect. A* **30**, 533 (1964).
[271] E. Vander Donckt, R. Dramaix, J. Nasielski, and C. Vogels, *Trans. Faraday Soc.* **65**, 3258 (1969).
[272] C. Parkanyi and R. Zahradnik, *Bull. Soc. Chim. Belg.* **73**, 57 (1964).
[273] G. Anthoine, G. Coppens, J. Nasielski, and E. Vander Donckt, *Bull. Soc. Chim. Belg.* **73**, 65 (1964).
[274] B. R. T. Keene and P. Tissington, *Rec. Trav. Chim.* **84**, 488 (1965).

rather than proton addition,[275] and the experimental value for phenanthridine ($E_{1/2} = -1.64$ V, in 95% DMF) fits precisely the regression line $E_{1/2}$ (V) $= 2.127k_{-1} - 0.555$ for a series of benzo- and dibenzopyridines, where $k_{-1}$ is the LCAO energy of the corresponding lowest unoccupied $\pi$ orbital.[272] Later investigators (using a silver reference electrode in an otherwise similar system) have described two waves, tentatively assigned to one-electron reductions; the half-wave potential of the first (less negative) wave correlates satisfactorily with the HMO energy of the lowest unoccupied orbital.[276] In aqueous solutions the first half-wave potential of phenanthridine shows the expected dependence on pH, but the effect of methyl groups at C-1 and C-10 is unusual in that $E_{1/2}$ becomes less negative, presumably because of steric strain. Methyl substitution at other positions produces the normal shift[276] to higher (negative) values.[274]

The first reduction wave for phenanthridine $N$-oxide in dimethylformamide (at 25°) appears at $-1.774$ (vs. s.c.e.) in good agreement with the HMO energy of the lowest vacant orbital[277]; a mechanism for the reduction has been proposed.[278] No esr signal could be detected during controlled-potential electrolysis at the appropriate potential, perhaps because of the relative instability of the anion radical. This behavior, which is paralleled by isoquinoline, contrasts sharply with that of the oxides of other polynuclear $N$-heteroaromatic systems (e.g., quinoline and acridine).[277]

## IV. Reactions of the Phenanthridine Nucleus

### A. Reactions at Nitrogen

A large number of quaternary phenanthridinium salts have been prepared in the search for biological activity[279] and many further examples have been reported since the last review, some obtained by

---

[275] P. H. Given, *J. Chem. Soc.* 2684 (1958).
[276] B. J. Tabner and Y. R. Yandle, *J. Chem. Soc. A* 381 (1968).
[277] T. Kubota, K. Nishikida, H. Miyazaki, K. Iwatani, and Y. Oishi, *J. Amer. Chem. Soc.* **90**, 5080 (1968).
[278] G. Anthoine, J. Nasielski, E. Vander Donckt, and N. Vanlautem, *Bull. Soc. Chim. Belg.* **76**, 230 (1967).
[279] This aspect of the chemistry of phenanthridinium salts has been well reviewed; see, e.g., B. A. Newton, *Advan. Chemother.* **1**, 35 (1964).

direct quaternization.[280] Kinetic measurements with azaphenanthrenes and azaanthracenes (Table I) have demonstrated the rate-controlling influence of *peri*-hydrogen atoms. Phenanthridine (9-azaphenanthrene) reacts with methyl iodide less than one-tenth as rapidly as isoquinoline but more than two hundred times faster than acridine (9-azaanthracene).[281]

The introduction of a methyl group at C-6 of phenanthridine is relatively more rate depressing than a similar substitution at C-2 in

TABLE I

SECOND-ORDER RATE CONSTANTS AND ARRHENIUS PARAMETERS FOR QUATERNIZATION WITH MeI IN NITROBENZENE[a]

| Compound | $10^4 k_2$ (25°) (liter mole$^{-1}$ sec$^{-1}$) | $E_A$ (kcal mole$^{-1}$) | $\log_{10} PZ$ |
|---|---|---|---|
| Pyridine | 3.6 | 13.9 | 6.72 |
| Isoquinoline | 5.0 | 13.5 | 6.60 |
| 1-Azaphenanthrene | 0.42 | 15.1 | 6.61 |
| 9-Azaphenanthrene | 0.41 | 14.7 | 6.41 |
| 9-Azaanthracene | 0.0017 | 16.8 | 5.65 |

[a] Adapted from Coppens and Nasielski.[281]

quinoline,[282] perhaps because of the buttressing effect of the *peri*-hydrogen atom at C-7 in the former.[102] Alkyl substitutions in the carbocyclic rings at positions other than C-1 and C-10 produce rate increases of the usual magnitude (Table II). The absence of similar activation in both 1- and 10-methylphenanthridine may reflect increased steric intervention by the *peri*-hydrogen owing to in-plane distortion of the nucleus, an effect which is partly alleviated in the nonplanar 1,10-dimethyl homolog.[102] Quaternizations with dimethyl

[280] See, e.g., T. I. Watkins (to Boots Pure Drug Co. Ltd.), *British Patent* 714,070, *Chem. Abstr.* **50**, 1982b; D. A. Peak, T. I. Watkins, J. S. Nicholson, J. L. Lowe, and C. I. Brodrick (to Boots Pure Drug Co. Ltd.), *British Patent* 818,257, *Chem. Abstr.* **54**, 8861i; F. F. Stephens (to Crookes Labs. Ltd.), *British Patent* 850,930, *Chem. Abstr.* **55**, 8437b; also T. I. Watkins, *J. Chem. Soc.* 1443 (1958) and B. L. Hollingsworth and V. Petrov, *J. Chem. Soc.* 3664 (1961).

[281] G. Coppens and J. Nasielski, *Bull. Soc. Chim. Belg.* **71**, 5 (1962); **68**, 187 (1959).

[282] J. Packer, J. Vaughan, and E. Wong, *J. Amer. Chem. Soc.* **80**, 905 (1958).

sulfate, normally in xylene or nitrobenzene, can also be carried out in refluxing nitromethane; anion exchange with the resulting methosulfate provides, for example, methobromides and methiodides without the inconvenience of sealed-tube reactions.[280, 283]

TABLE II

SECOND-ORDER RATE CONSTANTS FOR THE QUATERNIZATION OF PHENANTHRIDINE AND ITS HOMOLOGS WITH MeI IN $PhNO_2$[a]

| Substituent | $10^4 k_2$ (60°) | Substituent | $10^4 k_2$ (60°) |
|---|---|---|---|
| None | 5.2 | 6-Methyl | 0.2 |
| 1-Methyl | 5.0 | 3-Methyl | 7.4 |
| 10-Methyl | 4.9 | 8-Methyl | 6.8 |
| 3,8-Dimethyl | 10.1 | 1,10-Dimethyl | 7.2 |

[a] For all compounds except 6-methylphenanthridine, $E_A = 14.4$ ($\pm 0.3$) kcal mole$^{-1}$ (i.e., identical, within experimental error). In the case of 6-methylphenanthridine, $E_A = 16.3$ kcal mole$^{-1}$.

Despite the steric effect illustrated in Table II, the quaternization of 6-substituted phenanthridines offers preparative difficulty only where the substituent is very bulky (for example, naphthyl, 2,3-dimethoxyphenyl or 3,4,5-trimethoxyphenyl).[17] 6-Methylphenanthridine reacts almost quantitatively with dimethyl sulfate in refluxing xylene[284] or nitrobenzene[285] and the corresponding ethiodide has been prepared *en route* to cyanine dyes.[286] 6-Phenyl- and *p*-substituted phenylphenanthridines likewise form methosulfates without difficulty.[280] 6-Aminophenanthridinium methiodide is formed quantitatively by the action of methyl iodide on 6-aminophenanthridine at 125°, and the corresponding mono- and dimethylamino derivatives are likewise attacked at the ring nitrogen.[246]

Quaternizations with allyl halides have been described,[280] and high yields of salts of the type (**201**) are obtained by allowing phenanthridine to react with, for example, bromobutanone or chloracetaldoxime at room temperature in tetramethylene sulfone.[287]

[283] R. M. Acheson and G. J. F. Bond, *J. Chem. Soc.* 246 (1956).
[284] S. Sugasawa and H. Matsuo, *Chem. Pharm. Bull.* **6**, 601 (1958).
[285] R. M. Acheson, A. S. Bailey, and I. A. Selby, *J. Chem. Soc. C* 2066 (1967).
[286] I. I. Levkoev and V. V. Durmashkina, *Tr. Vses. Nauch.-Issled. Kinofotoinst.* 21 (1960); *Chem. Abstr.* **58**, 14154f (1963).
[287] C. K. Bradsher and R. W. L. Kimber, *J. Heterocycl. Chem.* **1**, 208 (1964).

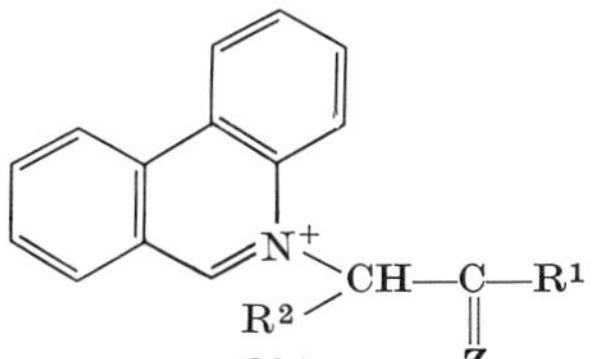

Sec. IV. A.]   PHENANTHRIDINES   381

(201)

R¹ and R² = H or Me; Z = O or NOH

7-Hydroxy-6-methyl-7,8,9,10-tetrahydrophenanthridine methiodide is formed directly by the action of methyl iodide on the corresponding base.[168] In an interesting reaction with acrylamide in methanolic hydrogen chloride, phenanthridine gives $N$-(2-carbamoylethyl)phenanthridinium chloride.[288]

An unsuccessful attempt has been made to prepare a borazaro compound by the action of boron trichloride on 6-benzylphenanthridine.[289]

$N$-Aminophenanthridinium salts can be obtained by the action of hydroxylamine-$O$-sulfonic acid on phenanthridine, followed, where required, by anion exchange. The salts readily give rise to the 1,3-dipolar system (202) which exists in equilibrium with its dimer at temperatures below 100°.[290]

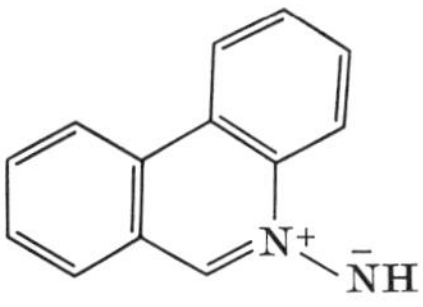

(202)

Phenanthridones and benzophenanthridones undergo ready alkylation on treatment with dialkyl sulfates in alkali.[113, 126] As in the case of other, similar, ambident nucleophiles, $N$-alkylation normally predominates, although with diethyl sulfate 2-nitrophenanthridone gives a significant amount (27%) of 6-ethoxy-2-nitrophenanthridine,[126] and, more recently, $O$-methylation has been observed in the reactions between both the methyl ester and the pyrrolidine amide of phenanthridone-4-carboxylic acid with methyl iodide in the presence

[288] R. Dowbenko, *J. Org. Chem.* **25**, 1123 (1960).
[289] M. J. S. Dewar, C. Kaneko, and M. K. Bhattacharjee, *J. Amer. Chem. Soc.* **84**, 4884 (1962).
[290] R. Huisgen, R. Grashey, and R. Krischke, *Tetrahedron Lett.* 387 (1962).

of sodium amide. In the latter case, none of the *N*-methylphenanthridone was isolated.[106] Good yields of *N*-methylphenanthridones are obtained by treating the preformed phenanthridone potassium salts with methyl iodide,[291] and the sodium salt of phenanthridone (from the lactam and sodamide) has been alkylated successfully with, for example, 2-dimethylaminoethyl chloride[292] and methyl bromoacetate.[245] The difficulty encountered in methylating 2,4-dimethylphenanthridone[99] is presumably steric in origin. *N*-Methylphenanthridone itself has been obtained in quantitative yield by the action of methyl iodide on a solution of phenanthridone containing an equivalent of thallous ethoxide in dimethylformamide.[293]

Prepared in tetrahydrofuran by the action of sodium hydride, the sodium salt (203) formed the *O*-benzoate (204) when treated with benzoyl chloride at −20°. A similar reaction at ambient temperature, followed by refluxing in toluene, gave (105). Clearly *O*-acylation is

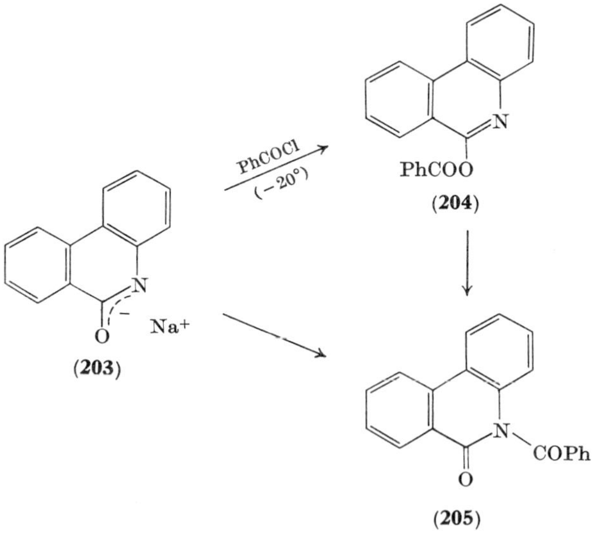

kinetically favored at lower temperatures, although (205) is thermodynamically the more stable isomer; after heating pure (204) in

[291] H. Gilman and J. Eisch, *J. Amer. Chem. Soc.* **79**, 5479 (1957).
[292] F. Hunzicker, H. Lauerer, and J. Schmutz, *Arzneim.-Forsch.* **13**, 324 (1963).
[293] E. C. Taylor, reported in "Reagents for Organic Synthesis" (by M. and L. F. Fieser), Vol. 2, p. 410. Wiley (Interscience) 1969.

methylene dichloride for some time, only (**205**) could be detected.[294] O→N thermal migrations of aryl substituents (under very vigorous conditions) have also been observed with 6-aryloxyphenanthridines.[129] The formation of 6-butylphenanthridine by the action of two equivalents of butyllithium on phenanthridone has been attributed to the preparation *in situ* of the *N*-lithium derivative shown[291] (Scheme 3).

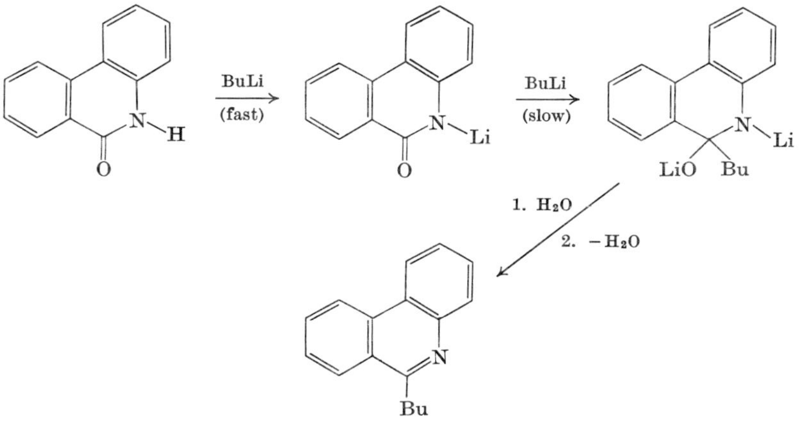

SCHEME 3

Phenanthridine *N*-oxides can be prepared in the usual way by direct oxidation with peracetic, perbenzoic, or permonophthalic acids.[77, 295-298] In aqueous media phenanthridones may be formed as by-products.[299] Hydrogen peroxide in acetic acid has been recommended, nevertheless, for the *N*-oxidation of 6-arylphenanthridines, although prolonged reaction times are needed for compounds bearing deactivating (e.g., nitro) substituents.[295] An attempt to prepare 3,8-dinitro-6-phenylphenanthridine-5-oxide by direct oxidation was unsuccessful, and other failures have also been reported[295]; however, 6-cyano- and 6-carboxamidophenanthridine both form *N*-oxides on heating with hydrogen peroxide–acetic acid at 100°[297] or peracetic

[294] D. Y. Curtin and J. H. Engelmann, *Tetrahedron Lett.* 3911 (1968).
[295] P. Mamalis and V. Petrow, *J. Chem. Soc.* 703 (1950).
[296] E. Hayashi and H. Ohki, *J. Pharm. Soc. Japan* **81**, 1033 (1961).
[297] E. Hayashi and Y. Hoshi, *J. Pharm. Soc. Japan* **87**, 651 (1967).
[298] T. M. Mishina and L. S. Efros, *Zh. Obshch. Khim.* **32**, 2217 (1962).
[299] K. Mitsuhashi, *J. Pharm. Soc. Japan.* **67**, 74 (1947).

acid in chloroform.[80] (Rather surprisingly, phenanthridine-6-carboxylic acid gives only phenanthridone and 6-hydroxyphenanthridine-5-oxide under the former conditions.[297]) A kinetic examination of the N-oxidation of a number of N-heteroaromatic bases with perbenzoic acid in aqueous dioxane has confirmed that this reaction is far less sensitive to steric factors than quaternization (Table III). Phenanthridine appears to react only slightly less rapidly than quinoline and isoquinoline, all three being markedly less reactive than acridine.[300]

TABLE III

RATES AND ACTIVATION ENERGIES FOR THE N-OXIDATION OF N-HETEROAROMATIC BASES WITH PERBENZOIC ACID

| Compound | $10^3 k_2$ (25°) (liter mole$^{-1}$ sec$^{-1}$) | $E_A$ (kcal mole$^{-1}$) |
|---|---|---|
| Pyridine | 5.3 | 14.6 |
| Quinoline | 5.9 | 14.4 |
| Isoquinoline | 5.8 | 14.5 |
| Phenanthridine | 3.3 | 14.7 |
| Acridine | 317 | 8.0 |

The major adducts formed when phenanthridine is allowed to react with dimethyl acetylenedicarboxylate[301] have been reformulated.[245, 283] An initial Michael-type reaction gives the zwitterion (**206**) and succeeding reactions depend on the nature of the solvent. Adduct (**207**) is formed in anhydrous methanol by the addition of a proton and methoxide ion, while in benzene nucleophilic attack on the carbonyl group of a second ester molecule and subsequent cyclization provides **208**. Alternatively, reaction at the triple bond of a second molecule of ester followed by ring closure of the new zwitterion gives **209**. Other products related to **207** arise if the methanol contains water.[283] 6-Methylphenanthridine with dimethyl acetylenedicarboxylate in benzene gives a mixture of the azepine (**210**) and tetramethyl 9a-methyl-9aH-dibenzo[a,c]quinolizine-6,7,8,9-tetracarboxylate. The

[300] J. Foucart, J. Nasielski, and E. Vander Donckt, *Bull. Soc. Chim. Belg.* **75**, 17 (1966).
[301] O. Diels and W. E. Thiele, *J. Prakt. Chem.* **156**, 195 (1940).

formation of **210** involves an ester group "migration" for which a plausible mechanism has been suggested. Similar additions with 6-ethylphenanthridine and diethyl acetylenedicarboxylate have been examined.[302]

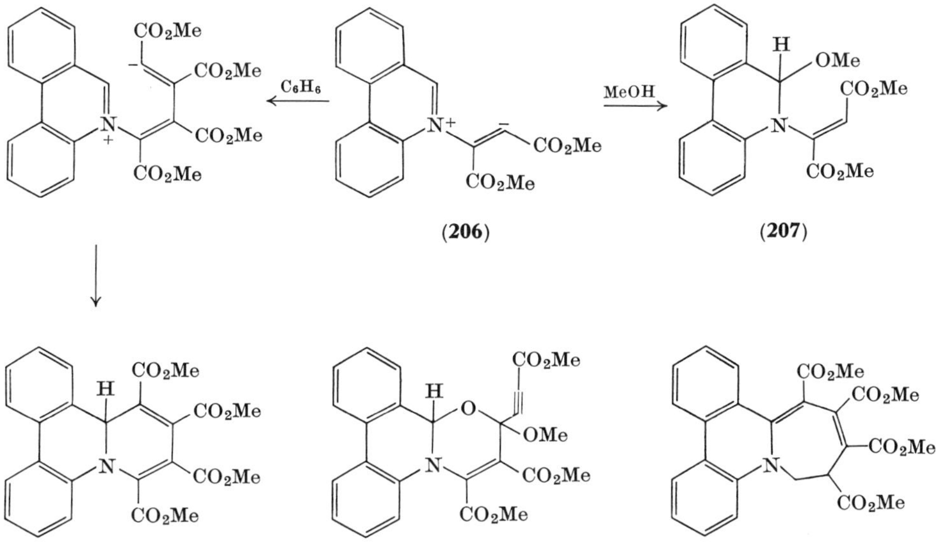

(206)    (207)

(209)    (208)    (210)

6-Methylphenanthridine and methyl propiolate give only the adduct (**211**) in rather poor yield. The corresponding reaction with phenanthridine itself is more complex, giving **212a** and **b** as well as smaller amounts of materials as yet unidentified.[303]

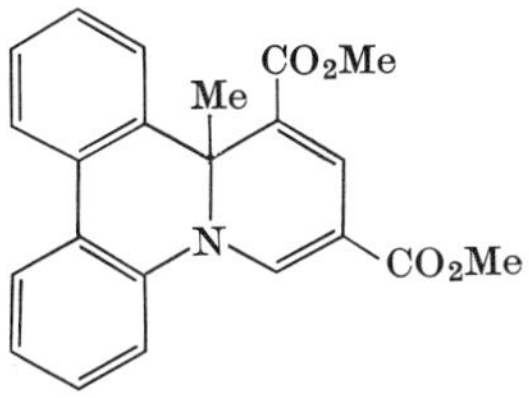

(211) 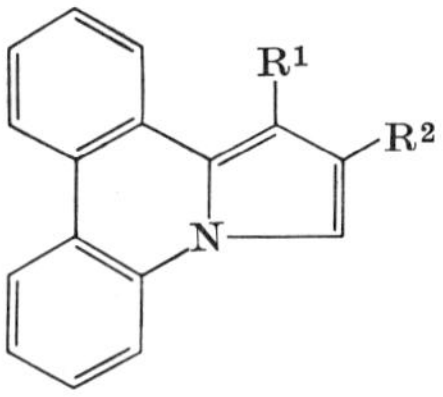   (212a) $R^1 = CO_2Me$, $R^2 = H$
(212b) $R^1 = CH_2CO_2Me$, $R^2 = CO_2Me$

[302] R. M. Acheson, J. M. F. Gagan, and D. R. Harrison, *J. Chem. Soc. C* 362 (1968).
[303] R. M. Acheson and M. S. Verlander, *J. Chem. Soc. C* 2311 (1969).

The structure **213** ($R = R^1 = H$, $R^2 = COCH_2COMe$) has been proposed for the adduct formed from phenanthridine and diketene,[304] but the corresponding product with dimethylketene has been shown to be **214** rather than the ketoamide **213** ($R = R^1 = R^2 = Me$).[305]

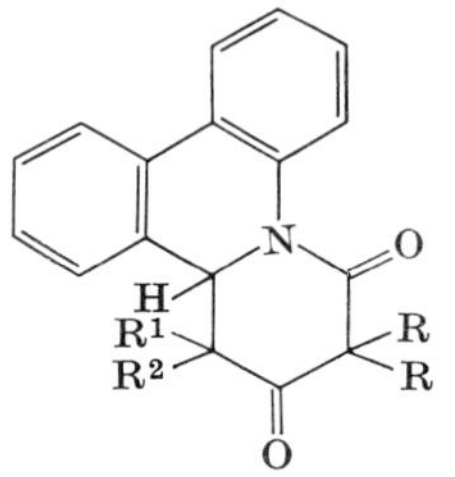

(213) 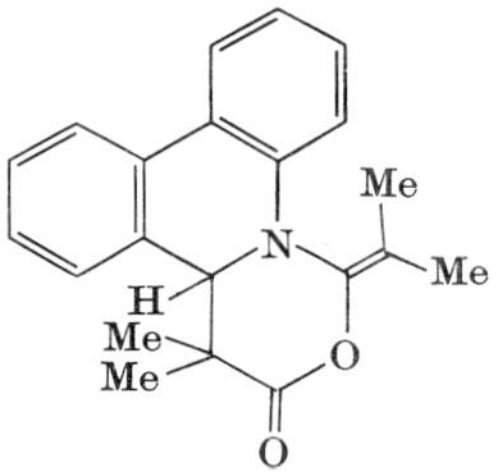   (214)

Several acetylenic esters have been shown to react with phenanthridine-5-oxide (or its 6-alkyl derivatives) forming adducts of the type (**215**).[306] Under necessarily more vigorous conditions (in dimethylformamide at 100°) the less reactive methyl phenylpropiolate combined with 6-methylphenanthridine-5-oxide to form 2-phenylpyrrolo[1,2-*f*]phenanthridine-3-carboxylate (**216**) directly, presumably via an intermediate of type (**215**).[285]

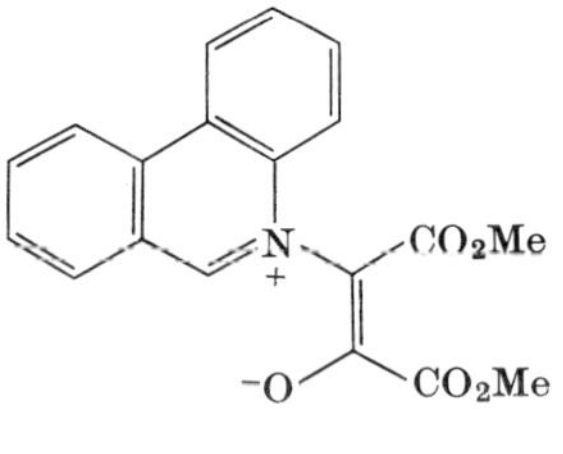

(215)   R = H or alkyl 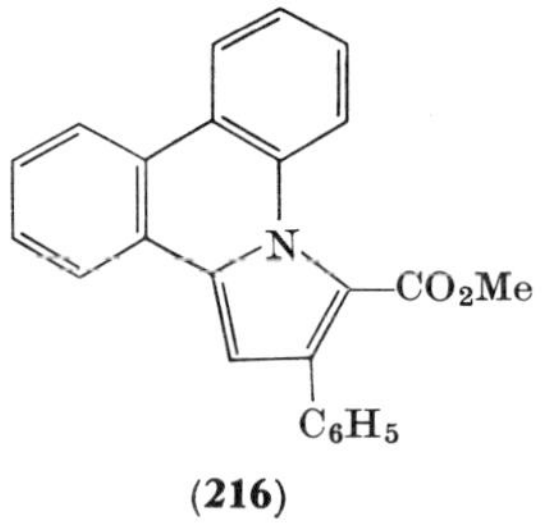      (216) 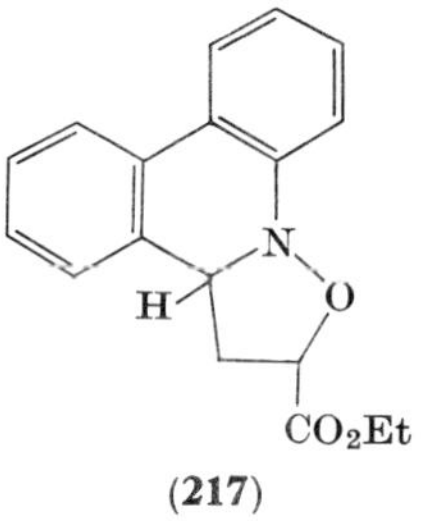      (217)

The reaction between phenanthridine-5-oxide and ethyl acrylate takes an entirely different course leading to the isoxazolidine (**217**), and then to ethyl 1-hydroxy-2-(6-phenanthridinyl)propionate.[307]

The formation of the adduct (**218**) from phenanthridine-5-oxide and dimethylketene in benzene is accompanied by deoxygenation

[304] T. Kato and Y. Yamamoto, *Chem. Pharm. Bull.* **14**, 752 (1966).
[305] R. N. Pratt, G. A. Taylor, and S. A. Proctor, *J. Chem. Soc. C* 1569 (1967).
[306] R. M. Acheson, A. S. Bailey, and I. A. Selby, *Chem. Commun.* 835 (1966).
[307] H. Seidl and R. Huisgen, *Tetrahedron Lett.* 2023 (1963).

to the parent base; phenanthridone is also formed. In chloroform containing traces of ethanol only phenanthridine, phenanthridone, and **219** are obtained.[308] Carbon dioxide is lost in the reaction between phenanthridine-5-oxide and phenyl isocyanate, which provides 6-anilinophenanthridine in good yield.[309]

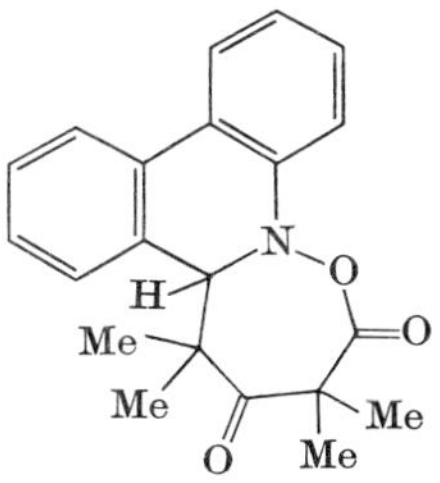

(218) 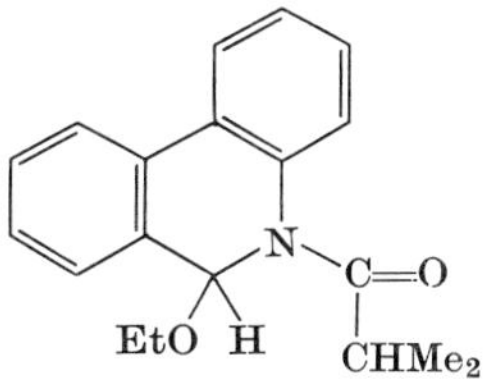    (219)

## B. Electrophilic Substitution Reactions at Ring Carbon Atoms

Previous reviewers have commented[4, 5] on the lack of information concerning the substitution reactions of phenanthridine. Although some progress has been made in this field, systematic studies are still much needed.

### 1. Acylation and Alkylation

The suggestion[5] that phenanthridine can probably be alkylated directly under suitably vigorous Friedel–Crafts conditions does not appear to have been tested experimentally. It receives support, however, from the successful conversion of the quaternary salts (**220**) ($R^1 = Me$, $R^2 = H$) and (**220**) ($R^1 = H$, $R^2 = Ph$) into the corresponding **221** by heating with polyphosphoric acid.[287] Not surprisingly, however, the related salt (**220**) ($R^1 = H$, $R^2 = m$-methoxyphenyl) forms **222** under similar conditions.[17] Direct alkylation of phenanthridine has been achieved photochemically (see p. 400).

Phenanthridone was recovered unchanged from an attempted acylation with acetyl chloride and aluminum chloride in tetrachloroethane.[291]

---

[308] R. N. Pratt and G. A. Taylor, *J. Chem. Soc. C* 1653 (1968).
[309] E. Hayashi, *J. Pharm. Soc. Japan* **81**, 1030 (1961).

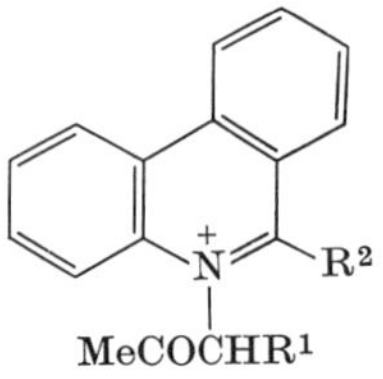

(220) 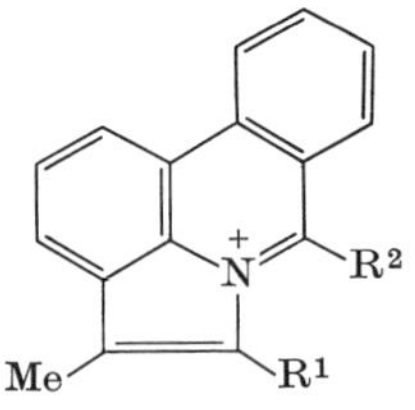 (221) 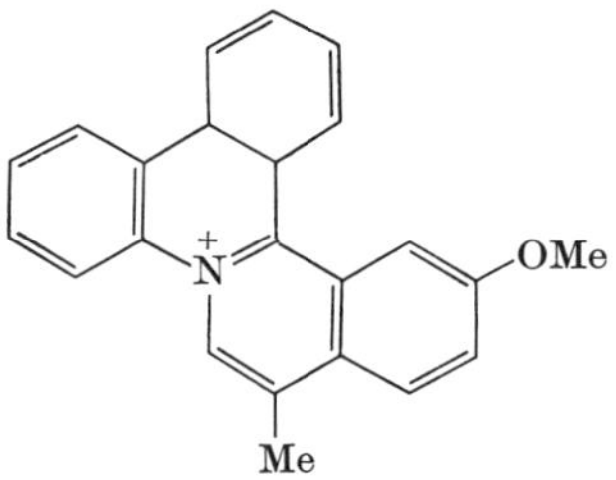 (222)

## 2. Halogenation

Phenanthridine has been brominated in solution using both $N$-bromosuccinimide in carbon tetrachloride[310] and bromine–silver sulfate in sulfuric acid.[311] Only 2-bromophenanthridine was isolated (in moderate yield) from the reaction in the organic solvent, but 2-, 4-, and 10-monobromophenanthridine were all obtained from the reaction in sulfuric acid, as well as small amounts of dibrominated derivatives and some unidentified material. The three monobromophenanthridines were obtained in the following order of abundance: $10 > 4 \gg 2$. Although these qualitative data probably do not precisely reflect the order of positional reactivity, it is noteworthy that the observed order agrees neither with MO predictions concerning substitution in the neutral molecule[225] (see p. 370) nor with the reactivity order qualitatively evaluated in nitration experiments (and which shows, in turn, some agreement with the order predicted for electrophilic substitution in the cation).[227] The pyrolysis of phenanthridinium bromide perbromide also gives 2-bromophenanthridine.[310]

It is uncertain whether the 2-bromophenanthridine-6-carboxylic acid obtained when **223** is treated with bromine in methylene dichloride (see p. 362) arises from attack on the azepinone or the phenanthridinecarboxylic acid, although the former seems more likely.

An attempt to demethylate 8-methoxyphenanthridine by heating with hydriodic acid (observed to contain some free iodine) led only to 8-hydroxy-3-iodophenanthridine; 8-hydroxyphenanthridine was itself iodinated under similar conditions.[117]

---

[310] H. Gilman and J. Eisch, *J. Amer. Chem. Soc.* **77**, 6379 (1955).
[311] G. S. Chandler, *Aust. J. Chem.* **22**, 1105 (1969).

(223)

An earlier report[51] that phenanthridone is brominated at C-2 on treatment with bromine in acetic acid has been confirmed[291]; chlorination (molecular chlorine) and iodination (iodine–iodate) in acetic acid give the corresponding 2-halogeno compounds in excellent yield.[291] The bromination of 2-acetamidophenanthridone has also been described, although the orientation was not established with certainty.[291]

Very recently, it has been shown that $N$-halosuccinimides in dimethylformamide convert phenanthridone and its 3,8-dihalo derivatives into the corresponding 2-halophenanthridones; under forcing conditions 2,4-dihalophenanthridones are formed in low yields. Although phenanthridone is chlorinated successfully with excess chlorine in acetic acid containing traces of ferric chloride, bromination does not occur with the milder reagent, aqueous hydrobromic acid in dimethyl sulfoxide.[101]

### 3. Nitration

Nitric acid alone is without effect on phenanthridine, but nitration occurs quite readily in sulfuric acid solution. The use of not more than one equivalent of nitric acid results in mononitration only, but a mixture of products is obtained and a quantitative recovery of the individual isomers has not been achieved. Valid assessments of reactivity are not possible from the available data, but the major products are certainly 1- and 10-nitrophenanthridine with smaller amounts of the 8-nitro isomers. To this extent the results are in agreement with localization energy calculations for the cation[227]; 2-, 3-, and 4-nitrophenanthridine have also been isolated, but in minor amounts which could not be determined accurately.[89]

It has been confirmed that, in line with qualitative prediction, the nitration of phenanthridone (with concentrated nitric acid) provides a mixture of 2- and 4-nitrophenanthridone; the isomer ratio is approximately 6:1. If C-2 is blocked by bromine or chlorine, nitration proceeds readily at C-4.[101] 2-Acetamidophenanthridone, on the other hand, appears to undergo nitration (in acetic acid) at either C-1 or C-3.[291] The nitration of 3-chloro- and 8-chlorophenanthridone leads to the corresponding 2-nitro compounds.[101]

As expected by analogy with quinoline, nitration of 7,8,9,10-tetrahydrophenanthridine (with mixed acids) provides a mixture of the 1- and 4-nitro derivatives.[312] Only the 1-nitro isomer was isolated when the tetrahydrophenanthridine nitrate was heated with a mixture of oleum and concentrated sulfuric acid.[171]

### 4. Sulfonation

The sulfonation of phenanthridine has not been reported. Phenanthridone forms phenanthridone-2-sulfonic acid virtually quantitatively on heating with concentrated sulfuric acid at 150°.[291]

## C. Nucleophilic Substitution Reactions

Although, predictably, C-6 of the phenanthridine ring is susceptible to nucleophilic attack, until recently only qualitative reactivity data were available. New examples of both direct substitutions (hydride ion extrusion) and nucleophilic replacements, notably of halogen atoms, have been reported since the last review.[5]

### 1. Alkylation

Phenanthridine is alkylated readily at C-6 by both Grignard reagents and organolithium compounds, offering convenient routes to 6-substituted derivatives. The first step, in both instances, clearly involves nucleophilic attack, and although dihydro intermediates have been isolated in a number of cases, aromatization occurs very readily (even during normal reactions) and the reactions are classified for convenience under the present heading. As usual in nucleophilic substitutions in azine systems, polynuclear compounds react more readily than pyridine (because of the increased delocalization of the negative charge), and with allylmagnesium bromide the reactivity order phenanthridine $\simeq$ acridine > quinoline $\simeq$ isoquinoline > pyridine has now been established.[313] The ease of reaction is shown in a typical example; phenanthridine with propylmagnesium bromide forms 6-propyl-5,6-dihydrophenanthridine which when heated with nitro-

[312] E. Hayashi and R. Goto, *J. Pharm. Soc. Japan* **85**, 645 (1965).
[313] H. Gilman, J. Eisch, and T. Soddy, *J. Amer. Chem. Soc.* **79**, 1245 (1957).

benzene gives 6-propylphenanthridine (84% overall).[314] A series of 6-alkylphenanthridines has been prepared by Japanese workers in essentially the same manner, but using picric acid to achieve the final dehydrogenation.[297] 6-Phenylphenanthridine also reacts readily with allylmagnesium bromide, giving 6-allyl-6-phenyl-5,6-dihydrophenanthridine.[315]

Ease of nucleophilic attack at C-6 has recently been demonstrated by the virtually quantitative conversion of phenanthridine into 6-methylphenanthridine by means of methylsulfinyl carbanion in dimethyl sulfoxide.[316]

In a reaction which is classified less easily, aralkylation with displacement of halogen takes place when 6-chlorophenanthridine and 6-methylphenanthridine are heated above the melting point; di- and tri-(6-phenanthridinyl)methane are formed in high yield.[317]

### 2. Amination and Substituted Amination

Phenanthridine is known to undergo direct amination readily with hydride extrusion (the Chichibabin reaction) and further examples have been reported.[22] More interesting is the preparation of 6-aminophenanthridine in high yield by the action of the sodium salt of $N,N$-dimethylhydrazine on phenanthridine in benzene. The adduct (**224**) (R = Me) loses dimethylamine on heating leaving the sodium salt of the 6-amino compound (**225**).[318]

(**224**)    (**225**)

Surprisingly, in the corresponding reaction with the sodium salt of hydrazine (**224**) (R = H) affords (**225**) in only very low yield; 5,6-dihydrophenanthridine is obtained almost quantitatively.[319]

[314] H. Gilman and J. Eisch, *J. Amer. Chem. Soc.* **79**, 4423 (1957).
[315] H. Gilman and J. Eisch, *J. Amer. Chem. Soc.* **79**, 2150 (1957).
[316] G. A. Russell and S. A. Weiner, *J. Org. Chem.* **31**, 248 (1966).
[317] G. Scheibe and H. J. Friedrich, *Chem. Ber.* **94**, 1336 (1961).
[318] Th. Kaufmann, H. Hacker, and H. Muller, *Chem. Ber.* **95**, 2485 (1962).
[319] Th. Kaufmann, H. Hacker, C. Kosel, and W. Schoeneck, *Angew. Chem.* **72**, 918 (1960).

6-Chlorophenanthridine reacts smoothly with methanolic ammonia and with both methylamine and dimethylamine to give the corresponding amines.[246] Rather surprisingly, 6-chloro-7,8-benzophenanthridine is reported to be unaffected by refluxing ethanolic ammonia.[320] Advantage has been taken of the ease of reaction between 6-chlorophenanthridine and primary aliphatic amines to prepare derivatives with antitumor activity, e.g., **226**[321]; other pharmacologically active products have been obtained from reactions with 1-substituted piperazines.[322, 323] Substituted aminations by means of

$$HN(CH_2)_3N{\diagup}^{C_3H_7}_{\diagdown (CH_2)_2OH}$$

(**226**)

α-amino-N-heteroaromatic compounds (e.g., 2-aminoquinoline) have been reported.[320] 6-Chloro-3,8-dinitrophenanthridine reacts with both aniline and p-nitroaniline at reflux temperature in the absence of a solvent.[324] It seems unlikely that such vigorous conditions are necessary and milder conditions have been used satisfactorily in the reactions between aniline and chlorobenzophenanthridines.[90] Related replacements have been reported in the synthesis of hydroderivatives related to chelidonine alkaloids.[325]

Kinetic data have been obtained for the reactions of a number of 6-chlorophenanthridines with piperidine, morpholine, and pyrrolidine; rates and activation parameters for the parent compound are given in Table IV.[102]

The data show 6-chlorophenanthridine to be more reactive than either 1-chloroisoquinoline or 2-chloroquinoline, the differences

---

[320] H. H. Credner, H. F. Friedrich, and G. Scheibe, *Chem. Ber.* **95**, 1881 (1962).
[321] R. M. Peck, A. P. O'Connell, and H. J. Creech, *J. Med. Pharm. Chem.* **9**, 217 (1966).
[322] A. S. Tomcufcik, P. F. Fabio, and A. M. Hoffmann (to American Cyanamid Co.), Belgian Patent 637,271; *Chem. Abstr.* **62**, 11830h (1965).
[323] H. J. Barber and D. H. Jones (to May and Baker Ltd.), S. African Patent 68/06,579; *Chem. Abstr.* **72**, 3505 (1970).
[324] M. Davis, *J. Chem. Soc.* 337 (1956).
[325] A. Bailey, R. Robinson, and R. S. Staunton, *J. Chem. Soc.* 2277 (1950).

## TABLE IV
### REACTION OF 6-CHLOROPHENANTHRIDINE WITH AMINES IN MeOH

| Nucleophile | $k_2$ (liter mole$^{-1}$ sec$^{-1}$) | Temperature (°C) | $E_A$ (kcal mole$^{-1}$) | $\Delta S^{\ddagger}$ (e.u.) |
|---|---|---|---|---|
| Piperidine | $6.70 \times 10^{-6}$ | 30.7 | | |
| | $1.68 \times 10^{-5}$ | 45.0 | 12.7 | $-42.6$ |
| | $3.22 \times 10^{-5}$ | 55.2 | | |
| Morpholine[a] | $3.73 \times 10^{-7}$ | 14.2 | | |
| | $8.27 \times 10^{-7}$ | 25.0 | 12.7 | $-45.9$ |
| | $1.86 \times 10^{-6}$ | 36.6 | | |
| Pyrrolidine | $9.86 \times 10^{-6}$ | 15.0 | | |
| | $2.67 \times 10^{-5}$ | 30.0 | 11.9 | $-42.4$ |
| | $6.97 \times 10^{-5}$ | 45.0 | | |

[a] Autocatalysis observed.

stemming from a decreased energy of activation in the case of the tricyclic system. Substituent effects are essentially in line with qualitative predictions based on simple electronic theory, except that 6-chloro-1,10-dimethylphenanthridine reacts with the three amines appreciably more readily than either the isomeric 3,8-dimethyl compound or 6-chlorophenanthridine itself, perhaps because of the release of steric strain in forming the transition state.

6-Chlorophenanthridine reacts with $p$-toluenesulfonyl hydrazide in chloroform in the presence of hydrogen chloride[213]: the resulting hydrazide (**227**) can be cleaved to the parent base without further purification (see p. 368).

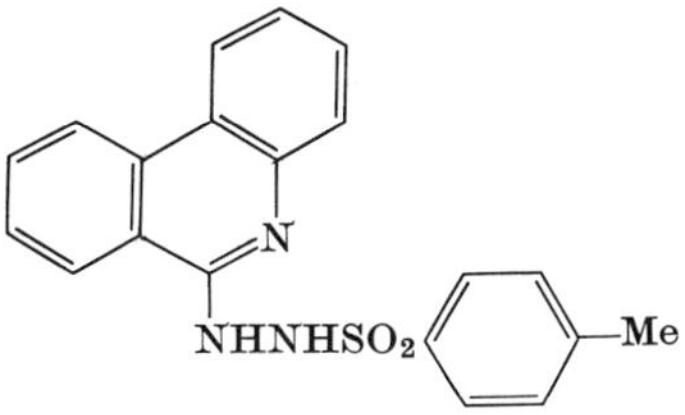

(**227**)

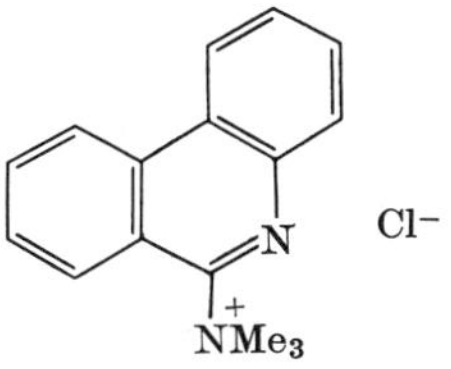

(**228**)

An attempt to repeat a reported preparation of the salt (**228**) from trimethylamine and 6-chlorophenanthridine failed; after metathesis only tetramethylammonium iodide could be isolated. A similar reaction took place with 4-methylmorpholine, which yielded 4,4-dimethylmorpholinium iodide.[326]

Aminative replacement in compounds other than those bearing halogen atoms are uncommon, but 6-aminophenanthridine is easily prepared by heating 6-phenoxyphenanthridine with urea[296] and 6-anilinophenanthridine has been obtained by the action of aniline on the 6-phenoxy compound.[268] The formation of 6-anilinophenanthridine from phenanthridine $N$-oxide and phenyl isocyanate probably proceeds via an initial 1,3-dipolar addition, followed by nucleophilic substitution and decarboxylation (Scheme 4).[309]

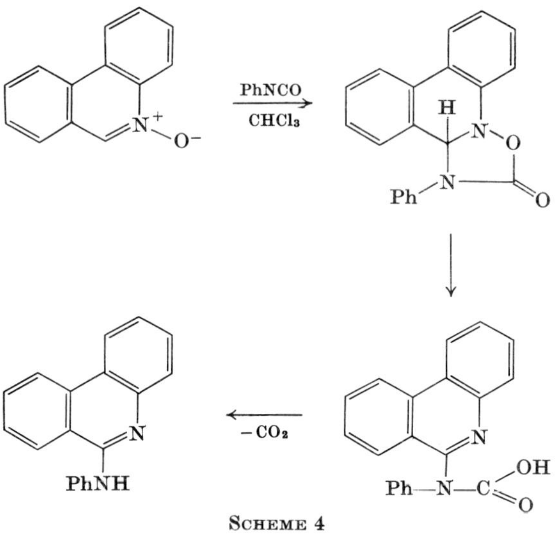

SCHEME 4

### 3. *Arylation*

The known reactivity of phenanthridine towards lithium aryls extends to the hindered $o$-tolyllithium, where the stability of the 6-$o$-tolyl-5,6-dihydrophenanthridine allows its isolation.[314] (6-Phenylphenanthridine forms 6,6-diphenyl-5,6-dihydrophenanthridine under similar conditions.[62, 314]) $p$-Biphenylyllithium reacts smoothly to give the dihydro adduct; oxidation *in situ* with boiling nitrobenzene gives the 6-biphenylylphenanthridine in good yield.[327]

---

[326] C. B. Reese, *J. Chem. Soc.* 899 (1958).
[327] J. J. Eisch and R. M. Thompson, *J. Org. Chem.* **27**, 4171 (1962).

## 4. Halogenation

The action of phosphorus halides on phenanthridones offers a convenient route to the corresponding 6-chloro- or bromophenanthridines, and further examples of this process have been reported since the last review.[5] A mixture of phosphorus tribromide and bromine converts phenanthridone (in toluene) into 6-bromophenanthridine in fair yield.[26] Phosphoryl chloride, alone[102, 111, 203, 324, 325] or in the presence of a base[90, 102, 105, 113, 213, 246] (normally a dialkylaniline) converts phenanthridones and benzophenanthridones into the corresponding 6-chloro compounds in good yield. Mixtures of phosphoryl chloride and phosphorus pentachloride have sometimes been used successfully,[26, 102, 126] although when 1,10-dimethylphenanthridone is heated with phosphorus pentachloride alone a dichloro derivative is formed.[102] (Other examples are known in which the action of phosphorus pentachloride on heterocyclic lactams is accompanied by additional nuclear chlorination.[328]) Difficulties encountered in preparing 6-chloro-1,10-dimethylphenanthridine using a mixture of phosphoryl chloride and phosphorus pentachloride in the usual manner arise from rapid hydrolysis after quenching; a non-aqueous reaction provides the required chloro compound in 30% yield.[102] The same mixture converts $N$-benzylphenanthridone into 6-chlorophenanthridine directly.[329] As expected, sulfuryl chloride[309] and phosphoryl chloride[295] both convert phenanthridine $N$-oxide into 6-chlorophenanthridine,[309] but with the latter reagent 6-phenylphenanthridine $N$-oxide forms 2-chloro-6-phenylphenanthridine. 6-Methyl-phenanthridine $N$-oxide forms a mixture of 6-chloromethylphenanthridine and, probably, 2-chloro-6-methylphenanthridine under similar conditions.[295]

## 5. Hydroxylation and Substituted Hydroxylation

Phenanthridine undergoes direct hydroxylation on heating with a mixture of prefused potassium hydroxide and barium oxide at 225°, giving phenanthridone in fair yield.[291] This appears to be the sole application of this reaction in the series, although it has been reported that 6-anilino- and 6-ethylaminophenanthridine also provide

---

[328] E. C. Taylor and P. Drenchko, *J. Amer. Chem. Soc.* **74**, 1101 (1952).

[329] A. Resplandy, A. Michailidis, and J. Brouard, *C. R. Acad. Sci. C* **269**, 781 (1969).

phenanthridone on fusing with solid potassium hydroxide at 380°–400°.[150] As noted above, 6-chlorophenanthridine is hydrolyzed only very slowly at room temperature in neutral or alkaline solution. The formation of phenanthridone is accelerated markedly by traces of acid; under pseudo-unimolecular conditions (in 20% aqueous dioxane, 0.005 $M$ with respect to sulfuric acid), the specific rate constant is $4.53 \times 10^{-7}$ sec$^{-1}$ at 45°. Cationization must be far from complete under these conditions and autocatalysis is observed.[102]

TABLE V

RATES AND ACTIVATION PARAMETERS FOR THE METHOXYDECHLORINATION OF 6-CHLOROPHENANTHRIDINES[a]

| Substituents | Solvent | Specific rate constant (liter mole$^{-1}$ sec$^{-1}$) | $E_A$ (kcal mole$^{-1}$) | $\Delta S^{\ddagger}$ (e.u.) |
|---|---|---|---|---|
| None | MeOH | $3.78 \times 10^{-4}$ | 18.8 | $-18.7$ |
| 3,8-Me$_2$ | MeOH | $1.30 \times 10^{-4}$ | 19.0 | $-18.6$ |
| 1,10-Me$_2$ | MeOH | $3.48 \times 10^{-3}$ | 15.9 | $-22.6$ |
| None | MeOH–40% dioxane | $6.02 \times 10^{-4}$ | 18.3 | $-18.1$ |
| 3,8-Br$_2$[b] | MeOH–40% dioxane | $2.62 \times 10^{-2}$ | 15.0 | $-19.5$ |
| 3,8-Cl$_2$[b] | MeOH–40% dioxane | $2.34 \times 10^{-2}$ | 15.0 | $-19.6$ |
| 3,8-Me$_2$ | MeOH–40% dioxane | $1.91 \times 10^{-4}$ | 19.0 | $-18.2$ |
| 1,10-Me$_2$ | MeOH–40% dioxane | $5.65 \times 10^{-3}$ | — | — |
| 3,8-(NO$_2$)$_2$[c] | MeOH–40% dioxane | $>1$ | — | — |

[a] At 45° unless otherwise stated.
[b] Measured at 36.6°.
[c] Measured at 25.0°.
[d] From rate measurements in the range 14.8°–55.2°.

Alkoxydechlorination occurs rapidly and smoothly and preparative examples abound.[90, 105, 111, 113, 126, 129, 246, 295] Quantitative data have been lacking until recently, but Table V shows rates and activation parameters for the methoxydechlorination of a number of chlorophenanthridines.[102] The reactivity of 6-chlorophenanthridine itself appears to be generally comparable with that of 9-chloroacridine,[330] and substituent effects are in line with qualitative prediction, with the exception of 6-chloro-1,10-dimethylphenanthridine which shows enhanced reactivity for steric reasons (see p. 393).

Good yields of 6-aryloxyphenanthridines are obtained by heating 6-chlorophenanthridine with the appropriate phenol at 150°–160°.[129, 309]

### 6. *Nitrile Formation*

Phenanthridine *N*-oxide forms 6-cyanophenanthridine *N*-oxide in fair yield when treated with aqueous potassium cyanide in the presence of potassium ferricyanide.[331]

### 7. *Thiation*

Little interest has been shown in sulfur-containing compounds other than the parent thione (**229**) which is formed in excellent yield when phenanthridone is heated in pyridine with phosphorus pentasulfide.[214] No data are available on the reactions of 6-chlorophenanthridine with mercaptans or thiophenols.

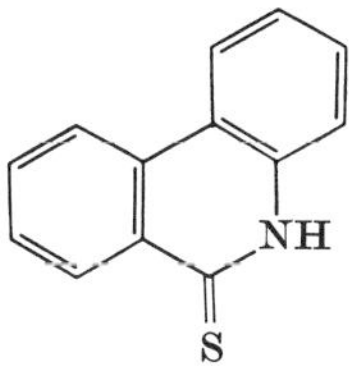

(**229**)

## D. OTHER REACTIONS

### 1. *Reduction*

The carbon–nitrogen double bond of phenanthridine can be reduced selectively by hydrogenation over Raney nickel, and attempted reductive dechlorination of 6-chloro derivatives in the presence of this catalyst normally results in the formation of the corresponding 5,6-dihydro compounds.[105] Hydrogenations over palladium catalysts are more successful.[203, 325] Desulfurization of phenanthridinthione

---

[330] G. Illuminati, G. Marino, and O. Piovesana, *Ric. Sci.* (*Parte II*, *Rend. Sez. A*) **4**, 437 (1964).
[331] Y. Kobayashi, I. Kumadaki, and H. Sato, *Chem. Pharm. Bull.* **18**, 861 (1970).

with Raney nickel is likewise complicated by the formation of dihydrophenanthridine[215] (although an almost quantitative conversion into phenanthridine in 1 : 1 ethanol–dimethylformamide has been reported[214]). Raney cobalt has been recommended for this reaction.[215]

A careful study of the reduction of phenanthridine with lithium, sodium, and magnesium–magnesium iodide has shown that radical-anion intermediates are involved. Thus, initial solutions of phenanthridine and sodium in tetrahydrofuran were found to give pronounced electron spin resonance signals indicating the presence of the radical anion (**230**) (R = H) and an esr signal was likewise obtained from similar solutions of 6-phenylphenanthridine, although the hyperfine splitting could not be analyzed in the case of **230** (R = Ph).

(**230**)   (**231**)   (**232**)

On refluxing in hydrocarbon solvents with sodium or magnesium–magnesium iodide, phenanthridine forms a mixture of 6,6′-biphenanthridyl, 5,6-dihydro-6,6′-biphenanthridyl, and 5,6-dihydrophenanthridine. By the action of lithium in tetrahydrofuran 5,6-dihydrophenanthridine and 5,5′,6,6′-tetrahydro-6,6′-biphenanthridyl were obtained. Chemical evidence in favor of **230** is provided by the fact that both 6-phenyl and 6-biphenylylphenanthridine give only the corresponding dihydro derivatives, with no trace of bimolecular reduction.[327]

Further examples of the reduction of the carbon–nitrogen link by means of lithium aluminum hydride have been reported,[90, 332] and sodium borohydride has been used for the same purpose in the reduction of a number of 6-alkyl-1,2,3,4,4a,10b-hexahydrophenanthridines.[42] Quaternary phenanthridinium salts form $N$-alkyldihydrophenanthridines on treatment with lithium aluminum hydride.[333]

[332] W. J. Van der Burg, I. L. Bonta, J. Delobelle, C. Ramon, and B. Vargaftig, *J. Med. Chem.* **13**, 35 (1970).
[333] E. Hoeft, A. Reiche, and H. Schultze, *Ann. Chem.* **697**, 181 (1966).

The conversion of phenanthridine into an unspecified octahydro-derivative by hydrogenation over a sodium metal–rubidium carbonate catalyst has been reported.[334] Hydrogenation of the tetrahydrophenanthridine (231) over platinum in acetic acid gave, rather surprisingly, the octahydrophenanthridine (232) with loss of methoxyl.[173]

## 2. Ozonization

Phenanthridine in methanol solution is unaffected by ozone, but in methylene dichloride quinoline-3,4-dicarboxylic acid (2%) and phenanthridone (23%) are formed; a large part of the remaining phenanthridine can be recovered unchanged.[335] The reaction thus proceeds much less readily than in the case of acridine. The mechanism by which the lactam is formed is not known with certainty, but an intermediate such as 233 has been proposed.[335] However, an initial electrophilic attack producing 234, followed by loss of oxygen and a proton shift, seems equally likely.

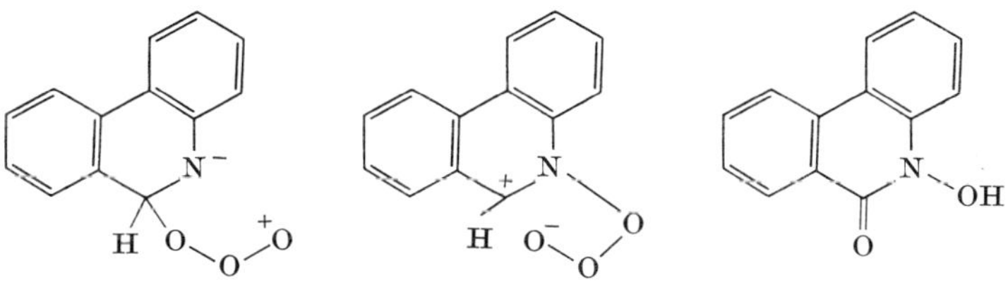

(233)   (234)   (235)

The absence of isoquinoline-3,4-dicarboxylic acid reflects the effective deactivation by the ring nitrogen atom of the adjacent ring toward electrophilic attack.

The ozonization of phenanthridine N-oxide proceeds more rapidly, as expected, yielding 77–80% 5-hydroxyphenanthridone (235) in methylene chloride and 56%, contaminated with phenanthridone and 2′-nitro-2-biphenylcarboxaldehyde, in methanol.[335]

[334] S. Friedman, M. L. Kaufman and I. Wender, *Am. Chem. Soc., Div. Fuel Chem. Preprints* **8**, 209 (1964).
[335] E. J. Moriconi and F. A. Spano, *J. Amer. Chem. Soc.* **86**, 38 (1964).

## 3. Photoalkylation

When a $5 \times 10^{-2}$ $M$ solution of phenanthridine in acidified ethanol is irradiated with ultraviolet light, conversion into 6-ethylphenanthridine (**236**) is essentially complete after 40 hours; prolonged exposure to sunlight achieves the same result. The mechanism shown below has been proposed for this (and related) photoalkylations.[336]

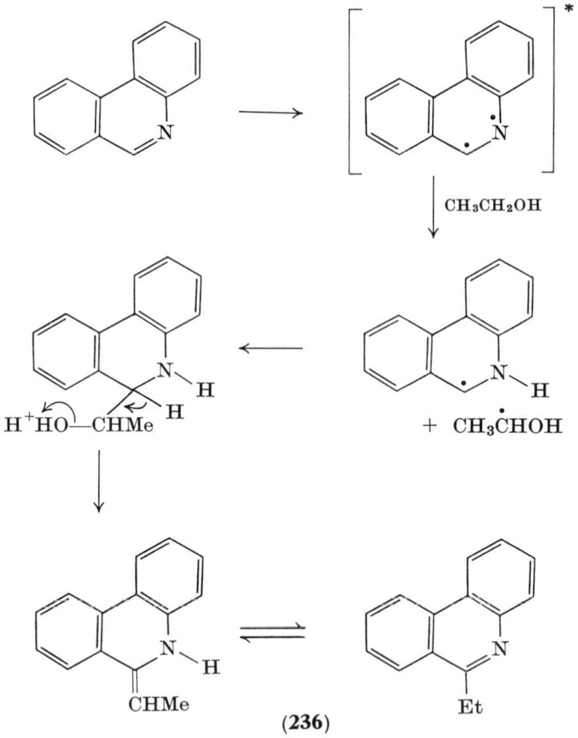

## 4. Oxidative Cleavage

Phenanthridines (and their 6-alkyl homologs) are oxidized to the corresponding phenanthridones by, for example, dichromate in acetic acid[14, 193, 313, 314] or acid permanganate.[310] Under more vigorous conditions the phenanthridones themselves undergo ring cleavage. Thus on heating with alkaline potassium permanganate 8-nitrophenanthridone forms 4-nitrophthalic acid; the 3-nitroisomer yields phthalic acid itself.[193]

---

[336] F. R. Stermitz, R. P. Seiber, and D. E. Nicodem, *J. Org. Chem.* **33**, 1136 (1968).

## 5. Electron-Impact Studies

Mass spectral data are available for relatively few phenanthridines. The salient features of the fragmentation pattern for phenanthridine itself are shown in Scheme 5 with relative abundances (%) in parentheses. Appropriate metastable peaks have been observed for each cleavage shown.[64, 337]

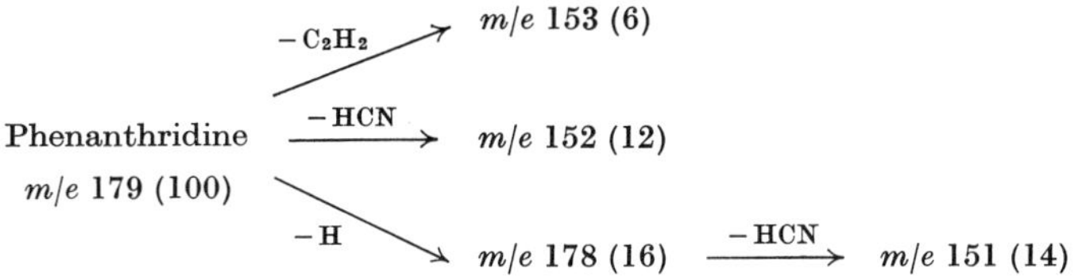

Phenanthridine $m/e$ 179 (100)
- $-C_2H_2 \rightarrow$ $m/e$ 153 (6)
- $-HCN \rightarrow$ $m/e$ 152 (12)
- $-H \rightarrow$ $m/e$ 178 (16) $\xrightarrow{-HCN}$ $m/e$ 151 (14)

SCHEME 5

The phenanthridone ring of oxyavicine (237) breaks down as shown.[338]

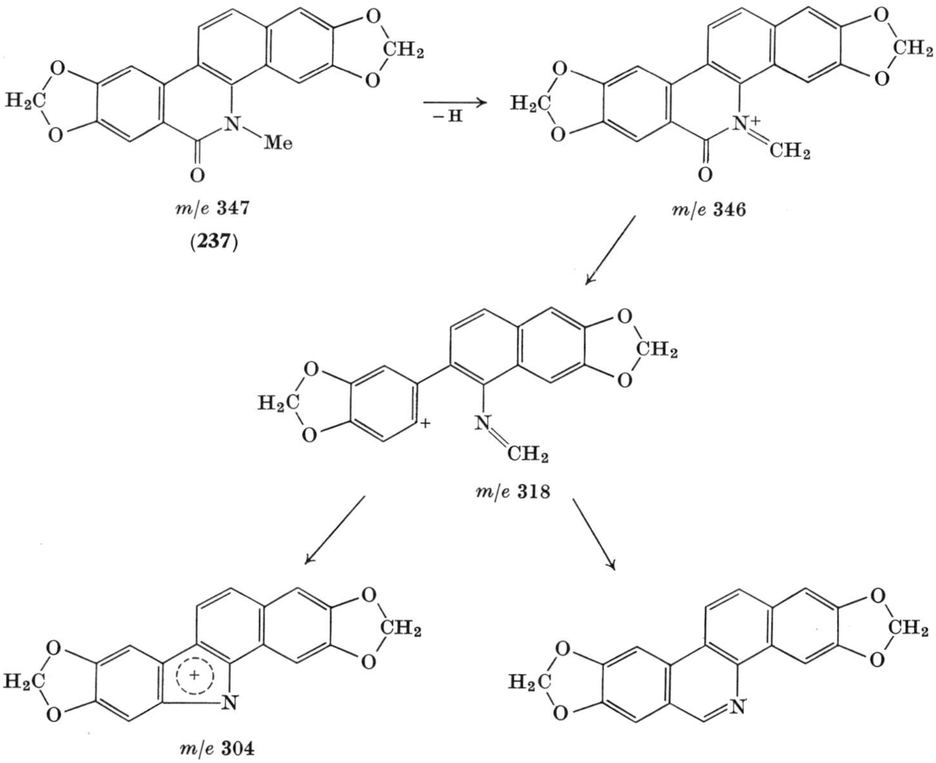

$m/e$ 347 (237)

$m/e$ 346

$m/e$ 318

$m/e$ 304

[337] J. F. Todd, private communication.
[338] M. Sainsbury, S. F. Dyke, and B. J. Moon, *J. Chem. Soc. C* 1797 (1970).

The spectrum of phenanthridine $N$-oxide is interesting in that the disintegration closely follows the pattern established for quinoline $N$-oxide in yielding M$^+$, (M–CO)$^+$, and (M–CO–HCN)$^+$ ions. In contrast, in isoquinoline $N$-oxide the loss of carbon monoxide and hydrogen cyanide occurs in the reverse order.[339] The spectrum of the 6-methylphenanthridinium enol betaine (**215**) (R = Me) shows a small parent ion (m/e 351) with the base peak at M/e 333 (M – 18)$^+$. This initial dehydration presumably parallels the easy chemical ring closure (to **216**) on vacuum sublimation; all the peaks in the spectrum of **216** are also present in that of **215** (R = Me). Further disintegration involves the stepwise loss of two ester groups. The analogous betaine (**215**) (R = H) shows a small parent peak (m/e 337), the loss of one ester group (M – 59)$^+$ and a base peak at (M – 158)$^+$, corresponding to loss of the 5-substituent.[285]

## V. The Properties of Functional Groups

These have not been widely studied, but simple substituents on the carbocyclic rings appear to behave normally in all respects. For this reason the treatment of such groups in this section is intended to be illustrative rather than exhaustive.

### A. Alkyl and Substituted Alkyl Groups

The acidity of protons attached to the α-carbon atoms of alkyl groups located at positions adjacent to pyridinelike nitrogen is well established. Alkyl groups at C-6 of the phenanthridine ring show pronounced reactivity of this type. The rates and activation parameters of ethoxide ion-catalyzed exchange between EtOD and 6-methylphenanthridine have been obtained: $k$ (at 35°) = 2.2 × 10$^{-5}$ sec$^{-1}$, $E_A$ = 20.9 kcal mole$^{-1}$, and log $A$ = 10.2. Exchange is thus appreciably more rapid than in, for example, 2-methylpyridine or 2-methylquinoline and a correlation has been claimed between these rates and the calculated electron densities on the ring carbon atoms carrying the methyl substituents. The corresponding values for uncatalyzed exchange with MeOD in the case of 6-methylphenanthridine are $k$ (at 75°) = 1.5 × 10$^{-5}$ sec$^{-1}$, $E_A$ = 13.0 kcal mole$^{-1}$, and

[339] O. Buchardt, A. M. Duffield, and R. H. Shapiro, *Tetrahedron* **24**, 3139 (1968).

log $A = 3.3$, while for the corresponding $N$-oxide $k$ (at 70°) $= 2.3 \times 10^{-5}$ sec$^{-1}$, $E_A = 13.9$ kcal mole$^{-1}$, and log $A = 4.2$. The (relatively small) decrease in going from the base to the $N$-oxide probably stems from the lower basic strength of the latter, since in the absence of base, exchange occurs via a partly cationized (alcohol-bonded) species. In the ethoxide ion-catalyzed reactions $N$-oxides exchange, in general, $10^2$ to $10^4$ times more rapidly than the bases, the activating influence of the $N$-oxide being shown in full where the exchanging species is the free base.[264]

When 6-methylphenanthridine is allowed to react with phenyllithium in ether, side-chain metalation takes place exclusively.[340] The lithiomethyl derivative so obtained forms 1,2-bis(6-phenanthridyl)ethane in good yield on treatment with oxygen.[341]

The activity of 6-alkyl substituents is well shown by the ease with which 6-methyl- and 6-ethylphenanthridinium-5-enol betaines (215) (R = Me and Et) undergo cyclodehydration to pyrrolophenanthridines (238) (R = H or Me) on heating, since the first step presumably involves proton transfer from the α-carbon atom of the alkyl substituent.[285]

6-Methylphenanthridine behaves as a nucleophile in its reactions with, for example, 2-chloroquinoline and 6-chlorophenanthridine, which give 6-(2'-quinolymethyl)- and 6-(6'-phenanthridylmethyl)-phenanthridine, respectively.[317] Although the methyl group appears to be less reactive than that of 9-methylacridine, 6-methylphenanthridine reacts smoothly with $p$-dimethylaminobenzaldehyde in the presence of piperidine, and cyanine dyes have been obtained from the corresponding ethiodide.[286] Coupling with $p$-nitrobenzenediazonium chloride gives the formazyl compound (239) and 6-methylphenanthridine-5-oxide behaves similarly, giving the $N$-oxide of 239. (The

---

[340] A. M. Jones, C. A. Russell, and S. Skidmore, *J. Chem. Soc. C* 2245 (1969).
[341] A. M. Jones and C. A. Russell, *J. Chem. Soc. C* 2246 (1969).

same compound can also be prepared by coupling *p*-nitrobenzenediazonium chloride with the phenylhydrazone of 6-formylphenanthridine-5-oxide, and its composition is thus established.[298] The failure of an attempt to hydroxydeaminate 3-amino-6-methylphenanthridine via the diazonium salt has been ascribed to "naphthoid reactivity" of the nucleus; it seems likely, in fact, that self-coupling occurs.[11]

Although 6-methylphenanthridine cannot be oxidized satisfactorily with neutral or acid permanganate,[245] Gopinath has shown that excess of selenium dioxide in boiling ethyl acetate converts the benzophenanthridine (**240**) (R = Me) into the corresponding acid (**240**)

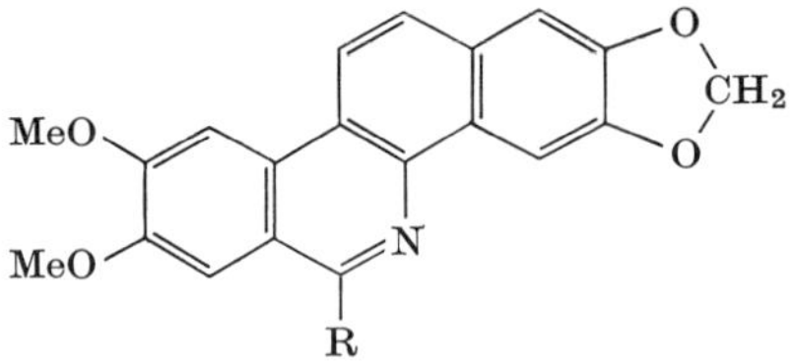

(**240**)

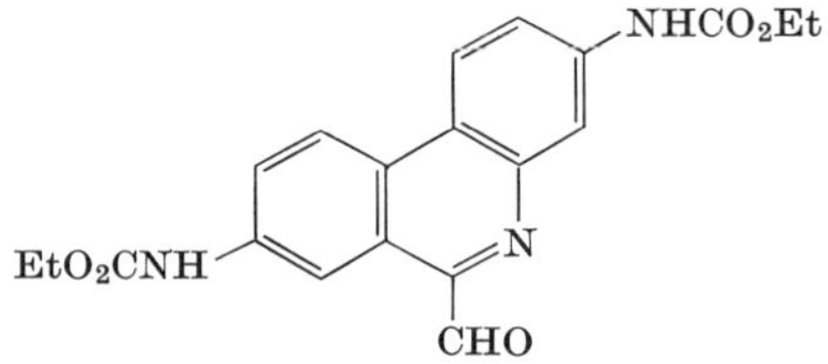

(**241**)

(R = COOH).[34] On the other hand, controlled conversion into the aldehyde is also possible. Thus aldehyde (**241**) was obtained by the action of selenium dioxide in boiling dioxane on 3,8-diethoxycarbonylamino-6-methylphenanthridine,[324] and 6-formylphenanthridine *N*-oxide has been obtained in good yield by oxidising the 6-methyl compound with selenium dioxide in pyridine.[298]

Indirect conversion of 6-methylphenanthridine to phenanthridine-6-carboxylic acid has been achieved by heating with benzaldehyde in the presence of zinc chloride and oxidizing the resulting styryl derivative with permanganate in acetone.[245]

As noted earlier, 6-alkyl substituents can be removed completely (yielding phenanthridones) by oxidation with acid dichromate.[14, 245, 313, 314]

Nucleophilic replacement (for example, by secondary amines) occur readily in 6-chloromethylphenanthridine.[295, 332]

### B. Amino and Substituted Amino Groups

Isolated examples of direct alkylation reactions have been reported. For example, 3-amino-5-ethylphenanthridone gives low yields of the corresponding 3-ethylamino and 3-diethylamino compounds when refluxed with ethyl iodide in aqueous sodium carbonate,[128] and the action of methanolic methyl iodide at 115° converts 3,8-diaminophenanthridone into the 3,8-bisdimethylaminobismethiodide.[324] Surprisingly, the same product is formed when either 3,8-diamino-6-$p$-aminoanilinophenanthridine or its triacetyl derivative is treated with methyl iodide under similar conditions.[324] Suitably activated heteroaromatic halogen compounds condense with amino groups at C-8 in quaternary phenanthridinium salts. Useful therapeutic activity has been claimed for the salt (242), prepared from 2-amino-6-chloro-3,4-dimethylpyrimidinium bromide and the appropriate diaminophenanthridinium methiodide.[342]

(242)  (243)

Under Skraup conditions, 4-aminophenanthridine forms the diazabenzphenanthrene (243) in poor yield.[178]

Acylation of amino groups on the carbocyclic rings proceeds normally[291, 324, 343]; in 3,8-diaminophenanthridinium salts, attack occurs more readily at the amino group at C-8.[324, 343] Both aminophenanthridines and aminophenanthridones condense readily with ethyl chloroformate in the presence of diethylaniline, and use has been made of the ethoxycarbonyl function as a protecting group.[39, 280, 324] Difficulty in regenerating the amino function by acid hydrolysis has been reported in one instance.[39]

$N$-Coupling with diazonium salts occurs fairly readily in phenanthridium salts bearing an amino group at C-8. Many diazoamino compounds have been so prepared in the search for trypanocides,[344–348]

[342] J. L. Lowe, W. Dickinson, and J. S. Nicholson (to Boots Pure Drug Co. Ltd.), British Patent 828,962.
[343] L. P. Walls, *J. Chem. Soc.* 3514 (1950).
[344] W. R. Wragg, K. Washbourn, K. N. Brown, and J. Hill, *Nature* **182**, 1005 (1958).
[345] S. S. Berg, *Nature* **188**, 1106 (1960).
[346] S. S. Berg, *J. Chem. Soc.* 3635 (1963).
[347] S. S. Berg, L. Bretherick, K. Washbourn, and W. R. Wragg, *J. Chem. Soc.* 4617, 4623 (1963).
[348] See also, e.g., W. R. Wragg and K. D. Washbourn (to May and Baker Ltd.), *British Patent* 834,231, *Chem. Abstr.* **54**, 24816i; S. S. Berg (to May and Baker Ltd.), *British Patent* 931,227, *Chem. Abstr.* **60**, 510c.

but the reaction is complicated by the formation of isomeric aminoazo compounds by electrophilic attack at carbon, probably C-7 or C-9.[345, 346]

Derived diazonium compounds, in general, undergo the usual interconversions although yields are often rather low. Reference has already been made (see p. 404) to an unsuccessful attempt to prepare 3-hydroxy-6-methylphenanthridine from the amine.[11] 1-Aminophenanthridone gives only impure 1-hydroxyphenanthridone on diazotization in sulfuric acid followed by heating,[127] but the method is more satisfactory in the case of 2-amino-$N$-methylphenanthridone.[126] Phenanthridones are usually more readily available than the corresponding 6-amino compounds, but as a structural proof phenanthridone itself has been obtained by the action of nitrous acid on 6-aminophenanthridine.[150]

3-Amino-8-bromophenanthridone can be deaminated smoothly by the action of cold hypophosphorus acid on the diazonium chloride.[122] 1-Bromophenanthridone is formed in moderate overall yield from the amino compound when the diazonium bromide–mercuric bromide complex is heated at 120° and a number of bromo- and chloro-phenanthridones have been obtained recently by Sandmeyer procedures.[26, 101]

The oxidation of 5-aminophenanthridone with lead tetraacetate in methylene chloride leads to benzocoumarin and phenanthridone rather than benzo[c]cinnoline (cf. the oxidation of 1-amino-3,4,5,6-tetraphenylpyridone[349]).

## C. Carbonyl, Carboxyl, and Related Groups

Few acylphenanthridines have been described since the earlier review[5] and additional information on reactivity is correspondingly meager. Phenanthridine-6-aldehyde is reported to undergo the benzoin reaction in the presence of 1,3-dimethylbenzimidazolium hydroxide[350]; an earlier attempt under conventional conditions was unsuccessful.[351] The hydrogen-bonded structure (**244**) is proposed for phenanthridoin.[350] Further synthetic use has been made of the easy permanganate oxidation of formyl groups at C-6[351]; aqueous

[349] C. W. Rees and M. Yelland, *Chem. Commun.* 377 (1969).
[350] E. Hayashi, T. Ishikawa, and M. Inaoka, *J. Pharm. Soc. Japan* **80**, 838 (1960).
[351] E. Ritchie, *Proc. Roy. Soc. N.S. Wales* **78**, 164 (1945).

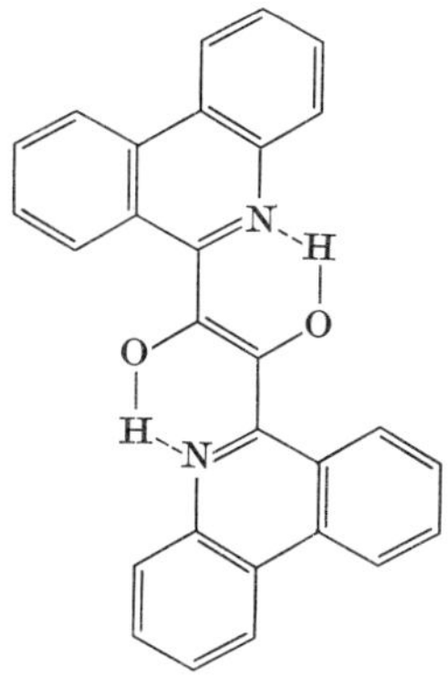

(244)

pyridine appears to be a satisfactory medium.[324] The known[351] resistance of 6-acylphenanthridines to electrophilic attack at ring nitrogen is confirmed by the reported failure of 6-benzoylphenanthridine to undergo $N$-oxidation with peracetic acid. Interestingly, however, an attempted synthesis by ring closure from **245** (R = COPh) fails because 6-benzoylphenanthridine $N$-oxide (**246**) undergoes nucleophilic attack by hydroxyl ion *in situ*.[78] The analogous 6-benzenesulfonylphenanthridine $N$-oxide (**245**) (R = SO$_2$Ph) forms the phenanthridone as shown, by nucleophilic attack in another sense.

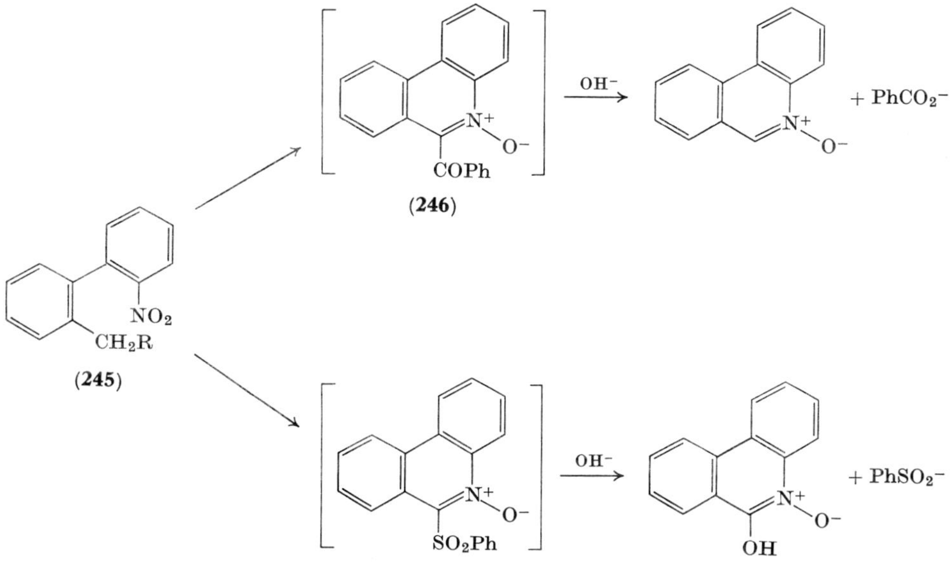

Synthetic and degradative use has been made in several instances of the easy decarboxylation of phenanthridine-6-carboxylic acids.[36, 80, 324] The corresponding $N$-oxide is decarboxylated by heating in aqueous sulfuric acid.[77] Phenanthridine-3,8-dicarboxylic acids are more resistant to decarboxylation, which can be achieved (in poor yield) by heating with copper powder in quinoline.[194] The usual carboxyl derivative interconversions (esterification, amide formation, ester and amide hydrolyses, etc.) proceed normally with both phenanthridine-6-carboxylic acid and its $N$-oxide,[77, 232, 352] although an unsuccessful attempt to prepare 6-acetylphenanthridine from the acid with methyllithium has been reported.[232]

The reactivity of the amide function of phenanthridine-6-carboxamide differs somewhat from that in the corresponding $N$-oxide. Although the former compound is smoothly converted to the amine under Hofmann conditions and forms the acid on treatment with nitrous acid, the amide $N$-oxide is virtually inert under the conditions of both reactions.[296]

1-Carboxyphenanthridone forms the aminophenanthridone in good yield on treatment with hydrazoic acid in sulfuric acid.[26]

## D. Halogen Atoms

The characteristic reactions of 6-halogenophenanthridines with nucleophiles have been discussed in Section IV,C (see p. 390), as have the synthetically useful reductive dechlorinations of 6-chloro derivatives (see p. 397). Little attention has been paid to the reactions of other halogenophenanthridines and -phenanthridones. 2-Bromophenanthridone (**247**) can be lithiated with two equivalents of

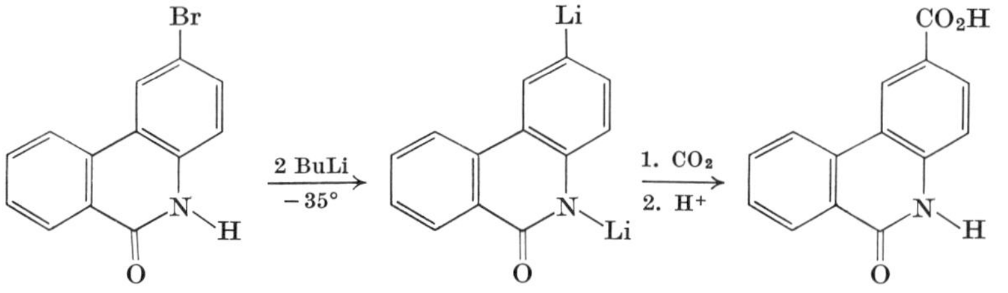

SCHEME 6

[352] C. W. Muth, G. A. Hall, and J. L. Essex, *Proc. W. Va. Acad. Sci.* **36**, 81 (1964); *Chem. Abstr.* **62**, 1527 (1965).

butyllithium at −35°; carbonation followed by partial decarboxylation in boiling acetic acid gives phenanthridone-2-carboxylic acid (**248**) (Scheme 6).[291]

### E. Hydroxyl and Substituted Hydroxyl Groups

Alkylations of 6-hydroxyphenanthridines (phenanthridones) have been dealt with earlier (see p. 381). Hydroxyl groups at positions other than C-6 have normal phenolic properties; for example, methylation (with diazomethane),[127] allylation (with allyl bromide and ethanolic potassium carbonate[11]), and acetylation (with acetic anhydride[126]) all appear to proceed normally. Methoxy groups are cleaved by boiling hydrochloric acid[11] and 3-allyloxy groups undergo an ortho Claisen rearrangement on heating at 195°.[11] Thus, 3-allyloxy-6-methylphenanthridine (**249**) forms an isomeric phenol to which the structure **250** has been assigned (although the compound was not obtained pure).

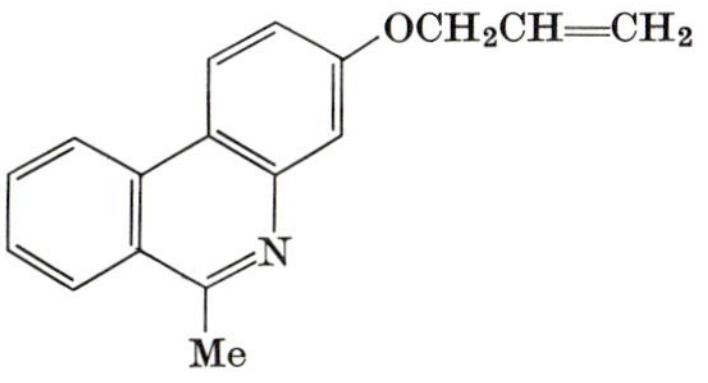

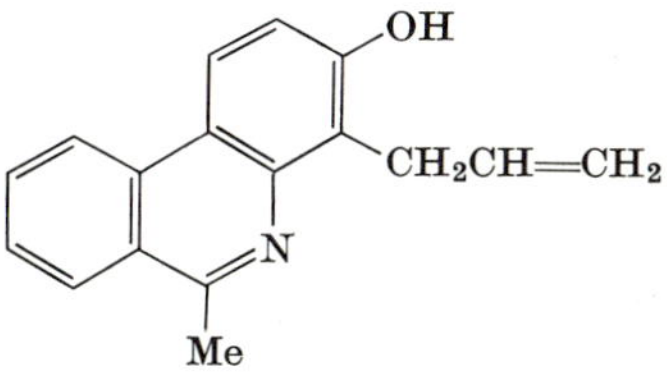

Mention has already been made (see p. 383) of the thermal O→N rearrangement of 6-aryloxyphenanthridines.[129]

The lithium aluminum hydride reduction of phenanthridones to dihydrophenanthridines is easily accomplished, and has considerable synthetic value[90, 134, 211, 213]; dehydrogenation to the parent bases occurs readily and, unless air is rigorously excluded, may be spontaneous. On treatment with lithium aluminum hydride, followed by acidification, N-methylphenanthridones form the corresponding quaternary salts.[211] N-Phenylphenanthridone gives, in contrast, 5-phenyl-5,6-dihydrophenanthridine.[333]

With an excess of methylmagnesium bromide the tetrahydrophenanthridone (**251**) forms **252** and the salt (**253**) in roughly equal amounts.[202]

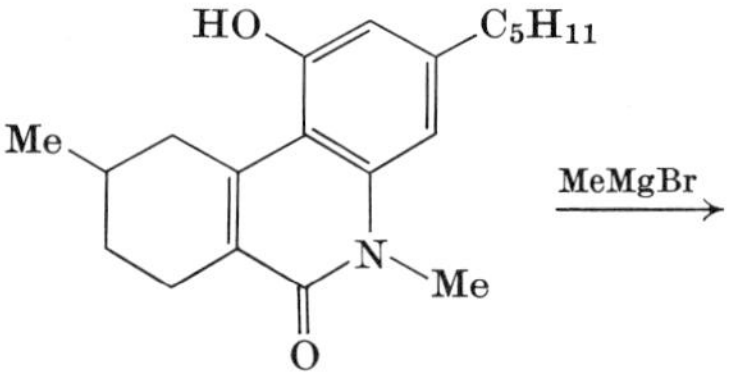

(251) → MeMgBr → (252) + (253)

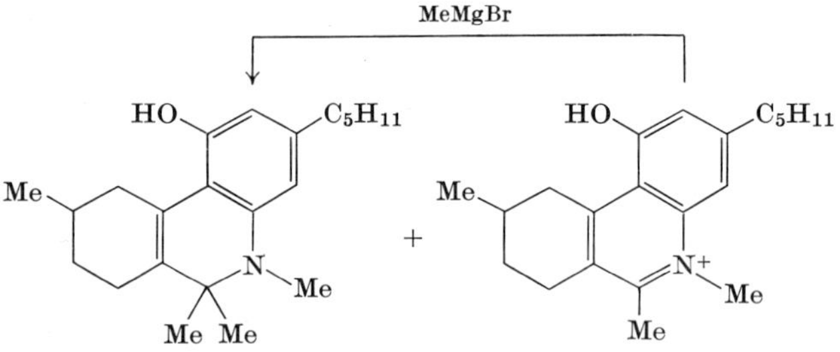

## F. Cyano Groups

6-Cyanophenanthridine forms the corresponding amide on treatment either with 90% sulfuric acid, or hydrogen peroxide in alkaline solution.[296] Complete hydrolysis, followed by decarboxylation to the parent phenanthridine, provides a ready method of structure determination.[80]

## G. N-Oxides

Phenanthridine N-oxides are smoothly deoxygenated by phosphorus tribromide in the usual manner.[353] Alternatively hydrogenation over Raney nickel can be used to obtain the parent base.[353] With ethereal lithium aluminum hydride, dihydrophenanthridine is formed.[354]

Further oxidation of the N-oxide with alkaline hydrogen peroxide gives 6-hydroxyphenanthridine-5-oxide.[353]

Dimethyl sulfate converts phenanthridine N-oxide into the methosulfate (254) and under Reissert conditions phenanthridone and 6-cyanophenanthridine are obtained.[353] [The latter compound is also obtained in good yield by treating the methosulfate (254) with potassium cyanide.[353]]

[353] E Hayashi and Y. Hotta, *J. Pharm. Soc. Japan* **80**, 834 (1960).
[354] W. Traber and H. Hubmann, *Helv. Chim. Acta* **43**, 265 (1960).

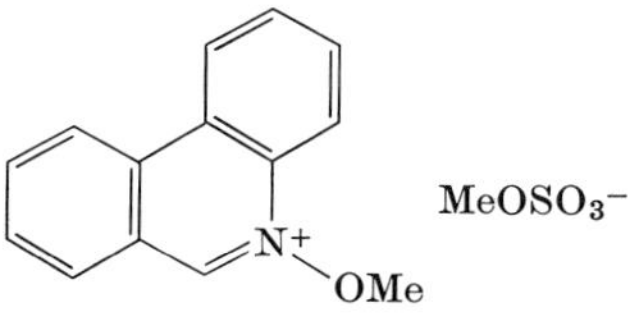

**(254)** MeOSO$_3^-$

Reference has been made previously to the conversion of phenanthridine *N*-oxides into 6-chlorophenanthridines by the action of sulfuryl chloride[309] or phosphoryl chloride.[295] Rearrangement to phenanthridone occurs on heating with acetic anhydride or *p*-toluenesulfonyl chloride.[353]

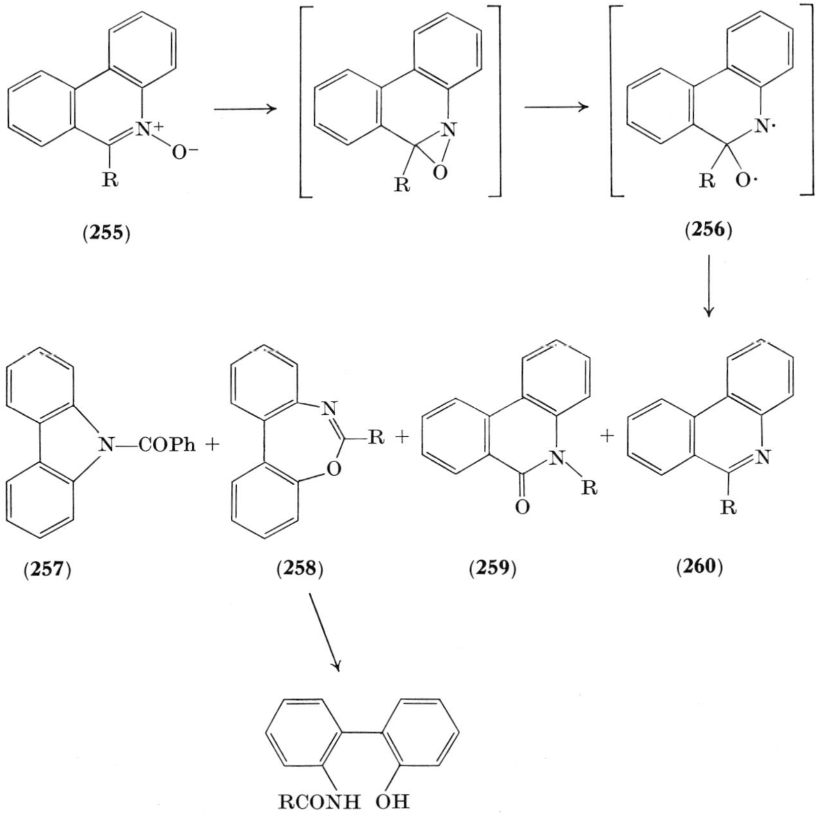

(255) → [ ] → [ (256) ]

↓

(257) N—COPh + (258) —R + (259) + (260)

↓

(261) RCONH OH

The photochemistry of phenanthridine $N$-oxides is typical of aromatic amine $N$-oxides.[355, 356] Rearrangement to phenanthridones is favored in protic solvents and good yields are obtained. In benzene, mixtures of products are formed; for example, 6-phenylphenanthridine 5-oxide (**255**) (R = Ph) gives five products (**257–261**) in the yields shown. The major product, 2-benzamido-2′-hydroxybiphenyl, arises from the hydrolysis of 6-phenyldibenz[$d,f$][1,3]oxazepine (**258**).[356] The isolation of 9-phenylcarbazole (**257**) is interesting since the compound cannot in this case[357] arise by the cyclodehydration of **261**. Support for the diradical intermediate (**256**) comes from the isolation of appreciable amount of 1,1,2,2-tetraphenylethane in the photolysis of 6-diphenylmethylphenanthridine 5-oxide (**255**) (R = Ph$_2$CH), and the total loss of optical activity observed when partly resolved (+) 6-($\alpha$-phenylpropyl)phenanthridine-5-oxide is rearranged photochemically in benzene or ethanol to 5-($\alpha$-phenylpropyl)phenanthridone.[356]

## H. Quaternary Salts

Quaternary phenanthridinium compounds are reduced to the corresponding $N$-alkyldihydrophenanthridines by hydrogenation over Adams catalyst. The reaction forms part of the modified Emde sequence[284] shown below.

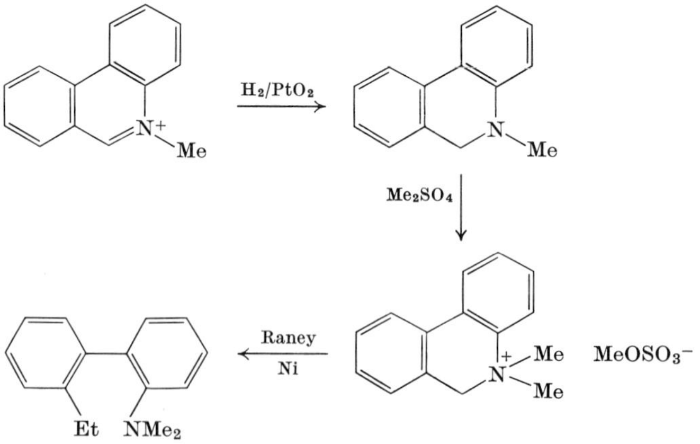

---

[355] M. Ishikawa, S. Yamada, H. Hotta, and C. Kaneko, *Chem. Pharm. Bull.* **14**, 1102 (1966).
[356] E. C. Taylor and G. G. Spence, *Chem. Commun.* 1037 (1968).
[357] *cf.* O. Buchardt, J. Becker, and C. Lohse, *Acta Chem. Scand.* **20**, 2467 (1966).

The oxidation of $N$-methylphenanthridinium hydroxide to $N$-methylphenanthridone by benzoquinone appears mechanistically similar to the older ferricyanide method. Use of the quinone is reported to offer no preparative advantage.[358]

The betaine (**262**) forms phenanthridine almost quantitatively on either heating with concentrated hydrochloric acid at 200° or refluxing with 20% aqueous potassium hydroxide.[283] Oxidation of **262** with alkaline hydrogen peroxide gives phenanthridone, whereas with alkaline ferricyanide $N$-(*trans*-1,2-dicarboxyvinyl)phenanthridone (**263**) is formed.[283]

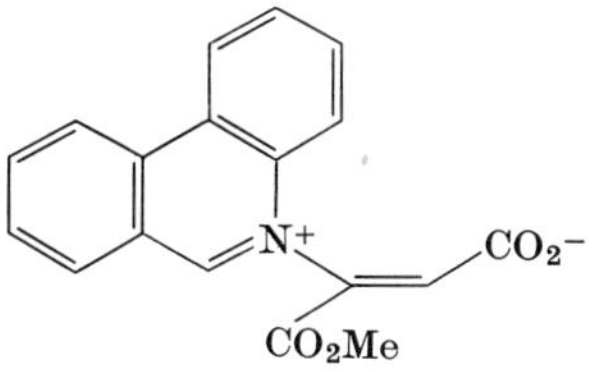

(**262**) 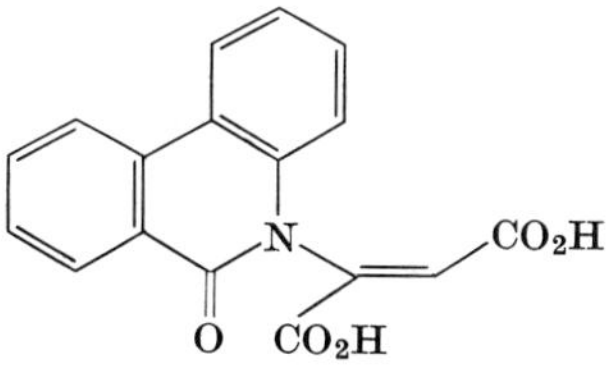   (**263**)

[358] T. Hass, *Acta Chem. Scand.* **18**, 1806 (1964).

# Author Index

Numbers in parentheses are reference numbers and indicate that an author's work is referred to, although his name is not cited in the text.

## A

Abe, K., 352
Abraham, N., 329, 407 (78)
Abraham, R. J., 239, 242
Abrahams, S. C., 213
Abramovitch, R. A., 11, 27, 326, 327 (62), 331, 346, 392 (62)
Acheson, R. M., 241, 373, 376, 377 (245), 380, 382 (245), 384 (245, 283), 385, 386 (285), 402 (285), 403 (285), 404 (245), 413 (283)
Acton, E. M., 28, 29 (119)
Adalberon, M., 42
Adam, W., 241
Adams, R., 282
Adkins, H., 327
Ahmad, Y., 331
Albert, A., 143, 146, 241, 281, 316 (3), 373, 377 (248)
Alder, R. W., 346, 367 (142)
Allen, K. A., 377, 394 (268)
Allinger, N. L., 362, 408 (194)
Alukar, R. H., 324
Amat di San Filippo, P., 253, 311 (109), 312, 313
Ames, G. R., 325
Amstutz, E. D., 240
Anand, N., 367
Anderson, D. G., 348
Anderson, D. J., 46, 47 (4), 75 (4)
Anderson, R., 237
Anderson, R. C., 363
Andrews, L. J., 301
Andrussow, K., 242
Anet, F. A. L., 336
Angelico, F., 290
Angelini, C., 328
Angell, C. L., 81

Anisimov, K. N., 41
Ankersmit, H. J., 316
Anthoine, G., 377, 378
Antipina, T. V., 10
Apprahanian, N. S., 252
Arakawa, K., 9
Aratani, T., 24, 27
Arauner, E., 252
Arcus, C. L., 337, 338, 340 (114), 364 (117), 388 (117), 395 (111), 396 (111)
Armstrong, A. T., 86
Armstrong, K. J., 286
Arndt, F., 162, 179 (2), 182 (1, 2), 183 (1, 2), 195 (2)
Arnold, D. R., 74
Aroyan, A. A., 284
Arthur, H. R., 320
Arumugan, N., 322, 345
Asatiani, L. P., 33
Asbury, R. L., 18
Ashton, B. W., 361
Aue, D. H., 47
Augenstein, L., 133
Austin, M. W., 256, 282, 283 (126)
Ayata, A., 46
Azogu, C. I., 11, 27

## B

Bac, N. V., 284
Bachmann, D. M., 295
Baciocchi, E., 247, 248 (76), 312
Bader, H., 364, 409 (202)
Badger, G. M., 319, 332, 348, 368, 371, 392 (90), 393 (213), 395 (213), 396 (90), 398 (90, 215), 408 (232), 409 (90)
Baggett, N., 34

Bailey, A., 392, 395 (325)
Bailey, A. S., 365, 380, 386 (285), 395 (203), 397 (203), 402 (285), 403 (285)
Bak, B., 237, 241 (12)
Baker, F. W., 301
Baker, R., 304
Balaban, J. E., 250
Banger, A., 319
Banks, R. E., 56, 57
Baranowicz, T., 353
Barben, I. K., 22, 25 (102)
Barber, H. J., 317, 393
Bardi, R., 162, 209 (5), 215
Barlin, G. B., 146
Barr, T. H., 10
Barton, D. H. R., 367
Basu, S., 373, 374, 377
Basu, U. P., 323
Battersby, A. R., 367
Bauer, K., 43
Bauer, W., 56, 76 (31)
Baumgarten, E., 374
Bavin, P. M. G., 336, 339
Becker, J., 412
Beckman, E., 334
Bedell, S. F., 284
Beer, R. J. S., 175, 180, 181, 186, 189 (32, 46, 61), 190 (46, 61), 194 (46, 50, 61), 206 (46), 214, 216, 225 (61), 226 (50, 61), 234 (46, 50, 61)
Behr, L. C., 283
Behringer, H., 176, 177 (38), 179, 182 (38), 183 (38), 189, 191 (64), 192 (38), 194 (38), 197, 200 (77), 201 (77), 224 (38), 225 (38, 54), 226 (38, 54, 64), 229 (64, 69, 82)
Belen'kii, L. I., 246, 290, 300 (300), 305
Belevskii, S. F., 318
Bell, F., 329
Bell, F. R., 316, 317 (3)
Belov, V. N., 359
Bender, D., 191, 229 (69)
Bendich, A., 101
Benford, G. A., 254
Benkeser, R. A., 6
Berde, H., 355
Berg, S. S., 405, 406 (345, 346)

Bergmann, E. D., 101, 102, 107 (93), 122 (98), 135, 136 (146), 137 (98)
Bergmann, F., 101, 107 (93), 130
Bergsson, G., 81
Berklava, I., 256
Berliner, E., 243, 249, 252, 253
Bernhauer, K., 84
Bernheimer, R., 310
Berry, D., 57
Berthier, G., 241
Berthod, H., 85, 86, 94 (59), 95 (59), 101, 107 (93), 121, 133, 143, 153
Bestmann, H. J., 62
Bezzi, S., 162, 209 (5), 214, 215
Bhattacharjee, M. K., 381
Bieber, T. I., 23, 24 (106)
Biel, J. H., 364, 409 (202)
Bignebat, J., 176, 177 (39), 180 (39), 181 (39), 187, 194 (62), 234 (62, 72)
Bindra, J. S., 367
Binks, J. H., 261
Birss, F. W., 240
Bisagni, M., 284
Bisnette, M. B., 43
Blackborow, V. R., 256, 282 (126), 283 (126)
Blackwood, J. E., 164
Blaschke, H., 47
Blicke, F. F., 284
Bloomfield, J. J., 30
Blout, E. R., 81
Bobbitt, J. M., 284
Boerwinkle, F. P., 58
Boettcher, F. P., 359
Bogaert-Verhoogen, D., 375, 376 (263)
Bohlmann, R., 172, 220 (22), 231 (22)
Boichard, J., 21, 22 (96, 97), 29 (96, 97), 32, 33 (96, 97), 35 (95, 96), 36 (98), 37 (96, 97), 39 (95, 96, 97, 98)
Bolton, E. S., 10
Bond, G. J. F., 380, 384 (283), 413 (283)
Bondi, A., 213
Bonta, I. L., 398, 404 (332)
Booth, H., 376
Borsche, W., 353
Boshard, H. H., 260
Bosshard, D. P., 236, 295 (1)

Bouillon, C., 189
Bourne, E. J., 258
Booth, D. J., 19
Bothorel, P., 111
Bourne, E. J., 34
Boyd, W. J., 290
Boyer, F., 357
Boyer, J. H., 54, 60, 283
Bozzini, S., 318
Bradfield, A. E., 247
Bradley, D. F., 151, 152 (167)
Bradley, E. le R., 344
Bradsher, C. K., 318, 380 (17), 387 (17, 287)
Braude, E. A., 273, 277, 329
Bredewey, C. J., 286
Bresinsky, E., 172, 220 (22), 231 (22)
Breslow, R., 48
Bretherick, L., 317, 405
Brewster, E. W., 9
Brimacombe, J. S., 34
Broadhead, G. D., 21, 28, 31 (99)
Brodrick, C. I., 379, 380 (280), 405 (280)
Brookes, P., 115, 119 (126), 121
Brophy, G. C., 286
Brouard, J., 395
Brouwer, D. M., 251, 252
Brown, D. N., 320, 405 (32)
Brown, D. J., 81
Brown, D. M., 115
Brown, E. I. G., 177, 187 (41), 198
Brown, G. M., 146, 155
Brown, H. C., 247, 262, 272, 273, 275, 276 (189), 279, 299 (185), 310
Brown, J., 48
Brown, K., 292
Brown, K. C., 175, 189 (32)
Brown, K. N., 405
Brown, N. M. D., 286
Brown, P. M., 327, 401 (64)
Brown, R. D., 244, 260, 262, 273 (171, 172, 173), 277 (171), 279 (171)
Brown, R. K., 290
Brownfield, R. B., 332, 333 (91)
Browning, C. H., 316, 317 (3)
Bryan, R. F., 82
Bublitz, D. E., 1

Buchanan, A. S., 262, 273 (171, 172, 173), 277 (171), 279 (171)
Buchardt, O., 402, 412
Buddenbaum, A. E., 310
Bürki, H., 145
Bugg, C. E., 82
Bullock, E., 242
Bunnett, J. F., 295
Bunting, J. W., 143
Burgard, A., 48
Burgess, E. M., 46
Burgot, J. L., 172
Burian, R., 260
Burton, A. G., 314
Burvill, M. I., 316 (3)
Burwell, J. T., 213
Butler, A. R., 245, 249, 271 (56), 298, 302, 303
Butler, J. M., 324
Butler, M. E., 50, 75 (15, 16)
Buu-Hoi, N. P., 284, 294, 295, 318

C

Cadogan, J. I. G., 330
Cagniant, D., 293
Cagniant, P., 293
Caillaud, G., 178
Caille, S. Y., 243
Caillet, J., 84, 153, 157
Cain, C. E., 9, 12, 14 (48), 37 (48)
Caldwell, A. G., 332, 337 (89), 389 (89)
Carithers, R., 46
Carmichael, J., 316, 317 (3)
Caroll, D. G., 86
Caronna, G., 332
Carpanelli, C., 241
Carr, R. P., 175, 180, 189 (32, 46), 190 (46), 194 (46), 206 (46), 234 (46)
Cartwright, D., 180, 181, 186, 189 (46, 61), 190 (46, 61), 194 (46, 50, 61), 206 (46), 214, 225 (61), 226 (50, 61), 234 (46, 50, 61)
Caserio, M. C., 264
Castro, A. J., 259
Catlin, W. E., 242, 299

Cava, M. P., 349
Chafetz, H., 363
Chakravorti, S. S., 323
Challenger, F., 284
Challis, B. C., 245
Chaldler, G. S., 319, 373(26), 374(26), 377(26), 388, 395(26), 406(26), 408(26)
Chané, J. P., 299
Chardonnens, L., 334, 341(99), 382(99)
Charton, M., 312
Chase, F. H., 356, 405(178)
Chaudhuri, J. N., 373
Chen, H. H., 104, 105, 135(99), 136(99)
Chiang, S. T., 328
Chiorboli, P., 241
Chirkova, V. V., 354
Chowdhury, M., 374
Christensen, B. E., 81
Christensen, D., 237, 241(12)
Chu, E., J.-H., 259
Chu, T. C., 259
Churanov, S. S., 9, 10(24)
Ciamician, G. L., 257
Ciana, A., 248
Clar, E., 371
Clark, L. B., 104, 105, 129, 135(99), 136(99), 138(141)
Classen, R. A., 348
Clauson-Kaas, N., 255
Claverie, P., 151, 153(168, 169)
Cleaver, C. S., 56
Clementi, S., 247, 249(77), 257, 258(142, 143), 259, 262(75), 266(142), 269(75), 270, 271(77), 273(75), 276, 277(192), 284(77), 285, 286(77), 287(77), 288(77), 289, 291(77), 293(143), 294(147), 300(143, 302), 305, 306(143), 307(143, 301, 302, 314), 308(302), 309, 311(143, 302), 314
Clesse, F., 170, 182(19), 188, 192(19)
Coffin, B., 327, 330(63), 332(63)
Coleman, K. J., 43
Collington, D. M., 344
Conley, R. T., 337
Cook, 345, 409(134)

Cooksey, M. R., 313
Coombs, M. M., 337, 338, 339, 340(114), 364(117), 388(117), 395(111), 396(111)
Coonradt, H. L., 247, 293
Cooper, W. D., 258
Coppens, G., 370, 371(226), 374, 377, 379
Cornforth, J. W., 281
Corriu, R. J. P., 243
Corwin, A. H., 251, 311(94)
Coulibaly, O., 174, 177(27), 180(27), 192(27), 198(27)
Coulson, C. A., 370
Cram, D. J., 49, 65, 70(12), 75(12)
Credner, H. H., 392
Creech, H. J., 392
Crick, F. H. C., 84
Cross, P. E., 353, 399(173)
Cuingnet, E., 13, 42(51)
Cumper, C. W., 369
Curphey, T. J., 6
Curtin, D. Y., 383
Cusachs, L. C., 86
Cusic, J. W., 21, 44(93)
Cymerman-Craig, J., 377, 394(268)

# D

Dabard, R., 16, 25, 32(73), 35(112, 113), 38(112, 113), 41(113)
Dalton, D. R., 346, 367(142)
Danilov, V. I., 122, 157
Dannies, P., 110
Danyashevskii, Ya. L., 263
DaRooge, M. A., 362, 408(194)
Das, B. P., 323
Das, M., 377
Das Gupta, A. C., 323
Daves, G. D., 359
Davey, W., 325
Davies, C. S., 43
Davies, D. W., 238, 239
Davis, J. C., Jr., 81
Davis, M., 392, 395(324), 404(324), 405(324), 407(324), 408(324)

Davison, A., 11
Davy, H., 191
Deans, F. B., 262, 263 (174), 269 (174), 273 (174)
DeBelder, A. N., 34
Decius, J. C., 81
Decombe, J., 357
De Dominicis, A. J., 295, 296 (275)
De Fabrizio, E. C. R., 253, 254 (108), 292 (107), 311 (109), 312 (109), 313 (109)
Defay, N., 375, 376 (263)
de Gauock, V., 369
de Heer, J., 240
De Joung, H. A. P., 238, 239
de la Mare, P. B. D., 236, 246
Dell'Erba, C., 240, 267 (22), 299
Delobelle, J., 398, 404 (332)
Del Re, G., 85, 96, 241
De Matte, M. L., 328
De Mayo, P., 368, 409 (211)
Demuynck, M., 191
Dennstedt, M., 257
Denton, D. A., 323
Dermer, O. C., 69
Derocque, J. L., 200
Desai, R. B., 349
DeVoe, H., 107, 118, 151, 155 (166)
Dewar, M. J. S., 240, 272, 298, 336, 370, 371, 381, 388 (227), 389 (227)
Dezelic, M., 259
Diamantis, A. A., 377, 394 (268)
Dickinson, W., 405
Diels, O., 384
Dille, K. L., 81
Dingwall, J. G., 167, 181, 183 (11), 193 (49), 194 (70), 196 (71), 203 (49), 224 (49), 225 (49), 226 (49), 227 (49), 230 (49), 231 (49), 232 (49), 233 (49), 234 (49)
Dixon, W. B., 237
Doak, K. W., 241, 311 (94)
Dobryak, V. A., 20
Doerr, F., 372
Dogliotti, L., 348
Donohue, J., 82, 102, 145, 213
Dorofeenko, G. N., 256
Dorsett, M. Y., 23, 24 (106)

Drake, J. W., 157
Dramaix, R., 377
Dran, R., 325
Drenchko, P., 395
Drobnik, J., 133
Drozd, V. N., 16, 18, 20, 37 (77, 78)
Drushel, H. V., 372
Dowbenko, R., 381
Duffield, A. M., 402
Duffin, H. C., 260
Duguay, G., 182
Dulenko, V. I., 256
Dunlop, A. P., 248, 264, 276 (80)
Durmashkina, V. V., 380, 403 (286)
Dyer, E., 107
Dyke, S. F., 346, 352, 401
Dzurilla, M., 26, 37 (116)

# E

Eaborn, C., 262, 269 (174), 273 (174), 277 (175), 286 (175), 291 (175), 300 (297), 303, 304, 308 (297)
Easton, D. B., 177, 189 (40), 191, 204 (40)
Eckert, R. J., 256
Edmiston, W. E., 363
Edwards, W. R., 256
Efraty, A., 43
Efros, L. S., 376, 383, 403 (264), 404 (298)
Egger, H., 14, 16 (62), 37, 41
Eguchi, S., 70
Eigendorf, G., 333
Eisch, J. J., 236, 244 (7), 316, 317, 373, 382, 383 (291), 387 (5, 291), 388, 389 (291), 390 (5, 291), 381, 394 (314), 395 (5, 291), 398 (327), 400 (310, 313, 314), 404 (313, 314), 405 (291), 406 (5), 409 (291)
Eldridge, E. M., 317
Elecko, P., 14, 35, 36
Elion, G. B., 119 (122), 121
Elkins, J. R., 328
Ellers, J. C., 329, 383 (77), 408 (77)
Elston, C. T., 325
El'tsov, A. V., 348
Elvidge, J. A., 238, 239

Engelhart, G., 7
Englemann, J. H., 383
Engelmann, T. R., 31
Ermili, A., 259
Essex, J. L., 408
Eugster, C. H., 236, 295 (1)
Evans, D. F., 371
Evans, J. V., 337, 338, 395 (111), 396 (111)
Everitt, P. M., 319

## F

Fabian, J., 241
Fabio, P. F., 392
Fahey, R. C., 249
Fairfull, A. E., 318
Fakstorp, J., 255
Fales, H. M., 367
Falk, H., 2, 3 (9), 4 (9), 5 (9), 25, 38, 43, 44
Falkenberg, J., 191, 197, 200 (77), 201 (77), 229 (69)
Fargher, R. G., 260
Farrar, M. W., 256, 284
Farris, R. E., Jr., 107
Favini, G., 153
Fawcett, J. S., 273, 277, 329
Fedotova, O. Y., 318
Felitte, A., 327
Fernandez-Alonso, J. I., 85
Ferorov, V. E., 27
Festal, D., 220
Feuer, B., 72
Feuer, B. I., 330, 331 (83)
Fields, M., 81
Fierens, P. J. C., 272
Fike, S., 282
Finar, I. L., 259
Fischer, E., 348
Fischer, E. O., 41, 42 (155)
Fischer-Hjalmars, I., 121
Fitzgerald, W. P., 6
Fletcher, T. L., 335, 337 (101), 389 (101), 390 (101), 406 (101)
Flood, S. H., 243, 255
Flores, M., 15, 44 (69)
Folmer, O. F., 329, 383 (77), 408 (77)
Foote, C., 324

Forsythe, P. P., 314
Foster, A. B., 34
Foucart, J., 384
Fourche, G., 111
Fournari, P., 295, 296 (273)
Fowler, F. W., 46, 53, 54 (22), 57, 58 (24), 66 (22), 67 (23), 70 (23), 73 (22)
Fox, J. J., 101
Fraenkel-Conrad, H., 115
Franck, B., 366
Franz, H., 44
Fraser, R. D., 81
Frazier, J., 81, 123 (10)
Freese, E., 122, 157
Frei, Y., 348
Frerichs, A. K., 9
Freure, B. T., 255
Fridinger, T. L., 51
Fried, M., 2, 3 (9), 4 (9), 5 (9), 18, 19 (81), 20 (81)
Friedler, H., 301
Friedman, B. S., 282
Friedman, S., 399
Friedrich, H. J., 369, 391, 392, 403 (317)
Fringuelli, F., 276, 299, 314
Fujita, H., 123
Fukuda, N., 151, 153 (169)
Furdik, M., 14, 16, 21, 26, 35, 36, 37 (116)
Furth, B., 330, 384 (80), 408 (80), 410 (80)
Furukawa, M., 299
Fusco, R., 281
Futterer, E., 181

## G

Gaertner, R., 284
Gagan, J. M. F., 385
Gait, R. J., 181, 194 (50), 216, 226 (50), 234 (50)
Gajewski, J. J., 48
Gal'bershtam, M. A., 262, 293, 303 (304), 308
Galt, R. H. B., 361, 362
Galust'yan, G. G., 257
Garbuglio, C., 162, 209 (5), 214
Gatlin, L., 81
Gautheron, B., 16, 25, 32 (73), 35 (112), 38 (112)

Gawlak, M., 328
Geerts-Evrard, F., 375, 376 (263)
Genel, F., 257, 258 (142), 259, 266 (142), 294 (147)
Gensler, W. J., 366
Gerasimenko, A. V., 18, 19
Germain, C. B., 294 (276), 296
Ghaisas, V. V., 284
Giacometti, G., 93
Gibson, M. S., 323, 347, 361
Giessner-Prettre, C., 85, 86, 94 (59), 95 (59), 110, 121, 133
Gilchrist, T. L., 46, 47 (4), 75 (4)
Gill, T. J., III, 44
Gillespie, R. J., 281
Gillet, C., 371
Gilman, H., 238, 254, 263, 316, 317, 382, 383 (291), 387 (5, 291), 388, 389 (291), 390 (5, 291), 391, 394 (314), 395 (5, 291), 400 (310, 313, 314), 404 (313, 314), 405 (291), 406 (5), 409 (291)
Ginman, R. F., 369
Given, P. H., 378
Gladys, C. L., 164
Gleicher, G. J., 370
Gleiter, R., 211
Goel, N. S., 151, 153 (169)
Gogan, N. J., 43
Gogte, V. N., 355
Gokel, G., 24
Gold, V., 245
Goldacre, R., 377
Gol'dfarb, Ya. L., 246, 256, 257, 263, 293 (141), 300 (300), 305
Goldman, G., 262
Goldstein, J. H., 81, 150
Golubeva, I. A., 31
Golubeva, S. K., 292
Gonda, T., 24, 27
Gopal, R., 360
Gopinath, K. W., 320, 404 (34), 408 (36)
Gordon, E., 282
Gordon, J. T., 351
Gore, P. H., 257, 258
Goto, R., 390
Govindachari, T. R., 319, 320, 322, 323, 345, 404 (34), 408 (36)

Gräbe, C., 316
Graham, P. J., 15, 21 (67), 28 (67), 29 (67), 32 (67)
Grandberg, I. I., 14, 35 (56), 39 (56), 281
Grandin, A., 199, 216 (80), 224 (80)
Grant, M. S., 290
Grashley, R., 381
Greenshields, J. B., 240
Grimison, A., 241, 251, 261 (98), 281
Grimm, A., 183, 225 (54), 226 (54)
Grimmett, M. R., 313
Grisdale, P. J., 298
Grom-Dursum, K., 259
Gronowitz, S., 236, 293 (2), 295 (2)
Gropper, H., 372
Gross, H., 259
Gross, M. L., 273 (239), 277 (239), 280 (239), 285, 286 (239), 291 (239), 293 (239)
Grovenstein, E., 252
Guanti, G., 240, 267 (22), 299
Gudriniece, E., 354
Guillouzo, G., 162, 167 (7), 208 (4), 219 (4), 230 (7), 231 (7), 232 (7), 233 (7)
Gundermann, K., 290
Gustafson, D. H., 9
Gverdtsiteli, I. M., 33
Gvodzeva, E. A., 300 (299), 305

H

Habib, M. J. A., 10
Hacker, H., 391
Hadlington, M., 23, 24 (107)
Hafner, K., 56, 76 (31)
Haines, A. H., 34
Hall, G. A., 408
Hall, J. E., 333, 356
Halls, C., 319
Halvarson, K., 245, 263 (57), 269 (57)
Ham, G. E., 69
Hammond, G., 270, 311 (183)
Handlar, K., 37
Hansch, C., 284
Hansen, J., 359
Hansen-Nygaard, L., 237, 241 (12)

Hari, K., 240, 267 (21)
Harper, E. T., 249, 250 (91)
Harris, D., 180, 186, 189 (46, 61), 190 (46, 61), 194 (46, 61), 206 (46), 225 (61), 226 (61), 234 (46, 61)
Harris, T. M., 182
Harrison, D. R., 385
Hartough, H. D., 247, 257, 284, 293, 294 (135)
Harvey, G. R., 58, 75 (40)
Hass, T., 413
Hassner, A., 46, 53, 54 (22), 57, 58, 66 (22), 67 (22), 70 (22), 73 (22)
Hata, K., 22, 28 (103), 35, 36
Hatch, M. J., 49, 65, 70 (12), 75 (12)
Hauser, C. R., 2, 6, 9, 12, 14 (48), 37 (48), 182
Havinga, E., 251, 252 (99, 100)
Hayashi, E., 383, 384 (297), 387, 390, 392 (297), 394 (296, 309), 395 (309), 397 (309), 406, 408 (296), 410 (296), 411 (309)
Hayes, K. J., 255
Heacock, R. A., 340, 345 (122), 406 (122)
Hemetsberger, H., 59
Hendrikson, J. B., 324, 346, 367 (142)
Hendry, J. B., 245, 249, 271 (56), 302
Henery-Logan, K. R., 51
Hennig, H., 25 (57, 58), 35 (57), 37 (57), 39 (57)
Hertz, H. G., 167
Hetzheim, A., 283
Hewlins, M. J. E., 115
Hey, D. H., 320, 328, 340, 341, 342, 343, 344 (123), 345 (122), 346, 347, 360, 366 (128), 367 (142), 375 (127), 381 (126), 383 (129), 395 (126), 396 (126, 129), 397 (129), 404 (128), 405 (32), 406 (122, 126, 127), 409 (126, 127, 129)
Heydenhauss, D., 26
Hill, E. A., 273 (239), 277 (239), 280 (239), 285, 286 (239), 289, 291 (239), 293 (239)
Hill, J., 405
Hill, M., 325

Hillers, S., 255, 256
Hinze, J., 94
Hirai, T., 299
Hisatome, M., 9, 10 (31), 11 (31)
Hoarau, J., 111
Hodge, P., 296
Hoeft, E., 398, 409 (333)
Hoergen, E., 81
Hoffman, R. A., 293
Hoffmann, A. M., 392
Hoffmann, K., 281
Hoffmann, P., 24
Hoffmann, R., 86, 211
Hoft, E., 259
Hoggett, J. G., 254, 255 (113)
Hollingsworth, B. L., 318, 321, 325 (14), 353, 355 (170, 171), 356, 379, 380 (280), 390 (171), 400 (14), 404 (14), 405 (39, 280)
Hollis, D. P., 81, 110
Holt, S. J., 317
Honig, B., 151, 152 (167)
Hoops, J. F., 364, 409 (202)
Hoppe, W., 369
Hondvik, A., 213, 214, 216, 217
Horning, E. C., 334
Hortmann, A. G., 73
Horton, C. A., 81, 115, 119 (124), 121
Hoshi, Y., 383, 384 (297), 392 (297)
Hotta, H., 412
Hotta, Y., 410
Howard, F. B., 81, 123 (10)
Howe, E. E., 290
Huang, E. C. Y., 50, 75 (15)
Hubert, A., 318, 369
Hubmann, H., 410
Huffman, J. W., 18, 31, 32 (127)
Hughes, E. D., 254
Huh, G., 48
Huisgen, R., 47, 62, 381, 386
Huisman, H. O., 358
Humffray, A. A., 262, 273 (171, 172, 173), 277 (171), 279 (171)
Hunzicker, F., 382
Huppatz, J. L., 319, 346, 368 (139), 373 (26), 374 (26), 377 (26), 395 (26), 406 (26), 408 (26)

# AUTHOR INDEX

## I

Ibraginova, A. B., 293
Iddon, B., 236
Illuminati, G., 248, 312, 397
Imamura, A., 123
Imoto, E., 240, 255, 263 (121), 267 (21), 273 (121), 299, 300 (121), 302 (121), 303 (121), 308 (121)
Inaoka, M., 406
Ingold, C. K., 254
Ishii, Y., 70
Ishikawa, M., 11, 12, 412
Ishikawa, T., 406
Isomura, K., 54, 55 (27), 59 (27), 70 (27)
Iwai, I., 333
Iwasaki, S., 11
Iwatani, K., 378

## J

Jackson, A. H., 290
Jackson, L. M., 229, 239
Jacobs, T. L., 281, 332, 333 (91)
Jacquignon, P., 318
Jaffé, H. H., 94, 242, 298
Jagow, R. H., 249
Jancke, H., 7
Janda, M., 295
Jardetzky, C., 81
Jardetzky, O., 81
Jardine, R. V., 290
Jarrett, A. D., 362, 400 (193)
Jeffrey, G. A., 213
Jennings, C. A., 31
Jensen, L. H., 82, 101, 145 (42), 146, 155
Johnson, C. D., 236, 244 (8), 314
Johnson, D. C., 292
Johnson, G. C., 247, 293
Johnson, J. R., 255
Johnson, P. L., 214, 216
Johnson, R. A., 333
Johnson, S. M., 214
Johnston, K. M., 321
Johnstone, R. A. W., 181, 194 (50), 208, 224 (92), 226 (50), 234 (50)
Jones, A. M., 403
Jones, B., 247
Jones, D. H., 392
Jones, E. R. H., 353, 399 (173)
Jones, G. H., 343
Jones, H. L., 242, 298
Jones, J. W., 115, 119 (123), 121
Jones, M., 262
Jones, R. A., 219, 236, 319, 373 (26), 374 (26), 377 (26), 395 (26), 406 (26), 408 (26)
Jorge, H., 277
Joshi, K. K., 43
Joshua, C. P., 348
Julshamn, K., 214

## K

Kaeppeler, W., 358
Kahlert, B., 284
Kaiser, G. V., 309
Kakuda, T., 282
Kalenda, N. W., 318
Kalinina, S. P., 36
Kalmus, A., 130
Kamrad, A. G., 245, 246 (54), 300 (299), 305
Kan, P. T., 11
Kandror, I. I., 262, 303 (304), 308
Kaneko, C., 381, 412
Kaplan, N. O., 110
Kataev, E. G., 293
Kato, H., 65, 68 (57), 71
Kato, T., 386
Katritzky, A. R., 219, 236, 244 (8), 292, 314
Kaufman, M. L., 399
Kauffmann, T., 359
Kaufmann, Th., 391
Kealy, T. J., 1
Keene, B. R. T., 319, 334 (22), 335, 336 (22), 337 (22), 338 (22), 368 (22, 102), 373 (22), 375 (22, 244), 376 (22, 102), 377 (244), 378 (274), 379 (102), 381 (113), 391 (22), 392 (102), 395 (102, 113), 396 (102, 113)
Kefurt, K., 295
Kehl, W. L., 213
Keith, L. H., 18
Kellogg, R. M., 249, 250 (91)
Kellom, D. B., 325

Kelly, R. B., 341
Kenner, G. W., 336
Kessar, S. V., 353, 360
Kharasch, N., 350
Khomutova, E. D., 282
Kieselack, P., 7
Kikuchi, K., 241
Kilimov, A. P., 372
Kim, G. J., 310
Kimber, R. W. L., 380, 387 (287)
King, R. B., 43
Kirby, G. W., 367
Kirby, J., 44
Kirkor, W., 353
Kiselov, B. I., 325
Kitaygonodskii, A. I., 153
Kitimura, T., 61
Kittle, P. A., 9
Kjøge, H. M., 217
Kleimann, H., 24
Kleinwächter, V., 133
Klemm, L. H., 321, 336, 337 (105), 395 (105), 396 (105), 397 (105)
Kliegman, J., 349
Klikorka, J., 37
Klingsberg, E., 173, 174 (26), 175 (26, 29), 176 (29), 177 (29), 178 (29), 181 (26), 183 (29), 192 (26, 29), 197, 200 (76), 201 (76), 202 (76), 203, 204, 207, 225 (26, 76), 226 (26), 227 (26), 229 (76), 232 (26)
Klusacek, H., 24
Knittel, D., 59
Knox, G. R., 10
Koboyashi, S., 54, 59 (28)
Koboyashi, Y., 397
Koblik, A. V., 259
Kobyak, N. K., 329
Kochetkov, N. K., 281, 282
Kochetkova, N. S., 32
Koda, Y., 299
Koenig, H., 63
Kokko, J. P., 81
Kolesnikov, G. S., 318
Kolobova, N. E., 41
Komarova, R. I., 30
Komppa, G., 284

Kon, S. K., 84
Kono, H., 22, 28 (103), 36
Konidze, A. A., 33
Korsakova, I. S., 293
Kontüm, K., 372
Kortum, G., 242
Kosak, A. I., 257, 284, 294 (135)
Kosel, C., 391
Kost, A. N., 14, 35 (56), 39 (56), 281
Kotrelev, V. N., 36
Kovac, S., 14
Kowanko, N., 368, 398 (215)
Kozlov, N. S., 325
Krasa, C., 25
Krasovitskii, B. M., 329
Kraut, J., 82
Krespan, C. G., 56
Krischke, R., 381
Kritskaya, I. I., 10
Krueger, W. E., 54, 60
Kruglyak, Yu. A., 122
Kubo, M., 299
Kubota, T., 372, 373 (243), 378
Kuhn, S. J., 243, 255
Kuhn, W. E., 363
Kumada, M., 11, 12
Kumadaki, I., 397
Kumar, A., 353
Kunii, T. L., 137, 140 (148)
Kunstmann, R., 62
Kupriyevich, V. A., 122
Kurabayashi, M., 333
Kuroda, H., 137, 140 (148)
Kurtz, D. W., 61
Kurzer, F., 283
Kurtney, J. P., 333
Kuznetsova, G. I., 36
Kyba, E. P., 326, 327 (62), 392 (62)
Kwiatkowski, J. S., 121, 123, 125
Kwietny-Govrin, H., 130
Kyogoku, Y., 81

## L

Labriola, R. A., 327
Lacan, M., 27
Lalor, F. J., 43
Lamatz, G., 310

Lambert, D. G., 252, 283 (104)
Lamy, J., 295
Land, A. H., 281
Lang, G., 199
Lansbury, P. T., 325, 326 (61), 335
Lauer, K., 248
Laurent, A., 70
Lauerer, H., 382
Laursen, R. A., 119 (125), 121
Laviron, E., 21, 25, 35 (113), 36 (98), 38 (113), 39 (98), 41 (113)
Lavit, D., 295
Lawesson, S.-O., 248
Lawley, P. D., 115, 119 (126), 121
Layer, R. W., 70
Lazzarone, M., 222
Leaver, D., 177, 187 (41), 189 (40), 191, 198, 204 (40)
Lecocq, J., 284
Lednicer, D., 2
Le Fèvre, R. J. W., 369
Lehner, H., 38
Lenk, C. T., 11
Lenz, G. R., 325
Leonard, J. A., 328, 341, 343, 360, 366 (128), 375 (127), 381 (126), 395 (126), 396 (126), 404 (128), 406 (126, 127), 409 (126, 127)
Leonard, N. J., 63, 64, 65 (55), 119 (125), 121
Leonteva, L. I., 6, 7 (17)
Le Roux, P., 336, 337 (106), 345, 382 (106)
Leshcheva, I. F., 24
Lester Smith, E., 84
Leung, F., 214
Levin, G., 130
Levine, R., 256, 284
Levkoev, I. I., 380, 403 (286)
Levy, L. A., 57
Lewett, G. L., 253, 283 (105)
Lewis, G. E., 348
Lewis, H. B., 350
Lielbriedis, I., 354
Lifschitz, Ch., 102, 122 (98), 137 (98)
Lifson, S., 151, 152 (167)
Lim, E. C., 372
Linda, P., 247, 248, 249 (77, 84), 250, 256, 257, 258 (139), 259, 261, 262 (75), 264, 265, 266, 267, 268, 269 (75), 270 (75), 271 (75, 77), 273 (75), 276, 277 (84, 130, 192), 279 (84, 90), 280 (84), 284 (77), 285 (77), 286 (77), 287 (77), 288 (77), 289 (77), 291 (77), 293 (90, 130), 296 (84), 300 (130), 305, 306, 307 (301, 314), 309, 314
Lindsay, J. K., 2, 14, 15 (55), 37 (55)
Lindsey, R. V., 15, 21 (67), 28 (67), 29 (67), 32 (67)
Lindwall, H. G., 284
Linfield, M. L., 329, 407 (78)
Lipinski, C. A., 273, 285
Lisitsyna, E. S., 22
Lister, J. H., 125
Litvinov, V. P., 257, 293 (141)
Lloyd, H. A., 334
Loader, C. E., 351
Loenig, K. L., 164
Loev, B., 15, 44 (69)
Löwdin, P. O., 82, 84 (46, 47), 93
Logothetis, A. L., 74
Loh, S. M., 319
Lohse, C., 412
Long, C. T., 318
Long, F. A., 245
Long, R. A. J., 341
Longuet-Higgins, H. C., 240, 370
Lord, G. H., 259
Lord, R. C., 81
Lorkowski, H. J., 7, 27, 28 (117)
Loudon, G. M., 273
Loudon, J. D., 330, 345, 361, 362, 409 (134)
Loven, J. M., 242
Lowe, J. L., 379, 380 (280), 405 (280)
Lozac'h, N., 167, 168, 170 (13), 173 (21), 174 (18, 20, 21), 176 (24), 177 (15, 20, 24, 30, 35, 36), 178 (28), 180 (27), 182 (12, 18, 19), 183 (12), 189 (25, 30), 192 (16, 17, 18, 19, 27, 35), 194, 195 (35), 196 (74), 197 (24), 198 (13, 20, 21, 27), 205 (74), 206, 219 (18), 221 (18, 30, 36), 222 (13), 228 (34), 229 (34), 230 (7, 13, 34, 90), 231 (7), 232 (7, 13, 30), 233 (7), 234 (7, 13)

Lubs, H. J., 366
Lucken, E. A., 338
Lukeš, R., 295
Lumbroso, H., 241
Lumme, P. O., 242
Luttringhaus, A., 176, 177 (37), 181

# M

McClellan, A. M., 241
McCloskey, P., 345, 409 (134)
McCullagh, L., 46
McGlynn, S. P., 86
Machus, F. F., 37, 42 (147)
Macintyre, W. M., 82, 145 (38)
McKee, R. L., 368
McKenzie, S., 181, 193 (49), 203 (49), 224 (49), 225(49), 226 (49), 227 (49), 230 (49), 231 (49), 232 (49), 233 (49), 234 (49)
McKinnon, D. M., 177, 187 (41), 189 (40), 204 (40)
McKreevoy, M. M., 240
McLeskey, J. J., III, 6
McMahon, J. E., 9
Madeiras, G., 81
Madelung, W., 290
Madinaveitia, J. L., 44
Maeda, K., 211
Magdesieva, N. N., 236, 245, 246 (59), 279 (59), 293 (4a)
Magnusson, R., 284
Maiorana, S., 58
Maitlis, P. M., 370, 388 (227), 289 (227)
Mallory, F. B., 348, 349 (153), 351
Maloney, J. H., 295
Mamalis, P., 383, 395 (295), 396 (295), 404 (295), 411 (295)
Mammi, M., 162, 209 (5), 214, 215
Manaresi, P., 241
Mandell, L., 81, 150
Mandolini, L., 247, 248 (76)
Mangini, A., 241
Manion, M., 273 (239), 277 (239), 280 (239), 285, 286 (239), 289 (239), 291 (239), 293 (239)

Manly, D. G., 240
Mann, L. T., Jr. 44
Mannschreck, A., 245
Marino, G., 247, 248, 249 (77, 84), 256, 257 (130), 258 (130, 139, 142, 143), 259, 261, 262 (75), 263 (72), 264, 265, 266 (72, 142), 267, 268 (182), 269 (75), 270 (75), 271 (75), 273 (72, 75), 276, 277 (84, 130, 192), 279 (84, 90), 280 (84), 284 (77), 285 (77), 286 (77), 287 (77), 288 (77), 289 (77), 291 (77), 293 (85, 90, 130, 143), 294 (147), 295 (85), 296 (84), 299 (85), 300 (85, 130, 143, 302), 301, 302 (72), 305, 306 (143), 307 (143, 302), 308 (302), 309, 311 (143, 302), 312, 314, 397
Marks, R. E., 338, 340 (114)
Marguarding, D., 24
Marquis, R., 254
Marr, G., 12, 19
Marsh, R. E., 82, 101
Martin, A. E., 368, 397 (214), 398 (214)
Martin, G., 260
Martin, M., 260
Martin, R. H., 272, 375, 376 (263)
Martin-Smith, M., 286
Marvel, C. S., 15, 30
Masamune, T., 329, 358 (75), 375, 376
Mashburn, T. A., Jr., 9, 14
Mason, S. F., 81, 82, 116, 119 (22), 122 (12), 125, 126 (12), 129 (22), 130 (12), 146 (22), 371, 373, 375
Masuda, G., 282
Matesich, M. A., 245
Matsuo, H., 380, 412 (284)
Matvelashvili, B. L., 318
Mautner, H. G., 81
Mayer, R., 245, 266 (53), 267 (53), 271 (53)
Maynard, J. C., 260
Meadow, M., 320
Mechtler, H., 5, 7 (11), 38 (11)
Meek, J. S., 54, 58 (24), 66
Mehlhorn, A., 241
Meimaroglou, C., 290
Meinwald, J., 47
Melander, L., 243, 245, 246, 263 (57), 269 (57)

Melzer, M. S., 6
Menez, C., 174, 177(30), 189(30), 221 (30), 232(30)
Meschke, R. W., 335
Mesley, R. J., 338
Metzger, H., 63
Metzger, J., 21, 36(98), 39(98)
Mez, H. C., 145
Mezentsova, N. N., 293
Michaels, A., 282
Michaels, R. J., 284
Michaelidis, A., 395
Michels, J. G., 255
Mikole, G. J., 54
Miles, H. T., 81, 123(10)
Miles, M. L., 182
Miller, S. A., 1, 284
Mills, B., 319, 356(30)
Mimura, K., 11
Minnier, C. E., 107
Mishima, H., 333
Mishina, T. M., 383, 404(298)
Miskel, J. J., 259
Mitra, S. S., 242
Mitsuhashi, K., 318, 345, 383, 404(11), 406(11), 409(11)
Mitsuhashi, T., 347
Miyazaki, H., 372, 373(243), 378
Mizoguchi, T., 46
Mlinkò, A., 255
Mndzhoyan, A. L., 284
Modest, E. J., 284
Modler, R., 60
Möckel, K., 283
Mohar, A., 9, 10, 14, 27(61)
Moller, F., 281
Mollier, Y., 168, 170(13), 173, 174(18, 20), 176(24, 25), 177(20, 24, 25, 27, 30, 31, 35, 36), 178, 180(27), 182(18), 189(25, 30), 195(35), 192(18, 27, 35), 197(24), 198(13, 20, 27, 31), 219(18), 220, 221(18, 30, 36), 222(13), 230(13), 232(13, 30), 234(13)
Momicchioli, F., 241
Monin, J., 21, 22(96), 29(96), 33(96), 35(96), 37(96), 39(96)
Montgomery, J. A., 83

Moodie, R. B., 254, 255(113)
Moon, B. J., 346, 401
Moore, G. J., 56, 57
Moore, H. W., 339, 376(121)
Moore, R. E., 12
Morgan, G. T., 316, 317
Morgan, M. J., 313
Moriconi, E. J., 399
Moriarty, R., 349
Morizur, J. P., 241
Moroe, M., 33
Morrey, D. P., 313
Morris, J. R., 241
Morrow, D. F., 50, 75(15, 16)
Mosby, W. L., 324, 389(51)
Moser, K. B., 318, 380(17), 387(17)
Most, E. E., 49
Motoyama, I., 22, 28(103), 35, 36
Motoyama, R., 240, 255, 263(121), 267(121), 273(121), 299, 300(121), 302, 303, 308
Moxi, T., 351
Moynahan, E. B., 3, 14, 17, 21(76), 23(76), 27(76), 28(94), 29(94), 35(52), 36(52), 44(52, 94)
Moynehan, T. M., 341, 342, 381(126), 383(129), 395(126), 396(126, 129), 397(129), 496(126), 409(126, 129)
Mowry, D. T., 284
Muller, A., 70
Muller, E., 84
Muller, H., 391
Muller, O., 84
Mulvaney, J. E., 30
Munns, A. R. L., 82
Munro, J. D., 10
Murakami, Y., 240, 267(21)
Murr, B. L., 310
Musante, C., 282
Muth, C. W., 328, 329, 383(77), 407(78), 408(77)
Muth, E. F., 64

N

Nachtwey, P., 162, 182(1), 183(1)
Nagakura, S., 121

Nagarajan, K., 319, 320, 322, 404 (34)
Nagata, Ch., 123
Nag-Chandhuri, J., 121
Nair, V., 51, 64, 74
Naito, T., 351
Nakane, R., 243
Nakano, A., 61
Nakayama, I., 282
Narasimhan, N. S., 324
Nasielski, J., 370, 371 (226), 372, 374, 377, 378, 379, 384
Nathan, C. C., 363
Natsubori, A., 243
Neber, P. W., 48
Neckers, D. C., 286
Nefedov, V. A., 20
Nefedova, M. N., 20
Neiman, Z., 101, 107 (93), 119 (120), 120
Nelhams, A., 23, 24 (107)
Nelson, K. L., 272
Nepras, M., 374
Nesmeyanov, A. N., 6, 7 (17), 8, 9, 10 (24), 16, 18, 19, 20, 23, 24, 25, 27, 28 (105), 30, 32, 35, 37 (77, 78), 41, 58
Neuse, E. W., 15
Newman, D., 331
Newton, B. A., 378
Newton, M. G., 214
Ng, Y. L., 320
Nguyen Kim Son, 170, 182 (19), 192 (19)
Nichols, R. W., 285
Nicholson, J. S., 379, 380 (280), 405 (280)
Nicodem, D. E., 400
Ninomiya, I., 351
Nishikida, K., 378
Nishimura, S., 240, 267 (21)
Nishiwaki, T., 61
Noland, W. E., 292
Nomura, T., 322
Norman, R. O. C., 236, 245
Nowacki, W., 145
Noyce, D. C., 291
Noyce, D. S., 273, 277, 282, 283, 285, 289, 309
Nozaki, H., 24, 27
Nübel, G., 137, 143
Nunn, A. J., 334
Nyburg, S. C., 214

## O

Oae, S., 212
O'Brien, C., 66, 67 (59)
Ochi, H., 9
Ochiai, E., 282
Ockenden, D. W., 318
O'Connell, A. P., 392
Östman, B., 245, 246, 255, 269 (58, 122), 271 (58), 273 (58, 122), 295, 296 (274)
Ogawa, J., 255, 263 (121), 273 (121), 300 (121), 302 (121), 303 (121), 308 (121)
Ogol'tsova, N. V., 348
Ogura, M., 12
Ohki, H., 383, 394 (296), 408 (296), 410 (296)
Ohta, M., 65, 68 (57), 71, 335
Ohuchi, S., 376
Oishi, Y., 378
Okada, K., 367
Okada, M., 54, 55 (27), 59 (27), 70 (27)
Okamoto, Y., 272, 299 (185)
Olah, G. A., 243, 255
Olsson, S., 245, 246, 269 (58), 271 (58), 273 (58)
Onda, M., 352
Orton, K. J. P., 247
Osborn, A. R., 373, 376 (247), 377 (247)
Osgerby, J. M., 2, 3 (6, 8), 8, 21, 24, 28 (108), 31 (99)
Ostroverkhov, V. G., 59
O'Sullivan, D. J., 43
Oswald, A., 290
Otsuji, Y., 299
Oyster, L., 327

## P

Pacault, A., 111
Packer, J., 379
Pacofsky, E. A., 329
Page, E. R., 44
Pai, B. R., 319, 322, 323
Pal, B. C., 81, 115, 119 (124), 121
Palkina, M. V., 293
Palmer, M. H., 241
Pan, H.-L., 335, 337 (101), 389 (101), 390 (101), 406 (101)

Pandit, U. K., 358
Pannier, R., 27, 28 (117)
Paquer, D., 191
Paquette, L. A., 241, 264 (37)
Parcell, R. F., 49
Pariser, R., 92, 93 (78), 94, 95 (82)
Parkanyi, C., 370, 371 (223), 377, 378 (272)
Parr, R. G., 92, 93 (78), 95 (82)
Parshall, G. W., 15, 21 (67), 28 (67), 29 (67), 32 (67)
Parthasarathy, P. C., 320, 408 (36)
Partridge, M. W., 347, 396 (150), 406 (150)
Pascal, Y., 241
Pasini, C., 248
Patel, H. R., 345
Patel, R. P., 345
Paterson, W., 324, 362 (53)
Patrick, J. B., 366
Patterson, A. M., 45
Paul, I. C., 214, 216
Pauling, I. L., 238
Pauling, L., 213
Pauly, C., 252
Pauly, H., 290
Paushkin, Y. M., 37, 42 (147)
Pauson, P. L., 1, 2, 3 (6, 8), 8, 9, 10 (25), 11 (25), 21, 24, 28 (108), 31 (99), 43
Pavlik, I., 37
Pawelkiewicz, J., 84
Peak, D. A., 379, 380 (280), 405 (280)
Peaston, W. C., 362, 363
Peck, R. M., 392
Pedersen, C. Th., 176, 228 (34), 229 (34), 230 (34)
Pelousek, H., 38
Perchel, G. G., 129, 138 (141)
Perevalova, E. G., 6, 7 (17), 9, 10 (24), 23, 25, 28 (105), 35
Pereyaslova, D. G., 329
Perkampus, H. H., 372, 374
Perkins, M. J., 343
Perrier, M., 200
Perrin, D. D., 143, 376
Peterlik, M., 10, 14, 15 (64), 38, 44
Peters, F. N., 248, 264, 276 (80)

Peterson, M. L., 15, 21 (67), 28 (67), 29 (67), 32 (67)
Peterson, W. R., Jr., 16
Petrarca, A. E., 164
Petrow, V., 318, 321, 325 (14), 353, 355 (170, 171), 356, 379, 380 (280), 383, 390 (171), 395 (295), 396 (295), 400 (14), 404 (14, 295), 405 (39, 280), 411 (295)
Pfau, M., 330, 384 (80), 408 (80), 410 (80)
Pfister-Guillouzo, G., 167, 170, 192 (16, 17), 230 (7), 231 (7), 232 (7), 233 (7), 234 (7)
Pfleiderer, W., 137, 138, 143
Phillips, J. N., 373, 377 (248)
Pictet, A., 316, 318, 369
Piers, K., 290
Pigenet, P., 241
Pignataro, S., 267, 268 (182), 309
Pilar, F. L., 241
Pinel, R., 168, 170 (13), 174 (18, 20), 177 (20, 30, 31), 182 (18), 189 (30), 192 (18, 30), 198 (13, 18, 31), 219 (18), 221 (18, 31), 222 (13), 230 (13), 232 (13, 31), 234 (13)
Piovesana, O., 397
Planchen, M., 272
Plesske, K., 41, 42 (155)
Plummer, L., 15
Plunkett, A. O., 373, 377 (245), 382 (245), 384 (245), 404 (245)
Poirer, Y., 170, 173 (21), 174 (21), 178 (28), 198 (21)
Pollack, P. I., 283
Pollak, M., 151, 153 (169)
Poltev, V. I., 133
Pople, J. A., 86, 92, 95, 99 (68), 100 (68, 69, 70)
Popp, F. D., 3, 14, 17, 21 (76), 22, 23 (76), 27 (76), 28 (94), 29 (94), 35 (52), 36 (52), 44 (52, 94)
Poppe, H., 369
Portail, C., 185, 186
Postnov, V. N., 20, 24
Potts, K. T., 283
Poulain, D., 13, 42 (51)
Povarov, L. S., 363, 368 (200)

Prasad, R. N., 119 (121), 121
Pratt, R. N., 386, 387
Price, C. C., 212
Pridham, J. B., 34
Prinzbach, H., 181
Procter, G. R., 324, 362 (53), 363
Proctor, S. A., 386
ProvóKluit, P., 251, 252 (99)
Prozorova, N. S., 10
Przhipalgovskaya, N. M., 359
Pruett, R. L., 9, 14
Pryde, C. A., 54, 58 (30), 76 (30)
Pullman, A., 84, 85 (50), 93, 94 (59), 95 (59), 97, 105, 113, 121, 122 (50), 133, 151 (63, 74), 152 (63), 153 (74), 155 (74), 157
Pullman, B., 84, 85, 86, 101, 102, 105, 107 (93), 110, 113, 121, 122 (98), 135 (76), 136 (146), 137 (98), 143, 151 (63, 74), 152 (63), 153 (74), 155 (74), 157, 158, 241
Parushothaman, K. K., 320, 404 (34)
Pusch, J., 162, 182 (1), 183 (1)
Pyman, F. L., 250, 260

## Q

Qazi, A. R., 43
Quilico, A., 281
Quiniou, H., 168, 170, 176, 177 (15, 39), 180 (39), 181 (39), 182 (19), 187, 188, 192 (19), 194 (62), 234 (62, 72)

## R

Rabb, D. J., 31, 32 (127)
Ramon, C., 398, 404 (332)
Ranneva, Yu. I., 245, 246 (54, 59), 279 (59)
Rao, S. T., 82
Rapic, V., 27
Rapoport, H., 242
Ratts, K. W., 58, 75 (40)
Rausch, M. D., 17, 33
Ravindranathan, T., 355, 357, 358 (180)

Rawlings, T. J., 198
Reddy, G. S., 150
Reed, R. I., 254
Rees, C. W., 46, 47 (4), 75 (4), 320, 328, 341, 343, 344, 347, 360, 366 (128), 375 (127), 381 (126), 395 (126), 396 (126), 404 (128), 405 (32), 406 (126, 127), 409 (126, 127)
Reese, C. B., 373, 377 (246), 380 (246), 392 (246), 394, 295 (246), 396 (246)
Reiche, A., 398, 409 (333)
Reid, D. H., 167, 181, 183 (11), 193 (49), 194 (70), 196 (71), 200, 202 (83), 203 (49), 224 (49), 225 (49), 226 (49), 227 (49), 230 (49, 83), 231 (49), 232 (49), 233 (49), 234 (49)
Reid, K. I. G., 216
Reid, W., 358
Reif, L., 246
Reimann, H., 179
Reimlinger, H., 74
Rein, R., 151, 153 (169)
Resplandy, A., 336, 337 (106), 345, 382 (106), 395
Restrup-Andersen, J., 237, 241 (12)
Reynolds, G. A., 175, 176 (33), 198 (33), 228 (33), 229 (33)
Reshetova, M. D., 25
Reutov, O. A., 8
Reynolds, J. W., 86
Richards, J. H., 6
Rickards, R. W., 296
Ridd, J., 236, 246
Ridd, J. H., 244, 251, 256, 260, 261 (98), 281 (96), 282 (126), 283 (126)
Rieche, A., 259
Rieke, R. D., 292
Rigatti, G., 209
Rigby, W., 368, 409 (211)
Rinehart, K. L., Jr., 1, 9
Ringertz, H., 82, 102
Rinkes, I. J., 254, 296
Risaliti, A., 318
Ritchie, E., 317, 319, 321, 406, 407 (351)
Robb, J. V. M., 316, 317 (3)
Robbins, R. F., 321, 327, 328, 330 (63), 332 (63)

Roberts, J. D., 264
Robertson, D. A., 73
Robertson, J. E., 10, 44 (33)
Robertson, P. W., 249
Robins, R. K., 101, 115, 119 (121, 123), 121, 140
Robinson, M. J. T., 336
Robinson, R., 392, 395 (325), 397 (325)
Robson, W., 290
Rockett, B. W., 6, 12, 19, 23, 24 (107), 43
Rodionov, V. M., 318
Rodionova, N. A., 19
Rodova, F. Z., 29
Roger, R., 284
Rogers, N. A. J., 353
Romanenko, V. I., 19
Romanskii, I. A., 246
Romers, C., 369, 370 (220)
Roothaan, C. C. J., 90
Rosenblum, M., 366
Ross, V., 110
Rossi, M., 94
Roth, S., 44
Rouessac, A., 181, 191, 193 (48)
Royan, C. A., 282
Rubbo, S. D., 316 (3)
Ruff, M., 176, 177 (38), 179, 182 (38), 183 (38), 192 (38), 194 (38), 224 (38), 225 (38), 226 (38)
Rush, J. E., 164
Rushworth, A., 19
Russell, C. A., 403
Russell, G. A., 391
Russell, J., 327, 401 (64)
Russell, P. J., 101
Rybinskaya, M. I., 58
Rynbrandt, R. H., 286

S

Sadovaya, N. K., 293
Sainsbury, M., 346, 352, 401
Salvemini, A., 240, 267 (22)
Sampson, R. J., 272
Sandhu, M. A., 9, 10 (25), 11 (25)
Sandström, J., 283

Sanesi, M., 202, 222 (84, 127)
Santini, S., 259, 293
Santry, D. P., 86, 92, 99 (68), 100 (68)
Sargent, M. V., 351
Sarma, R. S., 110
Sasse, W. H. F., 319, 346, 368 (139), 373 (26), 374 (26), 377 (26), 395 (26), 398 (215), 406 (26), 408 (26)
Satchell, D. P. N., 245
Sato, H., 397
Sato, S., 50, 54, 65, 67, 68, 71
Savelli, G., 299
Sazonova, N. S., 19, 30
Sazonova, V. A., 16, 18, 19, 20, 24, 27, 30, 37 (77, 78)
Schaap, A. P., 249
Schaaf, R. L., 11
Scheibe, G., 369, 391, 392, 403 (317)
Scheuner, G., 44
Scheuring, R., 176, 177 (37)
Schlesinger, A. H., 284
Schlessinger, R. H., 349
Schlögl, K., 2, 3 (9), 4 (9), 5 (9), 7 (11), 9, 10, 14, 15 (64), 16 (62), 18, 19 (81), 20 (81), 25 (68), 27 (61, 68), 28 (68), 37, 38 (11), 40 (63), 41, 43, 44
Schmidt, U., 176, 177 (37), 189, 191 (65), 226 (65), 234 (65)
Schmitt, H. F., 301
Schmitz-Dumont, O., 257
Schoeneck, W., 391
Schofield, K., 236, 244, 254, 255 (113), 294 (3), 316, 317 (4), 318, 319, 334, 337, 356 (30), 373, 376 (247), 377 (247), 318 (113), 387 (4), 395 (113), 396 (113)
Schmutz, J., 382
Schner, V. F., 359
Schottlander, M., 237
Schubert, H., 26
Schueller, K. E., 249
Schultze, H., 398, 409 (333)
Schwalbach, G., 245
Schweizer, M. P., 81, 110
Schwetlick, K., 245, 266 (53), 267 (53), 271 (53)
Scrowston, R. M., 236

Searles, S., 348
Seel, F., 43
Seelert, K., 63
Segal, G. A., 86, 92 (68), 99 (68), 100 (68, 69, 70)
Seiber, R. P., 400
Seidel, P., 290
Seidl, H., 386
Seidler, J. H., 332, 368, 392 (90), 393 (213), 395 (213), 396 (90), 398 (90), 409 (90, 213)
Seiler, H., 10
Selby, I. A., 376, 380, 386 (285), 402 (285), 403 (285)
Serebryanskaya, A. I., 245, 246 (59), 279 (59)
Sereda, I. P., 25
Shabica, A. C., 290
Shagalov, L. V., 318
Shani, A., 325
Shapiro, I. O., 245, 246 (54, 49), 279 (59), 300 (300), 305
Shapiro, R., 123, 125, 126 (133), 138 (133)
Shapiro, R. H., 402
Sharma, B. S., 213
Shatenshtein, A. I., 245, 246 (54), 269 (55, 299, 300), 271 (55), 286, 289, 291, 293 (55), 300 (55), 305 (55)
Shaw, J. T., 290
Shechter, H., 61
Sheehan, J. C., 359
Sheets, D. G., 284
Shemtova, M. R., 22
Sheppard, R. C., 239
Shepple, S. E., 310
Shiga, M., 22, 28 (103), 36
Shiina, K., 11
Shilov, E. A., 59
Shilovtseva, L. S., 23, 28 (105), 35
Shimanouchi, T., 81
Shimokawa, S., 376
Shiner, V. J., 310
Shioura, Y., 31, 32 (128), 33 (128)
Short, L. N., 373, 376 (247), 377 (247)
Short, W. F., 318
Shramko, O. V., 122
Shustorovich, E. M., 210

Shutte, L., 251, 252
Shvedov, V. I., 257, 293 (141)
Sidwick, N. V., 251
Siegel, A., 17, 33
Silverstein, R. M., 28, 29 (119)
Simonetta, M., 153
Singer, B., 115
Singh, B., 60, 61, 65 (46), 73 (46)
Singh, I., 353
Singh, M., 360
Singh, P., 82, 145 (38)
Skidmore, S., 403
Skorokhodova, T. A., 325
Slack, R., 281
Slater, R. A., 175, 180, 189 (32, 46), 190 (46), 194 (46), 206 (46), 234 (46)
Sletten, E., 82, 101, 145 (42), 146, 155, 214 216
Sletten, J., 82, 101, 145 (42), 146, 155, 206, 214, 216, 218
Sloan, A. D. B., 361
Slocum, D. W., 6, 31
Slough, W., 371, 373 (233)
Smalley, R. K., 323
Smart, J. C., 11
Smith, A., 290
Smith, B. V., 256, 282 (126), 283 (126)
Smith, G. F., 290
Smith, G. H., 10
Smith, H., 353, 381 (268)
Smith, L. R., 292
Smith, P., 6
Smith, P. A. S., 49, 325, 326 (60), 337
Smolinsky, G., 51, 52 (20), 54, 58 (30), 70 (20, 21), 72, 76 (30), 330
Smorgonskii, L. M., 256
Snipper, L. P., 295
Snyder, H. R., 290, 325, 339, 376 (121)
Sobell, H. M., 82
Soddy, T., 390, 400 (313), 404 (313)
Sokolov, S. D., 281
Solony, N., 240
Sommers, A. L., 372
Sousa, J., 348
Sowa, J. R., 259
Spaeth, E. C., 284, 294 (276), 296
Spano, F. A., 399

# AUTHOR INDEX

Sparks, M., 282
Speckamp, W. N., 358
Spence, G. G., 412
Spencer, M., 101, 112, 129 (95)
Sperber, V., 43
Speroni, G., 281
Sperry, J. A., 262, 277 (175), 286 (173), 291 (175)
Spinelli, D., 240, 267 (22), 299
Spitz, S. P., 325, 326 (61)
Sprague, J. M., 281
Sprecher, N., 374
Staab, H. A., 245
Stacey, M., 34, 258
Stasiewicz, M., 273 (239), 277 (239), 280 (239), 285, 286 (239), 289 (239), 291 (239), 293 (239)
Staskun, B., 319
Staunton, R. S., 392, 395 (325), 397 (325)
Stavaux, M., 168, 182 (12), 183 (12), 194, 195, 196 (74), 205 (74), 206, 230 (90)
Steinkopf, W., 247, 301
Stephens, F. F., 379, 380 (280), 405 (280)
Stephenson, E. F. M., 332
Stermitz, F. R., 400
Stern, E. S., 273
Sternhell, S., 229, 286
Stevens, M. F. G., 347, 396 (150), 406 (150)
Stewart, R. F., 82
Steyner, W., 14, 40 (63)
Stock, L. M., 247, 272, 273, 275, 276 (189), 279, 301
Stormer, R., 284
Stowe, G., 282
Stoyanovich, F. M., 246
Stradius, J., 255
Streitwieser, A., 246, 249, 272
Strojny, E. J., 329
Stromberg, V. L., 334
Stubbs, W. H., 43
Studzinskii, O. P., 348
Subbaswami, K. N., 357, 358 (180)
Suchy, J., 21, 26, 35, 36, 37 (116)
Sugasawa, S., 380, 412 (284)
Suginome, H., 329, 358 (75)
Sugiyama, N., 16, 31, 32 (128)
Suh, J. T., 9, 10 (32), 12, 44 (32, 50)

Sukhorukov, B. J., 133
Sundaralingam, M., 82
Sundararajan, V. N., 319, 323
Sundberg, R. J., 313
Suprunchuk, T., 31, 44 (126)
Suschitzky, H., 323, 361
Sutherland, R. G., 11, 27
Sutor, D. J., 82, 150 (34)
Suvorov, N. N., 318
Suzuki, H., 31, 32 (128), 33 (128)
Suzuki, S., 249
Sweet, R. M., 101
Swenton, J. S., 325
Sy, M., 294
Symon, J. D., 167, 183 (11), 200, 202 (83), 230 (83)
Szkrybalo, W., 362, 408 (194)
Szmuszkovicz, J., 284

## T

Tabner, B. J., 378
Takasugi, M., 329, 358 (75)
Tanaka, M., 121
Taniguchi, H., 54, 55 (27), 59 (27, 28), 70 (27)
Tarterat-Adalberon, M., 13, 42 (51)
Tatashina, T. I., 10
Taticchi, A., 276, 314
Tatlow, J. C., 258
Taylor, C. M. B., 336
Taylor, E. C., 318, 329, 330, 368, 382, 384 (80), 395, 397 (214), 398 (214), 408 (80), 410 (80), 412
Taylor, G. A., 386, 387
Taylor, J. D., 367
Taylor, R., 236, 245, 273, 277, 289, 304, 308
Taylor, W. I., 341
Tebboth, J. A., 1
Teitei, T., 16, 31, 32 (128), 33 (128), 365
Tennant, G., 330
Terent'ev, A. P., 290, 292
Terrier, F., 174, 176, 177 (30, 35, 36), 189 (30), 192 (35), 195 (35), 221 (30, 36), 231 (30)
Tertov, B. A., 259

Tertzakian, G., 331, 346
Teste, J., 180, 184 (45), 192 (45), 232 (45), 234 (45)
Theobald, R. S., 316, 317 (4), 334, 387 (4)
Thewalt, U. T., 82
Thiele, W. E., 384
Thomas, G. J., 81
Thomas, G. M., 367
Thomas, H. J., 84
Thomas, W. A., 239
Thompson, R. M., 394, 398 (327)
Thomson, B., 368, 393 (213), 395 (213), 409 (213)
Thomson, R. H., 327, 401 (64)
Thuillier, A., 185
Thyagarajan, B. S., 350
Tilak, B. D., 355, 357, 358 (180)
Timmons, C. J., 351
Timofeeva, L. M., 31
Tinland, B., 370, 388 (225)
Tinoco, I., Jr., 107, 118, 129, 138 (141), 151, 155 (166)
Tirouflet, J., 16, 21, 22 (96, 97), 25, 29 (96, 97), 32 (73), 33 (96, 97), 35 (95, 96, 113), 36 (98), 37 (96, 97), 38 (113), 39 (95, 96, 97, 98), 41 (113), 295, 296 (273), 299
Tishler, M., 290
Tissington, P., 319, 334 (22), 336 (22), 337 (22), 338 (22), 368 (22), 373 (22), 375 (22, 224), 376 (22), 377 (244), 378 (274), 391 (22)
Titov, A. I., 22
Todd, A. R., 343, 347
Todd, J. F., 401
Todd, M. J., 330
Tokizawa, M., 107
Tolgyesi, W. S., 243
Tollin, P., 82, 150 (34)
Toma, S., 16, 21, 26, 35, 36, 37 (116)
Tumcufcik, A. S., 392
Tomita, K. I., 82
Towne, E. B., 238
Townsend, L. B., 101, 140
Traber, W., 410
Tratt, K., 143
Traverso, G., 162, 167 (3), 182 (8), 183 (8, 52), 192 (3), 202, 203 (85), 209 (5), 215, 222 (84, 85)
Traynelis, V. J., 259
Trebaul, C., 180, 184 (45), 192 (45), 232 (45), 234 (45)
Treibs, A., 22
Treibs, W., 34
Tremaine, J. F., 1
Ts'o, P. O. P., 81, 110
Tsuboi, M., 81
Tsukervanik, I. P., 257
Tsunemi, H., 11, 12
Tsymbal, L. V., 292
Tupitsyn, I. F., 376, 403 (264)
Turner, E. E., 319
Turner, G. L., 335, 368 (102), 376 (102), 379 (102), 392 (102), 395 (102), 396 (102)
Turner, S., 239
Turpin, D. G., 340, 344 (123)
Tyson, F. T., 290
Tyurin, V. D., 23, 28 (105)

## U

Ubbelohde, A. R., 371, 373 (233)
Ugi, I., 24
Ullman, E. F., 60, 61, 65 (46), 73 (46)
Unverferth, K., 245, 266 (53), 267 (53), 271 (53)
Ushenko, I. K., 29
Ustynyuk, Yu. A., 6, 7 (17), 9, 10 (24), 35

## V

van den Hende, J. H., 203
Van der Burg, W. J., 398, 404 (332)
Vander Donckt, E., 371, 372, 374, 375, 377, 378, 384
Van der Vlugt, M. J., 251, 252 (100)
Van Duuren, B. L., 372
Vanlautem, N., 378
Van Sickle, D. E., 246
Van Thuijl, J., 369, 370 (220)
Van Zyl, G., 286
Vargaflig, B., 398, 404 (332)
Vas'Kovskii, V. E., 293

# AUTHOR INDEX

Vaughan, J., 379
Vaughan, J. D., 252, 253, 283 (104, 105)
Vaughan, V. L., 252, 253, 283 (104, 105)
Veillard, A., 84
Velardi, G., 290
Venters, K., 255
Vergoni, M., 307 (314), 314
Verlander, M. S., 385
Vialle, J., 172, 181, 185, 186, 189, 191, 193 (48), 199, 200, 216 (80), 224 (80)
Vilchevskaya, V. D., 32
Vinovskis, M., 366
Vishnyakova, T. P., 31, 37, 42 (147)
Viswanathan, N., 320, 408 (36)
Vlasova, I. D., 37, 42 (147)
Vogel, A. I., 369
Vogel, W., 242
Vogels, C., 377

## W

Wade, K., 193, 194 (70)
Walker, G. N., 322
Walker, I. S., 371, 408 (232)
Walker, J., 356
Walls, L. P., 316, 317, 332, 337 (89), 389 (89), 405
Walter, W., 167, 195
Wang, N., 366
Ward, S. D., 181, 194 (50), 208, 224 (92), 226 (50), 234 (50)
Warren, B. E., 213
Washbourn, K., 405
Washburne, S. S., 16
Watanabe, H., 35
Watkins, T. I., 379, 380 (280), 405 (280)
Watson, D. G., 82, 101, 150
Watson, J. D., 80, 84
Watts, W. E., 9, 10 (25), 11 (25)
Weber, D., 200, 229 (82)
Webster, B. R., 336
Wegerhoff, P., 334
Weidman, H., 59
Weiler-Feilchenfeld, H., 101, 107 (93), 119 (120), 120, 135, 136 (146)
Weiner, S. A., 391

Weinstein, J., 348
Weinstock, L. M., 283
Weisert, A., 321, 336, 337 (105), 395 (105), 396 (105), 397 (105)
Weissgerber, R., 284, 290
Wende, A., 27, 28 (117)
Wender, I., 399
Werkema, M. S., 82, 145 (38)
Westfall, P. A., 259
Westman, L. F., 9
Weston, A. W., 284
Wettermark, G., 348
Whaley, W. M., 320
Wheland, G. W., 238
White, R. F. M., 281
Whitman, G. M., 15, 21 (67), 28 (67), 29 (67), 32 (67)
Wiedenmann, R., 176, 177 (38), 182 (38), 183 (38), 189, 191 (64), 192 (38), 194 (38), 224 (38), 225 (38), 226 (38, 64), 229 (64, 69)
Weisner, K., 341, 363
Wihelm, O., 290
Wildman, W. C., 316, 367
Wiles, D. M., 31, 44 (126)
Willets, C. H., 81
Willson, C. D., 242
Winslow, E. C., 9
Witkop, B., 366
Woerner, F. P., 74
Wolf, L., 14, 25 (57, 58), 35 (57), 37 (57), 39 (57)
Wolf, V., 350
Wolfenden, R. V., 82, 113, 124 (111)
Wong, E., 379
Wood, C. S., 348, 349 (153)
Woods, C. S., 351
Wooldridge, K. R. H., 281
Wooton, W. C., 368
Worrall, R., 258
Worthing, C. R., 365, 395 (203), 397 (203)
Wotring, R. B., 329, 407 (78)
Wragg, W. R., 317, 405
Wright, G., 254
Wu, M. T., 245
Würmli, A., 334, 341 (99), 382 (99)
Wulff, J., 62

Wylie, A. G., 327, 401 (64)
Wynberg, H., 238, 239, 249, 250 (91)

## X

Xuong, N. D., 284, 294

## Y

Yakushima, T. A., 245, 269 (55), 271 (55), 293 (55), 300 (55), 305 (55)
Yamada, S., 46, 412
Yamakawa, K., 9, 10 (31), 11 (31), 33
Yamamoto, Y., 386
Yandle, Y. R., 378
Yang, N. C., 325
Yanovskaya, L. A., 290
Yelland, M., 406
Yokoyama, M., 329, 358 (75)
Yonan, P., 21, 44 (93)
Yonezwa, K., 352
Yoshimura, N., 324

Young, R. V., 263
Yu, J. M. H., 372
Yurchenko, O. P., 24
Yur'ev, Yu. K., 262, 293, 303 (304), 308

## Z

Zahradnik, R., 241, 370, 374, 377, 378 (272)
Zatsepina, N. N., 376, 403 (264)
Zauli, B., 241
Zhikhareva, K. D., 29
Ziegler, J. B., 290
Zimmer-Galler, R., 22
Zlotina, I. B., 41
Zollinger, H., 260, 261
Zolnikova, G. P., 19
Zvegintseva, L. N., 372
Zvyagintseva, E. N., 245, 246 (54), 269 (55), 271 (55), 293 (55), 300 (55, 300), 305
Zwanenburg, B., 63, 65 (55)

# Cumulative Index of Titles

## A

Acetylenecarboxylic acids, reactions with heterocyclic compounds, **1**, 125
Aminochromes, **5**, 205
Anthranils, **8**, 277
Aromatic quinolizines, **5**, 291
Aza analogs, of pyrimidine and purine bases, **1**, 189
Azines, reactivity with nucleophiles, **4**, 145
Azines, theoretical studies of, physicochemical properties and reactivity of, **5**, 69
Azinoazines, reactivity with nucleophiles, **4**, 145
1-Azirines, synthesis and reactions of, **13**, 45

## B

Benzisoxazoles, **8**, 277
Benzoazines, reactivity with nucleophiles, **4**, 145
Benzofuroxans, **10**, 1
Benzo[b]thiophene chemistry, recent advances in, **11**, 178

## C

Carbenes, reaction with heterocyclic compounds, **3**, 57
Carbolines, **3**, 79
Chemistry
    of benzo[b]thiophenes, **11**, 178
    of diazepines, **8**, 21
    of furans, **7**, 377
    of lactim ethers, **12**, 185
    of phenanthridines, **13**, 315
    of phenothiazines, **9**, 321
    of 1,3,4-thiadiazoles, **9**, 165
    of thiophenes, **1**, 1

Claisen rearrangements, in nitrogen heterocyclic systems, **8**, 143
Complex metal hydrides, reduction of nitrogen heterocycles with, **6**, 45
Covalent hydration, in heteroaromatic compounds, **4**, 1, 43
Cyclic enamines and imines, **6**, 147
Cyclic hydroxamic acids, **10**, 199
Cyclic peroxides, **8**, 165

## D

Development of the chemistry of furans, 1952–1963, **7**, 377
2,4-Dialkoxypyrimidines, Hilbert–Johnson reaction of, **8**, 115
Diazepines, chemistry of, **8**, 21
Diazomethane, reactions with heterocyclic compounds, **2**, 245
Diquinolylmethane, and its analogs, **7**, 153
1,2 and 1,3-Dithiolium ions, **7**, 39

## E

Electronic aspects of purine tautomerism, **13**, 77
Electronic structure of heterocyclic sulfur compounds, **5**, 1
Electrophilic substitutions of five-membered rings, **13**, 235

## F

Ferrocenes, heterocyclic, **13**, 1
Five-membered rings, electrophilic substitutions of, **13**, 235
Furan chemistry, development of the chemistry of (1952–1963) **7**, 377

## H

Halogenation of heterocyclic compounds, **7**, 1
Hammett equation, applications to heterocyclic compounds, **3**, 209
Hetarynes, **4**, 121
Heteroaromatic compounds, free-radical substitutions of, **2**, 131
  nitrogen, covalent hydration in, **4**, 1, 43
  prototropic tautomerism of, **1**, 311, 339; **2**, 1, 27
Heteroaromatic substitution, nucleophilic, **3**, 285
Heterocycles, photochemistry of, **11**, 1
Heterocyclic chemistry, literature of, **7**, 225
Heterocyclic compounds
  application of Hammett equation to, **3**, 209
  halogenation of, **7**, 1
  mass spectrometry of, **7**, 301
  quaternization of, **3**, 1
  reactions of, with carbenes, **3**, 57
  reactions of acetylenecarboxylic acids with, **1**, 125
  reactions of diazomethane with, **2**, 245
  sulfur, electronic structure of, **5**, 1
N-Heterocyclic compounds, electrolysis of, **12**, 213
Heterocyclic diazo compounds, **8**, 1
Heterocyclic ferrocenes, **13**, 1
Heterocyclic pseudo bases, **1**, 167
Heterocyclic syntheses, from nitrilium salts under acidic conditions, **6**, 95
Hilbert–Johnson reaction of 2,4-dialkoxypyrimidines, **8**, 115

## I

Imidazole chemistry, advances in, **12**, 103
Indole Grignard reagents, **10**, 43
Indoles, acid-catalyzed polymerization, **2**, 287
Indoxazenes, **8**, 277
Isoindoles, **10**, 113
Isothiazoles, **4**, 107
Isoxazole chemistry, recent developments in, **2**, 365

## L

Lactim ethers, chemistry of, **12**, 185
Literature of heterocyclic chemistry, **7**, 225

## M

Mass spectrometry of heterocyclic compounds, **7**, 301
Metal catalysts, action on pyridines, **2**, 179
Monoazaindoles, **9**, 27
Monocyclic sulfur-containing pyrones, **8**, 219

## N

Naphthyridines, **11**, 124
Nitrilium salts, heterocyclic syntheses involving, **6**, 95
Nitrogen heterocycles, reduction of, with complex metal hydrides, **6**, 45
Nitrogen heterocyclic systems, Claisen rearrangements in, **8**, 143
Nucleophiles, reactivity of azine derivatives with, **4**, 145
Nucleophilic heteroaromatic substitution, **3**, 285

## O

1, 3, 4-Oxadiazole chemistry, recent advances in, **7**, 183
1, 3-Oxazine derivatives, **2**, 311
Oxazolone chemistry, recent advances in, **4**, 75

## P

Pentazoles, **3**, 373
Peroxides, cyclic, **8**, 165

Phenanthridine chemistry, recent developments in, **13**, 315
Phenothiazines, chemistry of, **9**, 321
Phenoxazines, **8**, 83
Photochemistry of heterocycles, **11**, 1
Physicochemical aspects of purines, **6**, 1
Physicochemical properties
  of azines, **5**, 69
  of pyrroles, **11**, 383
3-Piperideines, **12**, 43
Prototropic tautomerism of heteroaromatic compounds, **1**, 311, 339; **2**, 1, 27
Pseudo bases, heterocyclic, **1**, 167
Purine bases, aza analogs of, **1**, 189
Purines
  physicochemical aspects of, **6**, 1
  tautomerism, electronic aspects of, **13**, 77
Pyrazole chemistry, progress in, **6**, 347
Pyridazines, **9**, 211
Pyridine, effect of substituents in, **6**, 229
Pyridines, action of metal catalysts on, **2**, 179
Pyridopyrimidines, **10**, 149
Pyrimidine bases, aza analogs of, **1**, 189
Pyrones, monocyclic sulfur-containing, **8**, 219
Pyrroles
  acid-catalyzed polymerization, of **2**, 287
  physicochemical properties of, **11**, 383
Pyrrolizidine chemistry, **5**, 315
Pyrrolopyridines, **9**, 27
Pyrylium salts, preparations, **10**, 241

## Q

Quaternization of heterocyclic compounds, **3**, 1

Quinazolines, **1**, 253
Quinolizines, aromatic, **5**, 291
Quinoxaline chemistry, recent advances in, **2**, 203
Quinuclidine chemistry, **11**, 473

## R

Reduction of nitrogen heterocycles with complex metal hydrides, **6**, 45
Reissert compounds, **9**, 1

## S

Selenazole chemistry, present state of, **2**, 343
Selenophene chemistry, advances in, **12**, 1
Synthesis and reactions of 1-azirines, **13**, 45

## T

Tautomerism, prototropic, of heteroaromatic compounds, **1**, 311, 339; **2**, 1, 27
1,2,3,6-Tetrahydropyridines, **12**, 43
Theoretical studies of physicochemical properties and reactivity of azines **5**, 69
1,2,4-Thiadiazoles, **5**, 119
1,2,5-Thiadiazoles, **9**, 107
1,3,4-Thiadiazoles, chemistry of, **9**, 165
1,2,3,4-Thiatriazoles, **3**, 263
Thiophenes, chemistry of, recent advances in, **1**, 1
Three-membered rings, with two hetero atoms, **2**, 83
1,6,6a$S^{IV}$-Trithiapentalenes, **13**, 161